AXIOMS, DEFINITIONS, AND THEOREMS

(1.2) $a - b = a + (-b)$

(1.3) $a + b = a(b^{-1})$, if $b \neq 0$

(1.4) $-(-a) = a$

(1.5) $-(a + b) = -a - b$

(1.6) $-(a - b) = -a + b$

(1.7) $a(0) = 0$ for every a

(1.8) $a(-b) = -(ab)$

(1.9) $(-a)(-b) = ab$

(1.10) $a/b = c$ if and only if $a = bc$, for $b \neq 0$

(1.11) $a \div 0$ is never defined

(1.12) $0 \div b = 0$, for $b \neq 0$

Cancellation laws:
If $a + c = b + c$, then $a = b$.
If $ac = bc$, then $a = b$, for $c \neq 0$.

(1.13) If $a < b$, then $a + c < b + c$

(1.14) If $a < b$ and $c > 0$, then $ac < bc$

(1.15) If $a < b$ and $c < 0$, then $ac > bc$

(1.16) If $0 < a < b$, then $1/a > 1/b$

(2.2) $a/a = 1$, if $a \neq 0$ (2.3) $a/1 = a$

(2.4) $\dfrac{ac}{bd} = \dfrac{a}{b} \cdot \dfrac{c}{d}$ (2.5) $\dfrac{a}{d} = a \cdot \dfrac{1}{d}$

(2.6) Dividend = (divisor)(quotient) + remainder

(2.7) $(ax + by)(cx + dy) = acx^2 + (ad + bc)xy + bdy^2$

(2.8) $(x + y)^2 = x^2 + 2xy + y^2$

(2.9) $(x - y)^2 = x^2 - 2xy + y^2$

(2.10) $(a + b)(a - b) = a^2 - b^2$

(2.11) $a^2 - b^2 = (a + b)(a - b)$

(2.12) $x^3 + y^3 = (x + y)(x^2 - xy + y^2)$

(2.13) $x^3 - y^3 = (x - y)(x^2 + xy + y^2)$

(3.1) If $ad = bc$, then $a/b = c/d$, and conversely,

(3.2) $\dfrac{a}{b} = \dfrac{ak}{bk} = \dfrac{a/d}{b/d}$

(3.3) $\dfrac{a}{b} = \dfrac{-a}{-b} = -\dfrac{-a}{b} = -\dfrac{a}{-b}$

(3.4) $\dfrac{a}{b} \cdot \dfrac{c}{d} = \dfrac{ac}{bd}$

(3.5) $\dfrac{a}{d} + \dfrac{c}{d} = \dfrac{a + c}{d}$

(4.1) $a^n = a \cdot a \cdots a$ to n factors

(4.2) $a^0 = 1$ if $a \neq 0$

(4.3) $a^{-n} = 1/a^n$ (4.4) $a^m a^n = a^{m+n}$

(4.5) $a^m/a^n = a^{m-n}$ (4.6) $(a^m)^n = a^{mn}$

(4.7) $(ab)^n = a^n b^n$ (4.8) $(a/b)^n = a^n/b^n$

(4.9) $\sqrt[n]{a} = b$
(for n a positive integer) means
$a = b^n$
i) if a and b are positive, n is any positive integer
or ii) if a and b are negative, n is an odd positive integer

(4.10) $\sqrt[n]{ab} = \sqrt[n]{a}\,\sqrt[n]{b}$

(4.11) $\sqrt[n]{a/b} = \sqrt[n]{a}/\sqrt[n]{b}$

(4.12) $\sqrt[m]{\sqrt[n]{a}} = \sqrt[mn]{a}$

(4.13) $a^{1/n} = \sqrt[n]{a}$

(continued on back cover)

McGraw-Hill Book Company

New York   St. Louis   San Francisco   Auckland   Bogotá   Hamburg

Johannesburg   London   Madrid   Mexico   Montreal   New Delhi

Panama   Paris   São Paulo   Singapore   Sydney   Tokyo   Toronto

Ninth Edition

COLLEGE ALGEBRA

Paul K. Rees

Professor of Mathematics
Louisiana State University

Fred W. Sparks

Late Professor of Mathematics
Texas Tech University

Charles Sparks Rees

Professor of Mathematics
University of New Orleans

COLLEGE ALGEBRA

234567890 DOCDOC 898765

ISBN 0-07-051735-5

This book was set in Times Roman by Monotype Composition Company, Inc.
The editors were Peter R. Devine and J. W. Maisel;
the production supervisor was Phil Galea.
The cover was designed by Joseph Gillians;
cover photograph by Paul Silverman.
The drawings were done by J & R Services, Inc.
R. R. Donnelley & Sons Company was printer and binder.

Library of Congress Cataloging in Publication Data

Rees, Paul Klein, date
    College algebra.

    Includes index.
    1. Algebra.   I. Sparks, Fred Winchell, date
II. Rees, Charles Sparks.   III. Title.
QA154.2.R44   1985   512.9   84-12527
ISBN 0-07-051735-5

This book is dedicated to the memory of Fred W. Sparks, who passed away on February 15, 1982, at the age of 90. He was an excellent teacher, a true friend, and an industrious coauthor. Fred first met Paul Rees in 1919, and they began working together on books in 1934, continuing until Fred retired from writing in 1974. At that time, Charles Sparks Rees became a coauthor. The three have written a total of about 60 books, not including foreign editions. Fred, we miss you.

Paul K. Rees
Charles Sparks Rees

CONTENTS

PREFACE

In writing the manuscript for the ninth edition of *College Algebra,* we have retained those features that have caused the book to be used by many thousands of students for over 45 years. Furthermore, we have made changes that we think will make this edition even more readable and teachable than any of its eight predecessors.

ORGANIZATION

We have used the current style of having one section for each exercise. However, we still follow our practice of putting the exercises one normal lesson apart.

The chapter on exponents and radicals has been moved from Chapter 5 to Chapter 4. The new Chapter 5 on equations and inequalities includes quadratic equations as well as linear and fractional material. Chapter 6 on graphs and functions has a more complete treatment of conics and also includes graphs of polynomials and rational functions. Translation and symmetry are also presented here. Exponential and logarithmic functions occur earlier in this edition, in Chapter 8. The last four chapters, on discrete mathematics, are shorter than the first twelve chapters. Keeping them separate makes it easier for the instructor to choose the material appropriate for her or his course.

TEACHING AIDS

We have, as usual, put the exercises a normal lesson apart, and almost all the problems are in groups of four similar ones. Most classes need be assigned only every fourth problem, but others may need more drill. Each exercise has a variety of graded problems which vary from the standard drill type to problems of a more challenging nature. About half the problems are new, and half are class-tested problems from the earlier editions. Many new problems have been included which deal directly with other disciplines such as anthropology, biology, chemistry, earth science, education, engineering, management, physics, political science, psychology, and sociology.

Each chapter, except for two short ones, finishes with a list of the key concepts in the chapter, the most important equations from that chapter, and a chapter review exercise.

Marginal notes highlight many features in the text and also the exercises.

There is a cumulative review exercise at the end of Chapter 4 which may be used either as a review or as a diagnostic tool for certain classes.

A list of important equations is printed on the endpapers for convenient reference.

EXERCISES

These have always been the hallmark of our books. We give enough routine problems to allow the student to *learn* the material, and we also have more challenging problems to keep up the students' interest and enthusiasm. There are 73 regular and 14 review exercises, with a total of about 4500 problems. In the back of the book, we have the answers to three-fourths the regular problems and all the review problems.

CALCULATORS

We have tried to adopt a realistic attitude toward calculators. First, we do not give detailed instructions on how to operate them because the variety is simply too great. Second, we do not give forced or artificial uses of a calculator. Third, we try to allow their use in the jobs they do best—actual calculation and as a table of function values.

INSTRUCTOR'S MANUAL

This booklet includes the answers to one-fourth of the problems in the regular exercises. There is also a list of examination questions for each chapter.

ACKNOWLEDGMENTS

We are appreciative of the suggestions made by people who have used earlier editions of the book and by those who read the manuscript for this edition. We have a better book than we would have had without these suggestions.

We also owe a special thanks to Mary Sparks Matthews for working each problem in the book. She is the daughter of the deceased senior author, Fred W. Sparks.

We began working with McGraw-Hill 46 years ago, and are proud to continue the tradition.

Paul K. Rees
Charles Sparks Rees

TO THE INSTRUCTOR

This book has 16 chapters and 73 sections, which is enough material to allow you to design a course specifically for your students depending on their backgrounds, abilities, and interests. In most situations, it will take about one class period of 50 minutes to cover each section and the problems at the end of that section. We have put the problems in groups of four similar ones so that it will be easy to give a reasonable assignment that covers the section. There are plenty of problems if needed for extra work.

The chapter titles with the normal number of non-quiz class periods are given below. For the first 12 chapters, we have given one more than the number of sections to allow for variability within the chapter and to allow time for the review exercise in the key concepts section. The last four chapters are shorter, and the extra day has not been included.

Paul K. Rees
Charles Sparks Rees

CHAPTER 1 · THE REAL NUMBER SYSTEM

1.1 The real numbers

In this book we shall be doing many things with numbers—adding, multi-plying, comparing sizes, etc. The first set of numbers we deal with is the set of *natural numbers* 1, 2, 3, 4, 5, 6, ... The three dots mean that there are an infinite number of natural numbers, of which 1984, 2001, and 64,000 are three examples. The natural numbers are also called the *counting numbers* or the *positive integers*. Note that 0 is not a positive integer.

Natural numbers

Counting numbers
Positive integers

One reason for looking closely at the natural numbers is that many of their properties will be seen in other situations in later sections and in later chapters. The familiarity gained now will be most helpful later.

Any two positive integers may be added or multiplied. If m and n are positive integers, their *sum* is written as $m + n$, whereas their *product* has several notations, including

Sum

Product

$$m \times n \qquad m \cdot n \qquad m(n) \qquad (m)(n) \qquad mn$$

For example, we have

$$16 + 3 = 19 \qquad \text{and} \qquad 16 \times 3 = 48$$
$$37 + 29 = 66 \qquad \text{and} \qquad 37 \cdot 29 = 1073$$
$$123 + 4567 = 4690 \qquad \text{and} \qquad (123)(4567) = 561{,}741$$

Note that the addition, and multiplication, of *two* positive integers is the thing which is defined. We only add three or more by repeatedly adding two. Thus we have $(15 + 11) + 9 = 26 + 9 = 35$ and $15 + (11 + 9) = 15 + 20 = 35$ and also $(4 + 19) + (43 + 26) = 23 + 69 = 92$. In Sec. 1.2, we will discuss the laws (associative, etc.) which govern this situation.

One way to classify the positive integers is as even and odd. A positive integer m is called *even* if $m = 2p$ for some positive integer p, that is, if m is a multiple of 2. Otherwise it is called *odd*. Thus 2, 4, 6, 8, 10, 12, 14, etc., are even, while 1, 3, 5, 7, 9, 11, 13, etc., are odd. In fact a positive

Even
Odd

integer is even precisely when its last digit is 2, 4, 6, 8, or 0. Hence the numbers 784 and 2,576,116 and 333,555,558 are even.

If m and n are natural numbers, then $m + n$ is even if m and n are both even or both odd, while $m \times n$ is even if either m or n (or both) is even. For example, $5 + 6$ is odd but 5×6 is even.

We may also consider the multiples of 3, namely, 3, 6, 9, 12, 15, etc.; the multiples of 4, namely, 4, 8, 12, 16, 20, etc.; the multiples of 5, namely, 5, 10, 15, 20, 25, etc.; and the multiples of any natural number, for example, 13, 26, 39, 52, 65, etc.

Factor   If m is a multiple of n, then we say n is a *factor* of m. For example, $65 = 13 \times 5$; hence 13 and 5 are factors of 65. Each natural number $n > 1$
Prime number can be classified as either a *prime number* if its only factors are 1 and n, or
Composite number as a *composite number* if it has at least one factor other than 1 or n.

Any even number (except 2) is composite since it has 2 as a factor. Other composite numbers include $15 = 5 \times 3$, $51 = 17 \times 3$, $511 = 7 \times 73$, $999 = 27 \times 37$, $1001 = 7 \times 11 \times 13$, and $4,294,967,297 = 641 \times 6,700,417$.

The number 1, for various reasons, is not usually considered to be a prime number. The first few prime numbers are 2, 3, 5, 7, 11, 13, 17, 19, 23, 29, and 31. Others are, for example, 257 and 65,537. It can be proved that there are an infinite number of primes.

The prime numbers are the building blocks of the natural numbers. The
Fundamental basis for this statement is the *fundamental theorem of arithmetic*. It states
theorem of that every natural number can be factored into prime factors, and that this
arithmetic can be done in only one way except for the order of the factors. For example, $105 = 3 \times 5 \times 7 = 5 \times 7 \times 3 = 7 \times 3 \times 5 = 5 \times 3 \times 7$, etc., but the only difference in these factorizations is the order in which we write down 3, 5, and 7. Also, $44 = 11 \times 2 \times 2 = 2 \times 11 \times 2 = 2 \times 2 \times 11$.

If m and n are natural numbers, the largest natural number which is a
Greatest factor of both m and n is called their *greatest common divisor* (written gcd),
common and it is denoted by the symbol (m, n). Thus $(24, 18) = 6$, since 6 is the
divisor largest number which is a factor of both 18 and 24. We may find the gcd of m and n by factoring m and n into prime factors, then taking each prime factor which occurs in both factorizations. Greatest common divisors are important when trying to factor the largest possible number out of each of several given numbers. See Example 1b.

● **Example 1**   Find the gcd of (a) 462 and 627, (b) 45 and 105.

Solution   (a)   Since $462 = 2 \times 3 \times 7 \times 11$ and $627 = 3 \times 11 \times 19$, we have

$$(462, 627) = 3 \times 11 = 33$$

(b)   Since $45 = 3 \times 3 \times 5$ and $105 = 3 \times 5 \times 7$, we have

$$(45, 105) = 3 \times 5 = 15$$

and furthermore

$$45 + 105 = 15(3 + 7)$$

Least
common
multiple

The *least common multiple* (written lcm) of m and n is the smallest natural number for which m and n are both factors. It is written $[m, n]$. Thus $[4, 6] = 12$, since 12 is the smallest number with both 4 and 6 as factors. We may find the lcm of m and n by factoring m and n into prime factors, then taking each factor which occurs in either factorization. Least common multiples, of numbers and expressions, are indispensable in adding fractions and in solving equations.

● **Example 2**   Find the lcm of (a) 20 and 36, (b) 77 and 143.

Solution   (a)   Since $20 = 2 \times 2 \times 5$ and $36 = 2 \times 2 \times 3 \times 3$, we have

$$[20, 36] = 2 \times 2 \times 3 \times 3 \times 5 = 180$$

(b)   Since $77 = 11 \times 7$, $143 = 11 \times 13$, we have

$$[77, 143] = 11 \times 7 \times 13 = 1001$$ ●

If we divide 27 by 4, we get a quotient of 6 and a remainder of 3. We may write either

$$
\begin{array}{r}
6 \\
4\overline{)27} \\
24 \\
\hline
3
\end{array}
\qquad \text{or} \qquad 27 = 4(6) + 3
$$

The same thing may be done if we divide any natural number m by another natural number n. We get a quotient q (a natural number) and a remainder r (either 0 or a natural number) such that

$$m = (n)(q) + r \qquad (1.1)$$

where r is between 0 and n, including 0 but not n.

● **Example 3**   Find the quotient and remainder if (a) $m = 111$, $n = 4$, (b) $m = 2880$, $n = 61$, (c) $m = 15{,}113$, $n = 119$.

Solution   Ordinary division gives the following quotients and remainders, which may be verified by multiplication.
(a) $111 = 4(27) + 3$, hence $q = 27$, $r = 3$.
(b) $2880 = 61(47) + 13$, hence $q = 47$, $r = 13$.
(c) $15{,}113 = 119(127) + 0$, hence $q = 127$, $r = 0$.   ●

In Example 3c, notice that because $r = 0$, the number 119 is a factor of 15,113.

● **Example 4**   Beginning with 108 and 30, we write several divisions, in each line using the previous n and r of Eq. (1.1).

$$
\begin{aligned}
108 &= (30)(3) + 18 \\
30 &= (18)(1) + 12 \\
18 &= (12)(1) + 6 \\
12 &= (6)(2) + 0
\end{aligned}
$$

Notice that the last nonzero remainder is 6, and 6 is the gcd of the original numbers 108 and 30. This will always be true.   ●

We will use Eq. (1.1) in a more general form in Chap. 2 when we divide one polynomial by another. It is also used in the factor theorem and the remainder theorem in Chap. 11.

Note   Notice that if $r = 0$ in (1.1), then $m = nq$, and hence n is a factor of m.

The Goldbach conjecture (not yet proved) says that any even natural number greater than 2 may be written as the sum of two primes. For example, $18 = 7 + 11 = 13 + 5$ and $50 = 47 + 3$ or $43 + 7$ or $31 + 19$.

n!   The symbol $n!$, read *n factorial*, means the product of all the natural numbers from 1 to n. Hence $3! = 3 \cdot 2 \cdot 1 = 6$ and $7! = 7 \cdot 6 \cdot 5 \cdot 4 \cdot 3 \cdot 2 \cdot 1 = 5040$. This notation will be used in the last four chapters of this book. One interesting way to use factorials is to produce a sequence of 99 consecutive natural numbers, each of which is composite. If we write

$$100! + 2, \ 100! + 3, \ 100! + 4, \ . \ . \ . \ , \ 100! + 99, \ 100! + 100$$

then the first number is divisible by 2, the second number is divisible by 3, the third by 4, etc., and the last one is divisible by 100. Incidentally, it has been shown by use of a computer that the numbers 396,733 and 396,833 are primes, but the 99 numbers between them are composite. (50! has 65 digits; so 100! is a *large* number.)

$\sqrt{n}$ When checking to see whether the natural number n is prime, we only need to divide n by the numbers from 1 to $\sqrt{n}$. The symbol $\sqrt{n}$ means the square root of n—we will discuss this and related topics in detail in Chap. 4. By definition, $(\sqrt{n})(\sqrt{n}) = n$. Thus $\sqrt{361} = 19$ since $(19)(19) = 361$. Also $\sqrt{212,521} = 461$. Note that $\sqrt{7}\sqrt{7} = 7$ exactly, but $\sqrt{7}$ is only approximated by 2.646 since $2.646 \times 2.646 = 7.001316$.

In the last few years, the study of prime numbers and of factoring natural numbers has been closely related to national security. The reason for this is that the most advanced ways of sending messages electronically by code involve the product of two large prime numbers. If you and your intended receiver both know a specific pair of large prime numbers, and if a third party does not know them, then you can send messages to the receiver safely. The third party may well intercept the electronic message, but decoding it will be impossible unless the two prime factors can be found.

Negative   Although the sum of any two natural numbers, or positive integers, is integers   also a positive integer, the same is not true for the difference of any two positive integers. For instance, $6 + 11$ is the positive integer 17, but $6 - 11$ is not a positive integer. Hence we define the set of *negative integers* as

$$\{-1, -2, -3, -4, -5, \ . \ . \ .\}$$

These numbers, are as important as positive integers, for example, in working with a checkbook, in measuring temperatures above and below freezing on the Celsius scale, and in recording weight gain or loss.

Integers We must also include 0, since $8 - 8$ is neither a positive nor a negative integer. The set of *integers* is then defined as

$$\{. . . , -4, -3, -2, -1, 0, 1, 2, 3, 4, . . .\}$$

which is composed of the positive integers, the negative integers, and 0.

Just as addition and subtraction forced us to include negative integers, multiplication and divison force us to expand a second time. Although $12 \div 4$ is an integer, $12 \div 5$ is not an integer since

$$12 = 4(3) \quad \text{and} \quad 12 = 5(2) + 2$$

Rational number We therefore define a *rational number* to be m/n or $m \div n$ where m and n are integers, as long as $n \neq 0$. Notice that a rational number is the ratio of two integers. It follows that $1/3$, $2/7$, $6/1$, $8/5$, and $-11/111$ are rational numbers.

The usual long-division process applied to $8/5$, $1/3$, and $2/7$ gives $8/5 = 1.6$, $1/3 = .333 \cdots$, and $2/7 = .285714 \cdots$. Each of these three decimals terminates or forms a repeating pattern. Clearly $.333 \cdots$ is a string of repeating 3s. Also, 1.6 is a terminating decimal. Finally $.285714 \cdots$ repeats every six digits.

Any rational number will form a repeating or terminating decimal. For example, $6/13$ will repeat after at most 12 digits, and $16/259$ will repeat after at most 258 digits.

By using infinite geometric progressions (see Chap. 15), it may be seen that any repeating nonterminating decimal is a rational number. In fact, there are a number of ways to show this. For instance if we let $x = 0.151515 \cdots$, then we have $100x = 15.151515 \cdots$. Subtracting gives $100x - x = 15$ since the decimal parts are identical and cancel out. Thus we get $99x = 15$ and $x = 15/99 = 5/33$.

Other examples of repeating nonterminating decimals are $0.314314 \cdots = 314/999$ and $0.5636363 \cdots = 31/55$.

Repeating decimals Combining these facts, we see that *the rational numbers are precisely the decimals* which either (1) terminate or else (2) repeat but do not terminate.

This brings us then to a third expansion of our number system. By
Irrational numbers definition, we say that *the irrational numbers are the nonrepeating and nonterminating decimals*. For example, $0.01001000100001 \cdots$ and $0.1234567891011121314 \cdots$ each has a definite pattern which is not repeating, and so each is irrational. It may be shown that π and $\sqrt{5}$ are also irrational. See Prob. 39 in Exercise 1.5 for a proof that $\sqrt{2}$ is irrational.

Real numbers The set of **real numbers** may now be defined as the set of decimals. Every real number is either rational or irrational, and every real number is positive, 0, or negative. Many other classifications are possible.

The real line or number line It is proved in more advanced mathematics that each real number can be associated with one and only one point on a *line* and each point on the line is associated with one and only one real number. Any point may be called 0. Then positive numbers are associated with points to the right of 0, and

FIGURE 1.1

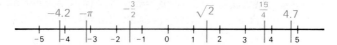

negative numbers with points to the left. See Fig. 1.1. There are an infinite number of points on the line, and also an infinite number of real numbers. The farther we go to the right, the larger the numbers become.

In addition to equality, it is also important to understand the concept of inequality of two real numbers. We know that \$5 is less than \$6, and 5 is to the left of 6 on the number line. We write "5 < 6" and read "5 is less $a < b$ than 6." This illustrates the fact that "$a < b$" means "a is to the left of b on the number line." This is the same as saying that $b - a$ is positive. We also write "$a > b$" when a is greater than b, or when a is to the right of b on the number line. See Fig. 1.1.

● **Example 5**
$$7 < 22 \qquad -17 < 41 \qquad -80 < -53$$
$$181 > 143 \qquad 0 > -5 \qquad -3 > -66$$

For any two real numbers a and b, we always have exactly one of these three conditions true:

$$a < b \qquad a = b \qquad a > b$$

The statements below are equivalent ways of saying that a is less than b:

$$a < b \qquad a - b < 0 \qquad b - a > 0 \qquad a \text{ is to the left of } b \qquad ●$$

If a is a real number, the symbol $|a|$ is used to denote the distance **Absolute** between 0 and a. This symbol, $|a|$, is read "the absolute value of a." Since **value** $|a|$ is simply the distance between two points, it is always positive or zero. In Fig. 1.2 we see that $|5| = 5$ and $|-5| = 5$ since 5 and -5 are both a distance of 5 from 0. Also $|6 - 1| = 5$ and $|1 - 6| = 5$. This illustrates the fact that

$|a - b|$ $|a - b| = $ the distance between a and b

In Sec. 1.2, where we formalize the properties of the real number system,

FIGURE 1.2

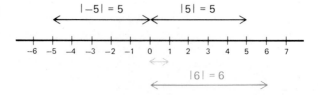

we will prove that $-(-a) = a$ for every real number a. With the help of this, we may write the algebraic definition of the absolute value of a as

$$|a| = \begin{cases} a & \text{if } a \geq 0 \\ -a & \text{if } a < 0 \end{cases}$$

Thus $|3| = 3$, $|-14.7| = -(-14.7) = 14.7$, and $|42 - 65| = |-23| = 23$.

The following properties of absolute value hold for all real numbers a and b:

$$|a| = |-a| \tag{1}$$

$$|ab| = |a| \cdot |b| \tag{2}$$

$$\left|\frac{a}{b}\right| = \frac{|a|}{|b|} \quad \text{if } b \neq 0 \tag{3}$$

$$|a + b| \leq |a| + |b| \tag{4}$$

$$|a - b| \geq |a| - |b| \tag{5}$$

$$-|a| \leq a \leq |a| \tag{6}$$

Triangle
inequality

We refer to (4) as the *triangle inequality*, and it is one of the most famous and useful inequalities in all of mathematics.

● **Example 6**   Verify (1) to (6) for $a = 8$ and $b = -4$.

Solution   (1) $|8| = 8$ and $|-8| = 8$

(2) $|(8)(-4)| = |-32| = 32$ and $|8| \cdot |-4| = 8(4) = 32$

(3) $\left|\dfrac{8}{-4}\right| = |-2| = 2$ and $\dfrac{|8|}{|-4|} = \dfrac{8}{4} = 2$

(4) $|8 + (-4)| = |4| = 4$ and $|8| + |-4| = 8 + 4 = 12$

(5) $|8 - (-4)| = |8 + 4| = |12| = 12$ and $|8| - |-4| = 8 - 4 = 4$

(6) $-|8| \leq 8 \leq |8|$, that is, $-8 \leq 8 \leq 8$ ●

In the triangle inequality, we will have equality if a and b are positive or both negative. If they have opposite signs, then we have a strict inequality.

Set
Element

One concept which is sometimes useful is the notion of a *set*. A set is a collection of things, called *elements*. The elements we use in this book will normally be either real numbers or points in the plane. There must be a criterion which enables us to decide whether the element a is in the set S:

$a \in S$ a belongs to the set S
$a \notin S$ a does not belong to S

We often use a capital letter to represent or stand for a set, and it can be described in either of two ways. In one method, we *list all elements* of the set and enclose them in braces { }. In the other method, we enclose a *descriptive phrase* of the elements in braces. For example, if W is the set of all names of days of the week, we may write

W = {Sunday, Monday, Tuesday, Wednesday, Thursday, Friday, Saturday}

or we may write

$$W = \{x \mid x \text{ is the name of a day of the week}\}$$

and read W is all x such that x is the name of a day of the week. The vertical line | is read "such that."

Equal sets   Two sets are *equal* if every element of each set is an element of the other set.

The equality of two sets does not require that their elements be arranged in the same order. Thus,

$$\{s, t, a, r\} = \{r, a, t, s\} = \{t, a, r, s\} = \{t, s, a, r\}$$

Subset   If each element of a set B is also an element of a set A, then B is a *subset* of A; furthermore, if in addition to each element of B being an element of
Proper subset A there are elements of A that are not elements of B, then B is a *proper subset* of A.

● **Example 7**   If $A = \{a, b, c, d, e\}$ and $B = \{a, c, e\}$, then B is a proper subset of A, since each element of B is also an element of A and there are elements of A that are not elements of B.

We indicate that B is a subset of A by writing $B \subseteq A$ and that B is a proper subset of A by the notation $B \subset A$.

If $B \subseteq A$, then each element of B is also an element of A; furthermore, if $A \subseteq B$, then each element of A is also an element of B. Consequently, we have

$A = B$ $A = B$ if and only if $A \subseteq B$ and $B \subseteq A$

It may happen that there are elements that are in both A and B without either A or B being a subset of the other. If that is the case, we say that those elements common to both A and B are their intersection.

Intersection   The *intersection* of two sets A and B is designated by $A \cap B$ and consists of those elements that belong to both A and B. The notation $A \cap B$ is read "the intersection of A and B." This definition may be succinctly stated as

$$A \cap B = \{x \mid x \in A \text{ and } x \in B\}$$ ●

Disjoint   If the intersection of two sets is empty (no elements in common), they
sets   are said to be *disjoint sets*. Thus, $A = \{2, 5, 8, 11\}$ and $B = \{1, 4, 7, 10\}$ are disjoint sets since they have no common elements.

Complement   The *complement* of the set B with respect to the set A is designated by
$A–B$ $A–B$ and $A–B = \{x \mid x \in A \text{ and } x \notin B\}$. If A is the universal set, we write
B' B' for $A–B$. Thus also $A–B = A \cap B'$.

Union

The set of elements that belong to S or to T or to both is designated by $S \cup T$, is called the *union* of S and T, and

$$S \cup T = \{x \mid x \in S \text{ or } x \in T \text{ or both}\}$$

● **Example 8**   If $A = \{2, 4, 6, 8, 10, 12\}$ and $B = \{3, 6, 9, 12\}$, then

$$A \cap B = \{6, 12\} \qquad\qquad A-B = \{2, 4, 8, 10\}$$
$$A \cup B = \{2, 3, 4, 6, 8, 9, 10, 12\} \qquad B-A = \{3, 9\}$$ ●

To recapitulate, the most important subsets of the real numbers that we have considered are

Natural numbers or
positive integers

$$\{1, 2, 3, 4, 5, \ldots\}$$

Integers

$$\{\ldots, -4, -3, -2, -1, 0, 1, 2, 3, 4, \ldots\}$$

Rational numbers

$\{m/n$ with m and n integers, $n \neq 0\}$ or
{decimals which either terminate or repeat}

Irrational numbers

{decimals which do not terminate or repeat}

Interval

Another important set is $\{x \mid a \leq x \leq b\}$, which is written $[a, b]$ and called the *closed interval* from a to b. The *open interval* is written $(a, b) = \{x \mid a < x < b\}$.

Exercise 1.1   A magic square is a square array of numbers in which every row and every column and both main diagonals have the same sum. Show that the following are magic squares:

1 8 3 4	**2** 10 11 6	**3** 16 5 4 9	**4** 40 30 33 27
1 5 9	5 9 13	2 11 14 7	49 11 56 14
6 7 2	12 7 8	13 8 1 12	32 38 25 35
		3 10 15 6	9 51 16 54

Verify the following products:

5 $1111 \times 1111 = 1{,}234{,}321$
6 $12{,}345 \times 11 = 135{,}795$ and $54{,}321 \times 11 = 597{,}531$ (note the pairs of reversed numbers)
7 $1{,}000{,}001 = 101 \times 9901$
8 $12 \times 483 = 5796$ (uses all digits from 1 to 9)

Verify these products, and notice that both members in each equation use precisely the same digits.

9 $8(473) = 3784$ **10** $15(93) = 1395$
11 $57(834) = 47{,}538$ **12** $86(251) = 21{,}586$

Show that the following numbers are prime by testing all integers from 1 to $\sqrt{n}$ as possible factors of n. Recall that $\sqrt{n} \cdot \sqrt{n} = n$.

13 $n = 73$ **14** $n = 97$ **15** $n = 127$ **16** $n = 163$

Factor these numbers into prime factors.

17 231   **18** 174   **19** 396   **20** 429

21 The number n is divisible by 4 if and only if its last two digits form a number divisible by 4. Test these numbers for divisibility by 4.
(a) 7124   (b) 88,554   (c) 12,345,678

22 The number n is divisible by 8 if and only if its last three digits form a number divisible by 8. Test these numbers for divisibility by 8.
(a) 12,474   (b) 66,666   (c) 765,432

23 The number n is divisible by 9 if and only if the sum of its digits is divisible by 9. Are these numbers divisible by 9?
(a) 67,463   (b) 2,323,233   (c) 98,577

24 A number is divisible by 11 whenever $x - y$ is a multiple of 11, where x is the sum of the first, third, fifth, etc., digits, and y is the sum of the second, fourth, sixth, etc., digits. Are these numbers divisible by 11?
(a) 21,825,364   (b) 90,939,497   (c) 7,436,912

The positive integer n is a prime if and only if $(n - 1)! + 1$ is a multiple of n. Verify this for the following values of n.

25 3   **26** 4   **27** 6   **28** 7

For greatest common divisors and least common multiples, it is always true that $(m, n) \cdot [m, n] = mn$. Verify this in Probs. 29 to 32.

29 $m = 8, n = 28$ **30** $m = 15, n = 33$
31 $m = 27, n = 63$ **32** $m = 42, n = 66$

Find the quotient and remainder.

33 $475 \div 7$ **34** $6821 \div 3$ **35** $4419 \div 13$ **36** $65,642 \div 4$
37 $6532 \div 481$ **38** $86,142 \div 47$
39 $10,032 \div 674$ **40** $91,119 \div 82$

In Probs. 41 to 44, let $A = \{1, 6/5, -5/6, -4, -\pi, 0, \sqrt{18}, 6.2, 6.222 \cdots\}$. Which numbers in A are:

41 Irrational   **42** Integers   **43** Negative numbers
44 Positive rationals

Perform the calculations in Probs. 45 to 48.

45 $|23| - |-21| + |-5 - 18|$ **46** $|64 - 31| - |-25 + 11| + |6|$
47 $|28| - |-20| + |8 + 1|$ **48** $|43 - 21| + |-6 - 8| - |7 - 5 + 3|$
49 $8 + |-4| - |6| - |-1|$ **50** $|2| - |6| - |2 - 6|$
51 $|3| - |4 + 3| + |4 + 2|$ **52** $|3 - 6| + |6 - 3| - |-4|$

Verify the triangle inequality [in Eq. (4)] for these values of a and b.

53 $a = 19, b = 3$ **54** $a = -16, b = -25$
55 $a = 27, b = -18$ **56** $a = -83, b = 44$

Find the sum and product of these numbers.

57 2.31 and 6.60   **58** 4.1 and 7.44   **59** 60.5 and 2.3   **60** 5.03 and 4.68

61 $\dfrac{(2n)!}{n!n!}$ is never a prime for any positive integer n. Verify this for $n = 4$ and $n = 5$.

62 Let P be the number of primes less than n. Show that P divides n for $n = 8$, $n = 30$, and $n = 33$.

63 If n is a positive integer, then $n^4/16$ always has a remainder of 0 or 1. Verify this for $n = 3$, $n = 4$, and $n = 5$. *Note:* $n^4 = n \cdot n \cdot n \cdot n$.

64 If the sum of all positive integer divisors of n (including 1 and n) is $2n$, we call n *perfect*. Show that $n = 6$ and $n = 28$ are perfect (496 is also perfect).

65 If a, b, and c are positive, then $a^2 + b^2 + c^2 + 3 \geq 2(a + b + c)$. Verify this for $a = 1.1$, $b = 1.2$, and $c = 1.3$. *Note:* $a^2 = a \cdot a$.

66 If x and y are positive, then $(x + y)(xy + 1) \geq 4xy$. Verify this for $x = 1.8$, $y = 1.9$.

67 If a, b, c, and d are positive, then $(a + b)(b + c)(c + d)(d + a) \geq 16abcd$. Verify this for $a = 3$, $b = 4$, $c = 5$, $d = 6$.

68 If $x + y = 2$, then $x^4 + y^4 \geq 2$. Verify this for $x = 0.8$, $y = 1.2$. *Note:* $x^4 = x \cdot x \cdot x \cdot x$.

69 If, x, y, and z are positive, and $x^2 + y^2 = z^2$, then $x^3 + y^3 < z^3$. Verify this for $x = 20$, $y = 21$, and $z = 29$. *Note:* $x^3 = x \cdot x \cdot x$.

70 Show that $|x - a| < d$ if and only if $a - d < x < a + d$.

71 Write $5 < x < 11$ in the form $|x - a| < b$. *Hint:* Try:

$$a = \frac{11 + 5}{2} \quad \text{and} \quad b = \frac{11 - 5}{2}$$

72 For any integer n, we have $n^3 - n$ always divisible by 6. Verify this for n equal to 7, 8, and 9.

73 Given a positive integer n, we will determine another one by using the following rule: If n is odd, we compute $3n + 1$, whereas if n is even, we compute $n/2$. Starting arbitrarily with $n = 6$ and using the rule repeatedly gives the numbers 6, 3, 10, 5, 16, 8, 4, 2, 1, 4, 2, 1 and the pattern 4, 2, 1 will keep on occurring. It is a long-standing conjecture that no matter which positive integer n we begin with, the pattern 4, 2, 1 will occur (and keep on occurring repeatedly). Try it yourself with your favorite positive integer n.

74 Consider the 10 numbers 0.06, 0.55, 0.77, 0.39, 0.96, 0.28, 0.64, 0.13, 0.88, 0.48, all in the interval (0, 1). Show by calculation that *each* of the following nine statements is true:

The first 2 numbers are in different halves of (0, 1).
The first 3 numbers are in different thirds of (0, 1).
The first 4 numbers are in different fourths of (0, 1).
The first 5 numbers are in different fifths of (0, 1).
. . . .
The first 10 numbers are in different tenths of (0, 1).

It is true (and fascinating to know) that the corresponding problem has a solution with 17 numbers, but there is no solution for 18 numbers (remember that *all* 17 statements would have to be true).

75 Show that the sum of the divisors of 220 (including 1 but not 220) is 284, and the sum of the divisors of 284 (including 1 but not 284) is 220.

1.2 Properties of the real numbers

In the first section of this chapter, we have used properties of real numbers such as multiplication, signs, and many more. In this section, we will present many of the axioms (assumptions) and theorems (consequences of these assumptions) of the real numbers. These are simply rules which tell us what we can and cannot do when working with numbers, and often how we may do some operation in the easiest and best way.

AXIOMS OF EQUALITY   E.1 The reflexive axiom: $a = a$.

E.2 The symmetric axiom: If $a = b$, then $b = a$.

E.3 The transitive axiom: If $a = b$ and $b = c$, then $a = c$.

E.4 The replacement axiom: If $a = b$, then a can be replaced by b in any mathematical statement without affecting the truth or falsity of the statement.

The first three axioms of equality need no detailed explanation here. The fourth will prove to be very useful in many situations, such as Example 1.

● **Example 1**   If we know that $2x + y = 11$ and $x = 3$, then we may replace x by 3 in the first equation. This gives $2(3) + y = 11$ or $6 + y = 11$; hence $y = 5$.
●

AXIOMS OF A FIELD   A.1 Closure axiom for addition: There exists a unique number $s \in R$ such that $a + b = s$.

A.2 Commutative axiom for addition: $a + b = b + a$.

A.3 Associative axiom for addition: $a + (b + c) = (a + b) + c$.

A.4 Identity element for addition: There exists an element $0 \in R$ such that $a + 0 = a$, for every $a \in R$.

A.5 Inverse element for addition: There exists a unique element $-a \in R$ such that $a + (-a) = 0$.

M.1 Closure axiom for multiplication: There exists a unique element $p \in R$ such that $a \times b = p$.

M.2 Commutative axiom for multiplication: $a \times b = b \times a$.

M.3 Associative axiom for multiplication: $a \times (b \times c) = (a \times b) \times c$.

M.4 Identity element for multiplication: There exists an element $1 \in R$ such that $a \times 1 = a$, for every $a \in R$.

M.5 Inverse element for multiplication: If $a \neq 0$, there exists an element $a^{-1} \in R$ such that $a \times a^{-1} = 1$.

AM.1 Distributive axiom: $a \times (b + c) = (a \times b) + (a \times c)$.

Field

A set S of elements is a *field* if there exist two binary operations $+$ and $\times$ (not necessarily the usual operations of addition and multiplication) such that all requirements of axioms A.1 to A.5, M.1 to M.5, and AM.1 are satisfied by the elements of S.

Reals are
a field

If $+$ and $\times$ denote the ordinary operations of addition and multiplication, then the elements of the set of *real numbers* R satisfy all the field axioms, and thus $(R, +, \times)$ is a field. Furthermore, since a rational number is a real number, axioms A.2 to A.5, M.2 to M.5, and AM.1 hold if a, b, and c are rational numbers. Moreover, since

$$\frac{a}{b} + \frac{c}{d} = \frac{ad + bc}{bd} \qquad \text{and} \qquad \frac{a}{b} \times \frac{c}{d} = \frac{ac}{bd}$$

Rationals
are a field

the sum and product of two rational numbers is rational. Hence the closure axioms A.1 and M.1 hold, and the set of *rational numbers* is therefore a field if we use ordinary addition and multiplication.

If, however, a is an integer other than 1, there is no integer k such that $a \times k = 1$. Hence the integer a has no inverse for $\times$ in the set of integers. Hence the set of integers is not a field.

The irrationals are not a field, since they do not contain 0.

In the field axioms, notice that we define the addition and multiplication of only two numbers at a time. Thus $1/3 + 2/7$ is a real number, and $(0.010101) \times (0.010010001 \cdots)$ is a real number. The commutative laws allow us to write the numbers in either order and get the same result.

If we want to add three numbers a, b, and c, we may first add a and b, then add c to the result, that is, $(a + b) + c$. We could also perform the addition $a + (b + c)$. The associative law assures us that the results are the same.

● **Example 2**

(a) $(2 + 5) + 9 = 7 + 9 = 16$ and $2 + (5 + 9) = 2 + 14 = 16$

(b) $(0.020202 \cdots + 0.121212 \cdots) + 0.353535 \cdots = 0.1414 \cdots + 0.3535 \cdots = 0.4949 \cdots$, and $0.0202 \cdots + (0.1212 \cdots + 0.3535 \cdots) = 0.0202 \cdots + 0.4747 \cdots = 0.4949 \cdots$

(c) $(6 \times 2) \times 5 = 12 \times 5 = 60$ and $6 \times (2 \times 5) = 6 \times 10 = 60$ ●

The commutative and associative laws combined enable us to say that when any two of the three numbers a, b, and c are added and then the third is added to this sum, the result is the same regardless of the way in which the first two numbers are chosen. Similar comments apply to multiplication. For instance,

$$(a + b) + c = b + (a + c)$$
$$c \times (a \times b) = (c \times b) \times a$$

The distributive axiom is the one which combines addition and multiplication. It will be used extensively in multiplying and in factoring not only sums of real numbers but also algebraic expressions later in the book. It may also be written as

$$(a + b) \times c = (a \times c) + (b \times c)$$

● **Example 3** $5(6 + 7) = 5(13) = 65$ and $(5 \times 6) + (5 \times 7) = 30 + 35 = 65$
$4t + 4u = 4(t + u)$
$(4 + 7) \times 3 = 11 \times 3 = 33$ and
$\qquad\qquad\qquad (4 \times 3) + (7 \times 3) = 12 + 21 = 33$ ●

We have discussed so far the closure, commutative, associative, and distributive axioms. Of the remaining four, which treat inverses and identities, special attention must be paid to M.5. It assures us of a

Note   multiplicative inverse for every real number with one exception. *The additive identity 0 has no multiplicative inverse.*

In a field, the operations defined are addition and multiplication. The other two fundamental operations, subtraction and division, for the real numbers are defined with the help of the inverse axioms A.5 and M.5. Specifically, we define

Subtraction $a - b = a + (-b)$ for any b (1.2)
Divison $a \div b = a \times (b^{-1})$ if $b \neq 0$ (1.3)

Note   Other notations for division are $a \div b$ or a/b or $a(1/b)$. Subtraction and division are defined for only two numbers at a time. Neither the commutative nor associative axiom holds for them. For example,

$8 - 5 = 3$ but $5 - 8 = -3$
$8 \div 4 = 8(4^{-1}) = 8(1/4) = 2$ but $4 \div 8 = 4(8^{-1}) = 4(1/8) = 4/8 = 1/2$
$15 - (7 - 3) = 15 - 4 = 11$ but $(15 - 7) - 3 = 8 - 3 = 5$
$(40 \div 20) \div 5 = 2 \div 5 = 2/5$ but $40 \div (20 \div 5) = 40 \div 4 = 10$

We will now present a number of results which help us calculate and deal with real numbers.

$$-(-a) = a \qquad\qquad\qquad\qquad (1.4)$$
$$-(a + b) = -a - b \qquad\qquad\qquad (1.5)$$
$$-(a - b) = -a + b \qquad\qquad\qquad (1.6)$$
$$a(0) = 0 \text{ for every } a \qquad\qquad (1.7)$$
$$a(-b) = -(ab) \qquad\qquad\qquad\quad (1.8)$$
$$(-a)(-b) = ab \qquad\qquad\qquad\quad (1.9)$$
$$a/b = c \text{ if and only if } a = bc \ (b \neq 0) \qquad (1.10)$$
$$a \div 0 \text{ is not defined for any } a \qquad (1.11)$$
$$0 \div b = 0 \text{ for every } b \neq 0 \qquad (1.12)$$

The proof of some of these will be given below. Some other proofs are requested in the exercise. We need first to give the cancellations laws for addition and multiplication.

If a and c are real numbers, then $a + c$ is a real number by closure of addition. Hence $a + c = a + c$ by E.1. Now if $a = b$, then by E.4 we have $a + c = b + c$. We have thus shown that

$$\text{If } a = b, \text{ then } a + c = b + c$$

Cancellation law for addition

The *cancellation law for addition* is the converse of this, namely,

$$\text{If } a + c = b + c, \text{ then } a = b$$

To prove it we first suppose $a + c = b + c$. Then

$$
\begin{aligned}
a + c &= b + c & &\text{given} \\
(a + c) + (-c) &= (b + c) + (-c) & &\text{adding } -c \\
a + (c - c) &= b + (c - c) & &\text{associative axiom} \\
a + 0 &= b + 0 & &c - c = c + (-c) = 0 \\
a &= b & &0 + x = x \text{ for all } x
\end{aligned}
$$

In a similar manner we may show that

$$\text{If } a = b, \text{ then } ac = bc$$

Cancellation law for multiplication

and also the *cancellation law for multiplication;* namely,

$$\text{If } ac = bc \text{ and } c \neq 0, \text{ then } a = b$$

We will now prove that $-(-a) = a$:

Proof of (1.4)

$$
\begin{aligned}
0 &= (-a) + [-(-a)] & &\text{additive inverse for } -a \\
0 &= a + (-a) = (-a) + a & &\text{commutative axiom} \\
(-a) + [-(-a)] &= (-a) + a & &\text{by E.3} \\
-(-a) &= a & &\text{cancellation law}
\end{aligned}
$$

Proof of (1.5)

To prove that $-(a + b) = -a - b$, we first write

$$0 = [a + b] + [-(a + b)]$$

and also

$$
\begin{aligned}
(a + b) + (-a - b) &= (a + b - a) - b \\
&= (a - a + b) - b = (0 + b) - b \\
&= b - b = 0
\end{aligned}
$$

Since both are equal to 0, we now have

$$
\begin{aligned}
(a + b) + [-(a + b)] &= (a + b) + (-a - b) \\
-(a + b) &= -a - b & &\text{cancellation law}
\end{aligned}
$$

Proof of (1.7)

We may show that any number multiplied by 0 is 0 by writing

$$
\begin{aligned}
0 &= 0 + 0 & &\text{definition of 0} \\
a(0) &= a(0 + 0) & &\text{multiplication by } a \\
a(0) + 0 &= a(0) + a(0) & &\text{distributive law and A.4} \\
0 &= a(0) & &\text{cancellation law for addition}
\end{aligned}
$$

Proof of (1.9)

To show that $(-a)(-b) = ab$, we first write

$$a(-b) + (-a)(-b) = (a - a)(-b) = 0(-b) = 0$$

and then

$$a(-b) + ab = a(-b + b) = a(0) = 0$$

Hence

$$
\begin{aligned}
a(-b) + (-a)(-b) &= a(-b) + ab & &\text{both are 0} \\
(-a)(-b) &= ab & &\text{cancellation law}
\end{aligned}
$$

We show, finally, why division by 0 is never defined. If it were defined and we let $b = 0$, then $a \div b = c$ would mean $a = bc$. We are looking for a value of c for which $a = 0(c)$ makes sense. Suppose first $a = 0$. Then

$$a = 0(c) \qquad \text{becomes} \qquad 0 = 0(c)$$

and the last equation is true for *every value* of c, since $0(c) = 0$. Suppose now that $a \neq 0$. Then we have

$$a = 0(c)$$

Never divide by 0

which is true for *no value* of c, since $a \neq 0$ but $0(c) = 0$. In either situation, we run into trouble. Hence we simply *never divide by 0*.

We know that 25 is larger than 19, and both are positive. This is written $25 > 19$ or $19 < 25$, and $25 > 0$ and $0 < 19$. We will now state one final set of axioms for the real numbers, the *order axioms*.

Order axioms

Closure: If $a > 0$ and $b > 0$, then $a + b > 0$ and $ab > 0$.
Trichotomy: For every a, either $a > 0$ or $-a > 0$ or $a = 0$.
Transitive law: If $a < b$ and $b < c$, then $a < c$.

The closure property says that the product of positive numbers is positive. Since $(-a)(-b) = ab$, the product of two negative numbers is also positive. However, the product of a positive and a negative number is negative, since $a(-b) = -(ab)$. To summarize:

Sign of a product

The product ab is *positive* if a and b have the *same* sign.
The product ab is *negative* if a and b have *opposite* signs.

Sign of a quotient

Since $a \div b = c$ means $a = bc$, we may use the above rule to show that the quotient a/b is positive (negative) if a and b have the same (opposite) sign.

● **Example 4**
$$4(5) = 20 \qquad 4(-5) = -20 \qquad (-4)(-5) = 20$$
$$6/(-3) = -2 \qquad (-6)/(-3) = 2 \qquad (-6)/3 = -2$$ ●

The rule for the sign of a sum, or the sign of a difference, is not quite so straightforward. First, recall that by definition (1.2),

$$a - b \text{ means } a + (-b)$$

and so we only need to give the rule for sums. For instance, $(8) - (5) = (8) + (-5)$ and $(-16) - (23) = (-16) + (-23)$ and $(5) - (-24) = (5) + (24)$.

Sign of a sum

To add two numbers with the same signs, we add their absolute values, then attach the common sign.

Sign of a difference

To add two numbers with opposite signs, we subtract their absolute values (larger minus smaller), then attach the sign of the original number which had the greater absolute value.

● **Example 5**
$$(16) + (27) = 16 + 27 = 43$$
$$(-28) + (-51) = -(28 + 51) = -79$$
$$(48) + (-27) = +(|48| - |-27|) = 48 - 27 = 21$$
$$(30) + (-62) = -(|-62| - |30|) = -(62 - 30) = -32$$ ●

When we write $a \leq b$, we mean either $a < b$ or $a = b$. Also $b \geq a$ means the same as $a \leq b$.

● **Example 6** $8 < 10$ $8 \leq 10$ $8 \leq 8$ $6 \geq -3$ and $-5 > -9$ ●

Note   Operating with inequality signs requires care. For example, it is *not* true that if $a < b$ and $c < d$, then $ac < bd$; for instance, $-5 < -2$ and $1 < 3$, but $(-5)(1) > (-2)(3)$, since $-5 > -6$.
Some rules for inequalities are given next.

If $a < b$, then $a + c < b + c$	(1.13)
If $a < b$ and $c > 0$, then $ac < bc$	(1.14)
If $a < b$ and $c < 0$, then $ac > bc$	(1.15)
If $0 < a < b$, then $1/a > 1/b$	(1.16)

In words, (1.15) says that the sense (direction) of an inequality is reversed if both members are multiplied by a negative number.

● **Example 7**   Since $5 < 9$, we have $5 + 10 < 9 + 10$, or $15 < 19$.
Since $6 < 11$ and $4 > 0$, we have $6(4) < 11(4)$, or $24 < 44$.
Since $-3 < 2$ and $-5 < 0$, then $-3(-5) > 2(-5)$, or $15 > -10$.
Since $0 < 2 < 3$, we have $1/2 > 1/3$. ●

The integers modulo *m*

One of the best ways to truly understand the field axioms is to use them in an unfamiliar situation. We shall actually write addition and multiplication tables in two cases, $m = 7$ and $m = 4$, but the same thing may be done with any positive integer m. It is proved in the theory of numbers that *the integers modulo m is a field if and only if m is a prime.*

We shall now discuss certain finite sets of real numbers and two operations, $\oplus$ and $\otimes$, that are not the usual operations of addition and multiplication. As an example, the remainder when 26 is divided by 7 is 5. This remainder 5 is called the *residue* of 26 modulo 7. Similarly, since $38 = (5 \times 7) + 3$, the residue of 38 modulo 7 is 3. In fact, any integer can be expressed in the form $7n + r$, where n is an integer and r is an element of the set

$$S = \{0, 1, 2, 3, 4, 5, 6\}$$

Residue set   Hence the residue of any specified integer modulo 7 is an element of the set S, and we call S the *complete residue set* modulo 7.

We will now define the operation $\oplus$ and $\otimes$ as follows: If $a \in S$ and $b \in S$, then

$a \oplus b$ $a \oplus b$ is the residue of $a + b$ modulo 7

$a \otimes b$ $a \otimes b$ is the residue of $a \times b$ modulo 7

Integers modulo 7   The set $S = \{0, 1, 2, 3, 4, 5, 6\}$, together with the addition $\oplus$ and the multiplication $\otimes$, is called the *integers modulo* 7. Notice that S has 7 elements.

● **Example 8**

$$3 \oplus 6 = 2 \text{ since } 3 + 6 = 9 = (1 \times 7) + 2$$
$$1 \oplus 3 = 4 \text{ since } 1 + 3 = 4 = (0 \times 7) + 4$$
$$5 \oplus 2 = 0 \text{ since } 5 + 2 = 7 = (1 \times 7) + 0$$
$$6 \oplus 0 = 6 \text{ since } 6 + 0 = 6 = (0 \times 6) + 6$$
$$6 \otimes 5 = 2 \text{ since } 6 \times 5 = 30 = (4 \times 7) + 2$$
$$5 \otimes 3 = 1 \text{ since } 5 \times 3 = 15 = (2 \times 7) + 1$$
$$6 \otimes 0 = 0 \text{ since } 6 \times 0 = 0 = (0 \times 7) + 0$$

●

We will refer to $a \oplus b$ as the *residue sum* of a and b and $a \otimes b$ as the *residue product*.

Residue sums modulo 7

$\oplus$	0	1	2	3	4	5	6
0	0	1	2	3	4	5	6
1	1	2	3	4	5	6	0
2	2	3	4	5	6	0	1
3	3	4	5	6	0	1	2
4	4	5	6	0	1	2	3
5	5	6	0	1	2	3	4
6	6	0	1	2	3	4	5

Residue products modulo 7

$\otimes$	0	1	2	3	4	5	6
0	0	0	0	0	0	0	0
1	0	1	2	3	4	5	6
2	0	2	4	6	1	3	5
3	0	3	6	2	5	1	4
4	0	4	1	5	2	6	3
5	0	5	3	1	6	4	2
6	0	6	5	4	3	2	1

S is a field   If we refer to the above tables, we see that 0 is the identity element for $\oplus$ and 1 is the identity element for $\otimes$, since if $a \in S$, then $a \oplus 0 = a$ and $a \otimes 1 = a$. Furthermore, since 0 appears in each line of the first table, each element of S has an inverse under $\oplus$. Since 1 appears in each line of the second table, except for the line of zeros, each nonzero element of S has an inverse element under the operation $\otimes$. The closure axioms hold, since for each two elements in S there exist in S a unique residue sum and a unique residue product. By use of methods illustrated in the following examples, it can be verified that all the other field axioms hold, and therefore S is a field with the operations $\oplus$ and $\otimes$.

● **Example 9**   (*a*) 3^{-1} is the number *a* for which $3 \otimes a = 1$. Since $3 \otimes 5 = 1$, then 3^{-1}
= 5. We may write $3^{-1} = 1/3 = 1 \div 3$.
(*b*) -3 is the number *b* such that $3 \oplus b = 0$. Since $3 \oplus 4 = 0$, we have
$-3 = 4$.   ●

● **Example 10**   Show that $(2 \oplus 4) \oplus 6 = 2 \oplus (4 \oplus 6)$.

Solution
$$(2 \oplus 4) \oplus 6 = 6 \oplus 6 = 5$$
$$2 \oplus (4 \oplus 6) = 2 \oplus 3 = 5 \qquad \text{since } 4 \oplus 6 = 3$$
●

● **Example 11**   Show that $(2 \otimes 4) \otimes 6 = 2 \otimes (4 \otimes 6)$.

Solution
$$(2 \otimes 4) \otimes 6 = 1 \otimes 6 \qquad \text{since } 2 \otimes 4 = 1$$
$$= 6$$
$$2 \otimes (4 \otimes 6) = 2 \otimes 3 \qquad \text{since } 4 \otimes 6 = 3$$
$$= 6$$
●

● **Example 12**   Show that $(2 \oplus 4) \otimes 6 = (2 \otimes 6) \oplus (4 \otimes 6)$.

Solution
$$(2 \oplus 4) \otimes 6 = 6 \otimes 6 = 1$$
$$(2 \otimes 6) \oplus (4 \otimes 6) = 5 \oplus 3 = 1$$
●

Integers modulo 4   The complete set of residues modulo 4 is
$$T = \{0, 1, 2, 3\}$$

We show below the tables of residue sums and residue products in *T*.

$\oplus$	0	1	2	3
0	0	1	2	3
1	1	2	3	0
2	2	3	0	1
3	3	0	1	2

$\otimes$	0	1	2	3
0	0	0	0	0
1	0	1	2	3
2	0	2	0	2
3	0	3	2	1

By use of the above tables, it can be verified that all the field axioms hold in the set *T* for $\oplus$ and $\otimes$ except axiom M.5. The number 1 does not appear in the fourth line of the second table, and therefore 2 does not have an *T* is not a field   inverse under $\otimes$. Hence *T is not a field*.

● **Example 13**   To illustrate the distributive law, we have
$$3 \otimes (2 \oplus 3) = 3 \otimes 1 = 3$$
and
$$(3 \otimes 2) \oplus (3 \otimes 3) = 2 \oplus 1 = 3$$
●

Exercise 1.2   Are the statements in Probs. 1 to 24 true or false?

1 The sum of two positive integers is a positive integer.
2 $8 - 3 = -3 + 8$ is an example of the commutative axiom.
3 $6 = 6$ is an example of the reflexive axiom.
4 $4(5 \cdot -6) = (4 \cdot 5)(-6)$ is an example of the associative axiom.
5 $8 \cdot 0 = 0$, by the additive identity axiom.
6 The integers contain a multiplicative identity.
7 $-\pi + 0 = -\pi$, by the additive inverse axiom.
8 By the associative axiom, $(1 + 2) + 3 = 1 \cdot (2 \cdot 3)$.
9 "If $x + 1 = y$ and $x = 3$, then $y = 4$" is true by the replacement axiom.
10 The odd integers have closure under addition.
11 $1 + (1^{-1}) = 0$ by the inverse axiom for addition.
12 $(x - 3) \cdot (x - 3)^{-1} = 1$ for all $x \neq 3$.
13 $\{0, 1\}$ is closed under multiplication.
14 $\{-1, 0, 1\}$ is closed under multiplication.
15 $\{x \mid 0 \leq x \leq 1\}$ is closed under multiplication.
16 $\{0, -1\}$ is closed under multiplication.
17 If $a > b$ and $b > c$ and $d > 0$, then $ad > dc$.
18 If $a > b$, then $a - c > b - c$.
19 $(5^{-1})^{-1} = -5$
20 $(-5)^{-1} = -(5^{-1})$
21 If $a + b = -c + a$, then $a = -c$.
22 If $a + b = 0$, then $-(-a) = -b$.
23 If $(a + c)b = ab + d$, then $bc = d$.
24 If $a + b > c + d$, then $a - c > d - b$.

Perform the following calculations:

25 $8 - 3(-6) + 1(-4)$ **26** $4(3 - 8) - 5(-1 - 2)$
27 $-(-8) - 8(1)$ **28** $4(-2)(1) + (-3)(-1)(-2)$
29 $4(6) - 4(3) + 4(-3)$ **30** $6(5 - 4) + 3(2 - 1) - 0(1 + 0)$
31 $1 + 0 - 1 - 0 + 1/0$ **32** $6 - 4 - 2 - (6 - 4 + 2)$

Prove the following statements:

33 $-(a - b) = -a + b$ **34** $a(-b) = -(ab)$
35 $0 \div b = 0$, if $b \neq 0$ **36** $(b^{-1})^{-1} = b$, if $b \neq 0$
37 If $ab = 0$ and $a \neq 0$, then $b = 0$.
38 If $a > b > 0$ and $c > d > 0$, then $ac > bd$.
39 If $a < b < 0$, then $a \cdot a > b \cdot b$.
40 If $0 < a < 1$, then $a \cdot a < a$.

List the numbers in order from smallest to largest in Probs. 41 to 44.

41 $6.2, 2.6, 12/5, \sqrt{2}, 6/2$ **42** $-4, 5, -6, 7, -8$
43 $|4 - 2|, |3 - 8|, -|-4|, -1 - 2, 1 - (-2)$
44 $-|2 - 5|, -(-2) - |-2|, -1.4, -\pi, 6 - |3|$

It is true for all positive numbers x, y, and z that

$$(x + y)(y + z)(z + x) \geq 8xyz$$

Verify this for the following values of x, y, and z:

45 $x = 2$, $y = 3$, $z = 4$ **46** $x = 1$, $y = 3$, $z = 5$

Verify $a^2 + b^2 + c^2 \geq |ab + bc + ca|$ in the following cases:

47 $a = 2$, $b = -3$, $c = -2$ **48** $a = 3$, $b = 4$, $c = 3$

Verify the inequality $(a + b + c)^2 \leq 3(a^2 + b^2 + c^2)$ for these cases:

49 $a = 2$, $b = 2$, $c = 1$ **50** $a = 4$, $b = 3$, $c = 2$

Verify the inequality $(ab + cd)^2 \leq (a^2 + c^2)(b^2 + d^2)$ in the following cases:

51 $a = 1$, $b = 2$, $c = 3$, $d = 2$ **52** $a = 3$, $b = 4$, $c = 3$, $d = 2$

Perform the calculations in Probs. 53 to 60 in the integers modulo 7.

53 $4 \oplus 3 \oplus 6 \oplus 2$ **54** $4 \otimes 3 \otimes 6 \otimes 2$ **55** $(4 \oplus 5) \otimes (6 \oplus 3)$
56 $(2 \oplus 1) \otimes (3 \oplus 4)$ **57** Find -5 and 5^{-1} **58** Find -6 and 6^{-1}.
59 Find $(-4)^{-1}$ and $-(4^{-1})$.   **60** Find $2/3$ $[= 2 \otimes (3^{-1})]$.

Perform the calculations in Probs. 61 to 69 in the integers modulo 4.

61 $3 \oplus 2 \oplus 1$ **62** $2 \otimes 2 \otimes 3$
63 $3 \otimes (2 \oplus 1)$ **64** $2 \oplus (3 \otimes 3)$
65 Find -3.   **66** Find -2.
67 Find 1^{-1}.   **68** Find 3^{-1}.
69 Verify that $(x \oplus y)^4 = x^4 \oplus y^4 \oplus 2x^2y^2$ for $x = 3$ and $y = 2$.

70 Show that the sum of two rational numbers is rational.
71 Show that the sum of a rational number and an irrational number is irrational. *Hint:* Let b be rational and c be irrational, and suppose that $b + c$ is rational; show that this contradicts Prob. 70 by considering $b + (-b - c)$.
72 (a) Show that the sum of two irrationals can be irrational by using $5 + \sqrt{2}$ and $-5 + \sqrt{2}$.
    (b) Show that the sum of two irrationals can be rational by using $7 - \sqrt{3}$ and $7 + \sqrt{3}$.

1.3   Positive integer exponents

We have used a^2 for the product $a \cdot a$ already. Furthermore, we will write

$$a^3 \text{ for } a \cdot a \cdot a \quad \text{and also} \quad a^4 \text{ for } a \cdot a \cdot a \cdot a$$

In general if n is a positive integer and a is a real number, we use the notation

$$a^n = a \cdot a \cdot a \cdots a \qquad \text{where there are } n \text{ factors of } a \qquad (1.17)$$

Base
Exponent

The number a is called the **base,** and the positive integer n is called the **exponent.** In particular $a^1 = a$.

● **Example 1**

$$5^3 = 5 \cdot 5 \cdot 5 = 125$$
$$3^5 = 3 \cdot 3 \cdot 3 \cdot 3 \cdot 3 = 243$$
$$(-2)^4 = (-2)(-2)(-2)(-2) = 16$$
$$(-7)^3 = (-7)(-7)(-7) = -343$$
$$(2a)^3 = (2a)(2a)(2a) = 8 \cdot a^3 = 8(a^3)$$

●

Notice the difference between $(2a)^3$ and $2a^3$. As in Example 1, $(2a)^3 = 8 \cdot a^3 = 8(a^3)$, whereas $2a^3 = 2 \cdot a^3 = 2(a^3)$.

Similarly the notation $-5a^4$ means $(-5)(a^4)$ and

$$(-5a)^4 = (-5a)(-5a)(-5a)(-5a) = 625a^4 = 625(a^4)$$

Our next example illustrates a general theorem.

● **Example 2**

$$a^3 \cdot a^5 = (a \cdot a \cdot a)(a \cdot a \cdot a \cdot a \cdot a)$$
$$= a \cdot a \cdot a \cdot a \cdot a \cdot a \cdot a \cdot a = a^8$$

●

We will now consider the product $a^m a^n$. By definition this product is equal to the product of m a's and n a's, and this is the product of $m + n$ a's. Hence we have

$$a^m a^n = a^{m+n} \qquad \text{where } m \text{ and } n \text{ are positive integers} \qquad (1.18)$$

In words, **we add exponents when the bases are the same.**

By the definition of the nth power of a number and the use of the commutative axiom, we can prove that

$$a^n b^n = (ab)^n \qquad \text{where } n \text{ is a positive integer} \qquad (1.19)$$

Proof

$$(ab)^n = ab \cdot ab \cdot ab \cdots ab \text{ to } n \text{ factors}$$
$$= (a \cdot a \cdot a \cdots a \text{ to } n \text{ factors})(b \cdot b \cdot b \cdots b \text{ to } n \text{ factors})$$
$$\qquad\qquad\qquad\qquad\qquad\qquad \text{by commutative axiom}$$
$$= a^n b^n \qquad \text{by definition of } n\text{th power}$$

In words, **we multiply bases when the exponents are the same.**

Note that the above two laws treat $a^m b^n$ in the two cases when the bases are equal ($a = b$) and when the exponents are equal ($m = n$).

Finally we prove that

$$(a^n)^m = a^{mn} \qquad \text{where } m \text{ and } n \text{ are positive integers} \qquad (1.20)$$

Proof

$$(a^n)^m = a^n \cdot a^n \cdot a^n \cdots a^n \text{ to } m \text{ factors}$$
$$= a^{n+n+n+\cdots+n} \text{ to } m \text{ addends of } n \qquad \text{by } (1.18)$$
$$= a^{mn}$$

In words, **for a power of a power, we multiply exponents.**

For division, we have a law similar to (1.19):

$$\left(\frac{a}{b}\right)^n = \frac{a^n}{b^n} \tag{1.21}$$

We will treat a^m/a^n for all cases of m and n in Chap. 2. For now, we will simply use the fact that

$$\frac{a^m}{a^n} = a^{m-n} \qquad \text{if } m > n \tag{1.22}$$

This follows by cancellation since, for example,

$$\frac{3^7}{3^5} = \frac{3 \cdot 3 \cdot 3 \cdot 3 \cdot 3 \cdot 3 \cdot 3}{3 \cdot 3 \cdot 3 \cdot 3 \cdot 3} = 3 \cdot 3 = 3^2 = 3^{7-5}$$

In Chap. 4, we will see that $a^0 = 1$ and $a^{-n} = 1/a^n$; these definitions will be made in order to be consistent with the laws we have in this section.
Furthermore using Eq. (1.19) gives

$$a^{m-n} \cdot a^n = a^{m-n+n} = a^m$$

which is clearly equivalent to Eq. (1.22) if we simply divide both sides by a^n.

We will illustrate the application of the above laws with several examples.

Example 3 $3^2 3^3 = 3^5 = 243$ by (1.18)

Example 4 $7^5/7^3 = 7^2 = 49$ by (1.22)

Example 5 $(2^3)^2 = 2^6 = 64$ by (1.20)

Example 6 $(2/3)^4 = 2^4/3^4 = 16/81$ by (1.21)

Example 7 $(2^3 3^2)^2 = 2^6 3^4 = (64)(81) = 5184$ by (1.19) and (1.20)

Example 8 $(3x^2 y^3)(4xy^5) = 3 \cdot 4x^2 xy^3 y^5$ by the commutative law

$\qquad\qquad\qquad\quad = 12x^3 y^8$ by (1.18)

Example 9 $\dfrac{15x^6 y^5}{3x^3 y^4} = 5x^3 y$ by (1.22)

Example 10 $(4x^2 y^3 z^7)^3 = 4^3 (x^2)^3 (y^3)^3 (z^7)^3$ by (1.19)

$\qquad\qquad\qquad\quad = 64x^6 y^9 z^{21}$ by (1.20)

Example 11 $\left(\dfrac{5x^5 y^8}{3z^4 w^2}\right)^3 = \dfrac{5^3 (x^5)^3 (y^8)^3}{3^3 (z^4)^3 (w^2)^3}$ by (1.21) and (1.19)

$\qquad\qquad\qquad\quad = \dfrac{125x^{15} y^{24}}{27z^{12} w^6}$ by (1.20)

Exercise 1.3   Simplification of exponential expressions

Perform the indicated operations.

1 $3^2 3^4$	**2** $2^3 2^4$	**3** $5^3 5^2$	**4** $4^3 4$
5 $7^4/7$	**6** $5^5/5^3$	**7** $8^2/8^0$	**8** $4^5/4^2$
9 $(3^2)^2$	**10** $(2^3)^2$	**11** $(3^2)^3$	**12** $(5^0)^5$
13 $(2/3)^5$	**14** $(5/2)^3$	**15** $(2/7)^2$	**16** $(3/4)^3$
17 $(2^3 3^2)^2$	**18** $(5^2 2^3)^3$	**19** $(3^4 5^2)^2$	**20** $(2^2 3^3)^3$
21 $(x^2)(-x^3)$	**22** $(x^3)(-x)^4$	**23** $(-x^4)(x^5)$	**24** $(-x^2)(-x^5)$
25 $(3x^2)(-2x^3)$	**26** $(-4x^3)(3x^4)$	**27** $(-2x^5)(-5x^3)$	**28** $(3x^3)(5x^5)$

29 $(2x^3y)(3x^4y^3)$ **30** $(-3x^2y^3)(5x^0y^2)$ **31** $(-6x^2y^2)(-3xy^3)$

32 $(7x^7y)(-3xy^3)$ **33** $(-x^2)^3$ **34** $(-x^3)^2$

35 $(-x^3)^4$ **36** $(-x^3)^3$ **37** $(2x)^4$

38 $(-3x^3)^3$ **39** $(-5x^4)^3$ **40** $(-5x^3)^4$

41 $(2x^2y)^5$ **42** $(-3x^3y)^3$ **43** $(-5x^4y^2)^2$

44 $(4x^2y^3)^4$ **45** $(3x^2y)(2xy^3)$ **46** $(-7x^2y^3)(2x^3y^4)$

47 $(2x^4y^2)(-3x^3y^5)$ **48** $(-2x^2y^5)(-3x^3y^0)$ **49** $\dfrac{12x^5y^4}{3x^2y^3}$

50 $\dfrac{27x^6y^7}{3x^3y^4}$ **51** $\dfrac{18x^4y^3}{3x^3y^2}$ **52** $\dfrac{24x^5y^6}{6x^3y^4}$

53 $(2a^2bc^3)^3$ **54** $(3a^3b^2c^4)^2$ **55** $(5ab^2c^4)^4$

56 $(7a^3b^2c^4)^3$ **57** $\left(\dfrac{a^3b^2}{2c^3d^4}\right)^2$ **58** $\left(\dfrac{a^4b^3c}{2d^2}\right)^3$

59 $\left(\dfrac{2x^4y^3}{3w^2z}\right)^2$ **60** $\left(\dfrac{3x^3y^5}{2p^2q^3}\right)^4$ **61** $(2a^2b^3)^2(3ab^0)^4$

62 $(3a^3b)^3(2a^2b^3)^2$ **63** $(5cd^2)^2(2c^2d^3)^3$ **64** $(3x^2y^3)^3(5x^2y)^4$

65 $\dfrac{2u^2v^3}{3w^4}\cdot\dfrac{6w^2u}{8u^2v^2}$ **66** $\dfrac{15b^2c^5}{16d^3}\cdot\dfrac{4b^4d^3}{5c^6}$ **67** $\dfrac{3x^2y^3}{7x^4z^2}\cdot\dfrac{28y^3}{33x^4z^3}$

68 $\dfrac{7b^4c^3}{8d^2}\cdot\dfrac{16b^3d^3}{21c^4d^5}$ **69** $\left(\dfrac{c^4d^3}{a^4}\right)^2\left(\dfrac{a^3}{c^2d}\right)^3$ **70** $\left(\dfrac{3a^2}{b^3c^0}\right)^4\left(\dfrac{bc^2}{9a^4}\right)^2$

71 $\left(\dfrac{4a^2b^5}{c^3}\right)^2\left(\dfrac{c^4}{12a^3b}\right)^3$ **72** $\left(\dfrac{6a^4}{b^2c^5}\right)^3\left(\dfrac{bc^2}{2a^3}\right)^4$ **73** $\dfrac{x^{3a+1}y^{a+3}}{x^{a+2}y^{a-1}}$

74 $\dfrac{(a^{b+3}d^{b+2})^3}{a^{3b}d^6}$ **75** $\dfrac{(a^{3n+1}b^{2n-1})^4}{a^{6n+4}b^{2n-5}}$ **76** $\dfrac{(a^{n+2}b^{2n+1})^p}{(a^{p+1}b^{2p-2})^n}$

Management   **77** If there are n levels of organization and the span of control is b, then the average time in days required to make a decision is

$$T = \frac{2b}{b-1} - \frac{2n}{b^n - 1}$$

Calculate T if $n = 4$ and $b = 3$.

Anthropology   **78** In the statistical studies of migration, the relatedness between any pair of populations in this system after t generations is

$$r = 1 - \left(\frac{Ny - 1}{N - 1}\right)^t$$

where $N = $ the number of populations and $0 < y < 1$. Find r for $N = 12$, $y = 1/4$, and $t = 8$.

Earth science   **79** In working with the atomic scattering factor for hydrogen, we need the expression

$$I = \frac{2ab}{(a^2 + b^2)^2}$$

Find I if $b = 3a$.

Psychology   **80** In a learning experiment with probability p in the nth trial, the average
and calculator   latency is

$$L = \frac{k}{1 - (1 - p)(1 - b)^{n-1}}$$

Calculate L if $k = 1.65$, $p = 0.11$, and $b = 0.10$ for $n = 4$ and then for $n = 12$.

Anthropology   **81** The diffusion of water into obsidian is governed by the equation $M^2 = Kt$, where M is the depth in micrometers, t is the time in years, and $K = 0.00354$. What is t if $M = 1.62$?
and calculator

1.4 Combining algebraic expressions

The expression πr^2 gives the area of a circle whose radius is r. The irrational number π has a well-known specific value, approximately 3.1415927, and we call π a constant. The symbol or letter r may theoretically take on any value which is a positive real number, and we call r a variable.

Constant   This illustrates the following definitions. If a symbol may take on only
Variable   one value, it is called a *constant*. If a symbol may take on more than one value (usually many values), it is called a *variable*. The result of applying the four fundamental operations of addition, subtraction, multiplication, and division to a collection of constants and variables is called an *algebraic*
Algebraic   *expression*. In Chap. 4 and later, we will also allow the extraction of roots.
expression   Special types of algebraic expressions called polynomials will be treated extensively in Chap. 2.

● **Example 1**   The following are algebraic expressions:

$$a + 3b \qquad 32a^2 - 5xy \qquad \frac{2p}{1 + 5q}$$

$$1 - \frac{v^2}{c^2} \qquad 4.6pv^2 \qquad (1 - x^3)^2(1 + x^2)^3$$

Value   If specific real numbers are substituted for the variables in an algebraic expression, the real number obtained is called the *value* of the expression (for those specific real numbers). For example, the value of

$$32a^2 - 5bc \qquad \text{when } a = 2, b = 4, c = 6$$

is found by substitution, and it is

$$32(2^2) - 5(4)(6) = 32(4) - 120 = 128 - 120 = 8$$

Note   Since algebraic expressions are symbols which represent real numbers, we may use any and all of the rules governing real numbers. In particular, the distributive law as well as $(-a)(-b) = ab$ and $-(a - b) = -a + b$ are very useful.

Inserting symbols of grouping   If a pair of grouping symbols is inserted in an expression after a *plus* sign, no changes in signs are necessary. If the grouping symbols are inserted immediately after a *minus* sign, the signs of all enclosed terms must be changed. For example,

$$x + y - z + w = x + (y - z + w)$$

and

$$a - b + c - d = a - (b - c + d)$$

● **Example 2**   All the following expressions are equal:

$$2x - 6y + 2p - 4q$$
$$2x - 6y - (-2p + 4q)$$
$$2x - 6y + (2p - 4q)$$
$$2x - (6y - 2p + 4q)$$
$$2(x - 3y + p - 2q)$$ ●

We will now show how to use additional signs of grouping to make the meaning of certain expressions clear and to indicate the order in which operations are to be performed. In addition to parentheses, we use the brackets, [ ], and the braces, { }, for these purposes.

Removing symbols of grouping   It is frequently desirable to remove the symbols of grouping from an expression, and we will explain and illustrate the procedure. If an expression that is enclosed in parentheses is preceded or followed by a factor, as in $x - 2y(3x - y + z)$, we apply the distributive law and replace $-2y(3x - y + z)$ by $-6xy + 2y^2 - 2yz$. Therefore, $x - 2y(3x - y + z) = x - 6xy + 2y^2 - 2yz$. Similarly, $a^2 + (a^3 - ab + b^2)2a = a^2 + 2a^4 - 2a^2b + 2ab^2$.

Since $-(n) = -1(n)$, the expression $x + y - (-2x^2 - y^2 + z^2) = x + y - 1(-2x^2 - y^2 + z^2) = x + y + 2x^2 + y^2 - z^2$ by use of the distributive law. Therefore, if an expression enclosed in parentheses is preceded by a **minus sign,** the parentheses can be removed if and only if

Note   the sign of each of the enclosed terms is **changed.** If an expression enclosed in parentheses is preceded by a plus sign, as in $a + (b + c - d)$, it is understood that the factor $+1$ precedes the parentheses but is not expressed. Since multiplying a number by 1 yields a product equal to the number, the parentheses can be removed from $a + (b + c - d)$ with no further changes, giving $a + b + c - d$.

Usually when braces or brackets or both appear together with parentheses in an expression, one or more sets of grouping symbols are enclosed in another set. When the symbols are removed from an expression of this type, it is advisable to *remove the innermost symbols first.* We will illustrate the procedure with the following examples.

Note

● **Example 3**   Remove the symbols of a grouping in

$$3a - [4b + 2(6a - b) - 3(a - 5b - 1) - 7]$$

Solution   We work first with the parentheses inside the brackets.

$$3a - [4b + 12a - 2b - 3a + 15b + 3 - 7]$$
$$= 3a - [17b + 9a - 4] \qquad \text{collecting terms}$$
$$= 3a - 17b - 9a + 4 \qquad \text{removing } [\ \ ]$$
$$= -6a - 17b + 4 \qquad \text{collecting terms} \qquad ●$$

● **Example 4** $$3x^2 - \{2x^2 - xy - [x(x - y) - y(2x - y)] + 4xy\} - 3y^2$$

Solution   We start with the given expression, indicate the successive steps in removing the symbols of operations, and explain each step at the right.

$$3x^2 - \{2x^2 - xy - [x(x - y) - y(2x - y)] + 4xy\} - 3y^2$$
$$\text{given expression}$$
$$= 3x^2 - \{2x^2 - xy - [x^2 - xy - 2xy + y^2] + 4xy\} - 3y^2$$
$$\text{applying the distributive law to the}$$
$$\text{expressions in the parentheses}$$
$$= 3x^2 - \{2x^2 - xy - [x^2 - 3xy + y^2] + 4xy\} - 3y^2$$
$$\text{adding similar terms in brackets}$$
$$= 3x^2 - \{2x^2 - xy - x^2 + 3xy - y^2 + 4xy\} - 3y^2$$
$$\text{since } -[x^2 - 3xy + y^2] = -x^2 + 3xy - y^2$$
$$= 3x^2 - \{x^2 + 6xy - y^2\} - 3y^2$$
$$\text{adding similar terms in braces}$$
$$= 3x^2 - x^2 - 6xy + y^2 - 3y^2$$
$$\text{since } -\{x^2 + 6xy - y^2\} = -x^2 - 6xy + y^2$$
$$= 2x^2 - 6xy - 2y^2 \qquad \text{adding similar terms} \qquad ●$$

Exercise 1.4   Find the value of the algebraic expressions in Probs. 1 to 8.

1 $3a + 14b - ab$ for $a = 5$ and $b = 2$

2 $\dfrac{4a + 1}{6b + a}$ for $a = -3$ and $b = 10$

3 $(16a + 5b) \div (1 - 4a)$ for $a = -1$ and $b = 5$

4 $3a + \dfrac{b}{1 + a}$ for $a = 2$ and $b = -12$

5 $\dfrac{x^2 + 1}{x + 1}$ for $x = 4$

6 $(x + 1)^2 - 2x$ for $x = -5$

7 $(x^3 + 1)^2 + \dfrac{1}{x}$ for $x = 2$

8 $(x + 4)^2 - (x + 2)^4$ for $x = -2$

Perform the indicated operations in Probs. 9 to 28.

9 $3ab - 4ab + 17ab$ 10 $6LS + 8LS - 9LS$

11 $-5alg + 15alg + 7alg - 9alg - 12alg$

12 $-8mt + 53mt - 40mt + 6mt - 16mt$

13 $2a + b + 6a - b - 4a + b$ 14 $x - 3y + 6x - 8y + 5y - 7x$

15 $y - 2x + 3x - 5x + y + 9y$ 16 $2y - b + 3y + 5b - y - 2b$

17 $(-2x)(-5)(3)(2y)$ 18 $(-3)(-x)(-2y)(8)$

19 $(-2)(-2)(-2)(-2)(-2)$ 20 $(4a)(3)(-6)(b)$

21 $-3(-2x + 5y) + 5(-x + 4y)$ 22 $4(a + 5b) - 3(2a - 3b)$

23 $x(-3a + 2b) - 4x(2a - 5b)$ 24 $-c(-m - 4n) - 3c(5m - 2n)$

25 $-5(3x + t) - 4(2x - 5t) + 2(-x - t)$

26 $4(-2n - 5b + 4a) - 3(-3n - 7b + 5a)$

27 $2x(3a + b) - y(a - b) + a(-x - 2y) - 3b(2x + y)$

28 $a(c - d) + b(d - c) - c(a - b) - d(-a - b)$

Complete the equations by inserting the correct expression inside the parentheses.

29 $6x - 5y + 3z = 6x + (\quad)$ 30 $4x + 3y - 2z = 4x - (\quad)$

31 $-2x + 3a + 5b = 3a - (\quad)$ 32 $4a + 2t + 3g = 3g - (\quad)$

33 $6x + 2y - 4 = -2(\quad)$ 34 $-2x + 6z - 8 = 2(\quad)$

35 $3a - 9b + 27 = -3(\quad)$ 36 $4a - 32x + 12 = -4(\quad)$

Remove the symbols of grouping in each of Probs. 37 to 52 and then combine similar terms.

37 $2a + [3a - 2(a - 2b) - 3a]$ 38 $3a + [5a - (2a - b)] + 3b$

39 $2a - [2c - 3b - (2a + 4c - 3b) - 2a] - (c - b)$

40 $5a - (2b - 3c) - [3a - 4b - (a - b - 2c) + c] - 2(a - b) + c$

41 $2\{2a - b[2a - c(2a - 1) + 2ac] - c\}$

42 $2a\{a^2 - a[2a - 3(a + 2) + 1] - a^2\}$

43 $4[3a - 2(a + 2b)] - 3\{a^2 - [3b + a(a - b)]\}$

44 $2a^3 - 5a\{a^2 + 3[3a - 4(a - 2) + 3] - a^2\}$

45 $2x + 2\{y - [4x - (z + 2y)] + z\} - 2y$

46 $3a - \{b - 2[c - 3b + 2(c - a) + b] + 2a\}$

47 $6d - 4e - \{2f + 2[-d + e - 2(d - f)] + e\} + e$

48 $2g - 3\{h - 4[i + 2(g - h + 2i) - g] + 2h\} + 3i$

49 $3 + x[-6 + x(4 + x)]$ 50 $-2 + x[5 - x(-7 + x)]$

51 $-1 + x\{6 + x[-4 + x(3 + x)]\}$ 52 $5 - x\{-5 + x[4 + x(1 - x)]\}$

1.5 Key concepts

Be sure you understand and can use the following important words and ideas.

Axiom	Irrational numbers
Theorem	Real numbers
Set	Absolute value
Subset	Triangle inequality
Equality of sets	Axioms of equality
Intersection	Closure, inverse, identity
Disjoint sets	Associative, commutative, distrib-
Complement	utive
Union	Subtraction
Universal set	Division
Ordered pair	Cancellation laws
Even	Order axioms
Odd	Integers modulo m
Factor	Variable
Prime	Constant
Composite	Factor
Fundamental theorem of arithmetic	Exponent
Greatest common divisor	Expression
Least common multiple	Coefficient
Natural numbers	Monomial, binomial, trinomial
Integers	Polynomial
Negative numbers	Similar terms
Rational numbers	

(1.1) $m = n \cdot q + r, 0 \leq r < n$

Axioms of equality
Field axioms
  Closure: $a + b$ and ab are real numbers.
  Associative laws: $(a + b) + c = a + (b + c)$ and $(ab)c = a(bc)$.
  Identity: $a + 0 = a$ and $a \cdot 1 = a$.
  Inverse: $a + (-a) = 0$ and $a(a^{-1}) = 1$.
  Commutative laws: $a + b = b + a$ and $ab = ba$.
  Distributive law: $a(b + c) = ab + ac$.

(1.2) $a - b = a + (-b)$
(1.3) $a \div b = a(b^{-1})$, if $b \neq 0$
(1.4) $-(-a) = a$
(1.5) $-(a + b) = -a - b$
(1.6) $-(a - b) = -a + b$
(1.7) $a(0) = 0$ for every a

(1.8) $a(-b) = -(ab)$
(1.9) $(-a)(-b) = ab$
(1.10) $a/b = c$ if and only if $a = bc$, for $b \neq 0$
(1.11) $a \div 0$ is never defined
(1.12) $0 \div b = 0$, for $b \neq 0$

Cancellation laws: If $a + c = b + c$, then $a = b$.
                   If $ac = bc$, then $a = b$, for $c \neq 0$.

Order axioms: closure, trichotomy, and transitivity

(1.13)   If $a < b$, then $a + c < b + c$
(1.14)   If $a < b$ and $c > 0$, then $ac < bc$
(1.15)   If $a < b$ and $c < 0$, then $ac > bc$
(1.16)   If $0 < a < b$, then $1/a > 1/b$
(1.17) $a^n = a \cdot a \cdot a \cdot \cdots \cdot a$ (1.20) $(a^n)^m = a^{mn}$
(1.18) $a^m \cdot a^n = a^{m+n}$ (1.21) $(a/b)^n = a^n/b^n$

(1.19) $(ab)^n = a^n \cdot b^n$ (1.22) $\dfrac{a^m}{a^n} = a^{m-n}$, for $m > n > 0$

Exercise 1.5

Let $A = \{1, 3, 5, 7, 9\}$, $B = \{2, 4, 6, 8, 10\}$, $C = \{1, 4, 7, 10\}$, $D = \{3, 4, 5, 6, 7\}$. Are the following statements true or false?

1 A and B are disjoint.   2 $D \subseteq A \cup B$
3 $B \cap D \subseteq A \cup C$ 4 $A \cup (B \cap C) = (A \cup B) \cap (A \cup C)$
5 $(B - C) \cup C = B$
6 $A \cup D = (A - D) \cup (D - A) \cup (A \cap D)$
7 Show that the sum along each row, each column,
and each main diagonal is the same. Note all
numbers are primes, except 1.

$$\begin{bmatrix} 67 & 13 & 31 \\ 1 & 37 & 73 \\ 43 & 61 & 7 \end{bmatrix}$$

Verify the following calculations:

8 $59 + 60 + 61 + 162 + 163 + 164 + 165 + 166 = 1000$
9 $123 + 45 - 67 + 8 - 9 = 100$
10 $1^3 + 3^3 + 5^3 + 7^3 = 2^4 + 2^5 + 2^6 + 2^7 + 2^8$
11 $4,210,124 \times 11 = 46,311,364$
12 In words, the distributive law says multiplication distributes over addition. If we ask whether addition distributes over multiplication, the equation is $a + (bc) = (a + b)(a + c)$. Show that this holds if and only if $a + b + c = 1$ or $a = 0$.
13 Factor 2970 and 3536.   14 Find the gcd and lcm of 980 and 126.
15 Find the quotient and remainder: $1440 \div 72$ and $1440 \div 71$.

Let $S = \{-\pi/\pi, 7.891011121314 \cdots, -71/79, 7.9, -0.434343 \cdots\}$. Write the following elements of S:

16 Irrationals   17 Integers   18 Smallest   19 Largest rational

20 Find the sum and product of 3.4 and 2.1.

21 Find the sum and product of 3.4 and 2.111 $\cdots$.

Are these statements true or false?

22 $6(8 - 8) = (5^{-1} \cdot 5) - 1$ **23** $3 + 3^{-1} = 0$

24 $3(-3) + 3(3) = 0$ **25** $4 + (-3) + (1)(1^{-1}) = 0$

26 $0.999 \cdots = 1$ **27** $\dfrac{x}{x + y}$ is a monomial.

28 If $a > b$ and $c < b$, then $-a < -c$.

29 $2^5 = 5^2$ in the integers modulo 7   **30** $[(6^{-1})^{-1}]^{-1} = 6^{-1}$

31 $6 \div 0 = 0 \times (1/6)$

Perform the indicated operations.

32 $7a(tr) + 4r(at) - 6t(ar)$ **33** $(2x^3)^3(4x^2)$

34 $\dfrac{(2x)^6(4x^2)^4}{(16x^4)^3}$ **35** $\dfrac{(6x^2)^3(2x^4)^5}{(3x^5)^3(4x)^4}$ **36** $-(-x + 2y) + x - (x - y)$

37 $2y - 3(-x - 3y) - (-5)(2x - 3y)$ **38** $-(-1)(3)(-2)(1^{-1})$

39 Try to follow all steps in the following proof that $\sqrt{2}$ is irrational. We will *assume* $\sqrt{2}$ is rational, and arrive at a contradiction. The contradiction forces us to conclude that $\sqrt{2}$ is not rational—hence it is irrational.

$\sqrt{2} = m/n$ with m and n positive integers whose gcd is 1
$2 = m^2/n^2$; hence $2n^2 = m^2$
m^2 even implies m even; hence $m = 2k$
$2n^2 = (2k)^2 = 4k^2$
$n^2 = 2k^2$; hence n is even
Thus m and n are both even, which contradicts the fact that their gcd is 1.

40 Show that $1.4 = 7/5 < \sqrt{2} < 99/70 \approx 1.4143$ by verifying that

$$(7/5)^2 < 2 \quad \text{and} \quad (99/70)^2 > 2$$

41 The number 381,654,729 is interesting because it is the only number with nine different nonzero digits with the following divisibility properties (*verify them*). Counting 3 as the first digit, 8 as the second, etc., then the first two digits form a number divisible by 2, the first three digits form a number divisible by 3, the first 4 are divisible by 4, . . ., up to and including 9.

CHAPTER 2·POLYNOMIALS

2.1   Sums and products of polynomials

Monomial

An expression involving only the product of a real number and variables to positive integral powers is called a *monomial*.

● **Example 1** The expressions $3d$, $2pr$, and $43x^3yz^5$ are monomials, but $a^2 + b$ and $2a/5xy$ are not monomials.   ●

Polynomial

The sum of a finite number of monomials is called a **polynomial.** The general polynomial in one variable of degree n is $a_nx^n + \cdots + a_1x + a_0$.

Binomial

Trinomial

A polynomial consisting of exactly two terms is a *binomial,* and a polynomial consisting of exactly three terms is a *trinomial.*

● **Example 2** Examples of polynomials are $2a^4 + a^3 - 4a^2 + 5a + 3$ and $2x^4y + 2x^3y - 4x^2y^2 + 2xy^3 - 4y^2$. However, $x^2/y + 3y^4$ is not a polynomial.   ●

Coefficient

If a monomial is expressed as the product of two or more symbols, each of the symbols is called the *coefficient* of the product of the others.

● **Example 3** In the monomial $3ab$, 3 is the coefficient of ab, a is the coefficient of $3b$, b is the coefficient of $3a$, and $3a$ is the coefficient of b. We call the 3 in $3ab$ the *numerical* coefficient. Usually when we refer to the coefficient in a monomial, we mean the numerical coefficient.   ●

Similar terms

Two monomials, or two terms, are called *similar* if they differ only in their numerical coefficients.

● **Example 4** The monomials $3a^2b$ and $-2a^2b$ are similar, and the terms in $4(3a/5b) + 2(3a/5b)$ are similar.   ●

The distributive axiom can be extended to cover situations in which the polynomial in the parentheses consists of more than two terms. For example,

$$(a + b + d)c = ac + bc + dc$$
$$p(q + r - s - t) = pq + pr - ps - pt$$

The distributive axiom enables us to express the sum of two or more similar monomials as a monomial. For example,

$$3ab + 2ab + 4ab = (3 + 2 + 4)ab = 9ab$$

This is just like saying "3 apples and 2 apples and 4 apples is 9 apples."

● **Example 5**

$$\begin{aligned}
7x^2y - 3x^2y - 2xy^2 + 8xy^2 &= 7 \cdot x^2y + (-3 \cdot x^2y) + (-2 \cdot xy^2) + 8 \cdot xy^2 \\
&= [7 + (-3)]x^2y + (-2 + 8)xy^2 \\
&= (7 - 3)x^2y + (-2 + 8)xy^2 \\
&= 4x^2y + 6xy^2
\end{aligned}$$

After some practice, the first and second steps in problems similar to Example 5 can be performed mentally, and thus we can proceed directly to the result in the third step. This is illustrated in Example 6.

● **Example 6**

$$\begin{aligned}
2a - 3a + 6b + 4b - 7c + 9c &= (2 - 3)a + (6 + 4)b + (-7 + 9)c \\
&= -a + 10b + 2c
\end{aligned}$$

Frequently, we are required to add two or more polynomials when at least one of them contains one or more terms that are not similar to any term in at least one of the others. The method for dealing with such situations is illustrated by Example 7, where we add similar terms together.

● **Example 7**   Add the polynomials $3x^2 + 4y^2 - 3xy + 7z^2$, $2x^2 + 4z^3$, and $4y^2 - 2z^2 - 2xy$.

Solution   We write the polynomials as shown below and perform the addition as indicated.

$$\begin{array}{l}
3x^2 + 4y^2 - 3xy + 7z^2 \\
2x^2 \qquad\qquad\qquad\quad + 4z^3 \\
\underline{\qquad 4y^2 - 2xy - 2z^2} \\
5x^2 + 8y^2 - 5xy + 5z^2 + 4z^3
\end{array}$$

● **Example 8**   Subtract $3x - 2y - 9z$ from $5x + 3y - 6z$.

Solution   The procedure here is to write the subtrahend below the minuend as indicated below, then mentally change the sign preceding each term of the subtrahend and proceed as in addition.

$$\begin{array}{ll}
5x + 3y - 6z & \text{minuend} \\
\underline{3x - 2y - 9z} & \text{subtrahend} \\
2x + 5y + 3z & \text{difference}
\end{array}$$

We may also write in this example,

$5x + 3y - 6z - (3x - 2y - 9z)$
$$= 5x + 3y - 6z - 3x + 2y + 9z = 2x + 5y + 3z$$

Products of monomials and polynomials

We employ the commutative, associative, and distributive axioms together with the law of signs and the law of exponents to obtain the product of two or more monomials, of a monomial and a polynomial, and of two polynomials. We shall illustrate the method with six examples.

● **Example 9** $3x^2y \cdot 4xy^2 \cdot 6x^3y^4 = 3 \cdot 4 \cdot 6 \cdot x^2 \cdot x \cdot x^3 \cdot y \cdot y^2 \cdot y^4$
$$= 72x^{2+1+3}y^{1+2+4} \qquad \text{commutative axiom}$$
$$= 72x^6y^7 \qquad \text{by (1.18)} \qquad ●$$

● **Example 10** $-4ab^2c^3 \cdot -2a^3b^4c \cdot 6a^2bc^5$
$$= -4 \cdot -2 \cdot 6 \cdot a \cdot a^3 \cdot a^2 \cdot b^2 \cdot b^4 \cdot b \cdot c^3 \cdot c \cdot c^5$$
$$= 48a^6b^7c^9 \qquad \text{by the law of signs and the law of exponents} \qquad ●$$

● **Example 11** $3ab(2a - 4b + 7a^2b) = 3ab(2a) - 3ab(4b) + 3ab(7a^2b)$
$$= 6a^2b - 12ab^2 + 21a^3b^2 \qquad \text{by the distributive axiom} \qquad ●$$

● **Example 12** $(3x^2y - 6xy^2 - 8y^3)(-5x^3y^2)$
$$= 3x^2y(-5x^3y^2) + (-6xy^2)(-5x^3y^2) + (-8y^3)(-5x^3y^2)$$
$$= -15x^5y^3 + 30x^4y^4 + 40x^3y^5 \qquad \text{by the distributive axiom} \qquad ●$$

The method for obtaining the product of two polynomials is illustrated in Examples 13 and 14.

● **Example 13**
$$(3x - 2y)(2x - 5y) = 3x(2x - 5y) - 2y(2x - 5y)$$
$$= 6x^2 - 15xy - 4xy + 10y^2$$
$$= 6x^2 - 19xy + 10y^2 \qquad ●$$

● **Example 14** To obtain the product $(-5x^2 + 2xy + 3y^2)(3x^3 - 6x^2y + 2xy^2 - 4y^3)$, we first consider the second factor as a single number, apply the distributive axiom, and then complete the computation as indicated.

$(-5x^2 + 2xy + 3y^2)(3x^3 - 6x^2y + 2xy^2 - 4y^3)$
$= (-5x^2)(3x^3 - 6x^2y + 2xy^2 - 4y^3) + (2xy)(3x^3 - 6x^2y + 2xy^2 - 4y^3)$
$\quad + (3y^2)(3x^3 - 6x^2y + 2xy^2 - 4y^3) \qquad \text{by the distributive axiom}$
$= -15x^5 + 30x^4y - 10x^3y^2 + 20x^2y^3 + 6x^4y - 12x^3y^2 + 4x^2y^3 - 8xy^4$
$\quad + 9x^3y^2 - 18x^2y^3 + 6xy^4 - 12y^5 \qquad \text{by the distributive axiom}$
$= -15x^5 + 30x^4y + 6x^4y - 10x^3y^2 - 12x^3y^2 + 9x^3y^2 + 20x^2y^3 + 4x^2y^3$
$\quad - 18x^2y^3 - 8xy^4 + 6xy^4 - 12y^5 \qquad \text{by the commutative axiom}$
$= -15x^5 + 36x^4y - 13x^3y^2 + 6x^2y^3 - 2xy^4 - 12y^5 \qquad ●$

The above process is usually abbreviated to the method illustrated below.

$$
\begin{array}{l}
3x^3 - 6x^2y + 2xy^2 - 4y^3 \\
- 5x^2 + 2xy + 3y^2 \\
\hline
-15x^5 + 30x^4y - 10x^3y^2 + 20x^2y^3 \\
+ 6x^4y - 12x^3y^2 + 4x^2y^3 - 8xy^4 \\
+ 9x^3y^2 - 18x^2y^3 + 6xy^4 - 12y^5 \\
\hline
-15x^5 + 36x^4y - 13x^3y^2 + 6x^2y^3 - 2xy^4 - 12y^5
\end{array}
$$

factors

multiplying by $-5x^2$
multiplying by $2xy$
multiplying by $3y^2$
adding similar terms

Exercise 2.1

Add the three expressions in each of Probs. 1 to 16.

1 $2a - 3b - 15c$
$3a + 4b + 17c$
$-4a - 5b + 2c$

2 $-3p + 14q - 6r$
$5p - 13q - 13r$
$4p - q + 18r$

3 $7x + 13y - 6z$
$-8x - 12y + 9z$
$3x - y + 2z$

4 $2a - 13p + 7x$
$3a + 15p - 18x$
$-4a + 4p + 13x$

5 $5ab - 13bc + 6ac$
$3ab + 17bc - 4ac$
$-7ab - 4bc - 3ac$

6 $9xy - 7xz - 4yz$
$3xy + 4xz - 13yz$
$-8xy + 5xz + 16yz$

7 $8pq + 5pr - 6qr$
$pq - 8pr + 7qr$
$-7pq + 3pr - qr$

8 $7bq + 18bz - 4qz$
$-9bq + 17bz - 3qz$
$2bq + bz + 7qz$

9 $12a + 13b - 4c, -13a - 16b + 3c, 5a - 4b + 10c$

10 $6x - 2y + 3z, -7x - 9y - 16z, 4x + 12y + 16z$

11 $5p + 7d - 3q, 3p - 18d - 16q, -7p + 13d + 19q$

12 $-7r + 5s - 4t, 8r - 6s + 7t, 2r + 3s - 2t$

13 $2ab^2 + 3ab + 5a^2b, 3ab^2 - 14ab - 3a^2b, 4ab^2 + 17ab - 8a^2b$

14 $4xy - 3xy^2 + 2xy^3, -7xy + 18xy^2 + 13xy^3, 2xy - 13xy^2 - 12xy^3$

15 $3p^2q + 4pq + 5pq^2, -7p^2q - 16pq + 13pq^2, 5p^2q + 14pq - 17pq^2$

16 $7r^2s + 12rs + 3s^2, -9r^2s - 17rs - 2s^2, 3r^2s + 5rs + s^2$

In each of Probs. 17 to 28, subtract the second number or expression from the first.

17 $18, 11$ **18** $81, 37$ **19** $28, -12$ **20** $-31, -42$

21 $2a + b - 3c, 3a + 2b + 4c$ **22** $7a - 5b + 18c, 6a - 6b + 19c$

23 $5x + 2y - 16z, 7x - 2y + 13z$ **24** $7a - 3k + 4p, 8a - 4k - 2p$

25 $3a^2 + 14b^2 - 5c^2, 2a^2 + 13b^2 - 6c^2$

26 $5x^2 - 17y^2 + 4z^2, 2x^2 - 18y^2 + 5z^2$

27 $9ax - 14ax^2 - 8a^2x, 2ax - 15ax^2 - 9a^2x$

28 $6p^2q^2 - 5pq^2 - 12pq, 6p^2q^2 - 7pq^2 - 13pq$

29 Subtract $2a + 3b - 2c$ from the sum of $a - b + c$ and $4a - 2b - 3c$.

30 Subtract $4x + 2y - 3z$ from the sum of $2x + 3y - 5z$ and $3x - 4y + 7z$.

31 Subtract the sum of $2a + 3p + 5x$ and $3a + 2p - 6x$ from $6a + 4p + x$.

32 Subtract the sum of $5x + 2y - 3a$ and $-2x + 3y - 4a$ from $3x + 4y - 8a$.

Find the indicated product in each of Probs. 33 to 68.

33 $(2x^2y^3)(-3xy^2)$ **34** $(5x^4y^3)(-3x^2y^2)$ **35** $(-7xy^4)(-2x^2y)$
36 $(-4x^2y^3)(-3x^5y^2)$ **37** $(2x^3)^2$ **38** $(3x^2)^4$
39 $(4x^5)^3$ **40** $(5x^2)^3$
41 $2x^2y(3xy^3 - 5x^2y^4)$ **42** $3x^3y(2xy^2 - 4x^2y)$
43 $4x^2y^3(2xy^3 - 3x^2y)$ **44** $5xy^4(2x^2y^3 - 5xy^2)$
45 $-2x^2y^3(3xy^2 - 2x^2y)$ **46** $-3xy^2(2x^2y^3 - 5x^3y)$
47 $5x^3y^4(2xy^2 - 4x^3y)$ **48** $7x^2y^4(3x^3y^2 - 2x^5y^3)$
49 $2x^2y(3y - 2x) - 3xy^2(2x - y)$ **50** $3xy(2x + 3x^2y) - 2x^2y(4 - xy)$
51 $5xy^3(2x^2y - 3xy^3) - 4x^2y(2xy^3 - 7y^5)$
52 $7x^2y(2xy^3 - 3x^2y^2) - 4xy^3(3x^2y - 5x^3)$
53 $(3x + 2y)(2x - 3y)$ **54** $(4x - 3y)(2x + 3y)$
55 $(5x - 3y)(3x - 2y)$ **56** $(7x - 4y)(2x - 5y)$
57 $(4x - 7)(3x - 4)$ **58** $(6x - 5)(3x - 8)$
59 $(5x + 4)(4x - 5)$ **60** $(2x - 7)(7x + 2)$
61 $(3x + 5)(2x^2 - 3x - 5)$ **62** $(4x + 1)(3x^2 + 4x - 1)$
63 $(4x^2 - 2x + 7)(2x^2 + 3x - 2)$ **64** $(5x^2 + 2x - 3)(x^2 - 3x - 3)$
65 $(2x^2 + 3xy - 3y^2)(x^2 - 3xy + 2y^2)$
66 $(2x^2 - xy + 3y^2)(3x^2 - xy - 2y^2)$
67 $(5x^2 + 2xy + y^2)(2x^2 - xy + 3y^2)$
68 $(x^2 - 3xy + 2y^2)(x^2 + 3xy - y^2)$

2.2   Division of two polynomials

As stated earlier, if $b \neq 0$, the quotient of a and b is expressed as $\dfrac{a}{b}$ or a/b or $a \div b$. This quotient is the unique number x such that $bx = a$. That is,

$$\text{If } b \neq 0, \text{ then } \frac{a}{b} = x \text{ if and only if } bx = a \tag{2.1}$$

Dividend
Divisor

The number a is the *dividend*, b is the *divisor*, and the procedure for computing x is called *division*.

Since $a \cdot 1 = a$, by definition of multiplicative identity, it follows that

$$\frac{a}{a} = 1 \qquad \text{provided } a \neq 0 \tag{2.2}$$

and also

$$\frac{a}{1} = a \tag{2.3}$$

We will now prove a theorem which is essential for the multiplication and division of fractions and the simplification of quotients.

○ **Theorem**

$$\frac{ac}{bd} = \frac{a}{b} \cdot \frac{c}{d} \qquad \text{if } b \neq 0, d \neq 0 \tag{2.4}$$

Proof To prove that $ac/bd = a/b \cdot c/d$, we will let $a/b = x$ and $c/d = y$, and thus have $a/b \cdot c/d = xy$. Furthermore, by the definition of a quotient, $a = bx$ and $c = dy$. Consequently $ac = bx \cdot dy = bd \cdot xy$ by the commutative and associative axioms. Hence $ac/bd = xy$. Therefore, $ac/bd = a/b \cdot c/d$, since each is equal to xy.

Furthermore, since $bd = db$, we have

$$\frac{ac}{bd} = \frac{ac}{db} = \frac{a}{d}\frac{c}{b} \qquad\qquad ○$$

This theorem can be extended to cover cases where there are more than two factors in the dividend, in the divisor, or in both. For example,

$$\frac{ace}{bdf} = \frac{a(ce)}{b(df)} = \frac{a}{b}\frac{ce}{df} = \frac{a}{b}\frac{c}{d}\frac{e}{f}$$

If in (2.4) we let $b = c = 1$ and apply the commutative axiom, we have

$$\frac{a}{d} = a \cdot \frac{1}{d} = \frac{1}{d} \cdot a \qquad \text{provided } d \neq 0 \tag{2.5}$$

In Sec. 1.3 we saw that $a^m/a^n = a^{m-n}$ for positive integers $m > n$. When $m \leq n$, we just recall the definition $a^n = a \cdot a \cdots a$ as a product, and then remove common factors from numerator and denominator using Eq. (2.2).

● **Example 1**

(i) $\dfrac{a^3}{a^5} = \dfrac{a \cdot a \cdot a}{a \cdot a \cdot a \cdot a \cdot a} = \dfrac{a}{a} \cdot \dfrac{a}{a} \cdot \dfrac{a}{a} \cdot \dfrac{1}{a^2} = 1 \cdot 1 \cdot 1 \cdot \dfrac{1}{a^2} = \dfrac{1}{a^2}$

(ii) $\dfrac{(2x)^3 3x^2}{6x^9} = \dfrac{2^3 \cdot x^3 \cdot 3 \cdot x^2}{2 \cdot 3 \cdot x^9} = \dfrac{2 \cdot 2^2}{2} \cdot \dfrac{3}{3} \cdot \dfrac{x^{3+2}}{x^9}$

$\qquad\qquad = 2^2 \cdot 1 \cdot \dfrac{x^5}{x^5 \cdot x^4} = 4 \cdot 1 \cdot \dfrac{1}{x^4} = \dfrac{4}{x^4}$ ●

Monomial divisors

We employ one or more of the theorems from (2.1) to (2.6) to obtain and simplify a quotient when the divisor is a monomial. The method is illustrated in the following examples.

● **Example 2**

$$6x^8y^6z^3 \div 3x^4y^3z^2 = \frac{6x^8y^6z^3}{3x^4y^3z^2}$$

$$= \frac{6}{3}\frac{x^8}{x^4}\frac{y^6}{y^3}\frac{z^3}{z^2} \qquad\qquad \text{by (2.4)}$$

$$= 2x^{8-4}y^{6-3}z^{3-2} \qquad \text{by law of exponents}$$

$$= 2x^4y^3z \qquad\qquad\qquad ●$$

If the dividend is a polynomial, we use the distributive axiom to write

$$\frac{a + b + c}{d} = \frac{1}{d}(a + b + c)$$

$$= \frac{a}{d} + \frac{b}{d} + \frac{c}{d} \qquad \text{by the distributive axiom}$$

● **Example 3** $(6x^6y^5 + 4x^5y^4 - 3x^4y^3 - 2x^3y^2) \div 3x^3y^2 = \dfrac{6x^6y^5 + 4x^5y^4 - 3x^4y^3 - 2x^3y^2}{3x^3y^2}$

$$= \frac{6x^6y^5}{3x^3y^2} + \frac{4x^5y^4}{3x^3y^2} - \frac{3x^4y^3}{3x^3y^2} - \frac{2x^3y^2}{3x^3y^2}$$

$$= 2x^3y^3 + \tfrac{4}{3}x^2y^2 - xy - \tfrac{2}{3} \qquad ●$$

The quotient of two polynomials

Recall from (1.1) that when dividing m by n, both positive integers, we have $m = n \cdot q + r$, and q is called the quotient and r the remainder. Thus $23 \div 5 = 4 + \tfrac{3}{5}$ may be written

$$23 = 5(4) + 3$$

Likewise $(6x^2 + 4)/3x = 6x^2/3x + 4/3x = 2x + (4/3x)$ may be written as

$$6x^2 + 4 = (3x)(2x) + 4$$

The following relationship is thus satisfied:

Division   **Dividend = (divisor)(quotient) + remainder**   (2.6)

Degree of a polynomial   The *degree of a polynomial* in any variable is the greatest exponent of that variable in the polynomial. For example, the polynomial $3x^4 + 2x^3y + 5x^2y^2 + xy^3$ is of degree 4 in x and 3 in y.

To divide one polynomial by another, we first arrange the terms in each polynomial so that the exponents of some letter that appears in each are in descending numerical order. Then we seek the quotient that is a polynomial, or possibly a monomial, that satisfies the relation (2.6), where the degree of the remainder in the variable chosen as the basis for the arrangement of terms is less than the degree of the divisor in that variable.

We will illustrate the procedure with an example.

● **Example 4**   Find the quotient and remainder obtained by dividing $6x^2 + 5x - 1$ by $2x - 1$.

Solution   Here the dividend is $6x^2 + 5x - 1$, the divisor is $2x - 1$, and we seek the quotient that satisfies the relation

$$6x^2 + 5x - 1 = (2x - 1)(\text{quotient}) + \text{remainder} \qquad (1)$$

Since the degree of the dividend is 2, the degree of the divisor is 1, and the degree of the remainder must be less than 1, it follows that the degree of the quotient is 1. Hence we write the quotient as $ax + b$, substitute this expression in (1), and get

$$6x^2 + 5x - 1 = (2x - 1)(ax + b) + \text{remainder} \qquad (2)$$

Now we can determine the values for a and b by the following procedure: We first perform the indicated multiplication in (2), and get

$$\begin{aligned}
6x^2 + 5x - 1 &= (2x - 1)ax + (2x - 1)b + \text{remainder} \\
&= x^2(2a) - ax + 2bx - b + \text{remainder} \qquad (3)
\end{aligned}$$

By inspection, we see that the only terms in the left and right members of (3) that involve x^2 are $6x^2$ and $2x(ax) = 2ax^2$, respectively. Hence $2ax^2 = 6x^2$, and it follows that $a = 3$. Now we substitute 3 for a in (3) and subtract $(2x - 1)3x = 6x^2 - 3x$ from each member and get

$$\begin{aligned}
8x - 1 &= (2x - 1)b + \text{remainder} \\
&= 2bx - b + \text{remainder} \qquad (4)
\end{aligned}$$

Again by inspection, we see that the only terms in (4) that involve x are $8x$ and $2bx$. Consequently, $2bx = 8x$, and therefore $b = 4$. Finally, we substitute 4 for b in (4) and subtract $(2x - 1)4 = 8x - 4$ from each member, and we get $3 = 0 + \text{remainder}$. Since the degree of 3 in x is 0, which is less than the degree of the divisor, 3 is the remainder. Hence we have

$$6x^2 + 5x - 1 = (2x - 1)(3x + 4) + 3$$

Therefore, the quotient is $3x + 4$ and the remainder is 3. ●

   The procedure above can be condensed in the usual long-division process shown below.

$$
\begin{array}{r}
\underline{3x + 4} \qquad \text{quotient} \\
2x - 1 \overline{\smash{\big)}\, 6x^2 + 5x - 1} \qquad \text{dividend} \\
\underline{6x^2 - 3x} \qquad\quad (2x - 1)3x \\
8x - 1 \qquad\quad \text{subtracting} \\
\underline{8x - 4} \qquad\quad (2x - 1)4 \\
3 \qquad\quad \text{remainder}
\end{array}
$$

   The above example illustrates the formal steps in the process of dividing one polynomial by another. These are stated below.

Dividing one
polynomial by another

1 Arrange the terms in the dividend and divisor in the order of descending powers of a letter that appears in each.
2 Divide the first term in the dividend by the first term in the divisor to get the first term in the quotient.

3 Multiply the divisor by the first term in the quotient and subtract the product from the dividend.

4 Treat the remainder obtained in step 3 as a new dividend, and repeat steps 2 and 3.

5 Continue this process until a remainder is obtained that is of lower degree than the divisor in the letter chosen in step 1 as the basis for the arrangement.

The computation can be checked by use of the relation (2.6). We will further illustrate the process by another example.

● **Example 5**   Divide $6x^4 - 6x^2 - 3 + 5x - x^3$ by $-2 + 2x^2 + x$.

Solution   We will arrange the terms in the dividend and divisor in the order of descending powers of x and proceed as indicated below.

$$
\begin{array}{ll}
\text{Divisor} \quad \underline{3x^2 - 2x + 1} & \text{quotient} \\
2x^2 + x - 2\,\overline{\big)\,6x^4 - x^3 - 6x^2 + 5x - 3} & \text{dividend} \\
\qquad\quad\underline{6x^4 + 3x^3 - 6x^2} & (2x^2 + x - 2)3x^2 \\
\qquad\qquad\qquad -4x^3 \qquad\quad +5x - 3 & \text{subtracting} \\
\qquad\qquad\qquad \underline{-4x^3 - 2x^2 + 4x} & (2x^2 + x - 2)(-2x) \\
\qquad\qquad\qquad\qquad\quad 2x^2 + x - 3 & \text{subtracting} \\
\qquad\qquad\qquad\qquad\quad \underline{2x^2 + x - 2} & (2x^2 + x - 2)(+1) \\
\qquad\qquad\qquad\qquad\qquad\qquad\quad -1 & \text{remainder} \quad ●
\end{array}
$$

Synthetic division

Dividing   A shorter and simpler method of procedure is available for use in division
by $x - r$ provided the divisor has the special form $x - r$. We will present this procedure by giving some examples. If we use the ordinary long-division method for dividing $2x^3 + x^2 - 18x - 7$ by $x - 3$, we have

$$
\begin{array}{l}
\underline{2x^2 + 7x + 3} \\
2x^3 + x^2 - 18x - 7\,\big)\,x - 3 \\
\underline{(2x^3) - 6x^2} \\
\qquad\quad 7x^2 - [18x] \\
\qquad\quad \underline{(7x^2) - 21x} \\
\qquad\qquad\qquad 3x - [7] \\
\qquad\qquad\qquad \underline{(3x) - 9} \\
\qquad\qquad\qquad\qquad 2
\end{array}
$$

Now let us examine this problem to see what can be omitted without interfering with the essential steps. In the first place, the division process requires that each term written in parentheses in the problem be exactly the same as the term just above it. Furthermore, the terms in brackets are the terms in the dividend written in a new position. If these two sets of terms are written only once, we have

$$\begin{array}{l} 2x^2 + 7x + 3 \\ \overline{2x^3 + x^2 - 18x - 7}\,\underline{|x - 3} \\ \underline{-\,6x^2} \\ \,7x^2 \\ \underline{-\,21x} \\ 3x \\ \underline{-\,9} \\ 2 \end{array}$$

We can save space by placing $-21x$ and -9 on the same line with $-6x^2$, and $3x$ and 2 on the same line with $7x^2$. Furthermore, it is not necessary to write the variable, since the problem tells us what it is. Hence, a shorter form of the work is

$$\begin{array}{l} 2 + 7 + 3 \\ \overline{2 + 1 - 18 - 7}\,\underline{|1 - 3} \\ \underline{-\,6 - 21 - 9} \\ 7 + 3 + 2 \end{array}$$

Note   Since the method we are developing applies only to division problems in which the divisor is $x - r$, it is not necessary for the coefficient of x, which is always 1, to appear. Moreover, in subtraction, we change the sign of the subtrahend and add. This latter change becomes automatic if we replace the -3 in the divisor by $+3$. Upon carrying out these suggestions, we have

$$\begin{array}{l} 2 + 7 + 3 \\ \overline{2 + 1 - 18 - 7}\,\underline{|3} \\ \underline{+\,6 + 21 + 9} \\ + 7 + 3 + 2 \end{array}$$

The final step in the process is to rewrite the 2 in the dividend as the first term in the third line. Then the first three terms in this line are the same as the coefficients in the quotient. Hence the latter can be omitted, and the problem becomes

$$\begin{array}{l} 2 + 1 - 18 - 7\,\underline{|3} \\ + 6 + 21 + 9 \\ \overline{2 + 7 + 3 + 2} \end{array}$$

Consequently, the essential steps in the process can be carried out mechanically as follows: Write the first 2 in the third line, multiply by 3, place the product 6 under 1, and add, obtaining 7; then multiply 7 by 3, obtaining 21, which is placed under -18 and added to it; finally, multiply the last sum by 3 and add the product to -7, getting 2. Hence the coefficients in the quotient are 2, +7, and +3 and the remainder is 2. We therefore have the following *rule for synthetic division*.

To divide a polynomial by $x - r$ synthetically:

Rules for synthetic division

1 Arrange the coefficients of the polynomial in order of descending powers of x, supplying zero as the coefficient of each missing power.
2 Replace the divisor $x - r$ by $+r$.
3 Bring down the coefficient of the largest power of x, multiply it by r, place the product beneath the coefficient of the second largest power of x, and add the product to that coefficient. Multiply this sum by r and place it beneath the coefficient of the next largest power of x. Continue this procedure until there is a product added to the constant term.
4 The last number in the third row is the remainder, which we will indicate by R, and the other numbers, reading from left to right, are the coefficients of the quotient, which is of degree one less than the given polynomial.

● **Example 6**   Determine the quotient and the remainder obtained by dividing $2x^4 + x^3 - 16x^2 + 18$ by $x + 2$ synthetically.

Solution   Since $x - r = x + 2$, we have $r = -2$. Upon writing the coefficients of the dividend in a line, supplying zero as the coefficient of the missing term in x, and carrying out the steps of synthetic division, we have

$$\begin{array}{r} 2 + 1 - 16 \quad\ 0 + 18\underline{|-2} \\ -4 + \ \ 6 + 20 - 40 \\ \hline 2 - 3 - 10 + 20 - 22 \end{array}$$

Hence, the quotient is $2x^3 - 3x^2 - 10x + 20$, and the remainder is $R = -22$.   ●

● **Example 7**   Use synthetic division to show that $x - 3$ is a factor of $x^3 - 2x^2 - x - 6$.

Solution   Using synthetic division gives

$$\begin{array}{r} 1 - 2 - 1 - 6\underline{|3} \\ 3 + 3 + 6 \\ \hline 1 + 1 + 2 \quad\ 0 \end{array}$$

so $R = 0$. Hence

$$x^3 - 2x^2 - x - 6 = (x - 3)(x^2 + x + 2) + 0$$
$$= (x - 3)(x^2 + x + 2)$$

which means that $x - 3$ is indeed a factor of $x^3 - 2x^2 - x - 6$.   ●

Exercise 2.2 division

Perform the division indicated in each of Probs. 1 to 20.

1 x^8/x^5 2 x^{11}/x^7 3 y^{10}/y^{17} 4 y^{21}/y^{27}

5 $x^3y^{15}/x^{12}y^{13}$ **6** x^7y^{19}/x^4y^{18} **7** x^8y^4/x^4y^8 **8** x^9y^{14}/x^7y^{13}

9 $48a^5b^4/16a^2b^6$ **10** $75a^{11}b^9/15a^7b^2$

11 $76a^7b^{22}/4a^3b^{22}$ **12** $60a^{17}b^9/12a^{17}b^6$

13 $\dfrac{21a^7 - 15a^4 - 6a^3}{3a^2}$ **14** $\dfrac{16a^9 - 24a^7 + 10a^4}{2a^6}$

15 $\dfrac{78x^7 + 66x^6 - 54x^4}{-6x^2}$ **16** $\dfrac{-21x^6 + 35x^5 - 14x^4}{-7x^5}$

17 $\dfrac{-35x^3y^4 + 42x^5y^3 - 14x^4y^5}{7x^2y^3}$ **18** $\dfrac{9x^7y^5 - 6x^4y^6 - 21x^5y^4}{3x^3y^{10}}$

19 $\dfrac{27x^5y^4 - 36x^6y^7 - 45x^4y^5}{9x^6y^4}$ **20** $\dfrac{39x^9y^2 - 65x^8y^7 - 91x^6y^4}{13x^5y}$

In Probs. 21 to 32, find the quotient and remainder obtained by dividing the first expression by the second.

21 $6x^3 - 11x^2 + 6x + 2,\ 2x^2 - 3x + 1$
22 $4x^3 + 2x^2 + 4x + 4,\ 2x^2 - 2x + 5$
23 $15x^3 - 41x^2 + 4x,\ 3x^2 - 7x - 2$
24 $20x^3 - 29x^2 + 17x - 2,\ 5x^2 - x + 3$
25 $6x^4 + 19x^3 + x^2 - 3x,\ 2x^3 + 5x^2 - 3x + 1$
26 $4x^4 + 4x^3 - 7x^2 + 4,\ 2x^3 + 3x^2 - 2x - 1$
27 $10x^4 - 19x^3 - 35x^2 + 3x + 3,\ 2x^3 - 5x^2 - 4x + 3$
28 $12x^4 - x^3 - 10x^2 - 17x - 2,\ 3x^3 + 2x^2 - x - 5$
29 $6x^4 - x^3 + 4x^2 + 5x + 1,\ 3x^2 + x - 2$
30 $5x^4 + 13x^3 + 13x^2 - 11x - 4,\ x^2 + 3x + 4$
31 $3x^4 - 4x^3 + 6x^2 + 4x + 1,\ x^2 - 2x + 3$
32 $8x^4 - 2x^3 - 19x^2 + 7x - 2,\ 2x^2 + x - 3$

In Probs. 33 to 44, find the quotient and remainder by dividing the first expression by the second synthetically.

33 $3x^2 - 11x + 8,\ x - 3$ **34** $2x^2 + 3x - 1,\ x - 1$
35 $3x^2 + 5x + 1,\ x + 2$ **36** $5x^2 + 22x + 1,\ x + 4$
37 $3x^3 + 10x^2 + 13x,\ x + 2$ **38** $2x^3 + 3x^2 - 8x - 2,\ x + 3$
39 $4x^3 - 22x^2 + 2,\ x - 5$ **40** $3x^3 - 10x^2 + 5x + 3,\ x - 3$
41 $2x^4 - 7x^3 + 10x^2 - 13x + 3,\ x - 2$
42 $4x^4 + 13x^3 + x^2 - 3x + 4,\ x + 3$
43 $3x^4 - 16x^3 + 21x^2 - 17x - 4,\ x - 4$
44 $3x^4 + 11x^3 - 22x^2 + 3,\ x + 5$

In Probs. 45 to 48, show that the remainder is 0 if the given polynomial is divided by $x^2 + x + 1$.

45 $x^{10} + x^5 + 1$ **46** $x^8 + x + 1$
47 $x^5 + 2x^4 + x^3 + 3x^2 + 2x + 3$ **48** $2x^4 + x^3 + 3x^2 + x + 2$

2.3 Special products

If the corresponding terms of two binomials are similar, as in $ax + by$ and $cx + dy$, we get the product by use of the distributive axiom. The procedure is explained below.

$(ax + by)(cx + dy)$
$$= ax(cx + dy) + by(cx + dy) \qquad \text{by the distributive axiom}$$
$$= acx^2 + adxy + bcxy + bdy^2 \qquad \text{distributive and commutative axiom}$$
$$= acx^2 + (ad + bc)xy + bdy^2$$

Hence we have

$$(ax + by)(cx + dy) = acx^2 + (ad + bc)xy + bdy^2 \qquad (2.7)$$

By observing the polynomial at the right of the equality sign in Eq. (2.7) we see that we obtain the product of two binomials with similar corresponding terms by performing the following steps:

Steps in obtaining the product of two binomials with similar terms

1 Multiply the first terms in the binomials to obtain the first term in the product.
2 Add the products obtained by multiplying the first term in each binomial by the second term in the other. This yields the second term in the product.
3 Multiply the second terms in the binomials to get the third term in the product.

FOIL

The terms on the right side of (2.7) may be remembered by the memory device called FOIL. This means that acx^2 is the product of the *first* terms ax and cx, $adxy$ is the product of the *outside* terms ax and dy, $bcxy$ is the product of the *inside* terms by and cx, and bdy^2 is the product of the *last* terms by and dy.

Taking $y = 1$ in (2.7) gives the formula in one variable:

$$(ax + b)(cx + d) = acx^2 + (ad + bc)x + bd \qquad (1)$$

Ordinarily, the computation required by these steps can be done mentally, and the result can be written with no intermediate steps. This fact is illustrated by the following example.

● **Example 1**   Obtain the product of $2x - 5y$ and $4x + 3y$.

Solution   The product is indicated below. To get it, we proceed as directed below the product, and then record the results in the position indicated by the flow lines.

$$(2x - 5y)(4x + 3y) = 8x^2 - 14xy - 15y^2$$

Get these products mentally:

1 $2x \cdot 4x = $
2 $(2x \cdot 3y) + (-5y \cdot 4x) = 6xy - 20xy = $
3 $-5y \cdot 3y = $

● **Example 2**   Find the product of $3x - 7$ and $4x - 1$.

Solution   By the FOIL method we have

$$\overset{\text{F}\qquad\quad\text{O}\qquad\quad\text{I}\qquad\quad\text{L}}{(3x - 7)(4x - 1) = (3x)(4x) + (3x)(-1) + (-7)(4x) + (-7)(-1)}$$
$$= 12x^2 - 3x - 28x + 7$$
$$= 12x^2 - 31x + 7$$

SQUARE OF THE SUM OR   The square of the sum of two numbers x and y is expressed as $(x + y)^2$. Since
DIFFERENCE OF TWO $(x + y)^2 = (x + y)(x + y)$, we can use Eq. (2.7) and get
NUMBERS

$$(x + y)^2 = (x + y)(x + y)$$
$$= x^2 + (xy + xy) + y^2 \qquad \text{by Eq. (2.7)}$$
$$= x^2 + 2xy + y^2$$

Consequently, $(x + y)^2 = x^2 + 2xy + y^2$ (2.8)

Similarly, $(x - y)^2 = x^2 - 2xy + y^2$ (2.9)

Square of the sum   Therefore, *the square of the sum (or of the difference) of two numbers is*
or difference   *the square of the first number, plus (or minus) twice the product of the two*
of two numbers   *numbers, plus the square of the second number.*

● **Example 3**   By use of Eqs. (2.8) and (2.9), obtain the square of $2a + 5b$ and the square
of $3x - 4y$.

Solution $$(2a + 5b)^2 = (2a)^2 + 2(2a)(5b) + (5b)^2 \qquad \text{by Eq. (2.8)}$$
$$= 4a^2 + 20ab + 25b^2$$
$$(3x - 4y)^2 = (3x)^2 - 2(3x)(4y) + (-4y)^2 \qquad \text{by Eq. (2.9)}$$
$$= 9x^2 - 24xy + 16y^2$$

In addition to $(a + b)^2$, we will consider now $(a + b + c)^2$ and $(a + b)^3$.
For the first we have, by (2.8) with $x = a + b$ and $y = c$,

$$[(a + b) + c][(a + b) + c] = (a + b)^2 + 2(a + b)c + c^2$$
$$= a^2 + 2ab + b^2 + 2ac + 2bc + c^2$$
$$= a^2 + b^2 + c^2 + 2ab + 2ac + 2bc$$

Square of a sum   Therefore, *the square of the sum of three numbers is the sum of the squares*
of the given numbers plus twice the sum of all possible products of two of
the given numbers.

The wording above generalizes to the square of the sum of four, or five,
or n, given numbers.

● **Example 4**   Find $(x + 2y - 3z)^2$.

Solution $$(x + 2y - 3z)^2 = x^2 + 4y^2 + 9z^2 + 4xy - 6xz - 12yz$$

The cube of the sum of two numbers is

Cube of the sum of two numbers

$$(a + b)^3 = (a + b)(a + b)^2$$
$$= a(a^2 + 2ab + b^2) + b(a^2 + 2ab + b^2)$$
$$= a^3 + 3a^2b + 3ab^2 + b^3$$

For $(a + b)^4$, $(a + b)^5$, etc., it is better to use the binomial theorem in Chap. 15 than it is to memorize more special formulas.

● **Example 5** Find $(2x - 5y)^3$.

Solution

$$(2x - 5y)^3 = [2x + (-5y)]^3$$
$$= (2x)^3 + 3(2x)^2(-5y) + 3(2x)(-5y)^2 + (-5y)^3$$
$$= 8x^3 - 60x^2y + 150xy^2 - 125y^3$$ ●

PRODUCT OF THE SUM AND DIFFERENCE OF THE SAME TWO NUMBERS

The product of the sum and the difference of the numbers a and b is expressed as $(a + b)(a - b)$. If we apply Eq. (2.7) to this product, we get

$$(a + b)(a - b) = a^2 + ab - ab - b^2$$
$$= a^2 - b^2$$

Consequently, $(a + b)(a - b) = a^2 - b^2$ (2.10)

Product of the sum and difference of the same two numbers

Therefore, *the product of the sum and the difference of the same two numbers is equal to the difference of their squares.*

We will illustrate the application of Eq. (2.10) with two examples.

● **Example 6** By use of Eq. (2.10), obtain the product of $3x + 5y$ and $3x - 5y$.

Solution

$$(3x + 5y)(3x - 5y) = (3x)^2 - (5y)^2 \quad \text{by Eq. (2.10)}$$
$$= 9x^2 - 25y^2$$ ●

● **Example 7**

$$(2x + y + a)(2x - y + a) = [(2x + a) + y] \cdot [(2x + a) - y]$$
$$= (2x + a)^2 - y^2$$
$$= 4x^2 + 4xa + a^2 - y^2$$ ●

Exercise 2.3 special products

Find the indicated products.

1 $(2x - 1)(x + 3)$

2 $(3b - 1)(b + 2)$

3 $(5a - 1)(a + 2)$

4 $(7c - 1)(c + 4)$

5 $(3x - 2)(2x + 3)$

6 $(5h + 3)(2h - 5)$

7 $(7i - 5)(3i + 4)$

8 $(4a - 7)(5a + 3)$

9 $(2r + 3s)(3r + 2s)$

10 $(7a + 2b)(6a + 5b)$

11 $(4x + 5y)(2x + 3y)$

12 $(8a + 3c)(2a + 5c)$

13 $(2x - 1)^2$

14 $(3i - j)^2$

15 $(3a - 4b)^2$

16 $(6m - 5n)^2$

17 $(5a - 2b)^2$

18 $(4x + 5y)^2$

19 $(7r - 3s)^2$

20 $(10h + 3k)^2$

21 $(2a^2 + 3b^3)^2$

22 $(4a^4 + 5b^3)^2$

23 $(3x^3 + 2y^4)^2$

24 $(5x^5 + 7y^4)^2$

25 $(x + 2y)^3$ **26** $(3x + 5y)^3$

27 $(2x + x^2)^3$ **28** $(3x^2 + x)^3$

29 $(5x - y)^3$ **30** $(6x - 5y)^3$

31 $(7x - 2)^3$ **32** $(5x - 4)^3$

33 $(x + 4)(x - 4)$ **34** $(x + 7)(x - 7)$

35 $(3x + 5)(3x - 5)$ **36** $(2x + 3)(2x - 3)$

37 $(2x - 5y)(2x + 5y)$ **38** $(3x + 4y)(3x - 4y)$

39 $(7x + 6y)(7x - 6y)$ **40** $(8x - 5y)(8x + 5y)$

41 $(2a^2 - 3b^2)(2a^2 + 3b^2)$ **42** $(2a^2 + 5b^2)(2a^2 - 5b^2)$

43 $(5a^2 + 7b^3)(5a^2 - 7b^3)$ **44** $(6a^3 + 7b^2)(6a^3 - 7b^2)$

45 $(a/3 - b/2)(a/3 + b/2)$ **46** $(3a/4 - 4b/3)(3a/4 + 4b/3)$

47 $(2a/3b - 3c/4d)(2a/3b + 3c/4d)$ **48** $(5a/4b + 2b/3a)(5a/4b - 2b/3a)$

49 $(x + y + z)^2 = [(x + y) + z]^2$ **50** $(2x - y + z)^2$

51 $(x - y + 3z)^2$ **52** $(2x + y - z)^2$

53 $(2x + y + z - 3w)^2 = [(2x + y) + (z - 3w]^2$

54 $(x - 2y + 2z + w)^2$ **55** $(x^3 + 2x^2 - 2x + 3)^2$

56 $(2x^3 + 3x^2 + 3x - 2)^2$ **57** $[2(x + y) + 3][3(x + y) + 2]$

58 $[5(2x - y) + 1][2(2x - y) - 3]$ **59** $[4(3x - 2y) + 5][3(3x - 2y) - 2]$

60 $[3(4x - 3y) + 2][5(4x - 3y) + 7]$

61 $[(x^2 - 2x) + 3][(x^2 - 2x) - 3]$

62 $[x^2 + (2x - 3)][x^2 - (2x - 3)]$

63 $(x^2 + 3x - 4)(x^2 + 3x + 4)$

64 $(x^2 - 3x + 4)(x^2 + 3x - 4)$

65 $[(x^3 + x) + (x^2 - 1)][(x^3 + x) - (x^2 - 1)]$

66 $[(x^2 + x) + (x^3 + 1)][(x^2 + x) - (x^3 + 1)]$

67 $[(2x^4 + x) + (x^3 - 2x^2)][(2x^4 + x) - (x^3 - 2x^2)]$

68 $[(x^5 - 3x) + (3x^3 - 1)][(x^5 - 3x) - (3x^3 - 1)]$

69 In a right triangle with sides a, b, and c, it can be shown that the radius of the inscribed circle is

$$r = \frac{ab}{a + b + c}$$

Show that r can be written in the simpler form

$$\frac{a + b - c}{2}$$

Hint: Multiply numerator and denominator by $a + b - c$, and use $a^2 + b^2 = c^2$, which is true in a right triangle with hypotenuse c.

2.4   Factoring: common factors and special binomials

Factored
Irreducible polynomial

We will consider only polynomials in which the numerical coefficients are integers. We say that a polynomial with integral coefficients is *factored* into prime factors if it is expressed as the product of two or more irreducible polynomials of the same type. A polynomial is *irreducible,* or *prime,* if it cannot be expressed as the product of two polynomials of lower degree and if the coefficients have no common factor.

Common factors

Common factor   If each term of a polynomial is divisible by the same term (often a monomial), this term is called a *common factor* of the terms of the polynomial. Such a polynomial can be factored by expressing it as the product of the common factor and the sum of the quotients obtained by dividing each term of the polynomial by the common factor. This procedure is justified by the *distributive axiom*. If either factor thus obtained is not prime, we continue factoring by use of one or more of the methods previously discussed. For example,

$$ab + ac - ad = a(b + c - d) \qquad \text{by the distributive law}$$

This method also can be applied to an expression that is the sum of two or more products which have a common factor.

● Example 1
$$\begin{aligned} a(x^2 + y^2) - a(x^2 - xy - y^2) &= a(x^2 + y^2 - x^2 + xy + y^2) \\ &= a(2y^2 + xy) \\ &= ay(2y + x) \end{aligned}$$

● Example 2
$$(a + b)(a - b) + 2(a + b) = (a + b)(a - b + 2)$$

● Example 3
$$\begin{aligned} (x - 1)(x + 2) - (x - 1)(2x - 3) &= (x - 1)[(x + 2) - (2x - 3)] \\ &= (x - 1)(x + 2 - 2x + 3) \\ &= (x - 1)(-x + 5) \end{aligned}$$

Factors of a binomial

THE DIFFERENCE OF
TWO SQUARES   If we interchange the members of Eq. (2.10), we obtain

$$a^2 - b^2 = (a + b)(a - b) \tag{2.11}$$

Consequently, we have the following rule for factoring the difference of the squares of two numbers:

Factors of the   *The difference of the squares of two numbers is equal to the product of*
difference of two   *the sum and the difference of the two numbers.*
squares

We will illustrate the application of this rule with the following example.

● Example 4   Factor $49a^2 - 16b^2$, $(a + 3b)^2 - 4$, and $x^2 - (y + z)^2$.

Solution
(1) $\begin{aligned}[t] 49a^2 - 16b^2 &= (7a)^2 - (4b)^2 \qquad &&\text{law exponents} \\ &= (7a + 4b)(7a - 4b) \qquad &&\text{by Eq. (2.11)} \end{aligned}$

(2) $\begin{aligned}[t] (a + 3b)^2 - 4 &= (a + 3b)^2 - 2^2 \\ &= (a + 3b + 2)(a + 3b - 2) \qquad &&\text{by Eq. (2.11)} \end{aligned}$

(3) $\begin{aligned}[t] x^2 - (y + z)^2 &= [x + (y + z)][x - (y + z)] \qquad &&\text{by Eq. (2.11)} \\ &= (x + y + z)(x - y - z) \end{aligned}$

Note   Note that the **sum of two squares** cannot in general be factored by using integer coefficients and real numbers.

THE SUM AND   The sum and difference of the cubes of two numbers can be expressed as
DIFFERENCE OF $x^3 + y^3$ and $x^3 - y^3$, respectively. If we divide $x^3 + y^3$ by $x + y$ by long
TWO CUBES   division, we get $x^2 - xy + y^2$ as the quotient. Hence,

$$x^3 + y^3 = (x + y)(x^2 - xy + y^2) \tag{2.12}$$

Similarly,

$$x^3 - y^3 = (x - y)(x^2 + xy + y^2) \tag{2.13}$$

Consequently, we have the following two rules:

Factors of the sum   *If a binomial is expressed as the sum of the cubes of two numbers, one*
or the difference of   *factor is the sum of the two numbers. The other factor is the square of the*
two cubes   *first number minus the product of the two numbers plus the square of the*
second number.

If a binomial is expressed as the difference of the cubes of two numbers,
one factor is the difference of the two numbers. The other factor is the
square of the first number plus the product of the two numbers plus the
square of the second number.

● Example 5   Factor $8x^3 + 27y^3$ and $27a^3 - 64b^6$.

Solution   (1) $8x^3 + 27y^3 = (2x)^3 + (3y)^3$
$= (2x + 3y)[(2x)^2 - (2x)(3y) + (3y)^2]$ by Eq. (2.12)
$= (2x + 3y)(4x^2 - 6xy + 9y^2)$

(2) $27a^3 - 64b^6 = (3a)^3 - (4b^2)^3$
$= (3a - 4b^2)[(3a)^2 + (3a)(4b^2) + (4b^2)^2]$ by Eq. (2.13)
$= (3a - 4b^2)(9a^2 + 12ab^2 + 16b^4)$ ●

Frequently, the factors obtained by use of Eqs. (2.11) to (2.13) can be further factored by a repeated application of one or more of these formulas.

● Example 6   Factor $x^8 - y^8$ and $a^{12} - 64b^6$.

Solution   (1) $x^8 - y^8 = (x^4)^2 - (y^4)^2$
$= (x^4 + y^4)(x^4 - y^4)$
$= (x^4 + y^4)(x^2 + y^2)(x^2 - y^2)$
$= (x^4 + y^4)(x^2 + y^2)(x + y)(x - y)$

(2) $a^{12} - 64b^6 = (a^4)^3 - (4b^2)^3$
$= (a^4 - 4b^2)[(a^4)^2 + (a^4)(4b^2) + (4b^2)^2]$
$= (a^2 - 2b)(a^2 + 2b)(a^8 + 4a^4b^2 + 16b^4)$ ●

Exercise 2.4   Remove the common factor in each of Probs. 1 to 12.
factoring

1 $3x + 9$ **2** $7x - 28$ **3** $5x + 30$
4 $6x - 18$ **5** $x^2 + 3x$ **6** $x^2 - 4x$
7 $x^3 + 6x^2$ **8** $x^3 - 3x^2$ **9** $x^3 - 5x^2 - 4x$
10 $4x^2 - 10x + 2$ **11** $6x^3 + 4x^2 + 8x$ **12** $6x^4 - 15x^3 + 21x^2$

Factor the expression in each of Probs. 13 to 72.

13 $a^2 - x^2$ **14** $9 - y^2$ **15** $x^2 - 25$
16 $y^2 - 36$ **17** $16y^2 - 25a^2$ **18** $64x^2 - 49y^2$
19 $16x^2 - 81y^2$ **20** $49y^2 - 36x^2$ **21** $p^3 - q^3$
22 $a^3 + b^3$ **23** $m^3 + n^3$ **24** $s^3 - t^3$
25 $x^3 - 8y^3$ **26** $27a^3 + b^3$ **27** $8a^3 + 125b^3$
28 $27x^3 - 512y^3$ **29** $a^2b^2 - 9c^2$ **30** $p^2q^2 - 4r^2$
31 $9a^2 - 4b^2c^2$ **32** $25x^2 - 36y^2z^2$ **33** $x^2 - y^4$
34 $x^6 - y^2$ **35** $x^6 - y^8$ **36** $x^4 - y^{10}$
37 $x^6 - y^3$ **38** $x^6 - y^9$ **39** $x^3 - y^9$
40 $x^{12} - y^9$ **41** $(x - 2y)^2 - 9z^2$ **42** $(3x - y)^2 - 4z^2$
43 $81x^2 - (4y - 5z)^2$ **44** $16a^2 - (3x - 2y)^2$ **45** $(x + y)^3 - 1$
46 $(2x - y)^3 - 8$ **47** $(3x + 2y)^3 + 27$ **48** $(x - 3y)^3 + 64$
49 $16x^4 - y^4$ **50** $x^4 - 81y^4$ **51** $81x^4 - 625y^4$
52 $16x^4 - 81y^4$ **53** $x^4 - y^8$ **54** $x^8 - 81$
55 $x^{12} - y^8$ **56** $x^8 - y^6$ **57** $x^4 - y^4$
58 $x^8 - y^8$ **59** $x^{16} - y^4$ **60** $x^8 - y^4$
61 $x^6 - y^6$ **62** $x^9 + y^9$ **63** $x^6 - 27$
64 $x^9 - y^9$ **65** $16x^4 - 1$ **66** $16 - 81x^4$
67 $625x^4 - 81y^8$ **68** $81x^8 - 256y^8$ **69** $x^5 + y^5$
70 $x^7 + y^7$ **71** $x^5 - y^5$ **72** $x^7 - y^7$

Hint: In Probs. 69 to 72, one factor is $x + y$ or $x - y$.

Suppose that $w^3 = 1$, but $w \neq 1$. Show that Probs. 73 to 76 are true.

73 $1 + w + w^2 = 0$. *Hint:* Factor $w^3 - 1 = 0$.
74 $x^3 + y^3 = (x + y)(x + wy)(x + w^2y)$
75 $(a + wb + w^2c)(a + w^2b + wc) = a^2 + b^2 + c^2 - ab - bc - ca$
76 $(1 - w)(1 - w^2)(1 - w^4)(1 - w^5) = 9$
77 The statements below are a proof of the following simplified version of Fermat's last theorem: If x, y, z, and n are positive integers with $x < y < z \leq n$, then $x^n + y^n \neq z^n$. Verify each step.

$$z^n - y^n = (z - y)(z^{n-1} + yz^{n-2} + y^2z^{n-3} + \cdots + y^{n-2}z + y^{n-1})$$
$$> (1)(x^{n-1} + x^{n-1} + x^{n-1} + \cdots + x^{n-1} + x^{n-1})$$
$$= n(x^{n-1})$$
$$> x^n$$

2.5 Factors of trinomials

Trinomials that are perfect squares

If a trinomial is the square of a binomial, we know by Eq. (2.8) that two of its terms are perfect squares and hence are positive and that the third term is plus or minus twice the product of the positive square roots of the two perfect squares. Furthermore, such a trinomial is the square of a binomial composed of the positive square roots of the perfect-square terms connected by the sign that precedes the other term. In other words, we have

$$a^2 + 2ab + b^2 = (a + b)^2$$
$$a^2 - 2ab + b^2 = (a - b)^2$$

● **Example 1** Factor   (1) $4x^2 - 12xy + 9y^2$,   (2) $9a^2 + 24ab + 16b^2$,   and (3) $(2a - 3b)^2 - 8(2a - 3b) + 16$.

Solution (1) Since $4x^2 = (2x)^2$, $9y^2 = (3y)^2$, and $12xy = 2(2x)(3y)$, we have

$$4x^2 - 12xy + 9y^2 = (2x - 3y)(2x - 3y) = (2x - 3y)^2$$

(2) $9a^2 + 24ab + 16b^2 = (3a + 4b)^2$

(3) $(2a - 3b)^2 - 8(2a - 3b) + 16 = [(2a - 3b) - 4]^2$ ●

Factors of a quadratic trinomial

A trinomial of the type $ax^2 + bxy + cy^2$, where a, b, and c stand for integers, is a *quadratic trinomial with integral coefficients*. In this section we will discuss methods for finding the two binomial factors of such a trinomial *if such factors exist*. Equation (2.7) with the members interchanged is

$$acx^2 + (ad + bc)xy + bdy^2 = (ax + by)(cx + dy) \qquad (2.7)$$

With $y = 1$, it is $acx^2 + (ad + bc)x + bd = (ax + b)(cx + d)$.   ·

● **Example 2** Factor $x^2 + 10x + 21$.

Solution We must have factors of the form

$$(x + \quad)(x + \quad)$$

because the 10 and 21 are preceded by + signs. The only possibilities are

$$(x + 21)(x + 1) \qquad \text{and} \qquad (x + 3)(x + 7)$$

and the one that works is $(x + 3)(x + 7)$.   ●

● **Example 3** Factor $x^2 - 7x + 12$.

Solution The factors must have the form

$$(x -)(x -)$$

because 12 has a + sign, but the middle term 7 has a − sign. The possible factors are

$$(x - 12)(x - 1) \qquad (x - 6)(x - 2) \qquad (x - 4)(x - 3)$$

and the one that works is $(x - 4)(x - 3)$. ●

Examples 2 and 3 illustrate that if $ax^2 + bx + c$ has $c > 0$, then the signs in the factors must be alike, and both are like the sign of b. That is, the factors have the form

$$(+)(+) \qquad \text{if } b > 0$$
$$(-)(-) \qquad \text{if } b < 0$$

If, however, $c < 0$, the form is $(+)(-)$, or $(-)(+)$.

● **Example 4** In factoring $3x^2 - 10xy - 8y^2$, we begin with

$$(3x)(x)$$

The signs must be opposite to give $-8y^2$, hence the form must be

$$(3x +)(x -) \qquad \text{or} \qquad (3x -)(x +)$$

Since the factors of 8 are 8, 4, 2, and 1, all possible factors are

$$(3x + 8y)(x - y) \qquad (3x - 8y)(x + y)$$
$$(3x + 4y)(x - 2y) \qquad (3x - 4y)(x + 2y)$$
$$(3x + 2y)(x - 4y) \qquad (3x - 2y)(x + 4y)$$
$$(3x + y)(x - 8y) \qquad (3x - y)(x + 8y)$$

The correct one is $(3x + 2y)(x - 4y) = 3x^2 - 10xy - 8y^2$. ●

Note Notice that also $3x^2 - 10xy - 8y^2 = (-3x - 2y)(-x + 4y)$, since $(-1) \times (-1) = 1$. Also, $3x^2 - 10x - 8 = (3x + 2)(x - 4)$, by taking $y = 1$.

It is often desirable to know whether a quadratic trinomial is factorable.

Is a trinomial factorable? In Chap. 5 we show that for a, b, and c integers, $ax^2 + bx + c$ is factorable with integer coefficients if and only if $b^2 - 4ac$ is a nonnegative perfect square.

● **Example 5** Is the trinomial factorable? (i) $7x^2 - 12x + 4$; (ii) $6x^2 - 13x + 6$.

Solution (i) $b^2 - 4ac = (-12)^2 - 4(7)(4) = 144 - 112 = 32$, which is not a perfect square; so it is not factorable.

(ii) $b^2 - 4ac = (-13)^2 - 4(6)(6) = 169 - 144 = 25 = 5^2$; so it is factorable. There are many possibilities to try—however, the middle signs must be the same because of the + sign in the constant term

+6. Further, they must both be minus because of the minus sign in $(-13x)$. The possibilities are

$$(6x - \quad)(x - \quad) \qquad \text{and} \qquad (3x - \quad)(2x - \quad)$$

The factors are $(2x - 3)(3x - 2) = 6x^2 - 13x + 6$. ●

● **Example 6** Factor $15x^2 + 11x - 12$.

Solution Some of the possibilities are $(15x - \quad)(x + \quad)$, $(5x + \quad)(3x - \quad)$, and $(5x - \quad)(3x + \quad)$. The actual factors are $(5x - 3)(3x + 4)$, and it requires a bit of trial and error to discover them.

Alternate solution To factor $15x^2 + 11x - 12$, first write

$$(15x + \quad)(15x - \quad)$$

using $15x$ in both factors. We now seek a and b for which $ab = (15) \times (-12) = -180$ and $a + b = 11$, where the 15, -12, and 11 come from the given trinomial. Trying 12 and -1, 13 and -2, 14 and -3, etc., gives $a = 20$ and $b = -9$. Now write

$$(15x + 20)(15x - 9)$$

and divide each term by its gcd:

$$\left(\frac{15x + 20}{5}\right)\left(\frac{15x - 9}{3}\right) = (3x + 4)(5x - 3)$$

which is the factorization. ●

For another alternate way to factor, see Prob. 68.

If a trinomial can be made a perfect square by *adding* a perfect square term, then the given trinomial can be factored as a difference of two squares.

● **Example 7**

$$\begin{aligned}
4x^4 + 8x^2y^2 + 9y^4 &= 4x^4 + 8x^2y^2 + 9y^4 + 4x^2y^2 - 4x^2y^2 \\
&= 4x^4 + 12x^2y^2 + 9y^4 - 4x^2y^2 \\
&= (2x^2 + 3y^2)^2 - (2xy)^2 \\
&= (2x^2 + 3y^2 + 2xy)(2x^2 + 3y^2 - 2xy) \\
&= (2x^2 + 3y^2 + 2xy)(2x - y)(x + 2y)
\end{aligned}$$

Exercise 2.5

Factor the expressions in Probs. 1 to 24.

1 $x^2 - 2x + 1$ **2** $x^2 - 6x + 9$ **3** $x^2 - 4x + 4$
4 $x^2 + 6x + 9$ **5** $2y^2 + 3y - 2$ **6** $3x^2 - 5x - 2$
7 $5h^2 + 2h - 3$ **8** $7p^2 + 2p - 5$ **9** $6r^2 + 13r - 5$
10 $8y^2 + 3y - 5$ **11** $9y^2 - 18y - 7$ **12** $9a^2 + 98a - 11$
13 $3a^2 + 12ab + 12b^2$ **14** $5x^2 + 18xy + 16y^2$
15 $18u^2 - 23uv + 7v^2$ **16** $24r^2 - 26rs + 5s^2$

17 $8h^2 + 5hk - 3k^2$ **18** $2c^2 + 7cd - 9d^2$
19 $3a^2 + 13ab - 10b^2$ **20** $5x^2 + 11xy - 12y^2$
21 $6p^2 - 13pq + 6q^2$ **22** $12y^2 + 17yz + 6z^2$
23 $15b^2 - 17bc + 4c^2$ **24** $16u^2 + 32uv + 15v^2$

Test the expression in each of Probs. 25 to 64 to find if it is factorable rationally in terms of the coefficients. Factor those that can be so factored, and give the value of $b^2 - 4ac$ for the others.

25 $3x^2 + 7x + 2$ **26** $2x^2 + 5x + 2$ **27** $6x^2 + 7x - 3$
28 $4x^2 - 19x + 12$ **29** $7x^2 - 10x + 3$ **30** $3x^2 + 13x - 10$
31 $4x^2 + 4x - 3$ **32** $5x^2 + 6x - 8$ **33** $2x^2 + 7x - 2$
34 $x^2 + 4x + 5$ **35** $5x^2 + 6x + 7$ **36** $5x^2 + 4x + 1$
37 $2x^2 + 9x - 9$ **38** $3x^2 + 8x + 2$ **39** $4x^2 + 9x - 5$
40 $3x^2 + 11x - 8$ **41** $x^2 + 4x - 5$ **42** $5x^2 + 4x - 1$
43 $2x^2 + 9x + 9$ **44** $4x^2 + 9x + 5$ **45** $2x^2 - xy - 28y^2$
46 $3x^2 - xy - 24y^2$ **47** $5x^2 - 9xy - 18y^2$
48 $15x^2 + 11xy - 12y^2$ **49** $20x^2 + 3xy - 35y^2$
50 $24x^2 + 26xy - 15y^2$ **51** $42x^2 + 5xy - 25y^2$
52 $45x^2 + 2xy - 15y^2$ **53** $30x^2 + 7xy - 49y^2$
54 $54x^2 - 3xy - 35y^2$ **55** $30x^2 + 37xy - 12y^2$
56 $30x^2 - xy - 99y^2$ **57** $77x^2 + 155xy - 48y^2$
58 $84x^2 - 23xy - 42y^2$ **59** $77x^2 + 13xy - 10y^2$
60 $60x^2 - 124xy + 63y^2$
61 $10(3x - 4y)^2 + 19(3x - 4y) + 6$
62 $12x^2 - x(2y - 3z) - 6(2y - 3z)^2$
63 $15(3x - 5y)^2 + 31(3x - 5y)z - 24z^2$
64 $12(5x - 2y)^2 - (5x - 2y)z - 20z^2$

Biology   **65** In the Wilcoxon test for unpaired data,

$$U = T - \frac{n(m + 1)}{2}$$

where T is total of the ranks, n is the number of cases in the group with the smaller mean, and m is the number of cases in the group with the larger mean. Find U if
(a) $m = 6$, $n = 8$, $T = 43$
(b) $m = n$ and $T = (n + 1)^2$

Psychology   **66** In the theory of learning in psychology, probabilities t and p and a constant b are related by

$$t = (1 - b)p$$

Show that this equation may be rewritten as

$$1 - t = (1 - b)(1 - p) + b$$

Psychology   **67** The number $\phi = (BC - AD)/(A + B)(C + D)$ measures the reproducibility of a test, or the degree to which one can reproduce a person's total response pattern from a knowledge of the total score and the order of difficulty of the items. A, B, C, and D are percentages with $A \leq B$ and $C \leq D$. (a) Find ϕ if $A = D$ and $B = C$ (b) Find ϕ if $A = 74$ percent, $B = 89$ percent, $D = 26$ percent, $C = 11$ percent

68 Verify each step in the following factorization; in step 2, we add and subtract $[\frac{1}{2}(\frac{7}{8})]^2$.

$$16x^2 + 14x - 15 = 16[x^2 + \tfrac{7}{8}x + \tfrac{49}{256} - \tfrac{15}{16} - \tfrac{49}{256}]$$
$$= 16[(x + \tfrac{7}{16})^2 - \tfrac{289}{256}] = 16[(x + \tfrac{7}{16})^2 - (\tfrac{17}{16})^2]$$
$$= 16[(x + \tfrac{7}{16} - \tfrac{17}{16})(x + \tfrac{7}{16} + \tfrac{17}{16})]$$
$$= (8)(2)(x - \tfrac{5}{8})(x + \tfrac{3}{2})$$
$$= (8x - 5)(2x + 3)$$

2.6 Further factoring

Grouping   Frequently, the terms in a polynomial can be *grouped* in such a way that each group has a *common factor*. We will illustrate the method with two examples.

● **Example 1**   Factor $ax + bx - ay - by$.

Solution   We notice that the first two terms have the common factor x and that the third and fourth terms have the common factor y. Hence we group the terms in this way, $(ax + bx) - (ay + by)$, and then proceed as indicated below.

$$ax + bx - ay - by = (ax + bx) - (ay + by) \quad \text{since } -(ay + by) =$$
$$= x(a + b) - y(a + b) \qquad -ay - by \text{ by the}$$
$$= (a + b)(x - y) \qquad\qquad \text{distributive axiom } ●$$

● **Example 2**   Factor $a^2 + ab - 2b^2 + 2a - 2b$.

Solution   Since $a^2 + ab - 2b^2 = (a + 2b)(a - b)$ and $2a - 2b = 2(a - b)$, we proceed as indicated below.

$$a^2 + ab - 2b^2 + 2a - 2b = (a^2 + ab - 2b^2) + (2a - 2b)$$
$$= (a + 2b)(a - b) + 2(a - b)$$
$$= (a - b)(a + 2b + 2)$$
$$\qquad\qquad a - b \text{ is the common factor } ●$$

Often, after the terms in a polynomial are suitably grouped, it becomes evident that the methods of previous sections can be applied.

● **Example 3**   Factor $4c^2 - a^2 + 2ab - b^2$.

Solution $4c^2 - a^2 + 2ab - b^2 = 4c^2 - (a^2 - 2ab + b^2)$ inserting parentheses
$$= 4c^2 - (a - b)^2 \qquad \text{square of a difference}$$
$$= (2c)^2 - (a - b)^2$$
$$= [2c + (a - b)][2c - (a - b)] \text{ difference of two}$$
$$= (2c + a - b)(2c - a + b) \qquad \text{squares} \qquad \bullet$$

In Chap. 11, we will show that $x - r$ is a factor of a given polynomial if and only if the value of the polynomial is zero when x is replaced by r. Synthetic division or long division by $x - r$ then yields the other factor.

● **Example 4**   Factor $x^3 + 3x^2 - 8x - 4$.

Solution   For now, we must resort to guessing the possible factors. Replacing x by 2 gives $8 + 12 - 16 - 4 = 0$, hence $x - 2$ is a factor.

Synthetic division
$$\begin{array}{r|rrrr} 2 & 1 & 3 & -8 & -4 \\ & & 2 & 10 & 4 \\ \hline & 1 & 5 & 2 & 0 \end{array}$$

Hence $x^3 + 3x^2 - 8x - 4 = (x - 2)(x^2 + 5x + 2)$. Note that $x^2 + 5x + 2$ is irreducible since $b^2 - 4ac = 5^2 - 4(1)(2) = 25 - 8 = 17$, which is not a perfect square.   ●

Exercise 2.6
grouping

Factor the expressions in Probs. 1 to 40.

1 $ab + a + b + 1$
2 $xy + 2x + y + 2$
3 $uv + 3u + 2v + 6$
4 $rs + 4r + 2s + 8$
5 $a^2 - ab + a - b$
6 $xy + x - 2y^2 - 2y$
7 $2c^2 + 4cd - 3c - 6d$
8 $2h^2 - 5hk + 4h - 10k$
9 $ac + ad + bc + bd$
10 $xy - y^2 + xz - yz$
11 $6rt + 2ru - 15st - 5su$
12 $6uv - 12ux - 5vw + 10wx$
13 $6a^2 - 4ac - 15ab + 10bc$
14 $4x^2 - 3xy - 24xz + 18yz$
15 $15h^2 - 9hk + 35hj - 21kj$
16 $6r^2 - 5rs + 18rt - 15st$
17 $a^2 - ab + ac - a + b - c$
18 $x^2 - xy - xz + x - y - z$
19 $r^2 - 2rs - rt - 3ru + 6su + 3tu$
20 $6bc - 9c^2 - 12cd - 8be + 12ce + 16de$
21 $x^2 - y^2 - x - y$
22 $a^2 - b^2 - a + b$
23 $a^2 - 2ab + b^2 - ac + bc$
24 $x^2 + 2xy + y^2 - xz - yz$
25 $x^3 - y^3 - x^2 + 2xy - y^2$
26 $a^2 + 2ab + b^2 - a^3 - b^3$
27 $x^2 + xy - 2y^2 - x^3 + y^3$
28 $2c^2 + 5cd - 3d^2 + 8c^3 - d^3$
29 $h^2 - 4hk + 4k^2 - 16$
30 $r^2 + 6rs + 9s^2 - 4t^2$
31 $25a^2 - b^2 + 4bd - 4d^2$
32 $9x^2 - 4y^2 - 12yz - 9z^2$
33 $x^2 - 4xy + 4y^2 - 4z^2 - 12zw - 9w^2$
34 $4a^2 - 4ab + b^2 - c^2 + 6cd - 9d^2$
35 $4r^2 + 12rs + 9s^2 - t^2 - 8tu - 16u^2$
36 $9h^2 - 6hi + i^2 - 16j^2 + 8jk - k^2$

37 $a^2 - 3ab + 3ac + 2b^2 - 3bc$ **38** $x^2 - 3xy + 2y^2 - xz + yz$
39 $4rs - 6s^2 + 17st - 6rt - 12t^2$ **40** $6ab - 12b^2 + 11bs - 4as - 2s^2$
41 $x^4 - (2x + 3)^2$ **42** $x^4 - (x + 6)^2$
43 $x^4 - (9x^2 - 12x + 4)$ **44** $x^4 - (4x^2 - 32x + 64)$

In Probs. 45 to 48, one factor is $x - 1$, $x - 2$, $x - 3$, $x - 4$, or $x - 5$.

45 $4x^3 - 16x^2 + 9x + 9$ **46** $6x^3 - 29x^2 - 6x + 5$
47 $x^4 - 4x^3 - x + 4$ **48** $x^4 - 4x^3 - 8x + 32$

2.7   Key concepts

Be certain that you understand and can use each of the following words, ideas, and indicated procedures.

Parentheses, brackets, braces	Polynomial
Synthetic division	Degree
Factorability	Special products
Monomial	

(2.1)   If $b \neq 0$, then $a/b = x$ if and only if $bx = a$

(2.2) $a/a = 1$, if $a \neq 0$

(2.3) $a/1 = a$

(2.4) $\dfrac{ac}{bd} = \dfrac{a}{b} \cdot \dfrac{c}{d}$

(2.5) $\dfrac{a}{d} = a \cdot \dfrac{1}{d}$

(2.6)   Dividend = (divisor)(quotient) + remainder

(2.7) $(ax + by)(cx + dy) = acx^2 + (ad + bc)xy + bdy^2$

(2.8) $(x + y)^2 = x^2 + 2xy + y^2$
(2.9) $(x - y)^2 = x^2 - 2xy + y^2$
(2.10) $(a + b)(a - b) = a^2 - b^2$
(2.11) $a^2 - b^2 = (a + b)(a - b)$
(2.12) $x^3 + y^3 = (x + y)(x^2 - xy + y^2)$
(2.13) $x^3 - y^3 = (x - y)(x^2 + xy + y^2)$

Exercise 2.7 review   Find the indicated sums, products, and quotients.

1 $(3a + 2b - 5c) - (2a + 6b - 4c) + (4a - 4b - c)$
2 $-(-6a + b + 5c) + (2a - 2b + 3c) - (4a - 3b + 7c)$
3 $3x - 2y + 4[-x + 3(y - 2x) - 3y] - 2x$
4 $3a + 2\{b - 4[a + 2(2a - b) + b] - a\} + 2b$

5 $(6x^3 + x^2 - 1) - (2x^3 - 3x^2 - 4) - (-4x^2 - 7x + 1)$

6 $3x + 2 + x\{4x + 3[2 - 5x(x + 1) - x] - 4x\} + 2 - x$

7 $2a - 3\{2a - 3[2a - 3(2a - 3) + 2a] - 3\}$

8 $\dfrac{(5x^2)^4(4x^4)^2}{(2x^8)^3}$ **9** $\dfrac{(11x^4)^0(12x^0)}{6x}$ **10** $4x^3(3x^2y - 6xy^4)$

11 $-x^8y^7(x^7y^{11} - x^6y^9)$ **12** $\dfrac{14x^8y^4 - 10x^6y^5}{2x^4y^4}$

13 $(4x - 3y)(2x + 5y)$ **14** $(2x - 3y)(3x + 2y)$

15 $(3x + 5y)^2$ **16** $(7x + 4y)^2$

17 $(2x + 7y)(2x - 7y)$ **18** $(3x^2 + 2y^3)(3x^2 - 2y^3)$

19 $(2x^2 - x + 3)(2x^2 + x - 3)$

20 $(x^3 - 3x^2 - 2x + 1)(x^3 - 3x^2 + 2x - 1)$

21 $(2x^2 - 3xy + y^2)(3x^2 + 2xy - y^2)$

22 $(5x^2 - 7xy - y^2)(x^2 + 3xy + 2y^2)$

23 Find the quotient and remainder if $8x^3 + 12x^2 - 14x + 7$ is divided by $4x^2 + 8x - 3$.

24 Find the quotient and remainder if $6x^4 + 5x^3 - x^2 + 14x + 1$ is divided by $2x^2 + 3x - 1$.

25 Use synthetic division to find remainder if $3x^4 - 2x^3 + 7x^2 - 4x - 50$ is divided by $x - 2$.

26 Use synthetic division to show $x - 4$ is a factor of $2x^3 - 7x^2 - 3x - 4$.

Factor the expression in each of Probs. 27 to 40 or state why it cannot be factored.

27 $7x - 21y$ **28** $3x^3 - 6x^2 + 9x$

29 $9 - 16x^2$ **30** $4x^2 - 12x + 9 - 25y^2$

31 $x^4 - 81y^4$ **32** $x^5 - y^{10}$

33 $x^2 - 14x + 49$ **34** $9x^2 - 24x + 16$

35 $12x^2 - 13x + 3$ **36** $12x^2 - 13x - 4$

37 $12x^2 - 13x - 3$ **38** $12x^2 - 13x$

39 $xy - 3x + 2y - 6$ **40** $x^3 - y^3 - x^2 + y^2$

41 Find $(a + b + c + d)^2$.

42 Show that the sum of two squares times the sum of two squares is a sum of two squares, by verifying that

$$(a^2 + b^2)(c^2 + d^2) = (ac - bd)^2 + (ad + bc)^2$$

Write $(3^2 + 4^2)(2^2 + 7^2)$ as a sum of two squares.

43 Simplify $\dfrac{(x + h)^3 - x^3}{h}$.

44 Find $(1 - r)(1 + r + r^2 + r^3 + r^4 + r^5 + r^6)$.

45 Factor $x^4 - (4x - 5)^2$.

46 Calculate $85^2 = (80 + 5)^2$, and $(73)(87) = (80 - 7)(80 + 7)$.

47 Verify $(a + b)^2 \geq 4ab$ for $a = 2$, $b = 5$ and for $a = 7$, $b = 8$.

48 If h is close to 0, then $(1 + h)(1 - h) = 1 - h^2 \approx 1$; hence $1/(1 + h)$ $\approx 1 - h$. Use this to approximate

$$\frac{1}{1.003} \quad \text{and} \quad \frac{4}{0.98} = 4\left[\frac{1}{1 + (-.02)}\right]$$

49 Write $x(x + 1)(x + 2)(x + 3) + 1$ in the form $(x^2 + ax + b)^2$ for some constants a and b.

50 Show that if $y = x + \dfrac{1}{x}$, then

(a) $y^2 - 2 = x^2 + \dfrac{1}{x^2}$

(b) $y^3 - 3y = x^3 + \dfrac{1}{x^3}$

(c) $y^4 - 4y^2 + 2 = x^4 + \dfrac{1}{x^4}$

51 Show that $(a - b)^3 + (b - c)^3 + (c - a)^3 = 3(a - b)(b - c)(c - a)$.

In Probs. 52 to 54, let $\emptyset = a^2 + b^2 + c^2 - ab - bc - ca$.

52 Show that $a^3 + b^3 + c^3 - 3abc = (a + b + c)\emptyset$.
53 Show that $(a - b)^5 + (b - c)^5 + (c - a)^5 = 5(a - b)(b - c)(c - a)\emptyset$.
54 Show that $(a - b)^7 + (b - c)^7 + (c - a)^7 = 7(a - b)(b - c)(c - a)\emptyset^2$.

CHAPTER 3 · FRACTIONS

3.1 Equivalent fractions

A *rational expression,* or a *fraction,* is a quotient of algebraic expressions (remember that division by zero is not defined). We will usually work with quotients of polynomials in this chapter. Since the values of these fractions are real numbers, our previous rules continue to hold, as well as a few new ones.

○ **Theorems**   For all real numbers a, b, c, and d with $b \neq 0$ and $d \neq 0$, we have

Equivalent fractions: $\dfrac{a}{b} = \dfrac{c}{d}$ if and only if $ad = bc$ (3.1)

Fundamental principle of fractions:

$$\text{For any } k \neq 0 \qquad \frac{a}{b} = \frac{ak}{bk} = \frac{a/d}{b/d} \qquad (3.2)$$

Signs of fractions: $\dfrac{a}{b} = \dfrac{-a}{-b} = -\dfrac{-a}{b} = -\dfrac{a}{-b}$

$$-\frac{a}{b} = \frac{-a}{b} = \frac{a}{-b} = -\frac{-a}{-b} \qquad (3.3)$$

Product of fractions: $\dfrac{a}{b} \cdot \dfrac{c}{d} = \dfrac{ac}{bd}$ (3.4)

Sum of fractions: $\dfrac{a}{d} + \dfrac{c}{d} = \dfrac{a+c}{d}$ (3.5)

To prove (3.1), suppose $a/b = c/d$. Multiplying each member of the equation by bd gives

$$\frac{a}{b} \cdot \frac{bd}{1} = \frac{a \cdot b \cdot d}{b \cdot 1 \cdot 1} = \frac{a}{1} \cdot \frac{b}{b} \cdot \frac{d}{1} = a \cdot 1 \cdot d = ad$$

and also

$$\frac{c}{d} \cdot \frac{bd}{1} = \frac{c \cdot b \cdot d}{d \cdot 1 \cdot 1} = \frac{c}{1} \cdot \frac{b}{1} \cdot \frac{d}{d} = c \cdot b \cdot 1 = bc$$

Thus if $a/b = c/d$, then $ad = bc$. The proof of the converse is similar: start with $ad = bc$ and divide each member by bd.

● **Example 1**
$$\frac{5}{8} = \frac{20}{32} \qquad \text{since } 5(32) = 8(20) = 160$$

$$\frac{12}{27} = \frac{28}{63} \qquad \text{since } 12(63) = 27(28) = 756$$

The proof of (3.2) follows from (3.1) since, for instance,

$$\frac{a}{b} = \frac{ak}{bk} \qquad \text{if and only if } a(bk) = b(ak)$$

and the last equation is true by the commutative law. The proof that

$$\frac{a}{b} = \frac{a/d}{b/d}$$

is similar. This fundamental principle of fractions is very useful in simplifying fractions and in getting common denominators.

● **Example 2**
$$\frac{3}{4} = \frac{3 \cdot 5}{4 \cdot 5} = \frac{15}{20}$$

● **Example 3**
$$\frac{x}{y} = \frac{x \cdot 2x^2y^3}{y \cdot 2x^2y^3} = \frac{2x^3y^3}{2x^2y^4}$$

● **Example 4**
$$\frac{24}{36} = \frac{24 \div 12}{36 \div 12} = \frac{2}{3}$$

● **Example 5**
$$\frac{15x^3y^7}{25x^5y^2} = \frac{15x^3y^7 \div 5x^3y^2}{25x^5y^2 \div 5x^3y^2}$$

$$= \frac{3x^0y^5}{5x^2y^0} \qquad \text{by law of exponents for division}$$

$$= \frac{3y^5}{5x^2} \qquad \text{since } x^0 = y^0 = 1$$

The proof of (3.3) again goes back to (3.1). We will only prove part of it:

$$\frac{a}{b} = \frac{-a}{-b} \qquad \text{if and only if} \quad a(-b) = b(-a)$$

and $\qquad a(-b) = -ab \qquad$ and $\qquad b(-a) = -ba = -ab$

Signs of fractions There are three signs associated with any fraction, and they are the sign preceding the fraction, the sign preceding the numerator, and the sign preceding the denominator. We have shown that if two of these signs in a given fraction are changed, the resulting fraction is equal to the given fraction.

● **Example 6**
$$\frac{-x}{y-x} = \frac{-(-x)}{-(y-x)} = \frac{x}{x-y}$$ ●

● **Example 7**
$$\frac{y^3 - x^3}{x - y} = -\frac{-(y^3 - x^3)}{x - y} = -\frac{x^3 - y^3}{x - y}$$ ●

Equation (3.4), on multiplying fractions, was shown in Sec. 2.2. As a result of it, we see how to divide two fractions.

$$\frac{a/b}{c/d} = \frac{(a/b)(d/c)}{(c/d)(d/c)} = \frac{ad/bc}{cd/cd} = \frac{ad}{bc} = \frac{a}{b} \cdot \frac{d}{c} \qquad (3.6)$$

Division of fractions Thus to divide by a fraction c/d, we *multiply by its reciprocal d/c.*

Equation (3.5), on adding fractions, is a direct result of the distributive law.

$$\frac{a}{d} + \frac{b}{d} = a\left(\frac{1}{d}\right) + b\left(\frac{1}{d}\right) = (a + b)\frac{1}{d} = \frac{a + b}{d}$$

Cancellation

A fraction is said to be in lowest terms if the numerator and denominator have no common factors except 1. We call such a fraction a *reduced fraction.* Consequently, to reduce a given fraction to lowest terms, we divide the numerator and denominator by each factor that is common to both and thus obtain the numerator and denominator, respectively, of the reduced fraction. In symbols, we have, for instance,

$$\frac{a \cdot p \cdot q}{b \cdot p \cdot q} = \frac{a}{b}$$

Note This process is called *canceling* the common factors. It is vital to cancel only *factors* that are in the numerator and denominator. Thus

$$\frac{a + b + 1}{a + 3} \text{ is not the same as } \frac{b + 1}{3}$$

because a is *not* a factor of numerator and denominator; it is added, not multiplied. However, it is true that

$$\frac{a(b + 1)}{3a} = \frac{b + 1}{3}$$

If the common factors are not clearly discernible, it is advisable to factor the members of the fraction before attempting the reduction. We will illustrate the procedure with two examples.

● **Example 8**   Reduce $35a^4b^2/42a^3b^3$ to lowest terms.

Solution   The common factor is $7a^3b^2$, hence

$$\frac{35a^4b^2}{42a^3b^3} = \frac{5a \cdot 7a^3b^2}{6b \cdot 7a^3b^2} = \frac{5a}{6b}$$

after canceling $7a^3b^2$ from the numerator and denominator.   ●

● **Example 9**   Reduce $(x^3 + x^2 - 6x)/(x^3 - 3x^2 + 2x)$ to lowest terms.

Solution   We will first factor the members of the given fraction and then proceed as indicated below.

$$\frac{x^3 + x^2 - 6x}{x^3 - 3x^2 + 2x} = \frac{x(x^2 + x - 6)}{x(x^2 - 3x + 2)} \qquad \text{common factor of } x$$

$$= \frac{x(x - 2)(x + 3)}{x(x - 2)(x - 1)} \qquad \text{factoring}$$

$$= \frac{x + 3}{x - 1} \qquad \text{canceling } x \text{ and } x - 2 \qquad ●$$

In many operations, such as adding fractions, it is desirable to convert a given fraction to another in which the denominator has a specified form. We accomplish this conversion by the application of Eq. (3.2), in which we multiply the members of the fraction by the number or expression necessary to produce the required denominator.

● **Example 10**   Convert $(a + b)/(a - b)$ to an equal fraction with $a^2 - b^2$ as the denominator.

Solution   Since $(a - b)(a + b) = a^2 - b^2$, we multiply the members of the given fraction by $a + b$ and get

$$\frac{a + b}{a - b} = \frac{(a + b)(a + b)}{(a - b)(a + b)} = \frac{a^2 + 2ab + b^2}{a^2 - b^2}$$
                                                                                              ●

The least common denominator

Least
common
denominator

The *least common denominator* of a set of fractions is the lowest common multiple of the denominators of the fractions in the set. It is abbreviated as lcd and is usually obtained in factored form. It must be divisible by every denominator in the set and must have no more factors than are needed to satisfy this requirement.

In obtaining the lcd, we begin by factoring each denominator into prime factors. Then the lcd is the product of the different prime factors of the denominators, each with an exponent that is equal to the greatest exponent of that prime factor in any given denominator.

● **Example 11**   Find the lcd if the denominators are $(x - 2)^4(x + 1)$, $(x - 2)(x + 1)^3 \times (x - 1)$, and $(x - 2)^2(x - 1)^2$.

Solution   The different factors are $x - 2$, $x + 1$, and $x - 1$ and their greatest exponents are 4, 3, and 2, respectively. Consequently, the lcd is $(x - 2)^4 \times (x + 1)^3(x - 1)^2$.   ●

● **Example 12**   Convert each of the fractions a/xy, $3b/x^3y$, and $2c/xy^2$ to an equal fraction with the lcd as its denominator.

Solution   The lcd is x^3y^2; hence, we multiply the members of the given fractions by $x^3y^2/xy = x^2y$, $x^3y^2/x^3y = y$, and $x^3y^2/xy^2 = x^2$, respectively, and get

$$\frac{ax^2y}{x^3y^2} \qquad \frac{3by}{x^3y^2} \qquad \text{and} \qquad \frac{2cx^2}{x^3y^2}$$
●

Exercise 3.1
Equivalent
fractions

Change the fraction in each of Probs. 1 to 24 to an equivalent fraction with the expression to the right of the comma as denominator.

1 $\dfrac{-2}{y - x}$, $x - y$

2 $\dfrac{a + b}{b - a}$, $a - b$

3 $\dfrac{2x - 3y + z}{x - z - y}$, $z + y - x$

4 $\dfrac{a - 3b + c}{-a - b + c}$, $a + b - c$

5 $\dfrac{a}{2b}$, $2ab$

6 $\dfrac{2x}{3y}$, $9xy$

7 $\dfrac{a + b}{a - b}$, $a^2 - b^2$

8 $\dfrac{x - 2}{x - 3}$, $x^2 - 9$

9 $\dfrac{3xy}{9x^2}$, $3x$

10 $\dfrac{12a^2b}{16ab^2}$, $4b$

11 $\dfrac{a^2 - b^2}{a^2 + 2ab + b^2}$, $a + b$

12 $\dfrac{x^2 - y^2}{x^3 + y^3}$, $x^2 - xy + y^2$

13 $\dfrac{a - 2b}{2a}$, $6a^2$

14 $\dfrac{2x - 3y}{3x}$, $12xy$

15 $\dfrac{4u - v}{-5v}$, $20v^2$

16 $\dfrac{5b + 3c}{4b}$, $-16b^2$

17 $\dfrac{x - 2y}{x - y}$, $x^2 - y^2$

18 $\dfrac{2a - b}{b - a}$, $a^2 - b^2$

19 $\dfrac{3h + 4k}{2h - k}$, $2h^2 + 3hk - 2k^2$

20 $\dfrac{p^2 - pq + q^2}{p - 2q}$, $p^2 - pq - 2q^2$

21 $\dfrac{5x + 2y}{x^2 + xy + y^2}$, $x^3 - y^3$

22 $\dfrac{a^2 - ab + b^2}{a - b}$, $a^3 - b^3$

23 $\dfrac{b^2 - bc - c^2}{b^2 - bc + c^2}$, $b^3 + c^3$

24 $\dfrac{r^2 - rt - t^2}{r^2 - rt + t^2}$, $r^4 + r^2t^2 + t^4$

Reduce the following fractions to lowest terms.

25 $\dfrac{x^2 + x - 6}{x^2 + 5x + 6}$

26 $\dfrac{a^2 + 4a + 3}{a^2 - a - 2}$

27 $\dfrac{2h^2 + 3h - 2}{3h^2 + 7h + 2}$

28 $\dfrac{3w^2 - 8w + 4}{2w^2 - w - 6}$

29 $\dfrac{(x - y)(2x^2 + xy - 6y^2)}{(x + 2y)(3x^2 - xy - 2y^2)}$

30 $\dfrac{(2a - b)(a^2 - ab - 6b^2)}{(a + 2b)(2a^2 + 3ab - 2b^2)}$

31 $\dfrac{(w + 2z)(6w^2 + 7wz - 3z^2)}{(3w - z)(2w^2 - wz - 6z^2)}$

32 $\dfrac{(2c - d)(6c^2 + 11cd + 3d^2)}{(3c + 2d)(6c^2 - cd - d^2)}$

33 $\dfrac{ms - 2mt - 2nt + ns}{2ms - mt - nt + 2ns}$

34 $\dfrac{ax - ay - 2by + 2bx}{ax + 2bx + 2by + ay}$

35 $\dfrac{3ah + 4bk - 2ak - 6bh}{2ah - 4bh + ak - 2bk}$

36 $\dfrac{xy + 3xz - 2wy - 6wz}{2xy + wy + 3wz + 6xz}$

37 $\dfrac{a^3 + b^3}{a^2 - b^2}$

38 $\dfrac{x^3 - y^3}{x^2 - y^2}$

39 $\dfrac{u^2 - v^2}{u^4 - v^4}$

40 $\dfrac{m^6 - n^6}{m^9 - n^9}$

41 $\dfrac{x^2 - y^2}{x^3 - y^3}$

42 $\dfrac{x^2 - y^2}{x^3 + y^3}$

43 $\dfrac{x^6 - y^6}{x^4 - y^4}$

44 $\dfrac{x^9 + y^9}{x^6 - y^6}$

45 $\dfrac{x + 1}{(x + 2)x + 1}$ **46** $\dfrac{x - 2}{(x - 1)x - 2}$

47 $\dfrac{x + 3}{(2x + 7)x + 3}$ **48** $\dfrac{x + 3}{(3x + 10)x + 3}$

49 $\dfrac{x - 1}{(2x - 1)x - 1}$ **50** $\dfrac{x + 2}{(x + 3)x + 2}$

51 $\dfrac{5x - 3}{x(5x - 3) + 5x - 3}$ **52** $\dfrac{3x - 2}{3x(x - 1) - 2(x - 1)}$

Convert each of the following sets of fractions to an equal set with a common denominator.

53 $\left\{\dfrac{3}{x}, \dfrac{2}{xy}, \dfrac{4}{x^2}\right\}$ **54** $\left\{\dfrac{1}{x^2}, \dfrac{2}{xy}, \dfrac{3}{y^2}\right\}$

55 $\left\{\dfrac{3}{x^2y}, \dfrac{-2}{xy^2}, \dfrac{1}{xy}\right\}$ **56** $\left\{\dfrac{5}{x^3y}, \dfrac{-4}{x^2y^2}, \dfrac{-3}{xy^3}\right\}$

57 $\left\{\dfrac{2x - y}{x - y}, \dfrac{x - 2y}{x + y}, \dfrac{x + 2y}{x^2 - y^2}\right\}$ **58** $\left\{\dfrac{x - y}{x^2 - xy + y^2}, \dfrac{x^2 + xy + y^2}{x + y}\right\}$

59 $\left\{\dfrac{x - y}{(x + y)(x - 2y)}, \dfrac{x + y}{(x - y)(x - 2y)}, \dfrac{x - 2y}{x^2 - y^2}\right\}$

60 $\left\{\dfrac{x - 3y}{(x - 2y)(x + y)}, \dfrac{2x - y}{(3x - y)(x + y)}, \dfrac{x - y}{(x - 2y)(3x - y)}\right\}$

Education **61** The mental age M and age in years C of a child are related by $M = CI/100$, if I is the child's IQ score. Find M for John if $C = 7.8$ and $I = 126$.

Management **62** If depreciation is taken according to the sum-of-the-digits method, then the fractions used each year of the total n years all have the same denominator: $1 + 2 + 3 + \cdots + n = n(n + 1)/2$. The numerators are successively $n, n - 1, n - 2, \ldots, 2, 1$. Calculate these fractions for $n = 8$.

Sociology **63** If the scores before and after a test are P and Q percent, the effectiveness index of the test is $E = (Q - P)/(100 - P)$. Find E if
(a) $P = 10$ and $Q = 20$
(b) $P = 80$ and $Q = 90$

Biology **64** If two dihybrids are crossed with complete domination, the phenotype ratios are 9/16, 3/16, 3/16, 1/16. How many of each type are expected out of a total of 112 individuals?

3.2 Multiplication and division of fractions

In Secs. 2.4 and 3.1, we saw how to multiply two fractions. This may be easily extended to three or more fractions:

$$\frac{a}{b} \cdot \frac{c}{d} \cdot \frac{e}{f} = \frac{ace}{bdf}$$

In words, we have

Product of fractions   *The product of two or more given fractions is a fraction whose numerator is the product of the numerators of the given fractions and whose denominator is the product of the given denominators.*

● **Example 1**   Obtain the product of

$$\frac{x^2}{y^3} \qquad \frac{x - y}{x + y} \qquad \text{and} \qquad \frac{x^2 + xy + y^2}{x^2 - 3xy + y^2}$$

Solution

$$\frac{x^2}{y^3} \cdot \frac{x - y}{x + y} \cdot \frac{x^2 + xy + y^2}{x^2 - 3xy + y^2} = \frac{x^2(x - y)(x^2 + xy + y^2)}{y^3(x + y)(x^2 - 3xy + y^2)}$$

When possible, the product should be reduced to lowest terms. For this reason, the members of the fractions should be factored, if they are reducible, before the product is formed. Then the factors that are common to the members of the product can be detected easily and should be canceled before forming the product.   ●

● **Example 2**   Obtain the product of

$$\frac{a^2 - 4b^2}{2a^2 - 7ab + 3b^2} \qquad \frac{6a - 3b}{2a + 4b} \qquad \text{and} \qquad \frac{a^2 - 4ab + 3b^2}{a^2 - ab - 2b^2}$$

Solution $\dfrac{a^2 - 4b^2}{2a^2 - 7ab + 3b^2} \cdot \dfrac{6a - 3b}{2a + 4b} \cdot \dfrac{a^2 - 4ab + 3b^2}{a^2 - ab - 2b^2}$

$$= \frac{(a - 2b)(a + 2b)}{(2a - b)(a - 3b)} \cdot \frac{3(2a - b)}{2(a + 2b)} \cdot \frac{(a - b)(a - 3b)}{(a + b)(a - 2b)} \qquad \text{factoring}$$

$$= \frac{3(a - 2b)(a + 2b)(2a - b)(a - 3b)(a - b)}{2(a - 2b)(a + 2b)(2a - b)(a - 3b)(a + b)} \qquad \begin{array}{l}\text{by the commutative}\\ \text{axiom and (3.4)}\end{array}$$

$$= \frac{3(a - b)}{2(a + b)} \qquad \begin{array}{l}\text{canceling}\\ (a - 2b)(a + 2b)\\ (a - 3b)(2a - b)\end{array}$$

●

● **Example 3**   Obtain the product of

$$\frac{x^2 - 3x + 2}{2x^2 + 3x - 2} \qquad \frac{2x^2 + 5x - 3}{x^2 - 1} \qquad \text{and} \qquad \frac{3x^2 + 6x}{2x - 4}$$

Solution $\dfrac{x^2 - 3x + 2}{2x^2 + 3x - 2} \cdot \dfrac{2x^2 + 5x - 3}{x^2 - 1} \cdot \dfrac{3x^2 + 6x}{2x - 4}$

$$= \frac{(x - 2)(x - 1)}{(2x - 1)(x + 2)} \cdot \frac{(2x - 1)(x + 3)}{(x - 1)(x + 1)} \cdot \frac{3x(x + 2)}{2(x - 2)} \qquad \text{factoring}$$

$$= \frac{3x(x + 3)}{2(x + 1)} \qquad\qquad\qquad\qquad\qquad\qquad \text{canceling} ●$$

We have seen in Sec. 3.1 that

$$\frac{a/b}{c/d} = \frac{a}{b} \cdot \frac{d}{c}$$

In words, we have

> *To obtain the quotient of two fractions, we multiply the numerator by the reciprocal of the denominator.*

● **Example 4**   Divide

$$\frac{x^2 - 3x + 2}{2x^2 - 7x + 3} \qquad \text{by} \qquad \frac{x^2 - x - 2}{2x^2 + 3x - 2}$$

Solution $\dfrac{x^2 - 3x + 2}{2x^2 - 7x + 3} \div \dfrac{x^2 - x - 2}{2x^2 + 3x - 2}$

$$= \frac{x^2 - 3x + 2}{2x^2 - 7x + 3} \cdot \frac{2x^2 + 3x - 2}{x^2 - x - 2} \qquad \text{multiplying by reciprocal}$$

$$= \frac{(x^2 - 3x + 2)(2x^2 + 3x - 2)}{(2x^2 - 7x + 3)(x^2 - x - 2)}$$

$$= \frac{(x - 2)(x - 1)(2x - 1)(x + 2)}{(2x - 1)(x - 3)(x - 2)(x + 1)} \qquad \text{factoring}$$

$$= \frac{(x - 1)(x + 2)}{(x - 3)(x + 1)} \qquad\qquad \text{canceling } (x - 2)(2x - 1) ●$$

Exercise 3.2   Perform the indicated operations in the following problems and reduce to
multiplication   lowest terms.
and division
of fractions   **1** $\dfrac{3a}{2b} \cdot \dfrac{4ab}{9c} \cdot \dfrac{3c^2}{4a^2}$ **2** $\dfrac{7y}{12x^2} \cdot \dfrac{10xy^2}{3z} \cdot \dfrac{6xz^2}{5y}$

3 $\dfrac{14u^2}{5v} \cdot \dfrac{10v^2}{21vw} \cdot \dfrac{9w^2}{8u^2v}$

4 $\dfrac{16rs}{7t^2} \cdot \dfrac{35t}{8r^2} \cdot \dfrac{4r}{15s^2}$

5 $\dfrac{25a^2b^2}{12c^2} \cdot \dfrac{36bc^3}{5a^3} \div \dfrac{15b^3}{7ac}$

6 $\dfrac{12pq}{r} \cdot \dfrac{35r^2p^3}{16q^2} \div \dfrac{7p^4}{24q^3}$

7 $\dfrac{28b^3}{15c^2d^2} \cdot \dfrac{18d^3}{35b^5} \div \dfrac{21d^4}{25b^2c^3}$

8 $\dfrac{24x^2y^3}{39z^2} \cdot \dfrac{25x^3z^3}{18y^4} \div \dfrac{15x^4}{26y^2z^2}$

9 $\dfrac{7a^2b^3c}{5x^3yz^5} \cdot \dfrac{15x^5y^2z^4}{28a^3b^4c}$

10 $\dfrac{12m^2n^3p^5}{25s^8t^2u^6} \cdot \dfrac{50s^6t^7u^5}{30\,mn^2p^3}$

11 $\dfrac{9a^2b^3}{4b^2c^4} \cdot \dfrac{20c^2d^4}{18a^3d^2} \cdot \dfrac{4b^3c^5}{5c^7d^3}$

12 $\dfrac{36a^2c^7}{5b^3d^8} \cdot \dfrac{15b^5d^4}{42a^3c^2} \cdot \dfrac{14a^4d^3}{18b^2c^3}$

13 $\dfrac{5a - 5b}{3a + 6b} \cdot \dfrac{a + 2b}{a - b}$

14 $\dfrac{4x - 2y}{5x + 10y} \cdot \dfrac{x^2 + 2xy}{2xy - y^2}$

15 $\dfrac{8h + 20k}{h^2 - 3hk} \cdot \dfrac{hk - 3k^2}{12h + 30k}$

16 $\dfrac{3wz - 7z^2}{w^2 + 5wz} \cdot \dfrac{3w + 15z}{12w - 28z}$

17 $\dfrac{4a^2 - b^2}{a + 2b} \div (4a^2 - 2ab)$

18 $\dfrac{x^2 - 9y^2}{2x} \div 3y(x + 3y)$

19 $(3b^2 + bc - 2c^2) \div \dfrac{9b^2 - 4c^2}{2c}$

20 $(20u^2 - 7uv - 6v^2) \div \dfrac{16u^2 - 9v^2}{3uv}$

21 $\dfrac{2a(a + b)^2}{3b^3} \cdot \dfrac{b^2(a - b)}{8a^3(a + b)} \cdot \dfrac{12a^2b^2}{a^2 - b^2}$

22 $\dfrac{3x^2y}{x + 2y} \cdot \dfrac{4y(x + 2y)^2}{3x(2x - y)} \cdot \dfrac{(2x - y)^2}{8xy^2(x + 2y)}$

23 $\dfrac{4c^2(3c - 5d)}{5d^3} \cdot \dfrac{15c^3d^4}{8(3c - 5d)(3c + 5d)} \cdot \dfrac{(3c + 5d)^2}{12c^4d^2}$

24 $\dfrac{24q^2(2p + 3q)^2}{5p^3} \cdot \dfrac{10p^5}{9q^3(4p^2 - 9q^2)} \cdot \dfrac{(2p - 3q)6q^4}{p^4}$

25 $\dfrac{x - y}{x + y} \cdot \dfrac{x^2 + xy}{x^2y^2 - xy^3} \cdot \dfrac{y}{x^2}$

26 $\dfrac{a^2b - ab^2}{a + b} \cdot \dfrac{a^2 - b^2}{ab^2 - b^3} \cdot \dfrac{b^2}{a^2}$

27 $\dfrac{p^2 - q^2}{pq^2} \cdot \dfrac{p^2}{pq + q^2} \cdot \dfrac{q^4}{p^2 - pq}$

28 $\dfrac{c^3 + dc^2}{c^2 + d^2} \cdot \dfrac{bc^2 + bd^2}{c^2 - d^2} \cdot \dfrac{c - d}{bc}$

29 $\dfrac{u^2 - 2u}{v^2 - v} \cdot \dfrac{uv^2 - uv}{u^2 - 4} \div \dfrac{u^2}{u + 2}$

30 $\dfrac{x^2y - xy}{y^2 - 1} \cdot \dfrac{y^3 + y^2}{x^3 - x^2} \div \dfrac{y^2}{y - 1}$

31 $\dfrac{a^3 - 3a^2}{b^2 + 2b} \cdot \dfrac{b^2 - 4}{a^2b - 3ab} \div \dfrac{ab - 2a}{b^2}$

32 $\dfrac{h^2 - 9}{k^2 - 9k} \cdot \dfrac{hk - 9h}{hk + 3k} \div \dfrac{h^3 - 3h^2}{k^3}$

33 $\dfrac{x^3 + y^3}{x^2 + 3xy + 2y^2} \cdot \dfrac{x^2 - xy - 6y^2}{x^2 - 2xy - 3y^2} \div \dfrac{x^2 - xy + y^2}{2x^2 + 2xy}$

34 $\dfrac{x^2 - xy - 2y^2}{x^3 - y^3} \cdot \dfrac{x^2 + xy + y^2}{x^2 - 4y^2} \div \dfrac{x^2 + 4xy + 3y^2}{x^2 + xy - 2y^2}$

35 $\dfrac{x^3 + 8y^3}{x^2 - 4y^2} \cdot \dfrac{x^2 - xy - 2y^2}{x^2 - 2xy + 4y^2} \div \dfrac{x^2 - 2xy - 3y^2}{x^2 - 3xy}$

36 $\dfrac{27x^3 - y^3}{3x^2 - 4xy + y^2} \cdot \dfrac{x^2 + 2xy - 3y^2}{9x^2 + 3xy + y^2} \div \dfrac{x^2 - 9y^2}{xy + 3y^2}$

37 $\dfrac{(x - 2)x + 1}{(x - 1)x - 2} \cdot \dfrac{(x - 2)x + (x - 2)}{x^2 - 1} \div \dfrac{x + 6}{x + 1}$

38 $\dfrac{x(x + 2) + 2(2x + 4)}{2x(x + 2) + (x + 3)} \cdot \dfrac{x^2 - (x + 2)}{x^2 - 4} \div \dfrac{x + 4}{x}$

39 $\dfrac{(x - 3)x - 4}{(x - 9)x + 20} \cdot \dfrac{(x + 3)x - 40}{(x^2 - 9) - 8x} \div \dfrac{x + 8}{x - 9}$

40 $\dfrac{(x + 1)x - 2}{(x + 1) - 4} \cdot \dfrac{(x - 2)x + 1(x - 2)}{(x - 2)(x - 1)} \div \dfrac{x + 2}{x - 3}$

41 $\dfrac{(x + 2)x - 8}{(x + 2) + 2} \cdot \dfrac{x(x - 3) + (x - 3)}{(x - 2)(x + 1)} \div \dfrac{x - 3}{x + 2}$

42 $\dfrac{(3x - 2)x - 1}{(x + 1)(3x + 1)} \cdot \dfrac{(x + 1)3x - 1(x + 1)}{3(x + 1)(x - 1) - 8x} \div \dfrac{x + 1}{x - 3}$

43 $\dfrac{(x - 3)x - 4}{x^2 - 16} \cdot \dfrac{(x + 4)x + 1(x + 4)}{(x + 1)(x - 3)} \div \dfrac{x + 3}{x - 3}$

44 $\dfrac{x(x - 1) - 6}{x(x - 1) - 20} \cdot \dfrac{x(x - 1) + (3x - 8)}{x^2 - 9} \div \dfrac{x - 2}{x - 5}$

3·3 Addition and subtraction of fractions

If two fractions have the same denominator, it is easy to add or subtract them by Eq (3.5).

$$\frac{a}{d} + \frac{b}{d} = \frac{a + b}{d} \qquad \frac{a}{d} - \frac{b}{d} = \frac{a - b}{d}$$

We simply keep the same denominator, and add or subtract the numerators.

● **Example 1**
$$\frac{4}{21} + \frac{2}{21} + \frac{8}{21} = \frac{14}{21} = \frac{2}{3}$$
●

● **Example 2**
$$\frac{3a}{2xy} + \frac{2a}{2xy} - \frac{b}{2xy} = \frac{3a + 2a - b}{2xy} = \frac{5a - b}{2xy}$$
●

If the denominators of the fractions to be added are not all equal, we begin by converting the given set of fractions to an equivalent set with the lcd as the denominator of each and then proceed as above.

● **Example 3**   Combine $1/2x + 1/5y + (8x - 2y)/20xy$ into a single fraction.

Solution   The lcd for the given fractions is $20xy$; hence, we multiply each member of the first fraction by $20xy/2x = 10y$ and each member of the second fraction by $20xy/5y = 4x$ and have

$$\frac{1}{2x} \cdot \frac{10y}{10y} + \frac{1}{5y} \cdot \frac{4x}{4x} + \frac{8x - 2y}{20xy} = \frac{1}{20xy}(10y + 4x + 8x - 2y)$$

$$= \frac{12x + 8y}{20xy} \quad \text{combining}$$

$$= \frac{3x + 2y}{5xy} \quad \text{reducing to lowest terms} \quad ●$$

● **Example 4**   Combine $2/(2x - y) + 3/(2x + y) - 9x/(4x^2 - y^2)y$ into a single fraction.

Solution   The factored form of the third denominator is $(2x - y)(2x + y)y$; furthermore, each of the other denominators is a factor of this one. Therefore, $(2x - y)$ $\times (2x + y)y$ is the lcd. Consequently, we multiply the members of the first fraction by $(2x - y)(2x + y)y/(2x - y) = (2x + y)y$ and the members of the second by $(2x - y)(2x + y)y/(2x + y) = (2x - y)y$ and have

$$\frac{2}{2x - y} \cdot \frac{y(2x + y)}{y(2x + y)} + \frac{3}{2x + y} \cdot \frac{y(2x - y)}{y(2x - y)} - \frac{9x}{(4x^2 - y^2)y}$$

$$= \frac{4xy + 2y^2 + 6xy - 3y^2 - 9x}{(4x^2 - y^2)y} = \frac{10xy - y^2 - 9x}{(4x^2 - y^2)y} \quad ●$$

● **Example 5**   Find the sum indicated by

$$\frac{x - 2y}{x^2 + xy - 2y^2} + \frac{y - x}{2x^2 - xy - y^2} + \frac{3y - x}{2x^2 + 5xy + 2y^2}$$

Solution   The lcd is $(x + 2y)(x - y)(2x + y)$ as seen after factoring the denominators. Now multiplying both members of each fraction by the quotient of this lcd and the denominator of the fraction that is being used, we have

$$\frac{(x - 2y)(2x + y) + (y - x)(x + 2y) + (3y - x)(x - y)}{(x + 2y)(x - y)(2x + y)}$$

$$= \frac{2x^2 - 3xy - 2y^2 - x^2 - xy + 2y^2 - x^2 + 4xy - 3y^2}{(x + 2y)(x - y)(2x + y)}$$

$$= \frac{-3y^2}{(x + 2y)(x - y)(2x + y)}$$

**Exercise 3.3
addition and
subtraction of
fractions**

Perform the additions and subtractions indicated in the following problems and then reduce to lowest terms.

1 $\dfrac{2x + 1}{6x + 1} + \dfrac{7x - 5}{6x + 1} + \dfrac{9x - 3}{6x + 1}$

2 $\dfrac{2x}{x^2 + 11} + \dfrac{5 - 3x}{x^2 + 11} - \dfrac{6 + 7x}{x^2 + 11}$

3 $\dfrac{3a + b}{d - 1} - \dfrac{4c - 3a}{d - 1} + \dfrac{6b - 5c}{d - 1}$

4 $\dfrac{4a + 5}{b + 13} - \dfrac{3a - b}{b + 13} - \dfrac{6 + 4b}{b + 13}$

5 $\dfrac{x + 1}{2} - \dfrac{x - 2}{3} + \dfrac{2x - 1}{9}$

6 $\dfrac{2x + 1}{4} - \dfrac{x - 2}{2} + \dfrac{x + 3}{3}$

7 $\dfrac{2x + 3}{3} - \dfrac{3x - 1}{2} + \dfrac{x - 4}{12}$

8 $\dfrac{x + 4}{2} - \dfrac{x - 1}{5} + \dfrac{2x + 5}{10}$

9 $\dfrac{2bc}{3a} - \dfrac{3ac}{2b} + \dfrac{5ba}{6c}$

10 $\dfrac{5bc}{4a} - \dfrac{4ac}{3b} + \dfrac{ab}{2c}$

11 $\dfrac{2a}{9bc} - \dfrac{c}{3ab} + \dfrac{3b}{2ac}$

12 $\dfrac{3a}{2bc} + \dfrac{2b}{ac} - \dfrac{5c}{3ab}$

13 $\dfrac{4b}{3a} - \dfrac{5a}{2b} + \dfrac{9a^2 - 8b^2}{6ab}$

14 $\dfrac{3a}{2b} - \dfrac{2b}{3a} - \dfrac{9a^2 - 4b^2}{6ab}$

15 $\dfrac{5a}{3b} - \dfrac{2b}{5a} + \dfrac{9b^2 - 22a^2}{15ab}$

16 $\dfrac{3b}{7a} - \dfrac{a}{2b} + \dfrac{3a^2 - 2b^2}{14ab}$

17 $\dfrac{2x - y}{3x + y} + \dfrac{5x}{2y}$

18 $\dfrac{3x + 4y}{2x + y} - \dfrac{2y}{x}$

19 $\dfrac{3x}{2y} - \dfrac{3x - 2y}{3x + 2y}$

20 $\dfrac{3x}{5y} - \dfrac{5x + 2y}{2x - y}$

21 $\dfrac{x + 2y}{x - y} - \dfrac{x - 2y}{x + y}$

22 $\dfrac{3x - 2y}{2x - y} - \dfrac{2x + y}{3x - y}$

23 $\dfrac{5x - 2y}{3x - 8y} + \dfrac{3x - 3y}{2x - 5y}$

24 $\dfrac{2x + 7y}{5x - 3y} + \dfrac{5x + 3y}{2x - 7y}$

25 $\dfrac{r + s}{rs} - \dfrac{1}{s} + \dfrac{s}{r(r - s)}$

26 $\dfrac{s}{r + s} - \dfrac{r^2}{s(r + s)} + \dfrac{2r}{s}$

27 $\dfrac{s}{r} - \dfrac{2rs}{r(r-2s)} + \dfrac{r}{r-2s}$

28 $\dfrac{r^2 - 3s^2}{r(r+3s)} + \dfrac{3s}{r} + \dfrac{2s}{r+3s}$

29 $\dfrac{6x}{x^2 - y^2} + \dfrac{2x}{y(x+y)} - \dfrac{3}{x-y}$

30 $\dfrac{2}{3x-2y} - \dfrac{3y}{x(3x+2y)} - \dfrac{8y}{9x^2 - 4y^2}$

31 $\dfrac{2x}{y(x+2y)} + \dfrac{4}{x-2y} + \dfrac{8x}{x^2 - 4y^2}$

32 $\dfrac{x}{x+3} - \dfrac{6x+6}{x^2 - 9} + \dfrac{x+1}{x-3}$

33 $\dfrac{3}{(x+2y)(x-y)} + \dfrac{2}{(x-2y)(x-y)} - \dfrac{4}{(x+2y)(x-2y)}$

34 $\dfrac{2}{(a+3b)(a+b)} - \dfrac{3}{(a+3b)(a-2b)} + \dfrac{5}{(a+b)(a-2b)}$

35 $\dfrac{4}{(a+5b)(a-4b)} + \dfrac{1}{(a-4b)(a-b)} + \dfrac{2}{(a+5b)(a-b)}$

36 $\dfrac{2}{(a+b)(a+2b)} - \dfrac{5}{(a-3b)(a+2b)} + \dfrac{4}{(a-3b)(a+b)}$

37 $\dfrac{4x+3y}{3x+2y} - \dfrac{2x+3y}{4x-3y} - \dfrac{3x-2y}{2x-3y}$

38 $\dfrac{5x+2y}{2x+5y} - \dfrac{3x+y}{x+3y} - \dfrac{2x-y}{x-2y}$

39 $\dfrac{2x-y}{x+3y} - \dfrac{x-3y}{2x+y} + \dfrac{3x+y}{3x-y}$

40 $\dfrac{x-3y}{x+2y} + \dfrac{x-2y}{4x+3y} - \dfrac{4x-3y}{x+3y}$

41 $\dfrac{x}{2x^2 + xy - 3y^2} + \dfrac{y}{2x^2 - 3xy + y^2} - \dfrac{x+y}{4x^2 + 4xy - 3y^2}$

42 $\dfrac{2x}{x^2 - 2xy - 3y^2} - \dfrac{y}{3x^2 + 4xy + y^2} + \dfrac{2x-y}{3x^2 - 8xy - 3y^2}$

43 $\dfrac{3x}{2x^2 + 3xy - 2y^2} + \dfrac{y}{x^2 - 4y^2} - \dfrac{2x+y}{2x^2 - 5xy + 2y^2}$

44 $\dfrac{x}{2x^2 + 7xy + 5y^2} + \dfrac{2y}{x^2 - y^2} + \dfrac{3x+2y}{2x^2 + 3xy - 5y^2}$

Verify the statements in Probs. 45 to 48.

45 $1 = \frac{1}{2} + \frac{1}{3} + \frac{1}{8} + \frac{1}{24}$

46 $1 = \frac{1}{2} + \frac{1}{4} + \frac{1}{5} + \frac{1}{20}$

47 $1 = \frac{1}{2} + \frac{1}{4} + \frac{1}{9} + \frac{1}{12} + \frac{1}{18}$

48 $1 = \frac{1}{3} + \frac{1}{5} + \frac{1}{6} + \frac{1}{7} + \frac{1}{10} + \frac{1}{30} + \frac{1}{42}$

Earth science   **49** In an orthorhombic crystal system

$$\frac{1}{d^2} = \frac{h^2}{a^2} + \frac{k^2}{b^2} + \frac{l^2}{c^2}$$

Find d if $h = k + p$, $l = k - p$, and $a = b = c$.

Anthropology   **50** In studying the class of similarity measures normally used in clustering problems, anthropologists use the inequality

$$\frac{r}{1+r} + \frac{s}{1+s} \geq \frac{t}{1+t}$$

valid whenever $r + s \geq t > 0$. Verify this inequality for $r = 7$, $s = 10$, and $t = 15$.

Manager   **51** The rule of 78 for charging interest for 1 year allows the lending institution to collect most of the interest early in the year. In the first month the allowable interest is 12/78, in the second month 11/78, in the third 10/78, etc., until 1/78 in the twelfth month. Why is the denominator always 78?

Earth science   **52** The rigidity modulus of a material is

$$M = \frac{E}{2(1+s)}$$

where E is Young's modulus and s is Poisson's ratio. Find the difference in the M values of two materials whose s values are $1.0s$ and $1.1s$.

53 If a and b are positive, then

$$\frac{a}{b} + \frac{b}{a} \geq 2$$

Verify this for $a = 14$ and $b = 15$.

54 If x and y are positive, then

$$(x + y)\left(\frac{1}{x} + \frac{1}{y}\right) \geq 4$$

Verify this for $x = 6$ and $y = 8$.

55 If a, b, and c are positive, then

$$\frac{ab}{c} + \frac{bc}{a} + \frac{ca}{b} \geq a + b + c$$

Verify this with $a = 3$, $b = 5$, $c = 6$.

56 If x, y, and z are positive, then

$$\frac{1}{x} + \frac{1}{y} + \frac{1}{z} \geq \frac{9}{x + y + z}$$

Verify this for $x = 4$, $y = 6$, and $z = 8$.

57 Show that

$$\frac{x^2}{x^2 + 1} \leq \frac{1}{2} \qquad \text{if } x = \frac{7}{8} \text{ and if } x = -\frac{4}{5}$$

58 Verify that

$$\frac{8}{x^2} - \frac{x}{4} \le \frac{3}{2} \quad \text{if } x = \frac{7}{3} \text{ and if } x = \frac{9}{4}$$

59 Show that

$$\frac{x^2 + x + 1}{x^2 + 1} \le \frac{3}{2} \quad \text{for } x = -2 \text{ and for } x = 3$$

60 Show that

$$\frac{x^2}{x^4 + 1} \le \frac{1}{2} \quad \text{for } x = 4 \text{ and for } x = -1$$

3·4 Complex fractions

Complex fraction   A *complex fraction* is a fraction in which at least one of the terms of one or both members is a fraction.

The following are examples of complex fractions:

$$\frac{3}{\frac{2}{3}} \qquad \frac{1 + \frac{x}{y}}{x + y} \qquad \frac{\frac{4x}{x + y} + \frac{2y}{x - y}}{3 - \frac{x^2 + y^2}{x^2 - y^2}}$$

First method   There are two methods for reducing a complex fraction to a simple
for reducing a   fraction. The first method consists of multiplying the numerator and
complex fraction   denominator of the complex fraction by the least common denominator
(lcd) of the denominators of the fractions that appear in it. This procedure
is justified by the fundamental principle of fractions shown by Eq. (3.2).
We will illustrate the procedure with an example.

● **Example 1**   Reduce

$$\frac{y - \frac{x^2}{y}}{\frac{y^2}{x} - x}$$

to a simple fraction.

Solution   The denominators of the fractions that occur in this complex fraction are x
and y. Hence the lcd is xy. Consequently, we multiply each member of the
complex fraction by xy and get

$$\frac{y - \dfrac{x^2}{y}}{\dfrac{y^2}{x} - x} = \frac{\left(y - \dfrac{x^2}{y}\right)xy}{\left(\dfrac{y^2}{x} - x\right)xy} \qquad \text{by Eq. (3.2)}$$

$$= \frac{xy^2 - x^3}{y^3 - x^2y}$$

$$= \frac{x(y^2 - x^2)}{y(y^2 - x^2)} \qquad \text{factoring}$$

$$= \frac{x}{y} \qquad \text{dividing the members by } y^2 - x^2 \qquad \bullet$$

Second method for reducing a complex fraction   The second method for simplifying a complex fraction consists of reducing the numerator and denominator separately to simple fractions and then performing the indicated division by multiplying by the reciprocal.

● **Example 2**   Reduce the complex fraction

$$\frac{\dfrac{x - y}{x + y} - \dfrac{x + y}{x - y}}{1 - \dfrac{x^2 - xy - y^2}{x^2 - y^2}}$$

to a simple fraction.

Solution $\dfrac{\dfrac{x - y}{x + y} - \dfrac{x + y}{x - y}}{1 - \dfrac{x^2 - xy - y^2}{x^2 - y^2}}$

$$= \frac{\dfrac{(x - y)^2 - (x + y)^2}{(x + y)(x - y)}}{\dfrac{x^2 - y^2 - (x^2 - xy - y^2)}{x^2 - y^2}} \qquad \begin{array}{l}\text{adding the fractions in the}\\ \text{numerator and in the denominator}\end{array}$$

$$= \frac{\dfrac{x^2 - 2xy + y^2 - x^2 - 2xy - y^2}{x^2 - y^2}}{\dfrac{x^2 - y^2 - x^2 + xy + y^2}{x^2 - y^2}} \qquad \text{performing the indicated operations}$$

$$= \frac{\dfrac{-4xy}{x^2 - y^2}}{\dfrac{xy}{x^2 - y^2}} \qquad \text{combining similar terms}$$

$$= \frac{-4xy}{x^2 - y^2} \cdot \frac{x^2 - y^2}{xy}$$

multiplying the members by the reciprocal of the denominator

$$= \frac{-4xy(x^2 - y^2)}{xy(x^2 - y^2)}$$

$$= -4$$

canceling $xy(x^2 - y^2)$

If either the numerator or denominator of a complex fraction is itself a complex fraction or if both are complex fractions, each should be reduced to a simple fraction as the first step in the simplification.

● **Example 3**   Reduce the complex fraction

$$\frac{1 + \dfrac{1}{1 + \dfrac{1}{x - 1}}}{1 - \dfrac{1}{1 - \dfrac{1}{x + 1}}}$$

to a simple fraction.

Solution $\dfrac{1 + \dfrac{1}{1 + \dfrac{1}{x - 1}}}{1 - \dfrac{1}{1 - \dfrac{1}{x + 1}}} = \dfrac{1 + \dfrac{x - 1}{x - 1 + 1}}{1 + \dfrac{x + 1}{x + 1 - 1}}$

multiplying both members of the complex fraction in the numerator by $(x - 1)$ and both members of the complex fraction in the denominator by $(x + 1)$

$$= \frac{1 + \dfrac{x - 1}{x}}{\dfrac{x + 1}{x}}$$

$$= \frac{x + x - 1}{x + 1}$$

multiplying both members by x

$$= \frac{2x - 1}{x + 1}$$

● **Exercise 3.4**
complex fractions

Reduce the following complex fractions to simple fractions:

1 $\dfrac{1 - \dfrac{1}{3}}{2 + \dfrac{2}{3}}$ **2** $\dfrac{4 + \dfrac{1}{2}}{3 - \dfrac{1}{3}}$ **3** $\dfrac{\dfrac{1}{2} - \dfrac{1}{3}}{\dfrac{3}{4} + \dfrac{7}{6}}$ **4** $\dfrac{\dfrac{2}{5} + \dfrac{1}{3}}{\dfrac{5}{6} - \dfrac{1}{2}}$ **5** $\dfrac{2 - \dfrac{1}{x}}{4 - \dfrac{1}{x^2}}$

6 $\dfrac{1 - \dfrac{9}{x^2}}{1 + \dfrac{3}{x}}$ **7** $\dfrac{x - \dfrac{16}{x}}{1 - \dfrac{4}{x}}$ **8** $\dfrac{2 + \dfrac{3}{x}}{4 - \dfrac{9}{x^2}}$ **9** $\dfrac{x^2 - \dfrac{4}{x^2}}{x + \dfrac{2}{x}}$ **10** $\dfrac{\dfrac{x}{3} - \dfrac{3}{x}}{\dfrac{1}{x} + \dfrac{2}{3x}}$

11 $\dfrac{\dfrac{x}{10} - \dfrac{1}{5}}{\dfrac{1}{2} - \dfrac{1}{x}}$ **12** $\dfrac{\dfrac{4}{x} - \dfrac{1}{2}}{\dfrac{3}{2x} + \dfrac{1}{x}}$ **13** $\dfrac{1 - \dfrac{2}{x} - \dfrac{15}{x^2}}{1 - \dfrac{1}{x} - \dfrac{12}{x^2}}$ **14** $\dfrac{3 - \dfrac{10}{x} + \dfrac{3}{x^2}}{3 + \dfrac{5}{x} - \dfrac{2}{x^2}}$

15 $\dfrac{6 + \dfrac{7}{x} + \dfrac{2}{x^2}}{2 + \dfrac{7}{x} + \dfrac{3}{x^2}}$ **16** $\dfrac{3 - \dfrac{11}{x} + \dfrac{6}{x^2}}{4 - \dfrac{13}{x} + \dfrac{3}{x^2}}$ **17** $\dfrac{1 + \dfrac{x}{x + y}}{1 - \dfrac{3x}{x - y}}$ **18** $\dfrac{2 - \dfrac{x}{2x + y}}{2 - \dfrac{5x}{x - y}}$

19 $\dfrac{1 + \dfrac{2x}{x + y}}{1 + \dfrac{x}{x + y}}$ **20** $\dfrac{2 - \dfrac{3x}{x - y}}{2 - \dfrac{3x}{2x + y}}$ **21** $\dfrac{a - 2 + \dfrac{a - 2}{a + 2}}{a - \dfrac{3a + 12}{a + 2}}$

22 $\dfrac{a + \dfrac{8a}{2a - 1}}{a + 2 - \dfrac{6}{2a + 3}}$ **23** $\dfrac{2a - \dfrac{3a + 4}{a - 2}}{a - \dfrac{10a + 4}{2a + 3}}$ **24** $\dfrac{a - \dfrac{a}{a + 2}}{a + \dfrac{1}{a + 2}}$

25 $\dfrac{\dfrac{p}{p + 3} + \dfrac{p}{p^2 - 9}}{\dfrac{1}{p - 3} + 1}$ **26** $\dfrac{\dfrac{2}{p + q} - \dfrac{1}{p - 2q}}{2 - \dfrac{p + q}{p - 2q}}$ **27** $\dfrac{\dfrac{1 - 2p}{1 + 3p} - \dfrac{1}{1 - p}}{1 - \dfrac{10}{1 + 3p}}$

28 $\dfrac{\dfrac{p + 2}{p - 2} - \dfrac{p}{p + 2}}{3 - \dfrac{4}{p + 2}}$ **29** $\dfrac{\dfrac{5}{x - 2} + \dfrac{3}{2x + 1}}{\dfrac{1 + 12x}{2 - x} - 1}$ **30** $\dfrac{\dfrac{2x + 3}{x + 2} - \dfrac{2x}{x + 1}}{\dfrac{x}{x + 2} - 1}$

31 $\dfrac{\dfrac{3}{2 - 3x} - \dfrac{2}{x + 1}}{1 - \dfrac{1 + 6x}{2 - 3x}}$ **32** $\dfrac{\dfrac{x}{x + 1} - \dfrac{x^2}{x^2 - 1}}{1 + \dfrac{1}{x - 1}}$ **33** $\dfrac{\dfrac{1}{x}}{1 - \dfrac{1}{1 + \dfrac{2x}{y}}}$

34 $\dfrac{x - \dfrac{x}{2 - \dfrac{1}{x}}}{y + \dfrac{y}{2x - 1}}$ **35** $\dfrac{u - \dfrac{u}{1 + \dfrac{v}{u}}}{w - \dfrac{w}{\dfrac{u}{v} + 1}}$ **36** $\dfrac{1 + \dfrac{1}{x}}{1 - \dfrac{1}{1 + \dfrac{2}{x - 4}}}$

Verify the fact that each of the following fractions is equal to 11/15.

37 $\dfrac{1}{1 + \frac{4}{11}}$ **38** $\dfrac{1}{1 + \dfrac{1}{2 + \frac{3}{4}}}$ **39** $\dfrac{1}{1 + \dfrac{1}{2 + \dfrac{1}{\frac{4}{3}}}}$ **40** $\dfrac{1}{1 + \dfrac{1}{2 + \dfrac{1}{1 + \frac{1}{3}}}}$

Political science   **41** In a model of negotiation, the equation

$$R = \dfrac{P - \dfrac{1}{m}}{Q - \dfrac{1}{m}}$$

occurs, where $P < Q$, P and Q measure goodness of fit, and m is a positive integer.
(a)   Express R without complex fractions.
(b)   Find R if $P = 0.271$, $Q = 0.547$, and $m = 5$.

Psychology   **42** The expected proportion of responses in a test compound is

$$\dfrac{m + \dfrac{n}{2}}{m + n + p}$$

Calculate this for (a) $m = 3, n = 8, p = 10$.
                  (b) $m = n = p$.

3.5 Key concepts

(3.1)   If $ad = bc$, then $a/b = c/d$; conversely, if $a/b = c/d$, then $ad = bc$
        for $b \neq 0$ and $d \neq 0$

(3.2) $\dfrac{a}{b} = \dfrac{ak}{bk} = \dfrac{a/d}{b/d}$ provided $b \neq 0, k \neq 0, d \neq 0$

(3.3) $\dfrac{a}{b} = \dfrac{-a}{-b} = -\dfrac{-a}{b} = -\dfrac{a}{-b}$ provided $b \neq 0$

(3.4) $\dfrac{a}{b} \cdot \dfrac{c}{d} = \dfrac{ac}{bd}$

(3.5) $\dfrac{a}{d} + \dfrac{c}{d} = \dfrac{a + c}{d}$

Least common denominator   Quotient of fractions
Product of fractions   Sum of fractions
Complex fraction   Cancellation

Exercise 3.5
review

Convert the fraction in each of Probs. 1 to 10 into an equal fraction with the number to the right of the comma as denominator.

1 y/x, xy **2** y/x, x^2 **3** yx/x^2, x **4** $xy/2y$, 2

5 $\dfrac{x - 2y}{3y - x}$, $x - 3y$ **6** $\dfrac{x - 3}{x + 4}$, $x^2 - 16$

7 $\dfrac{x + 4y}{x - 2y}$, $(x - 2y)(x + y)$ **8** $\dfrac{x - 3}{x + 1}$, $(x + 1)(x + 3)$

9 $\dfrac{(x + 5)(x - 2)}{(x - 3)(x + 5)}$, $x - 3$ **10** $\dfrac{2x^2 - 5x + 2}{x^2 + 2x - 8}$, $x + 4$

Reduce the fraction in each of Probs. 11 to 18 to lowest terms.

11 $\dfrac{(x - 2)(x - 3)}{(x + 2)(x - 3)}$ **12** $\dfrac{(2x + 3)(3x - 1)(x - 4)}{(x - 4)(3x + 1)(2x + 3)}$

13 $\dfrac{x^2 + 2xy - 8y^2}{x^2 - 5xy + 6y^2}$ **14** $\dfrac{(x + 2y)(x^2 + 4xy - 5y^2)}{(x - y)(x^2 + 8xy + 15y^2)}$

15 $\dfrac{xz + 2xw + yz + 2yw}{2xz - xw + 2yz - yw}$ **16** $\dfrac{x^3 + y^3}{x^2 - y^2}$

17 $\dfrac{(x + 3)(x - 4)}{-3(x + 4) + x(x + 2)}$ **18** $\dfrac{x - 1}{(2x - 1)x - 1}$

Convert each of the following sets of fractions to an equal set with a common denominator:

19 $\left\{ \dfrac{2x - 1}{x - 1}, \dfrac{x - 2}{x + 1}, \dfrac{x + 2}{x^2 - 1} \right\}$ **20** $\left\{ \dfrac{x - y}{x^2 - xy + y^2}, \dfrac{x^2 + xy + y^2}{x + y} \right\}$

Perform the multiplications and divisions in the following problems and reduce to lowest terms:

21 $\dfrac{2x^2 y^3}{3z^2 w} \cdot \dfrac{9z^4 w^2}{6x^4 y}$ **22** $\dfrac{14x^6 y^4}{5x^3 z^2} \div \dfrac{21x^3 y^2}{10wz^3}$

23 $(x^2 + 5x - 14) \div \dfrac{x^2 - 4}{5x}$ **24** $\dfrac{x^3 y + x^2 y^2}{x - y} \cdot \dfrac{x^2 - y^2}{xy^2 + y^3} \cdot \dfrac{y^3}{x}$

25 $\dfrac{x^2 + 2xy - 3y^2}{x^2 - 2xy - 15y^2} \cdot \dfrac{x^2 + xy - 30y^2}{x^2 + xy - 2y^2} \cdot \dfrac{x - 3y}{x + 6y}$

26 $\dfrac{(x + 2)x - 8}{(x + 2) + 2} \cdot \dfrac{x(x - 3) + (x - 3)}{(x - 2)(x + 1)} \div \dfrac{x - 3}{x + 2}$

Perform the indicated operations and reduce to lowest terms.

27 $\dfrac{7}{9} + \dfrac{8}{15} - \dfrac{1}{3}$ **28** $\dfrac{5z}{9xy} - \dfrac{y}{15xz} - \dfrac{3x}{5yz}$ **29** $\dfrac{x}{x+3} - \dfrac{6x+6}{x^2-9} + \dfrac{x+1}{x-3}$

30 $\dfrac{3}{(x+2y)(x-y)} + \dfrac{2}{(x-2y)(x-y)} - \dfrac{4}{(x+2y)(x-2y)}$

31 $\dfrac{2x+y}{x^2-3xy+2y^2} - \dfrac{x+4y}{x^2-4xy+3y^2} - \dfrac{x-7y}{x^2-5xy+6y^2}$

Reduce the following complex fractions to simple fractions:

32 $\dfrac{\dfrac{2}{7} - \dfrac{1}{3}}{\dfrac{12}{7} - \dfrac{7}{3}}$ **33** $\dfrac{\dfrac{1}{5} - \dfrac{5}{1}}{1 + \dfrac{1}{5}}$

34 $\dfrac{x - \dfrac{4}{x}}{1 - \dfrac{2}{x}}$ **35** $\dfrac{3 - \dfrac{10}{x} + \dfrac{3}{x^2}}{3 + \dfrac{5}{x} - \dfrac{2}{x^2}}$

36 $\dfrac{2 + \dfrac{3x}{y - 3x}}{3 - \dfrac{y + 9x}{y + 2x}}$ **37** $\dfrac{\dfrac{2}{x - 3y} + \dfrac{1}{x + y}}{\dfrac{3x - y}{x - 3y}}$

38 $\dfrac{2 + \dfrac{2}{x}}{2 - \dfrac{2}{2 + \dfrac{4}{x - 4}}}$ **39** $\dfrac{x - \dfrac{x^2}{x - 1}}{x + \dfrac{x^2 - 1}{1 + \dfrac{1}{x}}}$

40 Show that

$$1 + \dfrac{1}{2 + \frac{1}{2}} < \sqrt{2} < 1 + \dfrac{1}{2 + \dfrac{1}{2 + \frac{1}{2}}}$$

41 Show that if $\dfrac{1}{p} + \dfrac{1}{q} = 1$, then $p + q = pq$, $p \neq 0$, $q \neq 0$.

Biology   **42** A phenotype ratio of 3 to 1 is normal in fowls when considering the creeper gene which produces deformed legs. How many of each type are expected out of 228 fowls?

Earth science   **43** For two cells, the interplanar-spacing relation for the tetragonal system
and calculator   is

$$\frac{1}{d^2} = \frac{h^2 + k^2}{a^2} + \frac{l^2}{c^2}$$

Calculate d for $h = 255$, $k = 305$, $l = 290$, $a = 3.14$, and $c = 4.28$.

Management   **44** The total cost of orders is the ordering cost plus carrying cost, or

$$C = 75x + \frac{4680}{x}$$

The minimum value for C occurs for x nearly 8. Calculate C for
(a) $x = 5$ (b) $x = 8$ (c) $x = 10$

CHAPTER 4 · RATIONAL EXPONENTS AND RADICALS

Integer exponents

In Sec. 1.3, we defined a^n for positive integers n, and we also gave several properties satisfied by powers. We will now extend the definition of powers to allow n to be a negative integer or 0. Furthermore we will repeat the rules obeyed by powers because these same rules are satisfied whether the exponents are positive integers (Sec. 1.3), integers (this section), rational numbers (Sec. 4.3), or real numbers (Chap. 8).

○ **Definitions**

$$\overbrace{a^n = a \cdot a \cdots a}^{n \text{ factors}} \quad \text{if } n \text{ is any positive integer} \tag{4.1}$$

$$a^0 = 1 \quad \text{if } a \neq 0 \tag{4.2}$$

$$a^{-n} = \frac{1}{a^n} \quad \text{if } n \text{ is any positive integer, } a \neq 0 \tag{4.3}$$

●

○ **Theorems**

$$a^m a^n = a^{m+n} \tag{4.4}$$

$$a^m/a^n = a^{m-n} \qquad a \neq 0 \tag{4.5}$$

$$(a^m)^n = a^{mn} \tag{4.6}$$

$$(ab)^n = a^n b^n \tag{4.7}$$

$$(a/b)^n = a^n/b^n \qquad b \neq 0 \tag{4.8}$$

●

Note Once again, these laws hold for all real numbers as long as everything is defined. In particular, 0^0 is *not* defined, nor is division by zero.

We define $a^0 = 1$ in order that the laws above will hold. For instance, if we take $n = 0$ and $m = m$ in (4.4), then

$$a^m \cdot a^0 = a^{m+0}$$

which means that $\qquad a^m \cdot a^0 = a^m$

$a^0 = 1$ Now if $a \neq 0$, we may divide by a^m, giving $a^0 = 1$. Notice that we *defined* $a^0 = 1$, and we did so in order that the laws of exponents remain valid.

If we take $m = 0$, $n = n$ in (4.5), then we get

$$\frac{a^0}{a^n} = a^{0-n}$$

which means

$$\frac{1}{a^n} = a^{-n}$$

$a^{-n} = 1/a^n$ Again we *define* $a^{-n} = 1/a^n$ in order for the laws to be consistent. Notice that $a^{-n} = (a^{-1})^n = (1/a)^n$.

A complete proof of each of the theorems for integer exponents can be done by using mathematical induction (Chap. 13); however, we will give an argument only for (4.4) in the case where m and n are negative integers. In that case $-m$ and $-n$ are positive, so

$$a^m \cdot a^n = \frac{1}{a^{-m}} \cdot \frac{1}{a^{-n}} = \frac{1}{a^{-m-n}} = \frac{1}{a^{-(m+n)}} = a^{m+n}$$

The first equality is due to the definition of a negative exponent, the second to (4.4) for positive exponents, the third to algebra, and the fourth to the definition of negative exponents.

● **Example 1**

$$2^{-3} = \frac{1}{2^3} = \frac{1}{8} \quad \text{and} \quad 5^{-4} = \frac{1}{5^4} = \frac{1}{625}$$

$$3^{-5} \cdot 3^4 \cdot 3^3 = 3^{-5+4+3} = 3^2 = 9$$

$$\frac{5^3 \cdot 5^{-8}}{5^{-4} \cdot 5^{11} \cdot 5^{-2}} = \frac{5^{3-8}}{5^{-4+11-2}} = \frac{5^{-5}}{5^5} = \frac{1}{5^5 \cdot 5^5} = \frac{1}{5^{10}}$$

$$\frac{2^3 \cdot 4^{-2}}{4^5 \cdot 8^{-2}} = \frac{2^3(2^2)^{-2}}{(2^2)^5 \cdot (2^3)^{-2}} = \frac{2^3 \cdot 2^{-4}}{2^{10} \cdot 2^{-6}} = 2^{3-4-10+6} = 2^{-5} = \frac{1}{32}$$ ●

Note   The examples above, along with (4.3), illustrate that any *factor* in the numerator (denominator) of a fraction may be moved to the denominator (numerator) of the fraction provided that the sign of the exponent is changed.

● **Example 2** $\left(\dfrac{4a^3b^4}{3c^2d^3}\right)^4 \left(\dfrac{9c^5d}{2a^2b^5}\right)^2 = \dfrac{4^4(a^3)^4(b^4)^4}{3^4(c^2)^4(d^3)^4} \cdot \dfrac{9^2(c^5)^2d^2}{2^2(a^2)^2(b^5)^2}$ by (4.7) and (4.8)

$$= \frac{256a^{12}b^{16}}{81c^8d^{12}} \cdot \frac{81c^{10}d^2}{4a^4b^{10}} \quad \text{by (4.6)}$$

$$= \frac{64a^8b^6c^2}{d^{10}} \quad \text{by (4.5) and cancellation} \quad ●$$

Example 3   Convert $2^{-3}a^{-2}bc^{-1}/4^{-2}xy^{-3}z^4$ to an equivalent fraction in which all exponents are positive.

Solution

$$\frac{2^{-3}a^{-2}bc^{-1}}{4^{-2}xy^{-3}z^4} = \frac{\left(\dfrac{b}{2^3a^2c}\right)}{\left(\dfrac{xz^4}{4^2y^3}\right)} \qquad \text{by (4.3)}$$

$$= \frac{b}{2^3a^2c} \cdot \frac{4^2y^3}{xz^4} \qquad \text{invert and multiply}$$

$$= \frac{2by^3}{a^2cxz^4} \qquad \begin{array}{l}\text{cancellation and product}\\ \quad\text{of fractions}\end{array}$$

Example 4   Convert $2c^{-1}d^{-4}e^2/3c^{-3}d^{-2}e^{-1}$ to an equivalent fraction with only positive exponents.

Solution 1   In this fraction, we have c^{-1} in the numerator and c^{-3} in the denominator, and c^{-3} has the exponent with the greater absolute value. Also d^{-4} is in the numerator, and d^{-2} is in the denominator; here d^{-4} has the exponent with the greater absolute value. Furthermore, e^{-1} appears in the denominator. Hence if we multiply each member of the fraction by $c^3d^4e^1$, we will obtain an equivalent fraction in which all exponents are nonnegative. This procedure yields

$$\frac{2c^{-1}d^{-4}e^2}{3c^{-3}d^{-2}e^{-1}} = \frac{(2c^{-1}d^{-4}e^2)(c^3d^4e^1)}{(3c^{-3}d^{-2}e^{-1})(c^3d^4e^1)}$$

$$= \frac{2c^{-1+3}d^{-4+4}e^{2+1}}{3c^{-3+3}d^{-2+4}e^{-1+1}} \qquad \text{by Eq. (4.4)}$$

$$= \frac{2c^2d^0e^3}{3c^0d^2e^0}$$

$$= \frac{2c^2e^3}{3d^2} \qquad \text{since } d^0 = c^0 = e^0 = 1$$

Solution 2   Using the comment after Example 1 gives

$$\frac{2c^{-1}d^{-4}e^2}{3c^{-3}d^{-2}e^{-1}} = \frac{2c^3d^2e^2e^1}{3c^1d^4}$$

$$= \frac{2c^{3-1}e^{2+1}}{3d^{4-2}} = \frac{2c^2e^3}{3d^2}$$

Solution 3   Using (4.5) at once yields

$$\frac{2c^{-1}d^{-4}e^2}{3c^{-3}d^{-2}e^{-1}} = \frac{2}{3}c^{-1-(-3)}d^{-4-(-2)}e^{2-(-1)}$$

$$= \frac{2}{3}c^2 d^{-2} e^3$$

$$= \frac{2c^2 e^3}{3d^2} \qquad \bullet$$

The methods employed in the above examples can also be applied to fractions in which either or both of the members are the sum of two or more numbers. Remember: Only *factors* can be moved directly from numerator to denominator.

● **Example 5**   Convert the fraction $(4x^{-2} - 9y^{-2})/(3x + 2y)$ to an equal fraction in which all exponents are positive.

Solution   The terms with negative exponents that appear in this fraction are x^{-2} and y^{-2}. Consequently, we multiply each member of the fraction by $x^2 y^2$. Thus

$$\frac{4x^{-2} - 9y^{-2}}{3x + 2y} = \frac{(4x^{-2} - 9y^{-2})(x^2 y^2)}{(3x + 2y)(x^2 y^2)}$$

$$= \frac{4x^{-2+2}y^2 - 9x^2 y^{-2+2}}{(3x + 2y)x^2 y^2} \qquad \text{by distributive law and (4.4)}$$

$$= \frac{4x^0 y^2 - 9x^2 y^0}{(3x + 2y)x^2 y^2}$$

$$= \frac{4y^2 - 9x^2}{(3x + 2y)x^2 y^2} \qquad x^0 = y^0 = 1$$

$$= \frac{(2y + 3x)(2y - 3x)}{x^2 y^2 (3x + 2y)} \qquad \text{factoring}$$

$$= \frac{2y - 3x}{x^2 y^2} \qquad \begin{array}{l} \text{dividing the members} \\ \quad \text{by } (3x + 2y) \end{array} \qquad \bullet$$

Scientific notation   Scientific notation is useful in situations where very large or very small numbers are used. These occur not only in science but also in areas such as the national debt, corporate budgets, and the display on a calculator. A positive, real number N is written in scientific notation if $N = m \cdot 10^c$, where m is a real number between 1 and 10, including 1 but not 10 ($1 \leq m < 10$) and c is an integer. The integer c may be positive, negative, or zero.

● **Example 6**

N	m	Scientific notation
647.22	6.4722	$6.4722(10^2)$
0.008189	8.189	$8.189(10^{-3})$
5,000,000,000	5	$5(10^9)$
0.00000003303	3.303	$3.303(10^{-8})$

●

Calculators   Many calculators use scientific notation in the form "2.81643 17," for instance, which means $2.81643(10^{17})$. Thus the product (7.42 8)(4.06 11) is displayed as "3.01252 20," meaning $3.01252(10^{20})$.

In a number that represents an approximation, the digits known to be Significant digits   correct are called *significant digits*.

The digits 1, 2, 3, 4, 5, 6, 7, 8, and 9 are always significant if used in connection with a measurement, as are any zeros between any two of these digits. Zeros on the right may or may not be significant. Zeros on the left whose only function is as an aid in placing the decimal point are never significant. Such zeros occur between the decimal and the first nonzero digit in positive numbers less than 1.

● **Example 7**

N	Significant digits in N
238.71	2, 3, 8, 7, 1
2.051	2, 0, 5, 1
0.038	3, 8
0.0050	5 (last 0 may or may not be)

●

Note   Do not confuse *significant digits* with *decimal place accuracy*, i.e., the number of digits to the right of the decimal point. The number 84.35 has four significant digits but has two decimal places of accuracy. Also, the number 0.00251 has three significant digits but five decimal places of accuracy.

In computation involving approximate numbers we do not use a digit if its accuracy is doubtful. Eliminating such a digit or such digits is called Rounding off   *rounding off* a number. The procedure of rounding a number off to *n* significant digits consists of the following steps.

1 *Temporarily ignore the decimal point.*
2 *Consider only the significant digits, and discard all digits to the right of the nth one.*
3 *If the first digit discarded is 5, 6, 7, 8, or 9, increase the nth digit by one. Otherwise, leave it alone.*
4 *Put the decimal point where it should be. Add on any zeros necessary, or multiply by a power of 10, to make the number the proper magnitude.*

● **Example 8**

N	N rounded off to 3 significant digits
4.812	4.81
4.816	4.82
538,469	538,000 or 5.38(10⁵)
0.0024655	0.00247 or 2.47(10⁻³)

Note In calculating with approximate numbers, we must not be misled about the accuracy of our result. If $x = 3.15$ and $y = 86.2$, each to three significant digits, then

$$3.145 \le x < 3.155 \quad \text{and} \quad 86.15 \le y < 86.25$$

Hence

$$(3.145)(86.15) \le xy < (3.155)(86.25)$$
$$270.94175 \le xy < 272.11875$$

This shows that xy is close to 271 or 272. However, we may certainly *not* claim that $xy = (3.15)(86.2) = 271.530$ with all six digits significant. The commonly used rules are given below.

Products *When multiplying or dividing rounded numbers, round the answer to the least number of significant digits in any of the original numbers.*

Sums *When adding or subtracting rounded numbers, round the answer to the same decimal place as in the least accurate of the given numbers.*

● **Example 9** (a) $(48.2) \times (61.3) = 2954.66$, which should be rounded off to $2950 = 2.95(10^3)$.

(b) For $m = 1.81(10^{14})$, $M = 6.93(10^{15})$, $d = 2.98(10^9)$, we have

$$\frac{mM}{d^2} = \frac{(1.81)(6.93)10^{14+15}}{(2.98)^2(10^9)^2} = 1.41(10^{11}) \quad \text{after rounding}$$

(c) $1.2345 + 2.345 + 6.8 = 10.3795$, which should be rounded off to 10.4 since 6.8 has only one decimal place accuracy.

Exercise 4.1
operations with
integer
exponents

Find the value of the expression in each of Probs. 1 to 24.

1 4^{-2}	**2** 3^{-3}	**3** 5^{-1}	**4** 1^{-13}
5 $3^{-1}3^{-2}$	**6** $5^{-3}5^2$	**7** $1^{-5}1^{-8}$	**8** $4^{-5}4^3$
9 $7^{-2}/7^{-3}$	**10** $3^{-3}/3^{-1}$	**11** $11^{-2}/11^{-3}$	**12** $5^{-4}/5^{-2}$
13 $(3^{-1})^2$	**14** $(2^{-3})^{-2}$	**15** $(5^2)^{-2}$	**16** $(7^{-1})^3$
17 $(3^{-1}2)^{-1}$	**18** $(3^{-2}5^2)^{-2}$	**19** $(2^{-1}7^{-2})^2$	**20** $(5^{-2}3^3)^2$
21 $(3^{-2}/2^3)^2$	**22** $(3^{-2}/2^{-3})^{-3}$	**23** $(5/3^{-2})^{-3}$	**24** $(2^3/3^{-2})^3$

Write the expression in each of Probs. 25 to 32 without a denominator by using negative exponents if needed.

25 a^{-2}/b^2 **26** a^2/b^{-2} **27** a^2/b^2 **28** a^{-2}/b^{-2}

29 $2a^0b/a^{-1}b^2c^4$ **30** $3a^{-1}b^2/a^{-2}b^3c^{-1}$

31 $3a^{-2}b^{-1}/4^{-1}a^4b$ **32** $3^2a^{-4}b^{-2}/9^{-1}a^{-6}b^{-5}c^0$

In Probs. 33 to 68, simplify and express the results without zero or negative exponents.

33 $\dfrac{2^{-1}u^{-2}v^3w^1}{3^2u^0v^{-4}w^{-1}}$

34 $\dfrac{3^{-3}r^0s^{-2}t^{-4}}{6^{-2}r^{-3}t^{-5}s^3}$

35 $\dfrac{4^{-2}f^{-1}g^3h^0}{6^{-3}f^4g^{-2}h^{-3}}$

36 $\dfrac{4^{-3}r^0f^{-2}d^{-3}}{2^{-5}r^{-4}f^{-5}d^2}$

37 $\left(\dfrac{c^{-3}}{d^2}\right)^{-3}$

38 $\left(\dfrac{a^{-1}}{b^{-3}}\right)^2$

39 $\left(\dfrac{w^0}{z^3}\right)^{-2}$

40 $\left(\dfrac{c^4}{d^{-3}}\right)^{-3}$

41 $(m^{-3}n)^{-2}$

42 $(c^3d^{-2})^{-1}$

43 $(b^2r^3)^{-2}$

44 $(m^{-1}t^3)^{-3}$

45 $\left(\dfrac{3^{-3}m^2w^{-3}}{9^{-1}m^{-3}w^2}\right)^2$

46 $\left(\dfrac{2^3x^0y^{-5}}{8^2x^{-3}y^{-2}}\right)^{-3}$

47 $\left(\dfrac{6^{-3}b^{-2}z^{-4}}{2^{-6}b^4z^3}\right)^3$

48 $\left(\dfrac{6^{-2}r^0t^{-2}}{9^{-3}r^4t^{-1}}\right)^{-2}$

49 $\left(\dfrac{r^{-1}a^{-2}t^4}{r^0a^{-3}t^3}\right)^{-5}$

50 $\left(\dfrac{b^{-3}u^0m^4}{b^{-2}u^{-3}m^{-1}}\right)^4$

51 $\left(\dfrac{f^{-3}cd^5}{f^{-1}c^{-4}d^3}\right)^{-4}$

52 $\left(\dfrac{s^5u^{-3}m^{-2}}{s^3u^0m^4}\right)^5$

53 $\dfrac{x^{-1}+y^{-1}}{x^{-1}-y^{-1}}$

54 $\dfrac{a^{-2}-a^{-3}y^{-2}}{a^{-3}y^{-2}-y^{-2}}$

55 $\dfrac{a^{-1}b^{-2}-a^{-2}b^{-1}}{b^{-2}-a^{-2}}$

56 $\dfrac{w^{-2}z^{-3}+w^{-3}z^{-2}}{w^{-3}+z^{-3}}$

57 $\dfrac{y^{-2}-2x^{-1}y^{-1}-3x^{-2}}{x^{-1}y^{-2}-3x^{-2}y^{-1}}$

58 $\dfrac{y^{-2}-x^{-1}y^{-1}-6x^{-2}}{x^{-1}y^{-2}+2x^{-2}y^{-1}}$

59 $\dfrac{x^{-1}y^{-2}+x^{-2}y^{-1}}{y^{-2}-x^{-1}y^{-1}-2x^{-2}}$

60 $\dfrac{x^{-1}y^{-2}+3x^{-2}y^{-1}}{y^{-2}-9x^{-2}}$

61 $-2(x + 3)(x - 3)^{-4} + (x - 3)^{-3}$

62 $-(x - 2)^2(x - 3)^{-3} + (x - 3)^{-2}(x - 2)$

63 $-3(x - 3)^4(x - 5)^{-4} + 5(x - 3)^3(x - 5)^{-3}$

64 $-4(x + 3)^{-2}(x - 4)^{-5} - 3(x + 3)^{-3}(x - 4)^{-4}$

65 $-3(x + 5)(3x + 1)^{-2} + (3x + 1)^{-1}$

66 $-2(x + 1)^2(2x + 5)^{-3} + (x + 1)(2x + 5)^{-2}$

67 $2(2x - 3)^{-2}(3x + 4)^{-2} + 3(2x - 3)^{-3}(3x + 4)^{-1}$

68 $4(2x + 1)^{-1}(3x + 2)^{-2} + 3(2x + 1)^{-2}(3x + 2)^{-1}$

In Probs. 69 to 84, assume the given numbers are approximations. Perform the indicated operations.

69 $(6.92 \times 10^4)(1.13 \times 10^7)$

70 $(8.11 \times 10^{14})(3.26 \times 10^{16})$

71 $(5.53 \times 10^9)(1.72 \times 10^{12})$

72 $(6.88 \times 10^{23})(7.09 \times 10^{19})$

73 $(7.1 \times 10^{10})(8.42 \times 10^{15})$

74 $(8.2 \times 10^5)(9.411 \times 10^6)$

75 $(6.93 \times 10^{31})(1.071 \times 10^{27})$

76 $(4.45 \times 10^{18})(1.7921 \times 10^{22})$

77 $18.45 + 17.53$

78 $142.633 + 118.051$

79 $432.1 + 653.8$

80 $2.449 + 2.646$

81 $31.7 + 28.443$

82 $72.46 + 69.8$

83 $6.518 + 8.01$

84 $7.5689 + 4.71$

4.2 Radicals

For positive integers n, we have looked at the equation $a = b^n$ as defining the nth power of b. We will in this section use this same equation $a = b^n$ to define the nth root of a.

We are already familiar with square roots. For instance, $13^2 = 169$ may also be written as $13 = \sqrt{169}$. Although $(-13)^2$ is also 169, we use $\sqrt{169}$ only for $+13$, not for -13. Notice that -169 has no real square root because $b^2 \geq 0$ for every real number b. In general we define the square root of a by

Square root
$$\sqrt{a} = b \qquad \text{if and only if } a = b^2$$

for *positive* real numbers a and b. Also $\sqrt{0} = 0$. Thus for $a > 0$ we have $\sqrt{a} > 0$ by definition.

● Example 1 $\quad \sqrt{400} = +20$ (not -20, and not ± 20)

$\sqrt{207,936} = 456$ since $456 > 0$ and $456^2 = 207936$

$\sqrt{-55,555}$ is not defined as a real number since $-55,555 < 0$ ●

For cube roots, there is no problem with negative numbers. In fact for any real numbers a and b, we define the cube root of a by

Cube root
$$\sqrt[3]{a} = b \qquad \text{if and only if } a = b^3$$

● **Example 2** $\sqrt[3]{729} = 9$ since $729 = 9^3$

$\sqrt[3]{-125} = -5$ since $-125 = (-5)^3$

Note   If $a > 0$, both $\sqrt{a}$ and $\sqrt[3]{a}$ are defined and positive.

If $a < 0$, $\sqrt{a}$ is not defined but $\sqrt[3]{a}$ does exist and is negative.

In general, for any positive integer n and positive numbers a and b, we
*n*th root   define

$$\sqrt[n]{a} = b \qquad \text{if and only if } a = b^n \qquad\qquad (4.9)$$

We make the same definition for an odd positive integer n with negative
real numbers a and b. Furthermore we define $\sqrt[n]{0} = 0$ for any positive
integer n. $\sqrt{a}$ means $\sqrt[2]{a}$.

The *n*th root $\sqrt[n]{a}$ is called a *radical* of order n. Sometimes $\sqrt[n]{a}$ is called
the *principal nth root* of a to emphasize that it is defined to be positive if
$a > 0$. The number a is the *radicand*, $\sqrt[n]{}$ is the *radical sign*, n is the *index*,
and $\sqrt[n]{a}$ is called a *radical expression*, or the *n*th root of a.

● **Example 3** $\sqrt[4]{81} = 3$ since $3^4 = 81$ and $3 > 0$

$\sqrt[7]{-1} = -1$ since $(-1)^7 = -1$

$\sqrt[6]{1/64} = 1/2$ since $(1/2)^6 = 1/64$ and $1/2 > 0$

$\sqrt[5]{32/243} = 2/3$ since $(2/3)^5 = 32/243$

$\sqrt[8]{-2121}$ is not defined as a real number

It follows from the definitions that if $a > 0$ then

$$(\sqrt[n]{a})^n = a \qquad \text{and} \qquad \sqrt[n]{a^n} = a$$

Moreover, if $a < 0$ and n is odd, then the first equation is true, again by
definition.

We will now look at $\sqrt[n]{a^n}$ when $a < 0$. For n odd, its value is a. As an
example,

$$\sqrt[3]{(-4)^3} = \sqrt[3]{-64} = -4 \qquad \text{since } (-4)^3 = -64$$

However, for $a < 0$ and n even, the situation is different. For instance, $\sqrt{(-11)^2} \neq -11$ since $-11 < 0$, while the square root is positive. In fact

$$\sqrt{(-11)^2} = \sqrt{121} = +11 \qquad \text{not } -11$$

For any real number a,

$$\sqrt{a^2} = |a|$$

Notice that both members of this equation are positive. In addition, $\sqrt[n]{a^n} = |a|$ for n even.

● **Example 4** $\quad \sqrt{225} = 15 \qquad \sqrt{15^2} = |15| = 15 \qquad \sqrt{(-15)^2} = |-15| = 15$

$\sqrt[4]{(-7)^4} = |-7| = 7 \qquad \sqrt[6]{(-32)^6} = |-32| = 32$ ●

Remember that $(-32)^6 = 32^6$, whereas $-32^6 = -(32^6)$.

Just for convenience, to avoid the possibility of the absolute value notation, we will assume that

Note **Any variable under a radical sign is positive.**

With this assumption, it is true that

$$\sqrt[n]{a^n} = a \qquad \text{for any positive integer } n$$

There are three laws of radicals we will use extensively (for $a > 0$, $b > 0$):

LAWS OF RADICALS

$$\sqrt[n]{ab} = \sqrt[n]{a}\sqrt[n]{b} \tag{4.10}$$

$$\sqrt[n]{\frac{a}{b}} = \frac{\sqrt[n]{a}}{\sqrt[n]{b}} \tag{4.11}$$

$$\sqrt[m]{\sqrt[n]{a}} = \sqrt[mn]{a} \tag{4.12}$$

We will prove only the first of these. Letting $x = \sqrt[n]{a}$ and $y = \sqrt[n]{b}$ gives by definition $x^n = a$ and $y^n = b$. Thus by the laws of exponents,

$$(a)(b) = (x^n)(y^n) = (xy)^n$$

and so by definition of roots

$$\sqrt[n]{ab} = x \cdot y = \sqrt[n]{a}\sqrt[n]{b}$$

● **Example 5** (a) $\quad \sqrt{125} = \sqrt{25 \cdot 5} = \sqrt{5^2 \cdot 5} = \sqrt{5^2}\sqrt{5} = 5\sqrt{5}$

(b) $\quad \sqrt[3]{192} = \sqrt[3]{64 \cdot 3} = \sqrt[3]{4^3 \cdot 3} = \sqrt[3]{4^3} \cdot \sqrt[3]{3} = 4\sqrt[3]{3}$

(c) $\sqrt[4]{128a^6b^9} = \sqrt[4]{16a^4b^8(8a^2b)} = \sqrt[4]{(2ab^2)^4 8a^2b} = 2ab^2\sqrt[4]{8a^2b}$

(d) $\sqrt{8x^3y}\sqrt{6x^2y^5} = \sqrt{(8x^3y)(6x^2y^5)} = \sqrt{48x^5y^6} = \sqrt{16x^4y^6(3x)} = 4x^2y^3\sqrt{3x}$

(e) $\sqrt[3]{9a^5b^2}\sqrt[3]{81a^2b^7} = \sqrt[3]{729a^7b^9} = \sqrt[3]{3^6a^6b^9(a)} = 3^2a^2b^3\sqrt[3]{a} = 9a^2b^3\sqrt[3]{a}$ ●

INSERTION OF FACTORS INTO THE RADICAND Frequently, it is desirable to write a radical expression that has a coefficient as a single radical in which the coefficient is absorbed in the radicand; this is done below in part (a) by writing $2a = \sqrt{(2a)^2}$.

● **Example 6** (a) $2a\sqrt{4ab} = \sqrt{(2a)^2(4ab)} = \sqrt{4a^2 \cdot 4ab} = \sqrt{16a^3b}$ by (4.10)

(b) $(4x)\sqrt[3]{2y} = \sqrt[3]{(4x)^3}\sqrt[3]{2y} = \sqrt[3]{64x^3(2y)} = \sqrt[3]{128x^3y}$ by (4.10) ●

THE QUOTIENT OF TWO RADICALS OF THE SAME ORDER To get the quotient of two radicals of the *same order*, we use law (4.11). For example, $\sqrt[4]{a}/\sqrt[4]{b} = \sqrt[4]{a/b}$.

● **Example 7** (a) $\dfrac{\sqrt{128a^3b^5}}{\sqrt{2ab^2}} = \sqrt{\dfrac{128a^3b^5}{2ab^2}} = \sqrt{64a^2b^3} = \sqrt{(64a^2b^2)b} = 8ab\sqrt{b}$

(b) $\dfrac{\sqrt[3]{625x^{10}y^7z^{11}}}{\sqrt[3]{5x^2yz^4}} = \sqrt[3]{\dfrac{625x^{10}y^7z^{11}}{5x^2yz^4}} = \sqrt[3]{125x^8y^6z^7} = \sqrt[3]{(5x^2y^2z^2)^3x^2z}$

$= 5x^2y^2z^2\sqrt[3]{x^2z}$ ●

● **Example 8** In each of the following expressions, insert the integral factor at the left of the radical sign into the radicand and then arrange the given expressions in order of magnitude: $3\sqrt{3}, 2\sqrt{6}, 4\sqrt{2}$.

Solution
$$3\sqrt{3} = \sqrt{3^2 \cdot 3} = \sqrt{3^3} = \sqrt{27}$$
$$2\sqrt{6} = \sqrt{2^2 \cdot 6} = \sqrt{4 \cdot 6} = \sqrt{24}$$
$$4\sqrt{2} = \sqrt{4^2 \cdot 2} = \sqrt{16 \cdot 2} = \sqrt{32}$$

Since $24 < 27 < 32$, the desired arrangement is $2\sqrt{6}, 3\sqrt{3}, 4\sqrt{2}$. ●

The distributive law is used in adding two or more radical expressions if they have the same index and same radicand.

● **Example 9** Combine $\sqrt{108} + \sqrt{48} - \sqrt{3}$ into a single radical expression.

Solution
$\sqrt{108} + \sqrt{48} - \sqrt{3} = \sqrt{36 \cdot 3} + \sqrt{16 \cdot 3} - \sqrt{3}$ factoring

$= 6\sqrt{3} + 4\sqrt{3} - \sqrt{3}$ by Eq. (4.10)

$= (6 + 4 - 1)\sqrt{3}$ by distributive law

$= 9\sqrt{3}$ ●

● **Example 10** $\sqrt{8a^3b^3} + \sqrt[3]{ab} - \sqrt{2/ab} - \sqrt[3]{8a^4b^4} - \sqrt[4]{4a^2b^2}$.

Solution   In this problem, no two radicands are identical and we have radical expressions of order 2, 3, and 4. The radical expression $\sqrt[4]{4a^2b^2}$, however, can be converted to a lower order in this way: $\sqrt[4]{4a^2b^2} = \sqrt[4]{(2ab)^2}$ $= \sqrt{\sqrt{(2ab)^2}} = \sqrt{2ab}$. We therefore replace $\sqrt[4]{4a^2b^2}$ by $\sqrt{2ab}$ and then complete the solution as indicated below.

$$\sqrt{8a^3b^3} + \sqrt[3]{ab} - \sqrt{\frac{2}{ab}} - \sqrt[3]{8a^4b^4} - \sqrt[4]{4a^2b^2}$$

$$= \sqrt{(2ab)^2 2ab} + \sqrt[3]{ab} - \sqrt{\frac{2ab}{a^2b^2}} - \sqrt[3]{(2ab)^3 ab} - \sqrt{2ab}$$

factoring radicands and replacing $\dfrac{2}{ab}$ by $\dfrac{2ab}{a^2b^2}$

$$= 2ab\sqrt{2ab} + \sqrt[3]{ab} - \frac{1}{ab}\sqrt{2ab} - 2ab\sqrt[3]{ab} - \sqrt{2ab}$$

by Eqs. (4.10) and (4.11)

$$= 2ab\sqrt{2ab} - \frac{1}{ab}\sqrt{2ab} - \sqrt{2ab} + \sqrt[3]{ab} - 2ab\sqrt[3]{ab}$$

by the commutative axiom

$$= \left(2ab - \frac{1}{ab} - 1\right)\sqrt{2ab} + (1 - 2ab)\sqrt[3]{ab}$$

by the distributive axiom   ●

We obtain the product of two multinomials whose terms involve radical expressions by the distributive law and the laws of radicals. The method is illustrated in Example 11.

● **Example 11**   Find the product of $\sqrt{x} - 2\sqrt{xy} + 2\sqrt{y}$ and $\sqrt{x} + 2\sqrt{xy} - \sqrt{y}$.

Solution

$\sqrt{x} - 2\sqrt{xy} + 2\sqrt{y}$	multiplicand
$\sqrt{x} + 2\sqrt{xy} - \sqrt{y}$	multiplier
$x - 2x\sqrt{y} + 2\sqrt{xy}$	multiplying by $\sqrt{x}$
$2x\sqrt{y} \qquad -4xy + 4y\sqrt{x}$	multiplying by $2\sqrt{xy}$
$- \sqrt{xy} \qquad + 2y\sqrt{x} - 2y$	multiplying by $-\sqrt{y}$
$x \qquad + \sqrt{xy} - 4xy + 6y\sqrt{x} - 2y$	product

●

It is frequently desirable to convert a radical with a fractional radicand to a form in which no radical appears in the denominator. This process is called *rationalizing the denominator*. If the denominator of the radicand is a monomial, we use law (4.11) for the purpose. For example, to rationalize the denominator of $\sqrt{8a/3bc^3}$, we multiply the numerator and denominator of the radicand by the expression of lowest power that will convert the denominator into a perfect square. In this case, the expression is $3bc$; hence, we have

Rationalizing the denominator

$$\sqrt{\frac{8a}{3bc^3}} = \sqrt{\frac{8a(3bc)}{3bc^3(3bc)}} = \sqrt{\frac{2^2 6abc}{(3bc^2)^2}} = \frac{2\sqrt{6abc}}{3bc^2}$$

● **Example 12**

$$\sqrt[3]{\frac{4xy^2}{5x^5y^7}} = \sqrt[3]{\frac{4xy^2(25xy^2)}{5x^5y^7(25xy^2)}} = \sqrt[3]{\frac{100x^2yy^3}{(5x^2y^3)^3}} = \frac{y\sqrt[3]{100x^2y}}{5x^2y^3} = \frac{\sqrt[3]{100x^2y}}{5x^2y^2}$$ ●

RATIONALIZING BINOMIAL DENOMINATORS

If the denominator of a fraction is the sum or the difference of two terms at least one of which contains a radical expression of the second order, we rationalize the denominator by the method illustrated below. This simplifies the expression for further work or calculation.

● **Example 13**   Rationalize the denominator in $4/(\sqrt{5} - 1)$.

Solution   Since $(\sqrt{5} - 1)(\sqrt{5} + 1) = 5 - 1 = 4$, we multiply the members of the given fraction by $(\sqrt{5} + 1)$ and complete the problem as below.

$$\frac{4}{\sqrt{5} - 1} = \frac{4(\sqrt{5} + 1)}{(\sqrt{5} - 1)(\sqrt{5} + 1)}$$

$$= \frac{4(\sqrt{5} + 1)}{5 - 1}$$

$$= \frac{4(\sqrt{5} + 1)}{4}$$

$$= \sqrt{5} + 1$$ ●

● **Example 14**   Sometimes it is desirable to rationalize the numerator. In calculus, for instance, this is the case with the following expression.

$$\frac{\sqrt{x + h} - \sqrt{x}}{h} = \frac{(\sqrt{x + h} - \sqrt{x})(\sqrt{x + h} + \sqrt{x})}{h(\sqrt{x + h} + \sqrt{x})}$$

$$= \frac{(x + h) - x}{h(\sqrt{x + h} + \sqrt{x})} \qquad (a - b)(a + b) = a^2 - b^2$$

$$= \frac{h}{h(\sqrt{x + h} + \sqrt{x})} = \frac{1}{\sqrt{x + h} + \sqrt{x}}$$ ●

Exercise 4.2
radicals

Remove all possible factors from the radicands in Probs. 1 to 16.

1 $\sqrt{12}$ **2** $\sqrt{98}$ **3** $\sqrt{363}$ **4** $\sqrt{180}$

5 $\sqrt[3]{250}$ **6** $\sqrt[3]{40}$ **7** $\sqrt[4]{48}$ **8** $\sqrt[5]{486}$

9 $\sqrt{112a^9b^6}$ **10** $\sqrt{108a^0b^7}$ **11** $\sqrt{28a^4b^2}$ **12** $\sqrt{147a^8b^6}$

13 $\sqrt[3]{54x^3y^6}$ **14** $\sqrt[3]{375x^4y^9}$ **15** $\sqrt[4]{80x^6y^5}$ **16** $\sqrt[5]{160x^6y^{10}}$

In each of Probs. 17 to 28, combine into a single radical and then remove all possible factors from the radicand.

17 $\sqrt[3]{3}\sqrt[3]{432}$ **18** $\sqrt[4]{27}\sqrt[4]{15}$ **19** $\sqrt[5]{80}\sqrt[5]{6}$

20 $\sqrt[3]{18}\sqrt[3]{24}$ **21** $\sqrt{5x^2y}\sqrt{75xy^3}$ **22** $\sqrt{7x^3y^2}\sqrt{84xy}$

23 $\sqrt{18x^3y}\sqrt{6xy^2}$ **24** $\sqrt{75x^3y^2}\sqrt{3xy^3}$ **25** $\sqrt[3]{6x^2y^4}\sqrt[3]{9xy}$

26 $\sqrt[3]{8x^3y^5}\sqrt[3]{20x^2y^3}$ **27** $\sqrt[4]{27x^2y^3}\sqrt[4]{15x^3y}$ **28** $\sqrt[4]{54x^2y^5}\sqrt[4]{60x^3y^3}$

Express each of Probs. 29 to 36 as a single radical.

29 $\sqrt{\sqrt{2}}$ **30** $\sqrt[3]{\sqrt{5}}$ **31** $\sqrt{\sqrt[4]{3}}$ **32** $\sqrt[4]{\sqrt[3]{7}}$

33 $\sqrt{\sqrt[3]{a^2}}$ **34** $\sqrt[4]{\sqrt{a}}$ **35** $\sqrt[5]{\sqrt[4]{a^2}}$ **36** $\sqrt[5]{\sqrt[3]{a^4}}$

In Probs. 37 to 48, remove all possible factors from each radicand and rationalize each denominator; then make all combinations that can be made by additions and subtractions. Reduce the order of each radical if possible.

37 $\sqrt{3} - \sqrt{48} + \sqrt{12}$ **38** $\sqrt{5} - \sqrt{45} + \sqrt{80} + \sqrt{125}$

39 $\sqrt[3]{16} + \sqrt[3]{2} - \sqrt[3]{128}$ **40** $\sqrt[3]{40} + \sqrt[3]{135} - \sqrt[3]{5}$

41 $\sqrt{18} + \sqrt[3]{128} + \sqrt{8} - \sqrt[3]{54}$ **42** $\sqrt{27} + \sqrt[3]{54} - \sqrt{12} + \sqrt[3]{16}$

43 $\sqrt{50} - \sqrt[3]{375} + \sqrt{72} + \sqrt[3]{24}$ **44** $\sqrt[3]{81} + \sqrt[3]{256} - \sqrt[3]{24} - \sqrt[3]{108}$

45 $\sqrt{4a^2b} - \sqrt{25ab^2} + \sqrt{a^2b} + \sqrt{16ab^2}$

46 $\sqrt{2r^3t} - \sqrt{r^3t^2} + \sqrt{18rt^5} + \sqrt{r^5t^0}$

47 $\sqrt[3]{8c^5d} + \sqrt[3]{27c^4d^5} + \sqrt[3]{27c^2d^4} - \sqrt[3]{c^7d^5}$

48 $\sqrt[3]{3g^4m^2} - \sqrt[3]{8g^2m^4} - \sqrt[3]{24gm^8} + \sqrt[3]{27g^5m}$

Rationalize the denominator and then remove all possible factors from the radicand in Probs. 49 to 64.

49 $\sqrt{\dfrac{3x^5}{5y}}$ **50** $\sqrt{\dfrac{3x}{8y^3}}$ **51** $\sqrt{\dfrac{2x^3}{50y^5}}$ **52** $\sqrt{\dfrac{8x^7}{27y^3}}$

53 $\sqrt{\dfrac{243x^5y^6}{147xy^3}}$ **54** $\sqrt{\dfrac{24x^3y^6}{54xy}}$ **55** $\sqrt{\dfrac{7x^5y^4}{6x^3y}}$ **56** $\sqrt{\dfrac{64x^5y^3}{98xy}}$

57 $\sqrt[3]{\dfrac{27x^5y^2}{4x^2y^4}}$ **58** $\sqrt[3]{\dfrac{125x^4y}{18x^2y^2}}$ **59** $\sqrt[4]{\dfrac{8x^7y}{27x^2y^3}}$ **60** $\sqrt[5]{\dfrac{7xy^3}{16x^8y^6}}$

61 $\sqrt{\dfrac{147x^7y^2}{243x^2y^{-3}}}$ **62** $\sqrt{\dfrac{17x^5y^3}{6x^{-3}y^{-4}}}$ **63** $\sqrt{\dfrac{9x^3y^7}{98x^{-1}y^{-3}}}$ **64** $\sqrt{\dfrac{49x^5y^{-9}}{24x^{-3}y}}$

Rationalize the denominator in each of Probs. 65 to 76.

65 $\dfrac{2}{1-\sqrt{3}}$ **66** $\dfrac{7}{\sqrt{2}-3}$ **67** $\dfrac{3}{2+\sqrt{5}}$ **68** $\dfrac{1}{\sqrt{7}-2}$

69 $\dfrac{6}{\sqrt{5}-\sqrt{2}}$ **70** $\dfrac{2}{\sqrt{3}-\sqrt{7}}$ **71** $\dfrac{5}{\sqrt{10}-\sqrt{5}}$ **72** $\dfrac{12}{\sqrt{13}-\sqrt{7}}$

73 $\dfrac{2+\sqrt{3}}{\sqrt{5}-\sqrt{3}}$ **74** $\dfrac{1+2\sqrt{5}}{\sqrt{3}+\sqrt{5}}$ **75** $\dfrac{\sqrt{3}+2\sqrt{2}}{\sqrt{2}-\sqrt{3}}$ **76** $\dfrac{\sqrt{14}-2}{\sqrt{7}-\sqrt{2}}$

Physics **77** The approximate number of seconds for a complete oscillation of a pendulum of length L ft is given by

$$T = 2\pi\sqrt{\dfrac{L}{32.2}}$$

Find the time for a complete oscillation of a pendulum that is 32.2 in long.

Area of a triangle **78** The area of a triangle with sides of length a, b, and c is

$$K = \sqrt{s(s-a)(s-b)(s-c)}$$

where $a + b + c = 2s$. Find the area of a triangle with sides of length 8, 9, and 11 cm.

Value of e **79** If $f(x) = (1 + x)^{1/x}$, find a four-decimal-place approximation for $f(1)$, $f(0.1)$, $f(0.01)$, and $f(0.001)$. Compare each of them with 2.7183, which is an approximation to the limiting value of $(1 + x)^{1/x}$ for x very close to 0. This limiting value is called e.

Physics **80** Kepler's third law states that the square of the time required for a planet to make a circuit about the sun is a constant times the cube of the mean distance between the sun and the planet. Find the time in years for Mars to make a circuit about the sun if the mean distances of the Earth and Mars from the sun are 93 and 141 million miles, respectively.

Earth science   **81** The total reflecting power of a very thick crystal slab is

$$P = \frac{QA}{2m}$$

By what is P multiplied if Q is multiplied by $\sqrt{6}$, A by $\sqrt{7}$, and m by $\sqrt{8}$?

Earth science   **82** The velocity for compressional waves is $V_1 = \sqrt{(L + 2m)/p}$ and for shear waves is $V_2 = \sqrt{m/p}$. Find the ratio V_1/V_2 of compressional to shear velocity.

Political science   **83** In the theory of political development, the equation for the first-order partial correlation coefficient is

$$r = \frac{r_{13} - r_{12}r_{23}}{\sqrt{1 - r_{12}^2}\sqrt{1 - r_{23}^2}}$$

Calculate r if $r_{13} = r_{12} = r_{23} = a$.

Earth science   **84** The vertical gravitational attraction of a buried cylinder at a point off its axis is

$$g = K[\sqrt{L^2 + (x - r)^2} - \sqrt{L^2 + (x + r)^2}]$$

Express g by rationalizing the numerator.

In Probs. 85 to 88, use a calculator to find the roots to four significant digits.

85 $\sqrt{114.7}$ **86** $\sqrt[3]{278.4}$

87 $\sqrt[4]{788.1}$ **88** $\sqrt[5]{112.9}$

89 Find $\sqrt{2 + \sqrt{2}}$ and $\sqrt{2 + \sqrt{2 + \sqrt{2}}}$ and $\sqrt{2 + \sqrt{2 + \sqrt{2 + \sqrt{2}}}}$.

90 Rationalize the numerator in $\dfrac{\sqrt{x + 3 + h} - \sqrt{x + 3}}{h}$.

For $a^2 \geq b$ and $a > 0$, it is true that

$$\sqrt{a + \sqrt{b}} = \sqrt{\frac{a + \sqrt{a^2 - b}}{2}} + \sqrt{\frac{a - \sqrt{a^2 - b}}{2}}$$

91 Verify the above equation for $a = 3$ and $b = 2$.

92 Prove the above equation is true with the given conditions on a and b. *Hint:* Square the right-hand side, and simplify it to $a + \sqrt{b}$.

93 Arrange $\sqrt{2}$, $\sqrt[3]{3}$, and $\sqrt[4]{4}$ in order from largest to smallest. *Hint:* Use a calculator, or write each as a 12th root since 12 is the lcm of the indexes 2, 3, and 4.

94 Show $\dfrac{1}{\sqrt{m} + \sqrt{n} + \sqrt{p}} = \dfrac{(\sqrt{m} + \sqrt{n} - \sqrt{p})(m + n - p - 2\sqrt{mn})}{m^2 + n^2 + p^2 - 2mn - 2mp - 2np}$

95 Which of these numbers is real? $\sqrt{12}$, $\sqrt[4]{-6}$, $\sqrt[3]{-81}$, $\sqrt[5]{77}$.

96 Show that if $a > 1$ then $\sqrt{a + \dfrac{a}{a^2 - 1}} = (a)\sqrt{\dfrac{a}{a^2 - 1}}$. Show $\sqrt{6 + \dfrac{6}{35}} =$

$(6)\sqrt{\dfrac{6}{35}}$.

97 Show $\sqrt[3]{26 + 15\sqrt{3}} + \sqrt[3]{26 - 15\sqrt{3}} = 4$. Try $(2 + \sqrt{3})^3$.

98 Show $\sqrt[3]{26 + 15\sqrt{3}} - \sqrt[3]{26 - 15\sqrt{3}} = 2\sqrt{3}$. Try $(2 - \sqrt{3})^3$.

99 Show $\sqrt[3]{\sqrt{243} + \sqrt{242}} - \sqrt[3]{\sqrt{243} - \sqrt{242}} = 2\sqrt{2}$. Try $(\sqrt{2} + \sqrt{3})^3$.

100 Show $\sqrt[4]{17 + 12\sqrt{2}} = 1 + \sqrt{2}$. *Hint:* Calculate $(1 + \sqrt{2})^4$.

You can hear $\sqrt{2}$ **101** The frequency of any note is twice the frequency of the note one octave lower. On a piano each octave is broken into 12 steps (from one note to the adjacent one). Thus the frequency of any note is found by multiplying by $\sqrt[12]{2}$ the frequency of the note just lower. Hence starting at middle C and going successively to C#, D, D#, E, F, and F# multiplies the frequency of middle C by $(\sqrt[12]{2})^6 = \sqrt{2}$. Therefore playing C and F# simultaneously allows you to "hear" $\sqrt{2}$. For each pair of notes below, give both the exact value and a good rational approximation for the ratio of their frequencies (larger to smaller).

(a) C and E   (b) C and F   (c) C and G

4.3 Rational exponents

In this section we further extend the definition of a^n to include the case in which n is the quotient of two integers. We do this in such a way that the laws stated in Sec. 4.1 are valid for rational exponents.

In the law of exponents $(a^m)^n = a^{mn}$, we will take $m = 1/n$. Then since $mn = 1$, we get

$$(a^{1/n})^n = a$$

By definition of a radical, we also have

$$(\sqrt[n]{a})^n = a$$

and thus to be consistent with previous notation, we *define*

$$a^{1/n} = \sqrt[n]{a} \qquad\qquad (4.13)$$

● **Example 1** $289^{1/2} = \sqrt{289} = 17$

$(-243)^{1/5} = \sqrt[5]{-243} = -3$

$27^{1/3} = \sqrt[3]{27} = 3$

$(-16)^{1/4} = \sqrt[4]{-16}$, which is not defined   ●

For general rational exponents $m/n = m(1/n)$, we still want the laws of exponents to hold, and thus we require that

$$a^{m/n} = (a^{1/n})^m = (\sqrt[n]{a})^m$$

We want to show that this is equal to $(a^m)^{1/n}$, that is, that $(a^{1/n})^m$ is an nth root of a^m.

$$
\begin{aligned}
[(a^{1/n})^m]^n &= (a^{1/n})^{mn} && \text{by (4.6)}\\
&= (a^{1/n})^{nm} && \text{commutative law}\\
&= [(a^{1/n})^n]^m && \text{by (4.6)}\\
&= a^m && \text{definition of } a^{1/n}
\end{aligned}
$$

Now both $(a^m)^{1/n}$ and $(a^{1/n})^m$ are nth roots of a^m, and it can be shown that they have the same sign. Thus we have by definition

Rational exponents

$$a^{m/n} = (\sqrt[n]{a})^m = (a^{1/n})^m = (a^m)^{1/n} = \sqrt[n]{a^m} \qquad (4.14)$$

Note   In calculus it is especially important to be able to use Eq. (4.14) to write functions in the form $f(x) = x^{m/n}$, as this is the way in which differentiation and integration formulas are used.

The proofs that the laws of exponents stated in Sec. 4.1 hold for rational exponents as defined in (4.14) can be established, but we will not carry it out.

The processes involved in working with fractional exponents are illustrated in the following examples.

● **Example 2**   Evaluate $4^{1/2}, 8^{2/3}, (-32)^{3/5}$.

Solution
$$
\begin{aligned}
4^{1/2} &= \sqrt{4} = 2 && \text{by (4.13)}
\end{aligned}
$$

$$
\begin{aligned}
8^{2/3} &= (\sqrt[3]{8})^2 && \text{by (4.14)}\\
&= 2^2 = 4
\end{aligned}
$$

$$
\begin{aligned}
(-32)^{3/5} &= (\sqrt[5]{-32})^3 && \text{by (4.14)}\\
&= (-2)^3 = -8
\end{aligned}
$$
●

● **Example 3**   Express $3a^{1/2}b^{5/2}$ and $5a^{3/4}/b^{5/4}$ in radical form.

Solution

$$3a^{1/2}b^{5/2} = 3(ab^5)^{1/2} \qquad \text{by (4.6) and (4.14)}$$
$$= 3\sqrt{ab^5} \qquad \text{by (4.13)}$$

$$\frac{5a^{3/4}}{b^{5/4}} = 5\left(\frac{a^3}{b^5}\right)^{1/4} \qquad \text{by (4.8) with } n = \tfrac{1}{4}$$

$$= 5\sqrt[4]{\frac{a^3}{b^5} \cdot \frac{b^3}{b^3}} \qquad \text{by (4.13)}$$

$$= \frac{5}{b^2}\sqrt[4]{a^3 b^3}$$

● **Example 4**   Find the product of $3a^{1/2}$ and $2a^{2/3}$.

Solution

$$(3a^{1/2})(2a^{2/3}) = 6a^{1/2 + 2/3}$$
$$= 6a^{(3+4)/6}$$
$$= 6a^{7/6}$$

● **Example 5**   Find the quotient of $8x^{1/2}y^{5/6}$ and $5x^{1/4}y^{1/3}$.

Solution

$$8x^{1/2}y^{5/6} \div 5x^{1/4}y^{1/3} = \tfrac{8}{5}x^{1/2 - 1/4}y^{5/6 - 1/3} = \tfrac{8}{5}x^{1/4}y^{3/6}$$
$$= \tfrac{8}{5}x^{1/4}y^{1/2}$$

● **Example 6**   Combine $[(4x^4y^{3/4}z^2)/(9x^2y^{1/4}z)]^{1/2}$ wherever possible using the laws of exponents, and express the result without zero or negative exponents.

Solution

$$\left(\frac{4x^4y^{3/4}z^2}{9x^2y^{1/4}z}\right)^{1/2} = \left(\frac{4}{9}\frac{x^4}{x^2}\frac{y^{3/4}}{y^{1/4}}\frac{z^2}{z}\right)^{1/2}$$

$$= (\tfrac{4}{9}x^2y^{2/4}z)^{1/2}$$

$$= \tfrac{2}{3}xy^{1/4}z^{1/2}$$

● **Example 7**   Combine $[(4x^2y^{3/4}z^{1/6})/(32x^{-1}y^0z^{-5/6})]^{-1/3}$ wherever possible, and express the result without zero or negative exponents.

Solution

$$\left(\frac{4x^2y^{3/4}z^{1/6}}{32x^{-1}y^0z^{-5/6}}\right)^{-1/3} = \left(\frac{4}{32}\frac{x^2}{x^{-1}}\frac{y^{3/4}}{y^0}\frac{z^{1/6}}{z^{-5/6}}\right)^{-1/3}$$

$$= (\tfrac{1}{8}x^3y^{3/4}z)^{-1/3}$$

$$= \frac{1}{(\tfrac{1}{8}x^3y^{3/4}z)^{1/3}}$$

$$= \frac{2}{xy^{1/4}z^{1/3}}$$

**Exercise 4.3
rational
exponents**

Convert each number or expression in Probs. 1 to 36 to a form without exponents or radicals.

1 $81^{1/2}$ 2 $36^{1/2}$ 3 $125^{1/3}$ 4 $32^{1/5}$

5 $16^{3/4}$ 6 $64^{5/6}$ 7 $8^{5/3}$ 8 $25^{3/2}$

9 $9^{-1/2}$ 10 $27^{-1/3}$ 11 $32^{-4/5}$ 12 $16^{-3/4}$

13 $(9/16)^{3/2}$ 14 $(27/64)^{4/3}$ 15 $(32/243)^{3/5}$ 16 $(16/81)^{5/4}$

17 $(16/25)^{-1/2}$ 18 $(64/27)^{-2/3}$ 19 $(8/125)^{-2/3}$ 20 $(16/81)^{-3/4}$

21 $\sqrt{9^3}$ 22 $\sqrt[3]{27^2}$ 23 $\sqrt[6]{16^3}$ 24 $\sqrt[6]{27^2}$

25 $\sqrt{25b^2c^4}$ 26 $\sqrt{16a^4b^6}$ 27 $\sqrt{36c^6d^8}$ 28 $\sqrt{49m^0y^{10}}$

29 $\sqrt[3]{8m^6n^9}$ 30 $\sqrt[3]{27r^3s^9}$ 31 $\sqrt[4]{16x^8y^4}$ 32 $\sqrt[5]{32w^{10}z^{15}}$

33 $\sqrt{\dfrac{36a^4}{b^6c^8}}$ 34 $\sqrt[3]{\dfrac{m^6n^{12}}{8p^3}}$ 35 $\sqrt[4]{\dfrac{81s^8}{t^{12}v^0}}$ 36 $\sqrt[5]{\dfrac{32d^{10}}{g^{15}x^{25}}}$

Put the expression in each of Probs. 37 to 44 in radical form.

37 $a^{1/3}b^{2/3}$ 38 $a^{3/4}b^{1/2}$ 39 $a^{2/3}b^{3/5}$ 40 $a^{1/2}b^{3/5}$

41 $a^{2/3}y^{-1/5}$ 42 $a^{-1/2}b^{1/3}$ 43 $a^{-2/5}b^{-1/3}$ 44 $a^{-2/3}b^{3/2}$

Combine the expressions in each of Probs. 45 to 80 by use of the laws of exponents, and express each result without zero or negative exponents.

45 $(3x^{1/3})(2x^{1/2})$ 46 $(2x^{1/4})(5x^{1/3})$ 47 $(3x^{1/2})(4x^{1/5})$

48 $(7x^{1/6})(3x^{1/5})$ 49 $\dfrac{12x^{2/3}y^{1/2}}{3x^{1/3}y^{3/4}}$ 50 $\dfrac{6x^{3/5}y^{-1/3}}{2x^{2/5}y^{1/2}}$

51 $\dfrac{15x^{1/4}y^{-1/2}}{3x^{-1/4}y}$ 52 $\dfrac{27x^{4/5}y^{-1}}{9x^{4/3}y^0}$ 53 $(64x^{3/8}y^{-3/5})^{1/3}$

54 $(81a^{-4}b^{4/3})^{-1/4}$ 55 $(243^{-1}a^{5/3}b^{-5/7})^{1/5}$ 56 $(36x^{-2}y^{6/7})^{-1/2}$

57 $\left(\dfrac{8x^3y^{-4/3}}{27x^{-6}y}\right)^{-2/3}$ 58 $\left(\dfrac{x^{5/6}y^{-5/4}}{243x^0y^{-5/3}}\right)^{-1/5}$ 59 $\left(\dfrac{64x^{-1}y^6}{x^0y^{3/2}}\right)^{-1/6}$

60 $\left(\dfrac{36a^0b^3}{25a^{-1}b^{2/3}}\right)^{-1/2}$ 61 $(16x^6y^4)^{1/2}(27x^9y^6)^{-1/3}$

62 $(625a^4b^8)^{-1/4}(243a^{-10}b^5)^{1/5}$ 63 $(8a^6b^9)^{-2/3}(4a^4b^6)^{1/3}$

64 $(16x^{-4}y^2)^{1/2}(125^{-1}xy^{-3})^{1/3}$

65 $(x+1)(2x+3)^{-1/2}+(2x+3)^{1/2}$

66 $-(x+3)(5-2x)^{-1/2}+(5-2x)^{1/2}$

67 $3(3x + 2)(2x - 5)^{-1/4} + 6(2x - 5)^{3/4}$

68 $3(2x - 5)(3x + 4)^{-3/4} + 8(3x + 4)^{1/4}$

69 $3(2x - 1)^{2/3}(x + 1)^{-1/2} + 8(2x - 1)^{-1/3}(x + 1)^{1/2}$

70 $3(3x + 2)^{2/3}(x + 1)^{-1/2} + 4(3x + 2)^{-1/3}(x + 1)^{1/2}$

71 $3(2x + 3)^{1/4}(3x - 1)^{-1/2} + (3x - 1)^{1/2}(2x + 3)^{-3/4}$

72 $3(3x + 5)^{2/3}(4x - 3)^{-1/4} + 2(3x + 5)^{-1/3}(4x - 3)^{3/4}$

73 $\left(\dfrac{x^{1/(b-2)}}{x^{1/(b+2)}} \right)^{(b^2-4)/b}$
    74 $\left(\dfrac{x^{a+5b}}{x^{3b}} \right)^{a/(a+2b)}$

75 $\left(\dfrac{x^{a+3b}}{x^a} \right)^a \left(\dfrac{x^{a-2b}}{x^{-2b}} \right)^b$
    76 $\left(\dfrac{x^{a-5b}}{x^{-4b}} \right)^{b/(a^2-b^2)}$

Biology **77** For the number of pairs $N < 30$, the null hypothesis should be tested with the t distribution by using

$$t = \frac{r\sqrt{N - 2}}{\sqrt{1 - r^2}}$$

Find the value of t for $r = \frac{1}{3}\sqrt{5}$ and $N = 27$.

Biology **78** The error in a measurement in x-ray diffraction, in percent, is approximated by

$$e = \frac{201\sqrt{n + m}}{n - m}$$

Calculate e for $n = 2783$ and $m = 2117$.

Psychology **79** Stephen's power law in psychology states that

$$R = k(S - S_0)^c$$

Calculate the stimulus S in terms of the other variables.

Psychology and calculator **80** The altered behavior of a group of individuals (crowd psychology) will occur if the number of people present is at least

$$\frac{a\sqrt{2\pi(k^2 + s^2)}}{2A}$$

where all constants are positive. Calculate this number for $a = 6.83$, $k = 60$, $s = 63$, and $A = 0.215$.

Management and calculator **81** The production of pencils is governed by

$$Q = 650x^p y^{1-p}$$

where $0 < p < 1$, x is labor cost, and y is equipment value. Find Q if $x = 18$, $y = 75$ for (a) $p = 0.65$ (b) $p = 0.93$.

Earth science and calculator **82** Faust's statistical study of sedimentary-rock velocities shows that for sandstone and shale

$$V = 125.3(ZT)^{1/6}$$

where Z is depth in feet and T is age in years. Find V if $Z = 6000$ ft and $T = 12,860,000$ years.

Management and calculator **83** If T is the total cost of producing n units, and U is the cost of producing the ith unit, then by use of a learning curve it has been shown that

$$U = U_1 \cdot i^b \qquad \text{for } b = -0.152$$

$$T = \frac{U_1}{b + 1} \cdot n^{b+1}$$

(a) How much is production cost per unit decreased each time production is doubled?

(b) Find the total cost of producing 40 units if the first costs \$371.

Management and calculator **84** If earnings are $-300 + ax^n + b/x^m$, then the optimum value of x is

$$\left(\frac{bm}{an}\right)^{1/(m+n)} = \left(\frac{5m}{n}\right)^{1/(m+n)} \qquad \text{if } b = 5a$$

Calculate the optimum value if $m = 3$ and $n = 4$.

Complex numbers

In Sec. 4.2 we did not define $\sqrt{a}$ for $a < 0$ since no real number has a negative square. In this section we will extend our concept of number to include this case.

The equation $x^2 + 1 = 0$ has no *real* solution, so we define the *imaginary number* i as its solution; i.e., by definition

$$i^2 = -1 \qquad \text{or} \qquad i = \sqrt{-1}$$

This device allows us to solve not only the particular equation $x^2 + 1 = 0$ but also a great many more equations. We want the ordinary rules of arithmetic to apply to real numbers as well as i, hence we must deal with products of the form bi where b is real, and also with sums of the form

Complex numbers
$$a + bi \qquad \text{with } a \text{ and } b \text{ real}$$

Such numbers are called *complex numbers,* and they include as special cases *real* numbers (take $b = 0$) and *imaginary* numbers (take $a = 0$). To say that $a + bi = c + di$ means, by definition, that $a = c$ and $b = d$.

● **Example 1** These are complex numbers:

$$-1 + 7i \qquad 11 + 60i \qquad \sqrt{5} + \pi i \qquad 107 \qquad 33i \qquad 0 \qquad ●$$

We are now going to define the addition, subtraction, and multiplication of complex numbers. Later in the section, we will define division.

Addition

$$(a + bi) + (c + di) = (a + c) + (b + d)i \qquad (4.15)$$

Subtraction

$$(a + bi) - (c + di) = (a - c) + (b - d)i \qquad (4.16)$$

Multiplication

$$(a + bi)(c + di) = (ac - bd) + (ad + bc)i \qquad (4.17)$$

Instead of actually memorizing these definitions, it is easier to simply treat $a + bi$ as a first-degree polynomial with i replacing x. Addition and subtraction are then just like polynomial addition and subtraction, and multiplication is like polynomial multiplication remembering that $i^2 = -1$.

● **Example 2**   Add $3 + 4i$ and $6 + 7i$.

   Solution

$$(3 + 4i) + (6 + 7i) = 3 + 6 + 4i + 7i$$
$$= 9 + (4 + 7)i = 9 + 11i \qquad ●$$

● **Example 3**

$$8 + 3i - (5 - 2i) = 8 + 3i - 5 + 2i$$
$$= 8 - 5 + (3 + 2)i = 3 + 5i \qquad ●$$

To find the product of two complex numbers, we proceed as we would in obtaining the product of any other two binomials. Then use $i^2 = -1$.

● **Example 4**   Find the product of $2 + 3i$ and $5 + 7i$.

   Solution

$$(2 + 3i)(5 + 7i) = 2(5 + 7i) + 3i(5 + 7i)$$
$$= 10 + 14i + 15i + 21i^2$$
$$= 10 + 29i - 21 \qquad i^2 = -1$$
$$= -11 + 29i \qquad ●$$

Before finding the quotient of two complex numbers, we will find the product of the complex number $z = a + bi$ and its *conjugate* $\bar{z} = a - bi$. We have

Conjugate

$z\bar{z}$ is a real number

$$z\bar{z} = (a + bi)(a - bi)$$
$$= a^2 - abi + abi - b^2i^2$$
$$= a^2 + b^2$$

Consequently, we see that the product of any complex number and its conjugate is a *real number*. In fact, it is the sum of the squares of the real part and the coefficient of the imaginary part of the complex number. Thus, $(3 + 5i)(3 - 5i) = 3^2 + 5^2 = 34$.

Division

In order to find the quotient of two complex numbers, we **multiply the members of the fraction by the conjugate of the denominator,** and thereby obtain a real number in the denominator.

● **Example 5**
$$\frac{-11 + 29i}{2 + 3i} = \frac{-11 + 29i}{2 + 3i} \cdot \frac{2 - 3i}{2 - 3i}$$

$$= \frac{-22 + 33i + 58i - 87i^2}{4 + 9}$$

$$= \frac{65 + 91i}{13}$$

$$= 5 + 7i$$ ●

This is as it should be, since we found in Example 4 that

$$(2 + 3i)(5 + 7i) = -11 + 29i$$

It is customary to represent the complex number $a + bi$ by using the horizontal axis as the axis of reals and the vertical axis as the pure imaginary axis, as indicated in Fig. 4.1. If $z = a + bi$ is a complex number, then we represent the length of the line segment from the origin to z by $|z|$ and call

Absolute value it the *absolute value* of z. By use of the pythagorean theorem we have $|z| = \sqrt{a^2 + b^2}$. Thus $|5 - 12i| = \sqrt{5^2 + 12^2} = 13$.

FIGURE 4.1

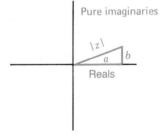

Pure imaginaries

$|z|$

a b

Reals

Exercise 4.4
complex numbers

Perform the operations indicated in Probs. 1 to 32.

1 $(3 + 5i) + (7 + 2i)$ **2** $(3 + 5i) - (-7 + 2i)$ **3** $(8 + 3i) + (1 + 4i)$

4 $(1 + 4i) + (8 - 3i)$ **5** $(2 - 5i) - (5 - 2i)$ **6** $\overline{(2 - 5i)} + (2 - 5i)$

7 $(1 + 4i) - \overline{(1 - 4i)}$ **8** $(1 + 4i) - (1 - 4i)$ **9** $(2 + 3i)(3 - 2i)$

10 $(2 + 3i)\overline{(3 - 2i)}$ **11** $(7 - i)\overline{(1 - 7i)}$ **12** $\overline{(7 - i)}(1 - 7i)$

13 $(3 + 4i)(3 - 4i)$ **14** $(3 - 4i)(2 + 3i)$ **15** $(5 - 2i)\overline{(5 + 2i)}$

16 $(5 + 2i)(5 - 2i)$ **17** $\dfrac{4 + i}{3 + 5i}$ **18** $\dfrac{3 + 5i}{4 - i}$

19 $\dfrac{3 - 2i}{2 - 3i}$ **20** $\dfrac{6 + 5i}{4 - 3i}$ **21** $\dfrac{12 + 5i}{2 + 3i}$

22 $\dfrac{13i}{3 + 2i}$

23 $\dfrac{14 + 48i}{7 - i}$

24 $\dfrac{14 - 48i}{7 + i}$

25 $\dfrac{2 + 3i}{2 - 3i} + \dfrac{4 - 3i}{4 + 3i}$

26 $\dfrac{4 + i}{4 - 3i} - \dfrac{4 - i}{4 + 3i}$

27 $\dfrac{3 + 4i}{2 - i} + \dfrac{3 - 4i}{2 + i}$

28 $\dfrac{3 + 2i}{1 + 5i} - \dfrac{1 + 5i}{3 + 2i}$

29 $|4 + 3i|$

30 $|3 - 4i|$

31 $|-5 + 12i|$

32 $|-7 - 24i|$

Verify the fact that $|z|^2 = z\bar{z}$ in each of Probs. 33 to 36.

33 $3 + 5i$ **34** $5 - 2i$ **35** $1 - 7i$ **36** $4 + 3i$

Prove the statement in each of Probs. 37 to 40 for $z = a + bi$ and $w = c + di$.

37 $\overline{z + w} = \bar{z} + \bar{w}$

38 $\overline{z - w} = \bar{z} - \bar{w}$

39 $\overline{zw} = \bar{z}\,\bar{w}$

40 $|zw| = |z| \cdot |w|$

For any two complex numbers $z = a + bi$ and $w = c + di$,

Triangle inequality

$$|z + w| \le |z| + |w|$$

Verify this in the specific cases in Probs. 41 to 44.

41 $z = 3 + 5i, w = 6 - 4i$

42 $z = 18 + 7i, w = -11 + i$

43 $z = 6 + 5i, w = -12 - 10i$

44 $z = -8 + 3i, w = 4 - 9i$

Verify the statements in Probs. 45 to 48 (n is a positive integer).

45 $i = i^5 = i^{13} = i^{4n+1}$

46 $i^2 = i^6 = i^{42} = i^{4n+2}$

47 $i^3 = i^7 = i^{99} = i^{4n+3}$

48 $i^4 = i^8 = i^{60} = i^{4n}$

Find x and y so that the statements in Probs. 49 to 56 are true. Use the fact that $a + bi = c + di$ if and only if $a = c$ and $b = d$.

49 $x + 2i + 2 = 5 + yi$

50 $x - 5i = 4 + 2i - yi$

51 $y + ix - 3i = 2 + 3i$

52 $5 + x + i = 2 + yi$

53 $(x + iy)(1 + 3i) = -1 + 7i$

54 $(x - 2iy)(3 + i) = 20$

55 $(x - iy)(3 - 5i) = -6 - 24i$

56 $(x - iy)(2 + 3i) = 4 + 6i$

57 Let $z = \dfrac{-1 + i\sqrt{3}}{2}$. Show that $z^3 = 1$ and $1 + z + z^2 = 0$.

4.5 Key concepts

Make sure you understand the following important words and ideas.

Exponent	Imaginary number
Power	Complex numbers
Base	Rationalize the denominator or
Index	numerator
Negative exponent	Conjugate
Rational exponent	Scientific notation
Radical	Rounding off
Principal root	Adding and multiplying approximate
Square root	numbers
Cube root	Laws of radicals
Radicand	

(4.1) $a^n = a \cdot a \cdots a$ to n factors   (4.2) $a^0 = 1$ if $a \neq 0$

(4.3) $a^{-n} = 1/a^n$ (4.4) $a^m a^n = a^{m+n}$

(4.5) $a^m/a^n = a^{m-n}$ (4.6) $(a^m)^n = a^{mn}$

(4.7) $(ab)^n = a^n b^n$ (4.8) $(a/b)^n = a^n/b^n$

(4.9) $\sqrt[n]{a} = b$ (for n a positive integer) means $a = b^n$
        (i) if a and b are positive, n is any positive integer
     or(ii) if a and b are negative, n is an odd positive integer

(4.10) $\sqrt[n]{ab} = \sqrt[n]{a}\sqrt[n]{b}$ (4.11) $\sqrt[n]{a/b} = \sqrt[n]{a}/\sqrt[n]{b}$

(4.12) $\sqrt[m]{\sqrt[n]{a}} = \sqrt[mn]{a}$ (4.13) $a^{1/n} = \sqrt[n]{a}$

(4.14) $a^{m/n} = (\sqrt[n]{a})^m = (a^{1/n})^m = (a^m)^{1/n} = \sqrt[n]{a^m}$

Exercise 4.5 review

Perform the indicated operations and simplify in Probs. 1 to 46.

1 $3^2 3^4$ **2** $6^5/6^3$ **3** $(2/3)^4$ **4** $(2^2)^3$

5 $(2^2 3^3)^3$ **6** $(5x^2 y^3)(3x^0 y^4)$ **7** $12x^4 y^5/3x^2 y^2$ **8** $(a^2 b^4)^2$

9 $(2a^3 b^2)^3$ **10** $(a^2 b^0)^4/(a^3 b^2)^2$ **11** $\left(\dfrac{a^3 b^2}{c^3}\right)^2 \left(\dfrac{a^2 c^4}{b}\right)^3$

12 $\left(\dfrac{8a^3 b^2}{c^4 d^3}\right)^3 \div \left(\dfrac{4a^4 b^3}{c^5 d^4}\right)^2$ **13** $\dfrac{(a^{2n-1} b^{n+3})^2}{a^{n-2} b^{n+1}}$ **14** 3^{-2}

15 $3^{-1}/3^{-3}$ **16** $(3^{-2})^2$ **17** $2^{-4} 2^3$

18 a^3/b^{-2} **19** $a^{-1} b^{-2}/a^2 b^2 c^3$ **20** $\dfrac{3a^{-1} b^{-2}}{2^{-1} a^3 b^{-5}}$

21 $\left(\dfrac{a^{-2}}{b^3}\right)^{-3}$

22 $125^{1/3}$

23 $0.64^{1/2}$

24 $\sqrt[6]{64}$

25 $\sqrt[4]{81}$

26 $\sqrt{25a^2b^6}$

27 $\sqrt[3]{\dfrac{8a^3}{b^6}}$

28 $\sqrt{\dfrac{676a^4}{b^2}}$

29 $(3x^{1/4})(2x^{1/3})$

30 $(5x^{1/3})(3x^{1/5})$

31 $(25x^2y^4)^{1/2}$

32 $(64a^{3/5}b^{-3})^{1/3}$

33 $(8x^3y^6)^{1/3}(9x^2y^6)^{-1/2}$

34 $\left(\dfrac{x^{2a+b}}{x^{a+b}}\right)^{1/a}$

35 $4(3x+2)^{1/2}(2x-3)^{-2/3} + 9(2x-3)^{1/3}(3x+2)^{-1/2}$

36 $\sqrt{27}$ **37** $\sqrt{80}$ **38** $\sqrt[3]{40}$ **39** $\sqrt[5]{96}$

40 $\sqrt{8}\sqrt{2}$ **41** $\sqrt[3]{18}\sqrt[3]{12}$ **42** $\sqrt[5]{a^2}\sqrt[5]{a^4}$ **43** $\sqrt[3]{20x^2y^4}\sqrt[3]{50xy^2}$

44 $\sqrt{15x^2y}\sqrt{6x^4y^3}$ **45** $\sqrt[3]{\dfrac{135x^{10}y^{-1}}{320x^{-2}y^5}}$ **46** $\dfrac{3-\sqrt{2}}{1+\sqrt{2}}$

Express the radicals in each of Probs. 47 to 50 as a single radical.

47 $\sqrt[3]{\sqrt{9}}$ **48** $\sqrt{\sqrt[3]{16}}$ **49** $\sqrt{\sqrt[3]{x^8}}$ **50** $\sqrt[3]{\sqrt{64}}$

Simplify the expressions in Probs. 51 to 58.

51 $\sqrt{2} + \sqrt{50} - \sqrt{72}$ **52** $\sqrt{9a} + \sqrt{4b} + \sqrt[3]{27a}$

53 $\sqrt[3]{ab^3} + \sqrt{4a^3b} + \sqrt{9a^3b^3}$ **54** $\sqrt{3ab} - \sqrt[4]{9a^2b^2} + \sqrt[6]{27a^3b^3}$

55 $(\sqrt{3} + \sqrt{7})(\sqrt{3} - \sqrt{7})$ **56** $(2\sqrt{2} + 3\sqrt{5})(3\sqrt{2} - \sqrt{5})$

57 $(\sqrt{u} + \sqrt{v})(u - \sqrt{uv} + v)$ **58** $(\sqrt{a} + \sqrt{b} + \sqrt{c})(\sqrt{a} - \sqrt{b} + \sqrt{c})$

Calculator   **59** Calculate $\sqrt{6 + \sqrt{6}}$ and $\{[(6 + \sqrt{6})^{1/2} + 6]^{1/2} + 6\}^{1/2}$.

60 If a and b are rational, then a^b may be rational or irrational. Verify that $(27/64)^{2/3} = 9/16$, which is rational, and $3^{1/2}$ is irrational.

61 If a and b are irrational, then a^b may be rational. The following argument, which you should verify, proves this without knowing whether $\sqrt{2}^{\sqrt{2}}$ is rational or irrational.

   (i) If $\sqrt{2}^{\sqrt{2}}$ is rational, then a^b is rational for the irrational numbers $a = \sqrt{2}$ and $b = \sqrt{2}$.

   (ii) If $\sqrt{2}^{\sqrt{2}}$ is irrational, then with $a = \sqrt{2}^{\sqrt{2}}$ and $b = \sqrt{2}$, both irrational, we have

$$a^b = (\sqrt{2}^{\sqrt{2}})^{\sqrt{2}} = \sqrt{2}^{(\sqrt{2}\cdot\sqrt{2})} = (\sqrt{2})^2 = 2$$

which is rational.

Physics   **62** The radium isotope Ra has a loss of intensity of 9.8 percent annually. If I_0 represents the intensity at a given time, find the intensity after 1 year, 2 years, n years.

Biology   **63** The probability of exactly 6 vestigial-winged flies out of a total of 12 is

$$\frac{12!}{6!6!}\left(\frac{1}{4}\right)^6\left(\frac{3}{4}\right)^6$$

Calculate this number if $12! = 12 \cdot 11 \cdot 10 \cdot 9 \cdot 8 \cdot 7 \cdot 6 \cdot 5 \cdot 4 \cdot 3 \cdot 2 \cdot 1$ and $6! = 6 \cdot 5 \cdot 4 \cdot 3 \cdot 2 \cdot 1$.

Psychology   **64** One of the learning curves in psychology is

$$y = 1 - (1 - p)(1 - b)^{n-1}$$

where $0 < p < 1$ and $0 < b < 1$. What happens to the values of y as n gets larger and larger?

Management and calculator   **65** In a simplified manufacturing situation, the earnings are

$$E = -200{,}000 + 60{,}000y - 81x + 42xy$$
$$+ 0.004x^2 - 5000y^2 - xy^2 - 0.002x^2y$$

Find E for $x = 8000$ and $y = 10$.

A radical inequality   **66** The following inequalities give an excellent approximation to $\sqrt{10}$. Verify each step.

$$3 < \sqrt{10} < 3.2 \qquad \textit{Hint: Square each term}$$
$$0 < \sqrt{10} - 3 < \tfrac{1}{5}$$
$$0 < (\sqrt{10} - 3)^4 < (\tfrac{1}{5})^4$$

and

$$(\sqrt{10} - 3)^4 = [(\sqrt{10} - 3)^2]^2 = (19 - 6\sqrt{10})^2 = 721 - 228\sqrt{10}$$

thus $$0 < 721 - 228\sqrt{10} < \tfrac{1}{625}$$

$$-721 < -228\sqrt{10} < -721 + \tfrac{1}{625}$$

$$721 > 228\sqrt{10} > 721 - \frac{1}{625} = \frac{450{,}624}{625}$$

$a > \sqrt{10} > b$ $$\frac{721}{228} > \sqrt{10} > \frac{450{,}624}{142{,}500} = \frac{112{,}656}{35{,}625}$$

Note that to six decimal places

b $$\frac{112{,}656}{35{,}625} = 3.162274$$

$\sqrt{10}$ $$\sqrt{10} = 3.162278$$

a $$\frac{721}{228} = 3.162281$$

67 Show that

$$\frac{1}{\sqrt[3]{4} + \sqrt[3]{6} + \sqrt[3]{9}} = \sqrt[3]{3} - \sqrt[3]{2}$$

68 Show $\sqrt[3]{9 + 4\sqrt{5}} + \sqrt[3]{9 - 4\sqrt{5}} = 3$. Try $\left(\dfrac{3 + \sqrt{5}}{2}\right)^3$.

69 Show $\sqrt[3]{19 + 9\sqrt{6}} + \sqrt[3]{19 - 9\sqrt{6}} = 2$. Try $(1 + \sqrt{6})^3$.

70 Show $\sqrt[3]{20 + 14\sqrt{2}} + \sqrt[3]{20 - 14\sqrt{2}} = 4$. Try $(2 + \sqrt{2})^3$.

71 Show $\sqrt[3]{6 + \sqrt{\dfrac{847}{27}}} + \sqrt[3]{6 - \sqrt{\dfrac{847}{27}}} = 3$.

Cumulative review exercise: Chaps. 1 to 4

The following exercise is a cumulative review exercise covering Chaps. 1 to 4. The answers to all these problems are in the back of the book.

1 Show that 460 is a composite number and give its prime factors.

2 What law is used in $21(13 + 4) = 21(13) + 21(4)$?

3 Why do we say that 17 is a prime number?

4 If $(2x - 3)(x + 5) = 0$, what is x?

5 What is the absolute value of $\sqrt{32} - \sqrt{64} + \sqrt{18} - \sqrt{8} - \sqrt{50}$?

6 Use the tests given in Probs. 10 and 11 of Exercise 1.1 to see if 28,271,463 is divisible by 3 and by 11.

7 Let A = the sum of the divisors of 42, not including 42, and B = the sum of the divisors of 24, not including 24. Show that $A - B = 42 - 24$.

8 Evaluate $(17.1)(26.2) + 23.7$ and $17.1(26.2 + 23.7)$.

9 Is $\{1, -1\}$ closed under multiplication? Under addition?

10 Which of the following are rational?
$\frac{2}{11}$, $3.141414 \cdots$ $2.121121112 \cdots$

11 Evaluate -3^4 and $(-3)^4$.

12 Evaluate $[(2)(3)]^4$ and $(5/7)^3$.

13 Evaluate 3^{-2}, 2^{-3}, and $(-2)^{-3}$.

14 Find the product of $\left(\dfrac{3a^4b^3}{2cd^0}\right)^2$ and $\left(\dfrac{4c^2d}{3a^2b^2}\right)^3$.

15 Write $4^{-1}c^{-1}a^{-2}t^{-3}/2^{-3}c^{-3}at^0$ without negative exponents.

16 Put 583, 0.0583, 50.83, and 0.00203 in scientific notation.

17 Round 73.64, 73.65, 0.07366, and 736,600 off to three significant digits.

18 Find $(28.7)(39.6)$ and $28.7/39.6$ and round off each to the proper number of digits.

19 Assume the numbers are approximations and evaluate $(3.1416)(2.718)$ and $3.1416 + 2.718$.

20 Remove the grouping symbols from $2a - 3\{a + 2a[3 - 4a(2a - 1) + 6a^2] - 2a\}$.

21 Find $(3x - 2y)(5x + 4y)$.

22 Find $(3x - 5y)(3x + 5y)$.

23 Find $(3x - 5y)^3$.

24 Find $(x - 2y + 3z)^2$.

25 Find $(2x - 5y + 7z)(2x + 5y - 7z)$.

26 By use of synthetic division, find the quotient and remainder if $3x^3 - 11x^2 + 14x + 11$ is divided by $x - 2$.

27 Find the product of $2x^2 - 5xy + y^2$ and $x^2 + 3xy - 2y^2$.

Factor the expression in each of Probs. 28 to 30.

28 $27a^3 - 8b^3$ **29** $(2a - b)^2 - 9$ **30** $6a^2 - ab - 15b^2$

31 Is $6x^2 - x - 14$ factorable into real factors with integer coefficients? What are the factors or why are there none?

32 Factor $4x^2 - 4y^2 + 12y - 9$.

33 Factor $x^4 + 3x^2 + 4$.

34 Reduce $72x^5y^3/27x^2y^4$ to lowest terms.

35 Reduce $\dfrac{x^2 - x - 2}{x - 1} \dfrac{6x^2 + x - 2}{3x^2 - 4x - 4} \dfrac{2x^2 + x - 3}{2x^2 + x - 1}$ to lowest terms.

36 Find the sum of $\dfrac{2x^2 + x - 6}{3x^2 + 8x - 3}$ and $\dfrac{2x^2 - 5x - 3}{2x^2 + 9x + 9}$.

37 Simplify $\dfrac{\dfrac{3}{2x + 3} - \dfrac{5}{x + 2}}{\dfrac{3x + 1}{x + 2} + 1}$.

38 Perform the indicated operations and reduce to lowest terms.

$$\frac{(x - 3)x - 4}{(x - 9)x + 20} \cdot \frac{(x + 3)x - 40}{(x^2 - 9) - 8x} \div \frac{x + 8}{x - 9}$$

Evaluate the expression in each of Probs. 39 to 42.

39 $\sqrt{(-3)^2}$ **40** $\sqrt[3]{(-3)^3}$ **41** $\sqrt{(-9)(-8)}$ **42** $\sqrt[4]{162/625}$

Remove all possible factors from the radicand in Probs. 43 and 44.

43 $\sqrt[3]{72x^5y^3z^2}$ **44** $\sqrt[5]{81x^3y^2}\sqrt[5]{21x^2y^9}$

45 Rationalize the denominator in $\dfrac{\sqrt{5} - 2\sqrt{3}}{\sqrt{5} - \sqrt{3}}$.

46 Express $\sqrt[3]{\sqrt{6a}}$ as a single radical.

Perform the indicated addition and subtraction in Probs. 47 to 49 and remove all possible factors from the radicands.

47 $\sqrt[3]{6} + \sqrt[3]{48} + \sqrt[3]{162} - \sqrt[3]{750}$ **48** $\sqrt{8a^3b^2} - \sqrt{18a^5} - \sqrt{8ab^2}$

49 $\sqrt[3]{375} + \sqrt{72} - \sqrt[3]{192} + \sqrt{98}$

50 Use the formula given in Prob. 78 of Exercise 4.2 to find the area of the triangle with sides 7, 13, and 18.

51 Evaluate $\sqrt{3 + \sqrt{3 + \sqrt{3}}}$ to two decimal places.

52 Evaluate $625^{1/2}, -125^{1/3}, -(125^{1/3}), -25^{1/2}, (-25)^{1/2}$.

53 Put $a^{2/5}b^{3/5}$ and $a^{-1/2}b^{1/3}$ in radical form.

54 Express $(625a^{-4}b^{4/3})^{-1/4}$ without negative exponents.

55 Express $-2(x + 2)^{1/2}(2x - 1)^{-1/3} + 5(x + 2)^{-1/2}(2x - 1)^{2/3}$ without negative exponents.

56 Simplify $[(x^{a^2 - b^2})^{1/(a + b)}]^{a/(a - b)}$.

57 Evaluate $(5m/n)^{1/(m + n)}$ for $m = 2$ and $n = 3$.

Perform the indicated operations in Probs. 58 to 63.

58 $(3 + 5i) + \overline{(2 - i)}$ **59** $3 + 5i - (2 - i)$

60 $(3 + 5i)(2 - i)$ **61** $(3 + 5i)/(2 - i)$

62 $|3 + 5i|$ **63** $\dfrac{3 + 5i}{2 - i} - \dfrac{2 + i}{3 - 5i}$

64 Find $2 + \dfrac{1}{4 + \dfrac{1}{3 + \frac{1}{5}}}$.

65 Show that $\sqrt{7 + 4\sqrt{3}} - \sqrt{7 - 4\sqrt{3}} = 2\sqrt{3}$. Try $(2 + \sqrt{3})^2$.

66 Show that $\sqrt[3]{7 + \sqrt{50}} + \sqrt[3]{7 - \sqrt{50}} = 2$. Try $(1 + \sqrt{2})^3$.

67 Show that $\sqrt[4]{193 + 132\sqrt{2}} + \sqrt[4]{193 - 132\sqrt{2}} = 6$. Try $(3 + \sqrt{2})^4$.

CHAPTER 5 · LINEAR AND QUADRATIC EQUATIONS AND INEQUALITIES

5.1 Linear equations

Equation

An **equation** is a statement that two expressions are equal. Later in the book we will discuss equations such as $2^x = 3$, $x^2 + 3y = 14$, and $x^8 = 6$. In this chapter we will consider

$$Ax + B = 0 \qquad \text{linear equations}$$

and

$$ax^2 + bx + c = 0 \qquad \text{quadratic equations}$$

The equation $6x - 5 = 2x + 7$ is true if x is replaced by 3 (both members have the value 13), but it is false if x is replaced by 4 (the values are 19 and

Solution
Root
Solution set

15). A number which makes the equation true is called a **solution** or a **root** of the equation. We say that a solution *satisfies* the equation. The set of all solutions of an equation is called the **solution set.**

Some equations are true for every permissible value of the unknown. For instance,

$$4x^2 - 25 - (2x + 5)(2x - 5) = 0$$

and

$$\frac{3}{x - 2} - \frac{2}{x - 2} = \frac{1}{x - 2}$$

Identity

are true for every real x, with the exception that $x = 2$ is not allowed in the second equation since division by zero is not defined. An **identity** is an

114

equation which is true for each permissible value. Every permissible real number is a solution of an identity.

If an equation is true for at least one value, but not all, it is called a **Conditional** **conditional equation.** The equation $x = 5$ is conditional, being true only if **equation** x is replaced by 5. The equation $6x - 17 = 2x + 3$ is also conditional, since it is true if x is replaced by 5 and false if x is replaced by -14.

The objective in solving an equation is to find all values for the variable that satisfy the equation. The simpler the equation is in form, the easier it is to solve. For example, we will consider the equations

$$7x - 45 = 5x - 43 \tag{1}$$

and $$2x = 2 \tag{2}$$

At this stage, the only way that we can find the root of (1) is to guess at a number and then substitute it for x and see if it satisfies the equation. In (2), however, it is obvious that the root is 1. Now if we substitute 1 for x in (1) and simplify, we get $-38 = -38$. Therefore, 1 is also a root of (1). We call attention to the fact that $7x - 45 = 5x - 43$ and $2x = 2$ are different statements. Each statement, however, is true if x is replaced by 1. Two equations of this type are said to be equivalent, and they illustrate the definition below.

Equivalent equations Two equations are **equivalent** if every root of each equation is also a root of the other equation.

If we employ the term "solution set," we can state the above definition in this way:

Two equations are *equivalent* if their solution sets are equal.

Operations that yield As we will presently see, we make extensive use of the concept of **equivalent equations** equivalent equations in the process of solving an equation. We will therefore consider the operations that can be performed on the members of a given equation to yield an equation that is equivalent to the given one.

Two principles from Chap. 1 are very useful in getting a succession of simpler equivalent equations. They are

If $a = b$, then $a + c = b + c$
If $a = b$, then $ac = bc$

In words, these state that the same real number can be added to or multiplied by both members of an equation. The equations

Equivalent $$a = b \quad \text{and} \quad a + c = b + c \tag{3}$$

are equivalent for every real number c. The equations

Equivalent if $c \neq 0$ $$a = b \quad \text{and} \quad ac = bc \tag{4}$$

are equivalent for every *nonzero* real number c.

● **Example 1** Solve the linear equation $ax + b = 0$ if $a \neq 0$.

Solution

$$(ax + b) + (-b) = -b \qquad \text{adding } -b$$

$$ax = -b \qquad b - b = 0$$

$$\frac{ax}{a} = -\frac{b}{a} \qquad \text{multiplying by } \frac{1}{a}$$

Solution of a linear
equation

$$x = -\frac{b}{a} \qquad\qquad\qquad (5.1)$$

●

In general, a linear equation has only one solution. We can check that $-b/a$ is in fact the solution by writing

$$ax + b = \frac{a(-b)}{a} + b = -b + b = 0$$

Instead of memorizing $-b/a$ as the solution, it is usually easier to actually go through the steps. The net effect in going from $ax + b = 0$ to the equivalent equation $ax = -b$ is to move the b to the other side of the
Transposing equation with its sign changed. This is called **transposing,** and it works for any term which is added. It also works for subtraction since subtracting k is merely adding $-k$.

● **Example 2**

$$5x - 6 = 9 \qquad \text{given}$$

$$5x = 9 + 6 \qquad \text{transposing 6}$$

$$5x = 15 \qquad \text{adding}$$

$$x = 3 \qquad \text{dividing by 5} \qquad ●$$

● **Example 3**

$$4x = 7x - 6 \qquad \text{given}$$

$$6 = 7x - 4x \qquad \text{transposing } 4x \text{ and } -6$$

$$6 = 3x \qquad \text{subtracting}$$

$$2 = x \qquad \text{dividing by 3} \qquad ●$$

● **Example 4**

$$6x - 7 = 2x + 1 \qquad \text{given}$$

$$6x - 2x = 1 + 7 \qquad \text{transposing}$$

$$4x = 8 \qquad \text{combining similar terms}$$

$$x = 2 \qquad \text{dividing by 4} \qquad ●$$

Note When multiplying both members of an equation by the same expression, we usually use the lcm of the denominators to get an equivalent equation with no fractions. We must be careful not to multiply by 0 in any form because the resulting equations may not be equivalent. Remember that dividing by p is just multiplying by $1/p$.

● **Example 5** Solve the equation

$$\tfrac{1}{2}x - \tfrac{2}{3} = \tfrac{3}{4}x + \tfrac{1}{12}$$

Since the lcm of the denominators is 12, and $12 \neq 0$, we will employ the theorem in (4) and multiply each member by 12; we get

$$12(\tfrac{1}{2}x - \tfrac{2}{3}) = 12(\tfrac{3}{4}x + \tfrac{1}{12})$$

$$6x - 8 = 9x + 1 \qquad \text{by the distributive axiom}$$

$$6x - 9x = 1 + 8 \qquad \text{transposing}$$

$$-3x = 9 \qquad \text{combining terms}$$

$$x = -3 \qquad \text{multiplying each member by } -\tfrac{1}{3} \qquad ●$$

Nonlinear equations   Some nonlinear equations are equivalent to linear equations.

● **Example 6** Solve the equation

$$\frac{x}{x+1} + \frac{5}{8} = \frac{5}{2(x+1)} + \frac{3}{4} \qquad (5)$$

Solution   The lcm of the denominators is $8(x+1)$.

$$8(x+1)\left(\frac{x}{x+1} + \frac{5}{8}\right) = 8(x+1)\left[\frac{5}{2(x+1)} + \frac{3}{4}\right] \qquad \text{multiplying by the lcm}$$

$$8x + 5(x+1) = 4(5) + 6(x+1) \qquad \text{canceling}$$

$$8x + 5x + 5 = 20 + 6x + 6 \qquad \text{the distributive axiom}$$

$$8x + 5x - 6x = 20 + 6 - 5 \qquad \text{transposing}$$

$$7x = 21 \qquad \text{combining similar terms}$$

$$x = 3 \qquad \text{multiplying each} \atop \text{member by } \tfrac{1}{7} \qquad ●$$

Since $x = 3$ does not make any denominator of (5) equal to zero, then $x = 3$ is a solution of (5), and it is the only one. In fact each member of (5) is $\tfrac{11}{8}$ when $x = 3$.

● **Example 7**

$$\frac{2x}{x-3} = 1 + \frac{6}{x-3}$$

$$2x = 1(x-3) + 6 \qquad \text{multiplying by } x - 3$$

$$2x = x - 3 + 6$$

$$2x - x = -3 + 6 \qquad \text{transposing } x$$

$$x = 3$$

However 3 is not a solution of the given equation since replacing x by 3 involves division by 0. The conclusion is that the first and last equations are *not* equivalent, and that the first equation has *no solution*. The first step, multiplying by $x - 3$, was in effect multiplying by 0.   ●

Remember: Do *not* multiply each member of an equation by 0 in any form.

● Example 8   Solve $\dfrac{2}{x-5} + \dfrac{5}{x-2} = \dfrac{7}{(x-5)(x-2)}$.

Solution   Multiplying by the lcm $(x-5)(x-2)$ gives

$$2(x-2) + 5(x-5) = 7$$
$$2x - 4 + 5x - 25 = 7$$
$$7x - 29 = 7$$
$$7x = 36$$
$$x = \tfrac{36}{7}$$

which is the solution since the value of the lcm $(x-5)(x-2)$ is not 0 for $x = \tfrac{36}{7}$.   ●

● Example 9   Solve $\dfrac{6}{2x-1} - \dfrac{1}{x-3} = \dfrac{2}{x+2}$.

Solution   Multiplying both members by the lcm $(2x-1)(x-3)(x+2)$ gives

$$6(x-3)(x+2) - (2x-1)(x+2) = 2(2x-1)(x-3)$$
$$6(x^2 - x - 6) - (2x^2 + 3x - 2) = 2(2x^2 - 7x + 3)$$
$$6x^2 - 6x - 36 - 2x^2 - 3x + 2 = 4x^2 - 14x + 6$$
$$-9x - 34 = -14x + 6 \qquad \text{combining terms}$$
$$5x = 40 \qquad \text{transposing}$$
$$x = 8 \qquad \text{dividing by 5}$$

The solution is $x = 8$ since it does not make the lcm be 0.   ●

Exercise 5.1   In Probs. 1 to 8, is the equation an identity or a conditional equation?

1 $x^2 - 4 = (x+2)(x-2)$ **2** $x^2 - 6 = (x+3)(x-2)$

3 $(2x-5)^2 = 4x^2 - 10x + 25$ **4** $x^3 + 8 = (x+2)(x^2 - 2x + 4)$

5 $\dfrac{x}{3} - \dfrac{x}{4} = \dfrac{x}{7}$ **6** $\dfrac{x}{3} - \dfrac{x}{4} = \dfrac{x}{12}$

7 $\dfrac{3x + 5x^2}{2} = (5x + 3)\dfrac{x}{2}$ **8** $\dfrac{x}{3} - \dfrac{3}{x} = \dfrac{x^2 - 3}{3x}$

In Probs. 9 to 12, state whether the equations are equivalent or not.

9 $7x - 2 = 4x + 3$
  $3x = 1$

10 $8x + 5 = 5x + 8$
  $3 = 3x$

11 $2x = 17$
  $x = 17 - 2$

12 $5 + 3x = 11$
  $3x = \frac{11}{5}$

Find the solution of the equation in each of Probs. 13 to 64.

13 $5x = 7x + 4$

14 $6x = 9x + 3$

15 $9x = 6x + 9$

16 $2x = 3x - 9$

17 $5(x + 2) = 3(x + 4)$

18 $4(x + 3) - 5(2x + 1) = 13$

19 $3(x + 5) + 2(3x - 7) = 10$

20 $7(2x - 3) - 3(x - 1) = 4$

21 $\dfrac{3x}{5} - 2 = \dfrac{1}{10}x + 3$

22 $\frac{5}{8}x + 6 = \frac{1}{4}x + 3$

23 $\frac{2}{3}x - 3 = \frac{1}{4}x + 2$

24 $\frac{7}{12}x + 3 = \frac{5}{6}x + 6$

25 $\dfrac{2x - 3}{3} = x - 3$

26 $\dfrac{3x + 4}{2} = 3x - 4$

27 $\dfrac{3x + 5}{4} = 2x - 5$

28 $\dfrac{5x - 7}{3} = 2x - 5$

29 $\dfrac{3x - 2}{5} + 3 = \dfrac{4x - 1}{3}$

30 $\dfrac{2x - 3}{3} + 1 = \dfrac{3x + 2}{5}$

31 $\dfrac{2x - 9}{3} - 2 = \dfrac{x - 3}{3}$

32 $\dfrac{5x + 1}{8} + 3 = \dfrac{3x + 1}{2}$

33 $\dfrac{3x + 7}{2} + 3x - 7 = \dfrac{2x + 3}{5}$

34 $\dfrac{3x + 10}{2} - x - 4 = \dfrac{3x + 6}{4}$

35 $\dfrac{5x + 7}{2} = \dfrac{3x + 5}{4} + 2x + 3$

36 $\dfrac{4x + 5}{5} = \dfrac{3x - 15}{2} + 2x - 5$

37 $\dfrac{2}{3x + 1} = \dfrac{1}{x}$

38 $\dfrac{2}{3x + 1} = \dfrac{5}{8x + 1}$

39 $\dfrac{11}{6x + 1} = \dfrac{2}{x + 1}$

40 $\dfrac{8}{5x - 4} = \dfrac{5}{3x - 1}$

41 $\dfrac{x + 1}{x - 2} = \dfrac{x - 1}{x - 3}$

42 $\dfrac{x + 1}{x - 5} = \dfrac{x + 4}{x - 4}$

43 $\dfrac{x + 5}{x - 1} = \dfrac{x + 2}{x - 2}$

44 $\dfrac{x + 4}{x - 3} = \dfrac{x + 2}{x - 4}$

45 $\dfrac{2x + 5}{4x + 1} = \dfrac{3x + 5}{6x - 1}$

46 $\dfrac{4x - 3}{2x - 3} = \dfrac{8x + 5}{4x + 1}$

47 $\dfrac{2x - 5}{4x - 1} = \dfrac{3x - 4}{6x + 9}$

48 $\dfrac{6x - 8}{9x + 8} = \dfrac{2x - 3}{3x + 2}$

49 $\dfrac{4}{x-2} - \dfrac{3}{x+1} = \dfrac{8}{(x-2)(x+1)}$

50 $\dfrac{1}{x+5} + \dfrac{1}{2x+9} = \dfrac{2}{(x+5)(2x+9)}$

51 $\dfrac{1}{2x+3} - \dfrac{3}{x-3} = \dfrac{3}{(2x+3)(x-3)}$

52 $\dfrac{2}{x+2} + \dfrac{1}{2x-1} = \dfrac{5}{(x+2)(2x-1)}$

53 $\dfrac{2}{x+1} + \dfrac{3}{2x-3} = \dfrac{6x+1}{2x^2-x-3}$

54 $\dfrac{5}{3x-1} - \dfrac{1}{5x-7} = \dfrac{11x-1}{15x^2-26x+7}$

55 $\dfrac{5}{2x+1} + \dfrac{4}{x-1} = \dfrac{12x+6}{2x^2-x-1}$

56 $\dfrac{9}{2x+3} - \dfrac{2}{x-1} = \dfrac{x+9}{2x^2+x-3}$ **57** $\dfrac{4}{2x-3} + \dfrac{5}{5x-4} = \dfrac{3}{x+2}$

58 $\dfrac{4}{3x-2} - \dfrac{1}{2x-3} = \dfrac{5}{6x+3}$ **59** $\dfrac{4}{3x-1} - \dfrac{3}{2x+3} = \dfrac{-1}{6x-24}$

60 $\dfrac{3}{x+3} - \dfrac{2}{2x-5} = \dfrac{6}{3x-13}$

61 $\dfrac{x+7}{(2x-3)(x+1)} = \dfrac{8x-9}{2x-3} - \dfrac{4x+9}{x+2}$

62 $\dfrac{3x+8}{(x+1)(x+3)} = \dfrac{x+3}{x+1} - \dfrac{2x+3}{2x+5}$

63 $\dfrac{5x+20}{(3x-5)(x+1)} = \dfrac{2x-7}{x-5} - \dfrac{6x-6}{3x-5}$

64 $\dfrac{15x-79}{(3x+1)(x-1)} = \dfrac{x-5}{x-1} - \dfrac{x-6}{x+3}$

In Probs. 65 to 68, find the value of b for which $x = 5$ is a solution of the equation.

65 $\dfrac{x-3}{b+2} + \dfrac{x-2}{b+6} = \dfrac{x}{b+4}$ **66** $\dfrac{x-3}{b+6} + \dfrac{2x-8}{2b-9} = \dfrac{x-2}{b-1}$

67 $\dfrac{x-3}{b+5} + \dfrac{x+1}{3b+5} = \dfrac{2x-6}{b+3}$ **68** $\dfrac{x-2}{3b+7} - \dfrac{x-4}{2b+18} = \dfrac{1}{2b-2}$

Find the solution of the equation in each of Probs. 69 to 76, for the letter given at the right of the comma.

69 $C = \dfrac{5}{9}(F - 32)$, F \qquad **70** $C = \dfrac{Ak}{4\pi d}$, d

71 $\dfrac{p}{q} = \dfrac{f}{q - f}$, f \qquad **72** $\dfrac{1}{p} + \dfrac{1}{q} = \dfrac{2}{R}$, q

73 $m = \dfrac{c(1 - p)}{1 - d}$, p \qquad **74** $I = \dfrac{Ne}{R + Nr}$, r

75 $S = \dfrac{a - ar^n}{1 - r}$, a \qquad **76** $M = \dfrac{L}{F}\left(\dfrac{25}{f} + 1\right)$, f

Solve the equation in Probs. 77 and 78 by adding the expressions in each member of the equation before multiplying by the lcd.

77 $\dfrac{1}{x + 3} - \dfrac{1}{x + 1} = \dfrac{1}{x + 4} - \dfrac{1}{x + 2}$

78 $\dfrac{1}{x + 5} - \dfrac{1}{x + 8} = \dfrac{1}{x + 3} - \dfrac{1}{x + 6}$

Anthropology **79** In the study of agricultural terraces in Ecuador, the following equations arise. Solve each of them for H.

(a) $z = \dfrac{WHx}{x^2 + W^2}$ \qquad (b) $H(W - x)^2 = W^2(H - 2z)$

(c) $\dfrac{H}{W} = \dfrac{H - y - z}{W - x}$

How thick is Scotch tape? **80** When tape is wrapped around a circular core, its length L, thickness T, inner radius a, and outer radius b are related by

$$L = \dfrac{\pi}{T}(b^2 - a^2)$$

Find the thickness if $L = 3290$ cm, $a = 1.78$ cm, and $b = 3.14$ cm. Use $\pi = 3.1416$.

Show that the equations in Probs. 81 to 84 have no solution.

81 $\dfrac{x}{x - 4} + 1 = \dfrac{4}{x - 4}$ \qquad **82** $\dfrac{2x}{x - 3} - 11 = \dfrac{6}{x - 3}$

83 $\dfrac{3x}{x + 4} = 7 - \dfrac{12}{x + 4}$ \qquad **84** $\dfrac{6x}{x + 7} = 10 - \dfrac{42}{x + 7}$

5.2   Applications

A stated problem is a description of a situation that involves both known and unknown quantities and that also involves certain relations between these quantities. If the problem is solvable by means of one equation, it must be possible to find two combinations of the quantities in the problem that are equal. Furthermore, at least one of the combinations must involve the variable.

The process of solving a stated problem by means of an equation is not always simple, and considerable practice is necessary before one becomes adept at problem solving. The following approach is suggested:

Solving stated problems

1 Read the problem carefully and study it until the situation is thoroughly understood.
2 Identify the quantities, both known and unknown, that are involved in the problem.
3 Select one of the unknowns and represent it by a letter, and then express the other unknowns in terms of this letter.
4 Search the problem for the information that tells what quantities, or combinations of them, are equal.
5 When the desired combinations are found, set them equal to each other, thus obtaining an equation.
6 Solve the equation thus obtained.
7 Check the solution *in the original problem.*

Note   Step 3 above is critical. We will now give several illustrations of translating English statements into algebraic equations.

(a)   "x is 25 more than y" or
      "x is greater than y by 25" or
      "the value of x decreased by 25 is y" or
      "y is 25 less than x"
      Equation: $x = 25 + y$ or $x - y = 25$

(b)   "the sum of x and y is 380"
      "y is 380 diminished by x"
      "the values of x and y total 380"
      Equation: $x + y = 380$ or $y = 380 - x$

(c)   "The sum of four consecutive integers is 178."
      Equation: $x + (x + 1) + (x + 2) + (x + 3) = 178$, where x is the smallest of the four integers.

(d)   "x is twice y" or "y is half of x"
      Equation: $x = 2y$ or $y = x/2$

(e)   "x is 17 less than twice y"
      Equation: $x = 2y - 17$

(f)   "Her age is 3 more than twice what it was 10 years ago."
      Equation: $x = 3 + 2(x - 10)$, where x is her age now.

(g)   "$106.50 is the interest earned on $1800 if part of it is invested at 5 percent and the rest at $6\frac{1}{2}$ percent."
Equation: $106.50 = 0.05x + 0.065(1800 - x)$, where x dollars are at 5 percent.

(h)   "A trip to town at 30 mi/h takes 12 min longer than the return trip at 48 mi/h."
Equation: $d/30 = d/48 + 12/60$, where d is the distance in miles of the trip.

Try a number to get a feel for the problem.

It is usually helpful to make a table of the given data, as in the examples below. Sometimes it helps to understand the problem if we simply select *any* number for the unknown, then try to calculate the various quantities using this specific number instead of, say, the variable x.

For instance in part (f) above, her current age is clearly more than 10, so suppose it is 19. Hence 10 years ago it was 9 (that is, 10 must be *subtracted* from her current age, not added to it). "Twice what it was 10 years ago" would then give 18 (that is, we double the 9, not the 19). Finally, "3 more than etc." means the 18 becomes 21. Therefore the numerical calculations

$$19 \qquad 9 \qquad 18 \qquad 21$$

lead in a natural way to

$$x \qquad x - 10 \qquad 2(x - 10) \qquad 3 + 2(x - 10)$$

PROBLEMS INVOLVING MOTION AT A UNIFORM VELOCITY

Problems that involve motion usually state a relation between the distances traveled, between the velocities (or speeds), or between the periods of time involved. The fundamental formula for use in solving such problems is

$$d = vt$$

where d represents the distance, v represents the velocity (or speed), and t represents the period of time. When this formula is used, d and v must be expressed in terms of the same linear unit and v and t must be expressed in the same unit of time. The formula can be solved for v and t to get the two additional formulas

$$v = \frac{d}{t} \qquad \text{and} \qquad t = \frac{d}{v}$$

● **Example 1**   A party of hunters made a trip of 380 km to a hunting lodge in 7 h. They traveled 4 h on a paved highway and the remainder of the time on a pasture road. If the average velocity through the pasture was 25 km/h less than that on the highway, find the average velocity and the distance traveled on each part of the trip.

Solution   We will first tabulate the data:

	Time, h	Velocity, km/h	Distance, km
On paved highway	4	x	$4x$
On pasture road	$7 - 4 = 3$	$x - 25$	$3(x - 25)$
Total	7		380

Quantities that are equal:

$$\text{Distance on highway} + \text{distance through pasture} = 380$$
$$4x \qquad + \qquad 3(x - 25) \qquad = 380$$

In the above table, the unknown quantities are expressed in terms of x and are printed in color, and these unknown quantities are the two velocities and the distance on each part of the trip. The known quantities are 380 km, the total distance; 7 h, the total time; 4 h, the time spent on the highway; and 25 km/h, the amount by which the velocity on the highway exceeds that through the pasture. The time spent on the pasture road was 7 h − 4 h = 3 h, and the total distance is equal to the sum of the distances traveled on each of the two parts.

If we let

$$x = \text{speed on the highway, km/h}$$

then

$$x - 25 = \text{speed through the pasture}$$

Furthermore,

$$4x = \text{distance traveled on the highway, in km}$$

$$3(x - 25) = \text{distance traveled through the pasture}$$

Hence, $4x + 3(x - 25) = 380$, the total distance in km

This is the desired equation, and we solve it below.

$4x + 3x - 75 = 380$	by the distributive axiom
$4x + 3x = 380 + 75$	adding 75 to each member
$7x = 455$	combining terms
$x = 65$	multiplying each member by $\frac{1}{7}$

Thus 65 km/h is the velocity on the highway, and 40 km/h is the velocity through the pasture, since $65 - 25 = 40$.

$$4 \times 65 = 260 \text{ km traveled on the highway}$$
$$3 \times 40 = 120 \text{ km traveled through the pasture}$$

Check $260 + 120 = 380.$

● **Example 2**   Three airports, A, B, and C, are located on a north-south line. B is 645 mi north of A, and C is 540 mi north of B. A pilot flew from A to B, delayed 2 h, and continued to C. The wind was blowing from the south at 15 mi/h during the first part of the trip, but during the delay it changed to the north with a velocity of 20 mi/h. If each flight required the same period of time, find the airspeed (i.e., the speed delivered by the propeller) of the plane.

Solution   We proceed as follows. Let

$$x = \text{airspeed, mi/h}$$

Then

$$x + 15 = \text{speed of the plane from } A \text{ to } B, \text{ mi/h}$$

and

$$x - 20 = \text{speed of the plane from } B \text{ to } C, \text{ mi/h}$$

Furthermore,

$$\frac{645}{x + 15} = \text{number of hours required for the first flight}$$

$$\frac{540}{x - 20} = \text{number of hours required for the second flight}$$

Hence, since these two periods of time are equal, we have

$$\frac{645}{x + 15} = \frac{540}{x - 20}$$

This is the required equation, and we solve it as follows:

$$(x - 20)(x + 15)\frac{645}{x + 15} = (x - 20)(x + 15)\frac{540}{x - 20}$$

multiplying each member by the
lcm of the denominators

$$(x - 20)645 = (x + 15)540$$

performing the indicated
multiplication

$$645x - 12{,}900 = 540x + 8100$$

by the distributive axiom

$$105x = 21{,}000 \qquad \text{combining terms}$$

$$x = 200 \qquad \text{multiplying each member by } \tfrac{1}{105}$$

Check $x + 15 = 200 + 15 = 215$, and this is the velocity of the plane during the first flight; $\frac{645}{215} = 3$, so the first flight required 3 h. Furthermore, $x - 20 = 200 - 20 = 180$ and $\frac{540}{180} = 3$; so the second flight required 3 h. Consequently, the airspeed of 200 mi/h satisfies the conditions of the problem.   ●

WORK PROBLEMS   Problems that involve the rate of doing certain things can often be solved by first finding the fractional part of the task done by each individual, or by each agent, in one unit of time and then finding a relation between these fractional parts. If this method is used, the unit 1 represents the entire job that is to be done.

● **Example 3**   A farmer can plow a field in 4 days by using a tractor. A hired hand can plow the same field in 6 days by using a smaller tractor. How many days will be required for the plowing if they work together?

Solution   We let

x = the number of days required to plow the field if they work together

Then $\dfrac{1}{x}$ = the part of the field plowed in 1 day by the two

Furthermore,

$\frac{1}{4}$ = the part of the field plowed in 1 day by the farmer

$\frac{1}{6}$ = the part of the field plowed in 1 day by the hired hand

Consequently, $$\frac{1}{4} + \frac{1}{6} = \frac{1}{x}$$

This is the required equation, and we solve it as follows:

$$12x \left(\frac{1}{4} + \frac{1}{6} \right) = 12x \frac{1}{x} \qquad \text{multiplying each member by the lcm, } 12x, \text{ of the denominators}$$

$$3x + 2x = 12 \qquad \text{by the distributive axiom}$$

$$5x = 12$$

$$x = 2\tfrac{2}{5} \qquad \text{days}$$

Check   If they plow the field in $2\frac{2}{5}$ days, then they complete $1/2\frac{2}{5} = \frac{5}{12}$ of it in 1 day. Furthermore, one plows one-sixth of it in 1 day, and the other plows one-fourth; thus:

$$\frac{1}{6} + \frac{1}{4} = \frac{2 + 3}{12} = \frac{5}{12}$$

●

● **Example 4**   If, in Example 3, the hired hand worked 1 day with the smaller machine and then was joined by the employer with the larger one, how many days were required for them to finish the plowing?

Solution   The hired hand plowed one-sixth of the field in 1 day, and hence five-sixths of it remained unplowed.

We let

x = the number of days required for the two to finish the job

Then

$\dfrac{x}{4}$ = the part plowed by the farmer, who plows one-fourth of the field in 1 day

$\dfrac{x}{6}$ = the part plowed by the hired hand

Therefore,

$$\frac{x}{4} + \frac{x}{6} = \frac{5}{6}$$

This is the required equation, and it is solved as follows:

$$12\left(\frac{x}{4} + \frac{x}{6}\right) = 12 \times \tfrac{5}{6} \qquad \text{multiplying each member by the lcm of the denominators}$$

$$3x + 2x = 10 \qquad \text{by the distributive axiom}$$

$$5x = 10$$

$$x = 2$$

Check   In 2 days, the farmer plowed $\tfrac{2}{4} = \tfrac{1}{2}$ of the field and the hired hand plowed $\tfrac{2}{6} = \tfrac{1}{3}$ of it, and

$$\frac{1}{2} + \frac{1}{3} = \frac{3 + 2}{6} = \frac{5}{6} \qquad \bullet$$

MIXTURE PROBLEMS   Many problems involve the combination of certain substances of known strengths, usually expressed in percentages, into a mixture of required strength in one of the substances. Others involve the mixing of certain commodities at specified prices. In such problems, it should be remembered that the total amount of any given element in a mixture is equal to the sum of the amounts of that element in the substances combined and that the value of any mixture is the sum of the values of the substances that are put together.

● **Example 5**   How many gallons of a liquid that is 74 percent alcohol must be combined with 5 gal of one that is 90 percent alcohol in order to obtain a mixture that is 84 percent alcohol?

Solution   If we let x represent the number of gallons of the first liquid and remember that 74 percent of x is $0.74x$, then the table on page 128, showing the data in the problem, is self-explanatory.

Since (the number of gallons of alcohol in the mixture) = (the number of gallons of alcohol in the first liquid) + (the number of gallons of alcohol in the second liquid), we have

$$0.74x + 4.5 = 0.84(x + 5)$$

the solution of which is given directly below the table.

	Number of gallons	Percentage of alcohol	Number of gallons of alcohol
First liquid	x	74	$0.74x$
Second liquid	5	90	$0.90 \times 5 = 4.5$
Mixture	$x + 5$	84	$0.84(x + 5)$

$$0.74x + 4.5 = 0.84x + 4.2 \qquad \text{by the distributive axiom}$$
$$-0.10x = -0.3 \qquad \text{combining terms}$$
$$x = 3 \qquad \text{multiplying each member by } -10$$

Hence the required number of gallons of the first mixture is 3.

Check
$$(0.74 \times 3) + 4.5 = 2.22 + 4.5 = 6.72$$
$$0.84 \times (3 + 5) = 0.84 \times 8 = 6.72$$

MISCELLANEOUS PROBLEMS

In addition to the three types of problems discussed above, there is a wide variety of problems that can be solved by means of equations. The fundamental approach to all of them is the same. It consists of finding two equal quantities, one or both of which involve the unknown. We will discuss three other kinds of problems, giving the general principle or formula to be used in solving each.

Many problems in physics and mechanics involve the lever. A lever is a rigid bar supported at a point, called the *fulcrum,* that is usually between the two ends of the bar. If two weights W_1 and W_2 at distances L_1 and L_2, respectively, from the fulcrum are balanced on a lever, then

$$W_1L_1 = W_2L_2$$

Furthermore, if a force F at a distance D from the fulcrum will just raise a weight R that is a distance d from the fulcrum, then

$$FD = Rd$$

In solving problems dealing with investments, the formula usually employed is

$$I = Prt$$

where P is the principal, or sum invested; I is the interest earned on the investment; r is the rate of interest, or earning per unit of time; and t is the total time the principal is invested.

Problems involving the digits in a number depend upon the place value of our number system. For example, if h is the hundreds digit in a three-place number, t is the tens digit, and u is the units digit, then $100h + 10t + u$ is the number. If the hundreds digit and the units digit are interchanged, then the number is $100u + 10t + h$.

There are, of course, many other formulas like these. Often, such formulas can be discovered by multiplying units and then canceling some of them. For example,

$$\text{Dollars} = \left(\frac{\text{dollars}}{\text{item}}\right)(\text{items})$$

shows that "income = price per item times the number of items."

Exercise 5.2
applications

In Probs. 1 to 8, give the original equation as well as the solution.

1 Find three consecutive integers whose sum is 75.

2 One fall the Hookings spent $224 outfitting their two children for school. If the clothes for the older child cost $1\frac{1}{3}$ the cost of those for the younger, how much did they spend for each child?

3 According to the 1980 census, the population of Mattville was 41,209. If this population was 5015 less than twice the population of Mattville in the 1970 census, what was the population increase in the 10 years?

4. Mr. Dixit jogged a total of 6600 yd in three nights. If each night he increased his distance 440 yd, how far did he jog on the first night?

5 The Kitchen family spent $625 buying a band instrument for each of their two chidren. If one instrument cost $195 more than the other, how much did each instrument cost?

6 The winning candidate for president of the freshman class received 2898 votes. If that was 210 more than half the votes cast, how many freshmen voted?

7 Fred noticed he had worked one-third of the problems in his math assignment and that when he had worked two more problems he would be halfway through the assignment. How many problems were in the assignment?

8 John agreed to work on his uncle's ranch 3 months one summer for $650 and a used car. At the end of 2 months he was needed at home; so his uncle paid him $200 and the car. What was the value of the car?

9 Fred has 316 more stamps in his collection than Bruce, and together they have 2736 stamps. How many stamps does each boy have?

10 Eight less than half the students in the freshman class have their own cars. If 258 cars are owned by freshmen, how many freshmen are there?

11 Ganesh figured that when he saved $21 more, he would have half the money saved for the camera he wanted. How much does the camera cost if he had already saved one-third of the amount?

12 One vacation season Jim earned $160 taking care of one neighbor's yard and feeding another neighbor's dog. How much did he earn doing the yard work if that amount was $40 more than he earned feeding the dog?

13 On a trip Jennifer noticed that her car averaged 21 mi/gal of gas except for the days she used the air conditioning, and then it averaged only 17 mi/gal. If she used 91 gal of gas to drive 1751 mi, on how many of those miles did she use the air conditioning?

14 A portion of $31,750 was invested at 9 percent, and the remainder at 10 percent. If the total income from the money is $3020, how much was invested at each rate?

15 The Billings spent $1488 on carpeting for their new home. The carpeting used in the living room cost $13/yd², and that used in the bedrooms cost $10/yd². If the bedroom area used 20 yd² more than the living room, how much did the Billings spend on each type of carpet?

16 At the beginning of the summer, Claude and Marilyn each earned $26.40/day from their summer jobs. After a time, Claude was assigned more responsibility and then earned $29.60/day. If they each worked 65 days and together earned a total of $3528, how long did Claude work at the higher rate?

17 Ellis collected $8200 in 1 year by renting two apartments. Find the rent charged for each if one rented for $50/month more than the other and if the more expensive one was vacant for 2 months.

18 A mountain resort that featured skiing in the winter was partially staffed by college students in the summer. One summer there were three times as many students employed as there were year-round employees. When September came, 40 of the students went back to school and 30 nonstudents were hired for the winter. If there were then twice as many nonstudents as students, how many people staffed the resort in the winter?

19 Fred is 3 years older than his sister Mary. In 7 years she will be six-sevenths of his age. How old are they?

20 The petty-cash drawer of a small office contained $16.25. If there were twice as many nickels as quarters and as many dimes as nickels and quarters combined, how many coins of each type were there?

21 Janice drove 30 mi to work each day and picked up a friend on the way. If she was able to average 40 mi/h on the trip, and drove 15 min longer with the friend than without him, how far did she live from her friend's house?

22 A boy rode his motorbike 20 min to a girl's home, and then the two drove in a car 30 min to a beach 35 mi from the boy's home. If the car speed was 10 mi/h faster than that of the motorbike, how fast did the car travel?

23 A bicycle club left the campus to ride to a park 24 mi away for an outing.

One member of the club left from the same place by car with picnic supplies $1\frac{1}{2}$ h later, traveled at a speed four times as fast, and arrived at the park at the same time as the cyclists. How fast did she drive?

24 A large plane left an airport at the same time as a small private plane that followed the same flight plan during the first hour of its flight. The speed of the large plane was five times that of the small one, and at the end of the hour, it was 500 mi ahead. What was the speed of the large plane?

25 Walt and Ramona left a stable for a ride to a mountain lookout. On the way up, they averaged 3 mi/h. On the way back along the same trail, they averaged $5\frac{1}{4}$ mi/h, and the return trip took $\frac{5}{7}$ h less time than the outward trip. How long was the entire horseback ride?

26 Jo Beth left her college town on a bus that traveled 60 mi/h. Three hours later, her father left home at a speed of 50 mi/h to meet the bus. If the college was 345 mi from Jo Beth's home and the father met the bus as it arrived at the station, how long did the father drive to meet the bus?

27 Two brothers took turns washing the family car on weekends. John could usually wash the car in 45 min, whereas Jim took 30 min for the job. One weekend they were in a hurry to go to a football game; so they worked together. How long did it take them?

28 Mrs. Windell spent 1 h addressing one-third of the family Christmas cards, and Mr. Windell spent $1\frac{1}{4}$ h addressing another third. Continuing the addressing together, how long did they take to finish the rest of the cards?

29 A three-man maintenance crew could clean a certain building in 4 h, whereas a four-man crew could do the job in 3 h. If one man of the four-man crew was an hour late, how long did the job take?

30 Three secretaries worked together typing a group of form letters. Jones could have done them alone in 2 h, Brown in 3 h, and Smith in 2 h. How long did the secretaries need to type the letters working together?

31 A chemical mixing tank can be filled by two hoses. One requires 42 min to fill the tank, and the other 30 min. If both hoses are used, how much time is needed to fill the tank?

32 Jean, Carol, and Linda were on a committee to compile and staple the pages of their club newsletter. Each girl could have done the job alone in 4 h. Jean started at 3:30 P.M., Carol came at 3:45, and Linda joined the work at 4. What time did they finish?

33 Dave, Joe, and Sabrina were assigned the job of cataloging the music for their school band. Dave could have completed the asignment in 2 h alone, Joe in 3 h, and Sabrina in 4 h. They started work together, but Joe left at the end of a half-hour and Sabrina at the end of an hour. How long did Dave work *alone* to finish the cataloging?

34 A swimming pool can be filled in 6 h and requires 9 h to drain. If the drain was accidently left open for 6 h while the pool was being filled, how long did filling the pool require?

35 The intake pipe to a reservoir is controlled by an automatic valve that closes when the reservoir is full and opens again when three-fourths of the water has been drained. The intake pipe can fill the reservoir in 6 h, and the outlet can drain it in 16 h. If the outlet is open continuously, how much time elapses between two consecutive times when the reservoir is full?

36 How many pounds of chocolates costing $1.60/lb may be mixed with 6 lb of chocolates costling $2/lb to produce a mixture that can be sold for $1.70/lb?

37 Gary paid $21.43 for a collection of seven records. Some were on sale for $3.94, and some for $1.89. How many of each type did he buy?

38 A contractor mixed two batches of concrete that were 9.3 and 11.3 percent cement to obtain 4500 lb of concrete that was 10.8 percent cement. How may pounds of each type of concrete was used?

39 The specifications for a shipment of gravel set as a minimum standard that 85 percent of the gravel should pass through a screen of a certain size. One load of 6 yd^3 tested at only 65 percent. How much gravel testing at 90 percent must be added to the 65 percent load to make it acceptable?

40 Only 5 percent of the area of a city could be developed into parks, whereas in the unincorporated area outside the city limits 25 percent of the area could be developed into parks. If the city covered an area of 300 mi^2, how many square miles of suburbs had to be annexed so the city could develop 10 percent of its total area as parks?

41 A chemist mixed 40 ml of 8 percent hydrochloric acid with 60 ml of 12 percent hydrochloric acid solution. She used a portion of this solution and replaced it with distilled water. If the new solution tested 5.2 percent hydrochloric acid, how much of the original mixture did she use?

42 A small plane was scheduled to fly from Los Angeles to San Francisco. The flight was against a head wind of 10 mi/h. Threat of mechanical failure forced the plane to turn back, and it returned to Los Angeles with a tail wind of 10 mi/h, landing $1\frac{1}{2}$ h after it had taken off. If the plane had a uniform airspeed of 150 mi/h, how far had it gone before turning back?

43 A group of tourists took a sightseeing bus trip of 240 mi and then boarded a plane which took them to their next stop 550 mi away. The average speed of the plane was 16.5 times that of the bus, and their travel time was 6 h and 50 min. Find the average bus speed and the average plane speed.

44 A boy rode his bicycle 15 mi with a tail wind of 8 mi/h, but in the same time rode only 3 mi of the return trip with the same wind now against him. How fast would he have traveled with no wind?

45 A rancher drove 40 km/h on a gravel road to the highway, on which she traveleu at 60 km/h until she reached a city that was 110 km from home. If the trip took 2 h, how far was the ranch from the highway?

46 Mr. Neff traveled 870 mi to attend a company conference. He drove his car 30 mi to an airport and flew the rest of the way. If his plane speed was 12 times that of the car and he flew 48 min longer than he drove, how long did he fly?

47 Gowdy invested $2400 in the common stock of one company and $1280 in the stock of another. The price per share of the second was four-fifths the price per share of the first. The next day the price of the more expensive stock advanced $1.50/share, the price of the other declined $0.75/share, and as a result, the value of his investment increased $30. Find the price per share of the more expensive stock.

48 Mrs. Johnson planned to spend $78 for fabric to make draperies. She found her fabric on sale at 20 percent less per yard than she expected and was able to buy her drapery fabric plus 4 extra yards for a bedspread for $83.20. How much fabric had she planned to buy, and what was the original cost per yard?

5.3 Quadratic equations

If a, b, and c are real numbers with $a \neq 0$, then

$$ax^2 + bx + c = 0$$

Quadratic equation   is called a *quadratic equation*. Examples of quadratic equations are

$$5x^2 + 3x - 1 = 0 \qquad\qquad 2x^2 + 5x + 6 = -x^2 + 3x$$
$$7x^2 - 4x = 2x + 1 \qquad\qquad \sqrt{2}x^2 + \tfrac{1}{3}x - \pi = 4 - \sqrt{3}x$$

If $ax^2 + bx + c$ can be factored, we may use the following result from Chap. 1 to solve the quadratic equation $ax^2 + bx + c = 0$.

Set each   *If p and q are real numbers with pq = 0, then either p = 0 or q = 0 or*
factor = 0   *both are zero.*

To solve by factoring, one member of the equation must be 0.

● **Example 1**   Solve the following quadratic equations by factoring:

(a) $12x^2 + 23x = -5$ (b) $8x^2 + 5x = 0$

Solution   (a) $12x^2 + 23x + 5 = 0$ transposing -5
$\quad\quad (3x + 5)(4x + 1) = 0$ factoring
$\quad\quad\quad 3x + 5 = 0 \qquad 4x + 1 = 0$
$\quad\quad\quad\quad x = -\tfrac{5}{3} \qquad\quad x = -\tfrac{1}{4}$

(b) $8x^2 + 5x = 0$
$\quad\quad x(8x + 5) = 0$ factoring
$\quad\quad\quad x = 0 \qquad 8x + 5 = 0$
$\quad\quad\quad\quad\quad\quad\quad\quad x = -\tfrac{5}{8}$

In this equation, it is tempting to some students to write

$$8x^2 = -5x \qquad \text{transposing } 5x$$

$$8x = -5 \qquad \text{``canceling'' the } x$$

$$x = -\tfrac{5}{8} \qquad \text{dividing by } 8$$

Note   In so doing, the root $x = 0$ is lost. The "canceling" of the x is in effect dividing by x. This is not allowed, since we may be dividing by 0. Note that the equations $8x^2 = -5x$ and $8x = -5$ are *not* equivalent; the solutions of the first one are 0 and $-\tfrac{5}{8}$, whereas the second one has only one solution, $-\tfrac{5}{8}$.

If $(x + d)^2 = e^2$, then we may transpose, factor, and write

$$(x + d)^2 - e^2 = 0 \qquad \text{difference of two squares}$$

$$(x + d - e)(x + d + e) = 0 \qquad \text{factoring}$$

$$x + d - e = 0 \quad \text{or} \quad x + d + e = 0 \qquad \text{solving}$$

$$\qquad x = -d + e \qquad\qquad x = -d - e \qquad\qquad (1)$$

$x = -d \pm e$

● **Example 2**   Solve $(x + 5)^2 = 23$.

$$(x + 5)^2 - (\sqrt{23})^2 = 0$$

$$(x + 5 - \sqrt{23})(x + 5 + \sqrt{23}) = 0$$

$$x = -5 \pm \sqrt{23} \qquad\qquad ●$$

Equations in this form are easy to solve. We will now show how to solve the general quadratic equation in the same way. Assume $a \neq 0$; otherwise we have a linear equation.

$$ax^2 + bx + c = 0 \qquad\qquad \text{given}$$

$$ax^2 + bx = -c \qquad\qquad \text{transpose } c$$

$$x^2 + \frac{b}{a}x = -\frac{c}{a} \qquad\qquad \text{divide by } a, a \neq 0$$

$$x^2 + \frac{b}{a}x + \left(\frac{b}{2a}\right)^2 = -\frac{c}{a} + \left(\frac{b}{2a}\right)^2 \qquad \text{adding } \left(\frac{b}{2a}\right)^2 \text{ to both sides}$$

Completing the square   The step just done in the left member is called "completing the square." It is accomplished by **adding the square of half the coefficient of x.** This allows the left member to be written as a square.

$$\left(x + \frac{b}{2a}\right)^2 = \frac{b^2}{4a^2} - \frac{c}{a}$$

$$\left(x + \frac{b}{2a}\right)^2 = \frac{b^2 - 4ac}{4a^2}$$

We now write the right member as a square also.

$$\left(x + \frac{b}{2a}\right)^2 = \left(\frac{\sqrt{b^2 - 4ac}}{2a}\right)^2$$

By Eq. (1), the solutions are thus

Quadratic formula
$$x = \frac{-b \pm \sqrt{b^2 - 4ac}}{2a} \qquad (5.2)$$

Try to factor first   As a general rule, we may solve a quadratic equation most simply and quickly if we first try to *factor* the equation. If it cannot be readily factored, the *quadratic formula* is the next best way. It works not only when a, b, and c are integers but when they are any real numbers. In fact it even works if they are complex numbers.

● **Example 3**   Solve the equation $3x^2 - 5x + 2 = 0$ by means of the quadratic formula.

Solution   To solve $3x^2 - 5x + 2 = 0$ by the quadratic formula, we see that $a = 3$, $b = -5$, and $c = 2$. Hence we substitute these values in (5.2) and get

$$x = \frac{-(-5) \pm \sqrt{(-5)^2 - 4(3)(2)}}{2(3)}$$

$$= \frac{5 \pm \sqrt{25 - 24}}{6} = \frac{5 \pm 1}{6} = \frac{6}{6} \text{ or } \frac{4}{6}$$

Factoring gives
$(x - 1)(3x - 2) = 0,$
so $x = 1$ **or** $\frac{2}{3}$.
$$= 1 \text{ or } \tfrac{2}{3}$$ ●

● **Example 4**   Solve the equation $4x^2 = 8x - 5$ by means of the quadratic formula.

Solution   The first step in solving the given equation is to convert it to the equivalent equation in standard form. It is $4x^2 - 8x + 5 = 0$. Hence we see that $a = 4$, $b = -8$, and $c = 5$. If we substitute these values in (5.2), we get

$$x = \frac{-(-8) \pm \sqrt{(-8)^2 - 4(4)(5)}}{2(4)}$$

$$= \frac{8 \pm \sqrt{64 - 80}}{8} = \frac{8 \pm \sqrt{-16}}{8}$$

$$= \frac{8 \pm 4i}{8} = \tfrac{1}{2}(2 \pm i)$$

Hence the solutions are $\tfrac{1}{2}(2 + i)$ and $\tfrac{1}{2}(2 - i)$.   ●

We will check to see if $x = (2 + i)/2$ is a root by substitution. Recall from Chap. 1 that $i = \sqrt{-1}$ and $i^2 = -1$.

$$4x^2 = \frac{4(2 + i)^2}{4} = (2 + i)^2 = 4 + 4i + i^2 = 3 + 4i$$

$$8x - 5 = \frac{8(2 + i)}{2} - 5 = 4(2 + i) - 5 = 8 + 4i - 5 = 3 + 4i$$

If it is properly applied, the quadratic equation will always give the two solutions to $ax^2 + bx + c = 0$.

● **Example 5**   Solve $2x^2 + 2x = 1$.

Solution $2x^2 + 2x - 1 = 0$ transposing

$$x = \frac{-2 \pm \sqrt{4 - 4(2)(-1)}}{2(2)}$$ quadratic formula with $a = 2, b = 2, c = -1$

$$= \frac{-2 \pm \sqrt{12}}{4} = \frac{-2 \pm 2\sqrt{3}}{4} = \frac{-1 \pm \sqrt{3}}{2}$$

●

● **Example 6**   Solve $x^2 + 4ix - 5 = 0$.

Solution   Using $a = 1, b = 4i, c = -5$ gives

$$x = \frac{-4i \pm \sqrt{(4i)^2 - 4(1)(-5)}}{2(1)}$$

$$= \frac{-4i \pm \sqrt{-16 + 20}}{2}$$

$$= \frac{-4i \pm 2}{2} = -2i \pm 1$$

●

Completing the square   Completing the square can be used directly to solve a quadratic equation. However it is most useful in working with circles, parabolas, ellipses, and hyperbolas (Chap. 7), in solving polynomial equations of degree 4 (Chap. 11), and in finding the maximum or minimum value of a quadratic expression (coming right up). We write

$$ax^2 + bx + c = a\left[x^2 + \frac{b}{a}x + \frac{c}{a}\right]$$

$$= a\left[x^2 + \frac{b}{a}x + \left(\frac{b}{2a}\right)^2 + \frac{c}{a} - \left(\frac{b}{2a}\right)^2\right]$$

$$= a\left[\left(x + \frac{b}{2a}\right)^2 + \frac{4ac - b^2}{4a^2}\right] \qquad (2)$$

The quadatic $ax^2 + bx + c$ has a maximum if $a < 0$ and minimum if $a > 0$. From (2), this extreme value occurs for

Maximum or minimum

$$x = -\frac{b}{2a}$$

and we find the extreme value by replacing x by $-b/2a$. This extreme value, by (2), is

$$a\left(0 + \frac{4ac - b^2}{4a^2}\right) = \frac{4ac - b^2}{4a}$$

● **Example 7**   The minimum value for $2x^2 - 8x + 3$ occurs where

$$x = \frac{-b}{2a} = \frac{-(-8)}{2(2)} = \frac{8}{4} = 2$$

The actual minimum value is

$$2(2^2) - 8(2) + 3 = 8 - 16 + 3 = -5$$ ●

In the quadratic formula, $b^2 - 4ac = D$ occurs under the radical sign.
Discriminant   It is called the **discriminant,** and its sign tells us about the roots of $ax^2 + bx + c = 0$. If a, b, and c are real numbers, then we see from (5.2) that:

Nature of the roots   (1)   If $D = 0$, then the roots are real, and both are equal to $-b/2a$.
(2)   If $D > 0$, then the roots are real and unequal.
(3)   If $D < 0$, then the roots are imaginary, and are complex conjugates of each other.

Suppose a, b, and c are rational. It follows from (5.2) that x is rational if and only if $\sqrt{D}$ is rational. This is equivalent to D being the square of a rational number. Hence the following statement is true.

Rational roots   (4)   If $D > 0$ and a, b, c are rational, then the roots are rational if and only if D is a perfect square.

● **Example 8**

Equation	D = discriminant	Nature of the roots
$4x^2 - 4\sqrt{5}x + 5 = 0$	$D = 80 - 80 = 0$	Real, both $= -\dfrac{b}{2a} = \dfrac{\sqrt{5}}{2}$
$5x^2 + 2x - 9 = 0$	$D = 184 > 0$	Real, irrational, and unequal
$\sqrt{2}x^2 + 3x + \sqrt{5} = 0$	$D = 9 - 4\sqrt{10} < 0$	Imaginary
$3x^2 - 7x - 6 = 0$	$D = 121 = 11^2 > 0$	Rational and unequal

●

If in (5.2) we let r represent the root $(-b + \sqrt{b^2 - 4ac})/2a$ and s represent the root $(-b - \sqrt{b^2 - 4ac})/2a$, then

$$r = \frac{-b + \sqrt{D}}{2a} \quad \text{and} \quad s = \frac{-b - \sqrt{D}}{2a}$$

we can show that the sum and the product of the roots of a quadratic equation are simple combinations of the coefficients in the equation. For example, the sum of the two roots is

$$r + s = \frac{-b + \sqrt{D}}{2a} + \frac{-b - \sqrt{D}}{2a} = \frac{-2b}{2a} = \frac{-b}{a}$$

Furthermore, the product is

$$rs = \left(\frac{-b + \sqrt{D}}{2a}\right) \cdot \left(\frac{-b - \sqrt{D}}{2a}\right) = \frac{(-b)^2 - (\sqrt{D})^2}{(2a)^2}$$

$$= \frac{b^2 - D}{4a^2} = \frac{b^2 - (b^2 - 4ac)}{4a^2} = \frac{4ac}{4a^2} = \frac{c}{a}$$

Therefore,

Sum of the roots

$$r + s = \frac{-b}{a} \tag{5.3}$$

Product of the roots

$$rs = \frac{c}{a} \tag{5.4}$$

● **Example 9**

Equation	Sum of roots	Product of roots
$x^2 - 3x + 2 = 0$	$-\dfrac{b}{a} = \dfrac{-(-3)}{1} = 3$	$\dfrac{c}{a} = \dfrac{2}{1} = 2$
$2x^2 + 8x - 5 = 0$	$-\dfrac{b}{a} = \dfrac{-8}{2} = -4$	$\dfrac{c}{a} = \dfrac{-5}{2}$
$\sqrt{2}x^2 + 5x - \sqrt{8} = 0$	$-\dfrac{b}{a} = \dfrac{-5}{\sqrt{2}}$	$\dfrac{c}{a} = \dfrac{-\sqrt{8}}{\sqrt{2}} = -\sqrt{4} = -2$

● **Example 10**   Find a quadratic equation with roots $\frac{4}{3}$ and $-\frac{7}{5}$.

Solution   If $x = \frac{4}{3}$, then $3x - 4 = 0$, while if $x = -\frac{7}{5}$ then $5x + 7 = 0$. Thus

$$0 = (3x - 4)(5x + 7) = 15x^2 + x - 28$$

Exercise 5.3 Solve the following quadratic equations.

1 $x^2 - 4 = 0$

2 $9x^2 - 1 = 0$

3 $9x^2 - 25 = 0$

4 $121x^2 - 169 = 0$

5 $x^2 + 6x = 0$

6 $x^2 = -5x$

7 $5x^2 = 10x$

8 $4x^2 - 9x = 0$

9 $x^2 + 4 = 0$

10 $x^2 = -9$

11 $4x^2 = -25$

12 $9x^2 + 16 = 0$

13 $x^2 - 5x + 4 = 0$

14 $x^2 - 9x + 20 = 0$

15 $x^2 - x - 12 = 0$

16 $x^2 - 2x - 24 = 0$

17 $x(x - 5) = -6$

18 $x(x - 2) = 3$

19 $x(x + 7) = -12$

20 $x(x - 3) = 4$

21 $(2x - 1)(3x - 5) = -1$

22 $(3x - 1)(4x - 1) = 13$

23 $(5x - 12)(2x + 1) = -18$

24 $(3x - 2)(x - 3) = -4$

25 $10x^2 - 3 = 13x$

26 $12x^2 + 16x = 3$

27 $29x + 10 = 21x^2$

28 $18x^2 = 10 - 3x$

29 $5x^2 - 17x + 6 = 0$

30 $3x^2 - 7x + 2 = 0$

31 $7x^2 - 17x + 6 = 0$

32 $2x^2 - 9x + 4 = 0$

33 $x^2 - 2x - 2 = 0$

34 $x^2 - 4x - 1 = 0$

35 $x^2 - 6x + 7 = 0$

36 $x^2 - 10x + 18 = 0$

37 $4x^2 - 8x + 1 = 0$

38 $9x^2 - 18x + 7 = 0$

39 $9x^2 - 12x + 1 = 0$

40 $2x^2 - 10x + 11 = 0$

41 $x^2 - 8x - 5 = 0$

42 $x^2 + 4x - 6 = 0$

43 $2x^2 - 6x - 7 = 0$

44 $6 = 3x^2 - 2x$

45 $x^2 - 4x + 13 = 0$

46 $x^2 - 6x + 13 = 0$

47 $x^2 - 8x + 20 = 0$

48 $x^2 - 10x + 34 = 0$

49 $2x^2 - 6x + 5 = 0$

50 $9x^2 - 12x + 5 = 0$

51 $25x^2 - 30x + 13 = 0$

52 $4x^2 - 20x + 41 = 0$

Calculate the discriminant, determine the nature of the roots, and find their sum and product.

53 $x^2 - 5x + 6 = 0$

54 $12x^2 - 5x - 2 = 0$

55 $x^2 - 6x + 9 = 0$

56 $9x^2 + 12x + 4 = 0$

57 $x^2 - 6x - 9 = 0$

58 $4x^2 + 2x - 5 = 0$

59 $4x^2 + 2x + 5 = 0$

60 $x^2 - 2x + 9 = 0$

Find the quadratic equation that has the given solutions.

61 $2, 1$

62 $-\frac{2}{3}, \frac{4}{5}$

63 $2 + \sqrt{5}, 2 - \sqrt{5}$

64 $5 - 2i, 5 + 2i$

Find the maximum or minimum value of each of the following:

65 $2x^2 - 5x - 8$

66 $-5x^2 - x + 8$

67 $-6x^2 + 4x - 5$

68 $7x^2 + 3x + 6$

69 $2x^2 + 6x - 1$

70 $3x^2 - 3x + 5$

71 $-x^2 + 5x + 4$

72 $-3x^2 + 1$

Management   **73** If earnings are $E = -0.0025x^2 + 27x - 66,000$, find the number of units x which produce maximum earnings.

Management   **74** A certain debt will be repaid after n months where

$$416 = \frac{n}{2}[2(11) + (n - 1)2]$$

How many months will repayment take?

Chemistry   **75** The heat of vaporization in calories per mole is given by
and
calculator
$$\Delta H_v = A + BT + CT^2$$

For hexafluorobenzene, $A = 12,587.5$, $B = -10.3365$, and $C = -1.0917$. Find ΔH_v for $T = 25°C$.

Chemistry   **76** The equation $2\pi r(V_2 - V_3) + \pi r^2 pgh = 0$ occurs in physical chemistry in connection with free energy of a liquid at equilibrium. Solve it for r.

Anthropology **77** In the study of intermarriage of two groups in population stability, the equation

$$x^2 - (B + C)x + BC - AD = 0$$

arises. (a) Find its solutions. (b) Show that both roots are real if A, B, C, D are all positive.

Biology   **78** The equation $4x^2 - 2x - 1 = 0$ is connected with Mendelian heredity. Find its roots.

Chemistry   **79** The ionization constant K_V, volume V, and dissociation α are connected by

$$K_V = \frac{4\alpha^2}{(1 - \alpha)V}$$

Find α if $K_V = 1$ and $V = 10$.

Chemistry   **80** The equation $K = x^2/(a - x)(b - x)$ is used in connection with equilibrium in liquid flow. Solve for x if $a = b = 1$ and $K = 4$.

5.4 Equations which lead to quadratic equations

The unknown in a quadratic equation may be any quantity if it appears only to the second and first power. The equation

$$at^2 + bt + c = 0$$

is in quadratic form, and we may solve it for both values of t. If t is an expression involving x, we then solve the resulting two equations for x.

● **Example 1**   Solve the equation $(x^2 - 3x)^2 - 2(x^2 - 3x) - 8 = 0$ for x.

Solution   The given equation is a quadratic provided we think of $t = x^2 - 3x$ as the unknown. We will do that and solve $t^2 - 2t - 8 = 0$ by use of the quadratic formula. Thus,

$$t = \frac{-(-2) \pm \sqrt{(-2)^2 - 4(1)(-8)}}{2(1)}$$

$$= \frac{2 \pm 6}{2} = 4, -2$$

We now find the desired values of x by solving $x^2 - 3x = 4$ and $x^2 - 3x = -2$. From the former, we get $x = 4$ and -1, and from the latter, we find that $x = 2$ and 1. Therefore, $4, -1, 2$, and 1 are the solutions of the given equation.

● **Example 2**   Solve the equation $x^4 - 5x^2 - 36 = 0$ for x.

Solution   The given equation is a quadratic provided we consider $t = x^2$ as the unknown since the equation is then $t^2 - 5t - 36 = 0$. Hence

$$(t - 9)(t + 4) = 0 \qquad \text{factoring}$$

$$t = x^2 = 9, -4$$

$$x = \pm 3, \pm 2i$$

Radical   A *radical equation* is one in which either or both members contain a
equations   radical that has an unknown in the radicand. If the radicals are of the second order, we can solve the equation by the method explained in this section. The method depends upon the following theorem:

Any root of the equation $\qquad f(x) = g(x)$ $\qquad\qquad$ (1)
is also a root of $\qquad\qquad f^2(x) = g^2(x)$

To prove this, we simply multiply left and right members of $f(x) = g(x)$ and $f(x) = g(x)$, which gives

$$[f(x)]^2 = [g(x)]^2 \qquad\qquad (2)$$

Note   It is not true that the equations (1) and (2) are always equivalent. For example, -1 and 3 are both roots of $(x - 1)^2 = 2^2$, but only 3 is a root of $x - 1 = 2$. Any root of (2) which is not a root of (1) is called an *extraneous*
Extraneous root   *root.*
Roots must be checked   As a result of the above, *we must always check each possible solution of a radical equation in the original equation.*
Sometimes it is obvious from the form of an equation that it has no roots. Such is the case with $\sqrt{x + 3} + \sqrt{x - 2} = -5$, since the left side is positive and the right member is negative.
Solving a radical   If an equation contains three or fewer radicals and all radicals are of the
equation   second order, first isolate the most complicated radical. Then square, collect like terms, and repeat the process as necessary.

● **Example 3** Solve $x + 1 = \sqrt{4x + 9}$.

   Solution
$$(x + 1)^2 = (\sqrt{4x + 9})^2 \qquad \text{squaring}$$
$$x^2 + 2x + 1 = 4x + 9$$
$$x^2 - 2x - 8 = 0 \qquad \text{collecting like terms}$$
$$(x - 4)(x + 2) = 0 \qquad \text{factoring}$$
$$x = 4 \text{ or } x = -2 \qquad \text{possible solutions}$$

To check, we first use $x = 4$:

$$x + 1 = 4 + 1 = 5 \quad \text{and} \quad \sqrt{4x + 9} = \sqrt{16 + 9} = \sqrt{25} = 5$$

and so $x = 4$ is a solution. If, however, we take $x = -2$:

$$x + 1 = -2 + 1 = -1 \quad \text{and} \quad \sqrt{4x + 9} = \sqrt{-8 + 9} = 1$$

and so $x = -2$ is not a solution.   ●

● **Example 4** Solve the equation $\sqrt{5x - 11} - \sqrt{x - 3} = 4$.

   Solution
$$\sqrt{5x - 11} = \sqrt{x - 3} + 4 \qquad \text{isolating } \sqrt{5x - 11}$$
$$5x - 11 = x - 3 + 8\sqrt{x - 3} + 16 \qquad \text{equating squares of the}$$
$$\text{members}$$
$$4x - 24 = 8\sqrt{x - 3} \qquad \text{isolating } 8\sqrt{x - 3}$$
$$x - 6 = 2\sqrt{x - 3} \qquad \text{dividing each member by 4}$$
$$x^2 - 12x + 36 = 4(x - 3) \qquad \text{squaring each member}$$
$$x^2 - 12x + 36 = 4x - 12$$
$$x^2 - 16x + 48 = 0 \qquad \text{combining similar terms}$$
$$(x - 12)(x - 4) = 0 \qquad \text{factoring left member}$$
$$x = 12, 4$$

   Check
$$\sqrt{60 - 11} - \sqrt{12 - 3} = \sqrt{49} - \sqrt{9} \qquad \text{substituting 12 for } x \text{ in the left}$$
$$= 7 - 3 = 4 \qquad \text{member of the given equation}$$
$$\sqrt{20 - 11} - \sqrt{4 - 3} = \sqrt{9} - \sqrt{1} \qquad \text{substituting 4 for } x \text{ in the left}$$
$$= 3 - 1 = 2 \qquad \text{member of the given equation}$$

Consequently, since the right member of the given equation is not 2, 4 is not a root, and the only solution of the given equation is 12.   ●

● **Example 5** Solve the equation $\sqrt{x + 1} + \sqrt{2x + 3} - \sqrt{8x + 1} = 0$.

   Solution
$$\sqrt{x + 1} + \sqrt{2x + 3} - \sqrt{8x + 1} = 0 \qquad \text{the given equation}$$
$$\sqrt{x + 1} + \sqrt{2x + 3} = \sqrt{8x + 1} \qquad \text{isolating } \sqrt{8x + 1}$$

$$x + 1 + 2\sqrt{(x + 1)(2x + 3)} + 2x + 3 \qquad \text{squaring each member}$$

$$= 8x + 1$$

$$2\sqrt{(x + 1)(2x + 3)} \qquad \text{isolating } 2\sqrt{(x + 1)(2x + 3)}$$

$$= 8x + 1 - x - 1 - 2x - 3$$

$$2\sqrt{2x^2 + 5x + 3} = 5x - 3 \qquad \text{simplifying the radicand and combining similar terms}$$

$$4(2x^2 + 5x + 3) = 25x^2 - 30x + 9 \qquad \text{squaring each member}$$

$$17x^2 - 50x - 3 = 0 \qquad \text{combining similar terms}$$

$$x = \frac{50 \pm \sqrt{2500 + 204}}{34} \qquad \text{by the quadratic formula}$$

$$= \frac{50 \pm \sqrt{2704}}{34} = \frac{50 \pm 52}{34}$$

$$= \frac{102}{34} \quad \text{and} \quad -\frac{2}{34}$$

$$= 3 \quad \text{and} \quad -\frac{1}{17}$$

Checking will show that $x = 3$ is a root of the original equation and that $x = -\frac{1}{17}$ is not. ●

OTHER EQUATIONS

● **Example 6** $$\frac{1}{2x - 5} + \frac{1}{x - 1} = 1$$

$$1(x - 1) + 1(2x - 5) = (x - 1)(2x - 5) \qquad \text{multiplying by LCD}$$

$$3x - 6 = 2x^2 - 7x + 5$$

$$0 = 2x^2 - 10x + 11 \qquad \text{combining similar terms}$$

$$x = \frac{10 \pm \sqrt{100 - 4(2)(11)}}{4} \qquad \text{solution by quadratic formula}$$

$$= \frac{10 \pm \sqrt{12}}{4} = \frac{10 \pm 2\sqrt{3}}{4}$$

$$= \frac{5 + \sqrt{3}}{2}, \frac{5 - \sqrt{3}}{2}$$

●

Note   We must be sure that the solutions do not cause division by zero in the given equation since that is not an allowable operation.

● **Example 7** $|2x - 5| + |2x - 3| = 4.$

Solution $|2x - 5| = 4 - |2x - 3|$ isolating $|2x - 5|$
$4x^2 - 20x + 25$ squaring; $t^2 = |t|^2$
$\quad = 16 - 8|2x - 3| + 4x^2 - 12x + 9$
$-8x = -8|2x - 3|$ combining similar terms
$x = |2x - 3|$
$x^2 = 4x^2 - 12x + 9$ squaring
$0 = 3x^2 - 12x + 9$ combining similar terms
$0 = x^2 - 4x + 3 = (x - 3)(x - 1)$ factoring

$x = 3$ or $x = 1$ ●

Both these roots check.

● **Example 8**   Solve the following equation.

Solution
$$\frac{2x + 1}{3x + 4} = \frac{3x - 1}{9x - 8}$$ given

$(2x + 1)(9x - 8) = (3x - 1)(3x + 4)$ multiplying by the LCD
$18x^2 - 7x - 8 = 9x^2 + 9x - 4$
$9x^2 - 16x - 4 = 0$ combining similar terms
$(9x + 2)(x - 2) = 0$
$\qquad x = -\frac{2}{9}$ and $x = 2$ solution by factoring   ●

We must make sure that the solutions do not result in division by zero in
the given equation.

**Exercise 5.4
equations that
lead to quadratics**   Solve the following equations for x.

1 $x^4 - 13x^2 + 36 = 0$ **2** $x^4 - 20x^2 + 64 = 0$
3 $x^4 - 25x^2 + 144 = 0$ **4** $9x^4 - 37x^2 + 4 = 0$
5 $4x^{-4} - 25x^{-2} + 36 = 0$ **6** $36x^{-4} - 289x^{-2} + 400 = 0$
7 $4x^{-4} - 13x^{-2} + 9 = 0$ **8** $100x^{-4} - 229x^{-2} + 9 = 0$
9 $100x^{4/3} - 409x^{2/3} + 36 = 0$ **10** $x^{4/3} - 13x^{2/3} + 36 = 0$
11 $4x^{4/3} - 25x^{2/3} + 36 = 0$ **12** $x^{4/3} - 34x^{2/3} + 225 = 0$
13 $(x^2 + 3)^2 - 19(x^2 + 3) + 84 = 0$
14 $(2x^2 - 7)^2 + 4(2x^2 - 7) - 5 = 0$
15 $(x^2 + x)^2 - 18(x^2 + x) + 72 = 0$
16 $(2x^2 - 5x)^2 - (2x^2 - 5x) = 6$

17 $\left(\dfrac{2x - 1}{x + 3}\right)^2 - 4\left(\dfrac{2x - 1}{x + 3}\right) + 3 = 0$

18 $\left(\dfrac{3x + 2}{2x + 1}\right)^2 - 3\left(\dfrac{3x + 2}{2x + 1}\right) + 2 = 0$

19 $\dfrac{x - 3}{x + 1} - 1 - 2\left(\dfrac{x + 1}{x - 3}\right) = 0$

Hint: Multiply through by $(x - 3)/(x + 1)$ or its reciprocal.

20 $3\left(\dfrac{2x + 3}{3x - 9}\right) - 4 + \dfrac{3x - 9}{2x + 3} = 0$ **21** $\left(x + \dfrac{1}{x}\right)^2 - 3\left(x + \dfrac{1}{x}\right) - 10 = 0$

22 $\left(x - \dfrac{1}{x}\right)^2 + x - \dfrac{1}{x} - 12 = 0$ **23** $x - 4\sqrt{x} + 3 = 0$

24 $4x - 16\sqrt{x} + 15 = 0$ **25** $\sqrt{2x + 3} = \sqrt{3x}$

26 $\sqrt{7x + 11} = \sqrt{1 - 3x}$ **27** $x - 2 = \sqrt{3x - 2}$

28 $\sqrt{5x^2 - 6x + 9} = 2x + 3$ **29** $\sqrt{2x + 3} = \sqrt{4x - 1} + 1$

30 $\sqrt{x} + \sqrt{5} = \sqrt{x + 5}$ **31** $\sqrt{x^2 - 3x} = \sqrt{x^2 + 2x - 8} - 2$

32 $\sqrt{2x^2 + 3x - 5} + \sqrt{2x^2 - x - 2} = 5$

33 $\sqrt{2x + 3} + \sqrt{x - 2} = \sqrt{5x + 1}$ **34** $\sqrt{5x + 1} - \sqrt{2x} = \sqrt{3x + 1}$

35 $\sqrt{7x + 2} - \sqrt{3x - 2} = \sqrt{6x - 8}$

36 $\sqrt{9x + 7} + \sqrt{2x - 1} = \sqrt{9x + 16}$

37 $\dfrac{2x + 1}{3x - 2} = \dfrac{5x - 1}{4x + 2}$ **38** $\dfrac{3x - 1}{2x + 3} = \dfrac{7x - 5}{5x + 3}$

39 $\dfrac{3x + 2}{2x} = \dfrac{5}{1 - 2x}$ **40** $\dfrac{3x + 4}{4x + 1} = \dfrac{6x - 5}{2x + 7}$

41 $\dfrac{2}{x + 2} + \dfrac{3}{4x - 2} = 1$ **42** $\dfrac{4}{3x - 1} + \dfrac{5}{3x + 1} = 1$

43 $\dfrac{x^2 - 2x - 3}{(x - 1)(x - 2)} + \dfrac{3x + 1}{x - 1} = 5$ **44** $\dfrac{x^2 - 3x + 4}{(x + 1)(x - 3)} - \dfrac{5x - 3}{x + 1} = -3$

45 $|x + 1| = 3 - 2x$ **46** $|x + 1| = |3 - 2x|$

47 $|x - 2| = 2x - 7$ **48** $|x - 2| = |2x - 7|$

5.5 Applications

Many problems in various areas lead to quadratic equations. The same basic procedures apply here as in solving linear equations—see Sec. 5.2. Some of the formulas which help in translating the problem from English to an equation are:

1 Distance = rate multiplied by time

$$d = rt \quad \text{or} \quad \text{kilometers} = \frac{\text{kilometers}}{\text{hour}} \text{ (hours)}$$

2 Work done = rate of working multiplied by time worked

$$w = rt \quad \text{or} \quad \text{number of cars produced} = \frac{\text{cars}}{\text{day}} \text{ (days)}$$

3 Interest = principal times rate times time

$$I = Prt \quad \text{or} \quad \text{dollars} = \text{(dollars)} \frac{\text{percent}}{\text{year}} \text{ (years)}$$

4 Revenue = cost per item times number of items

$$R = CI \quad \text{or} \quad \text{dollars} = \frac{\text{dollars}}{\text{item}} \text{ (items)}$$

5 Area of a rectangle = length times width

$$A = lw \quad \text{or} \quad \text{square meters} = \text{(meters)(meters)}$$

6 $W_1L_1 = W_2L_2$ in a fulcrum, or (weight)(length) is a constant

7 $100h + 10t + u$ is the number with h hundreds, t tens, and u units

● **Example 1** (a) If the sum of two numbers is 40, then we may let one of them be x. The other one will be $40 - x$.

(b) A pecan orchard yields 120 kg of pecans/tree when there are 8 trees/acre. Every extra tree per acre decreases production per tree by 10 kg. Since

$$\frac{\text{kg of pecans}}{\text{tree}} \cdot \frac{\text{trees}}{\text{acre}} = \frac{\text{kg of pecans}}{\text{acre}}$$

the production with 8 trees is $(120)8 = 960$ kg/acre. If x trees are added, then there are $8 + x$ trees and production is

$$(8 + x)(120 - 10x)$$

(c) Two cars A and B on perpendicular roads are going 80 and 70 km/h, and they are 50 and 20 km from the intersection, respectively. After t h, car A has gone $80t$ km toward the intersection and car B has gone $70t$ km away from it, and so their new positions are $A' = 50 - 80t$ and $B' = 20 + 70t$ as shown in Fig. 5.1. By the pythagorean theorem, the distance d between them is then $\sqrt{(50 - 80t)^2 + (20 + 70t)^2}$ km.

●

FIGURE 5.1

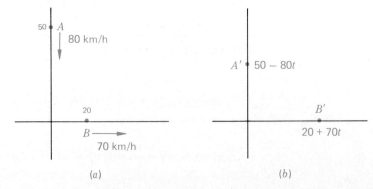

(a)   (b)

FIGURE 5.2

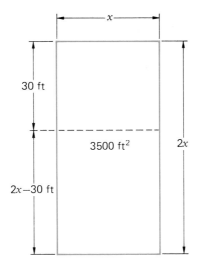

● **Example 2**   A rectangular building whose depth is twice its frontage is divided into two parts by a partition that is 30 ft from and parallel to the front wall. If the rear portion of the building contains 3500 ft², find the dimensions of the building.

Solution   In problems of this type, it is advisable to draw a diagram such as Fig. 5.2. In doing so, we find it best to let

$$x = \text{frontage of the building, ft}$$

Then $\qquad\qquad 2x = \text{depth of the building, ft}$

and $\qquad 2x - 30 = \text{length of the rear portion, ft}$

Since the area of a rectangle is equal to the product of its length and its width, the area of the rear portion of the building is $x(2x - 30)$ ft². Furthermore, since we know that this area is 3500 ft², we have

$$x(2x - 30) = 3500 \qquad\qquad (1)$$

We solve (1) as follows:

$$2x^2 - 30x = 3500 \qquad \text{performing the indicated multiplication}$$
$$2x^2 - 30x - 3500 = 0 \qquad \text{adding } -3500 \text{ to each member}$$
$$(x - 50)(2x + 70) = 0 \qquad \text{factoring the left member}$$
$$x - 50 = 0$$
$$x = 50$$
$$2x + 70 = 0$$
$$x = -35$$

Hence the solutions of (1) are 50, −35. Since, however, the dimensions of the building are positive, we reject −35; and we have

$$x = 50 \qquad \text{frontage in feet}$$
$$2x = 100 \qquad \text{depth in feet}$$

Check $2x - 30 = 100 - 30 = 70$ and $50 \text{ ft} \times 70 \text{ ft} = 3500 \text{ ft}^2.$ ●

● **Example 3**   The periods of time required for two painters to paint a square yard of floor differ by 1 min. Together, they can paint 27 yd² in 1 h. How long does it take each to paint 1 yd²?

Solution   In this problem, the numbers in our equation will be expressed in units of time. Therefore, we let

x = the number of minutes required by the faster painter to paint 1 yd²

Then $x + 1$ = the number of minutes required by the other. Consequently,

$$\frac{1}{x} = \text{fractional part of 1 yd}^2 \text{ painted by the first person in 1 min}$$

and

$$\frac{1}{x + 1} = \text{fractional part of 1 yd}^2 \text{ painted by the other person in 1 min}$$

Consequently,

$$\frac{1}{x} + \frac{1}{x + 1} = \text{fractional part of 1 yd}^2 \text{ painted by both painters in 1 min}$$

Since however, together they painted 27 yd² in 60 min, they covered $\frac{27}{60} = \frac{9}{20}$ yd² in 1 min. Therefore,

$$\frac{1}{x} + \frac{1}{x + 1} = \frac{9}{20} \tag{2}$$

Equation (2) is the desired equation, and we solve it as follows:

$(20x + 20) + 20x = 9x^2 + 9x$ multiplying each member of (2) by $20x(x + 1)$

$-9x^2 + 31x + 20 = 0$ adding $-9x^2 - 9x$ to each member

The solutions of the latter equation are $-\frac{5}{9}$, 4. We reject $-\frac{5}{9}$, however, since a negative time has no meaning in this problem. Therefore, we have

$x = 4$ number of minutes required by faster painter to paint 1 yd²
$x + 1 = 5$ number of minutes required by slower painter

Check   5 min − 4 min = 1 min.   ●

● **Example 4**   In Example 1(c), when are the cars closest together?

Solution   Instead of minimizing the distance d between them, we minimize d^2, since these minimums will occur at the same time and d^2 does not involve a square root. We have

$d^2 = (50 - 80t)^2 + (20 + 70t)^2$
$\quad = 100[(5 - 8t)^2 + (2 + 7t)^2]$ factoring out $10^2 = 100$
$\quad = 100[25 - 80t + 64t^2 + 4 + 28t + 49t^2]$
$\quad = 100[113t^2 - 52t + 29]$ collecting terms

The minimum occurs for

$$t = -\frac{b}{2a} = -\frac{-52}{2(113)} = \frac{26}{113} \text{ h}$$

or almost 14 min since $\frac{26}{113}(60) = 13.8$.   ●

Exercise 5.5   In Probs. 1 to 12, write a quadratic equation which will solve the problem,
applications   and then solve the equation.

1 Find two consecutive integers whose product exceeds their sum by 19.
2 The tens digit of a certain number is 4 more than the units digit. The
   sum of the squares of the two digits is 26. Find the number.
3 Find two numbers that differ by 8 and whose product is 273.
4 Divide 67 into two parts whose product is 1120.
5 If the length of the side of a square is increased by 6 units, the area is
   multiplied by 4. Find the original side length.
6 The product of a positive even integer and the reciprocal of the next
   larger positive even integer equals the reciprocal of the first integer. Find
   that integer.
7 The sum of a number and its reciprocal is $\frac{25}{12}$. Find the number.
8 Two numbers differ by 9, and the sum of their reciprocals is $\frac{5}{12}$. Find
   the numbers.
9 Wesley purchased some shares of stock for $1560. Later, when the price
   had gone up $24/share, he sold all but 10 of them for $1520. How many
   shares had he bought?
10 Audrey drives 10 miles, then increases her speed by 10 mi/h and drives
   another 25 miles. Find her original speed if she drove for 45 min.
11 Two brothers washed the family car in 24 min. Previously, when they
   each had washed the car alone, it had been found the younger boy took
   20 min longer to do the job than the older boy. How long did the older
   boy take to wash the car?
12 Mr. Billings received $72 in dividends one year from some stock. Mrs.
   Billings received $50 from some other stock which was worth $200 less
   and whose rate was 1 percent less. Find the higher rate.

Solve Probs. 13 to 24.

13 Find two consecutive integers whose product exceeds their sum by 11.
14 The difference between the square of a positive number and 7 times the
   number is 18. Find the number.
15 The tens digit of a certain number is 3 more than the units digit. The
   sum of the squares of the two digits is 117. Find the number.
16 Find a negative number such that the sum of its square and 5 times the
   number is 14.
17 Find two numbers that differ by 4 and whose product is 525.
18 Find two numbers that differ by 7 and whose product is 228.

19 The area of a triangle is 30 ft². Find the base and altitude if the latter exceeds the former by 7 ft.

20 Find the dimensions of a rectangle whose perimeter is 56 ft and whose area is 192 ft².

21 To make a straight sidewalk from his house to his garage, a man used 54 ft of forming, into which he poured 24 ft³ of concrete to form a slab 4 in thick. What were the dimensions of the sidewalk?

22 To cover one wall of a room, 63 ft of wallpaper 2 ft wide was required. If the wall contained no windows or doors and if the width exceeded the height by 5 ft, find the dimensions.

23 A farmer rode around his rectangular farm inspecting fences from a jeep. He noticed that his jeep speedometer measured the trip as 3.75 mi. If the area of the farm was 560 acres, what were the lengths of its sides? *Note:* 1 mi² = 640 acres.

24 A man purchased 4¼ lb of grass seed to plant a rectangular lawn. Find the dimensions of the area to be planted if the length exceeds the width by 35 ft and if a pound of seed is required for 1000 ft².

In Probs. 25 to 36, use quadratic expressions to find the specified maximum or minimum.

Physics **25** The height in meters of a ball thrown upwards is $-5t^2 + 18t + 12$ after t s. How high does the ball go?

Physics **26** An arrow shot into the air will be $-5t^2 + 40t + 2$ m high after t s. What story of a tall building will the arrow reach if each story is 3 m high?

Geometry **27** What is the maximum area which can be enclosed in a rectangular field if 400 m of fencing is available? See Fig. 5.3.

FIGURE 5.3

Prob. 27   Prob. 28

Geometry **28** What is the largest rectangular field which can be divided into two smaller rectangular fields by 400 m of fencing? (See Prob. 63, Exercise 5.8 for a similar problem.)

Marketing **29** A box with no top is to be made from a rectangular piece of metal with perimeter 500 cm by cutting out squares 16 cm on a side from each of the four corners, then folding up the sides. What is the maximum possible volume of the box?

Construction **30** A piece of a long gutter is to be made from a rectangular piece of metal 32 cm by 170 cm by folding up two sides. What is the maximum capacity of the gutter?

Economics **31** If a club operator charges $4 admission, he has an average of 240 people per night. He finds that the average attendance decreases by 12 for each 25 cents increase. What price will maximize revenue?

Management   **32** If a manufacturer charges $40 for a radio, the sales average 2400 per week. The sales decrease by 120 for each $2.50 increase in price. What price will maximize revenue?

Farming   **33** A farmer has 15,000 kg potatoes in the ground. For every day he waits to harvest, he gains 175 kg but loses an average of 1 percent weight due to spoilage. When should he harvest so as to get the most kilograms?

Economics   A demand function $D(x)$ gives a price per unit at which x items can be sold. A revenue function $R(x) = x \cdot D(x)$ gives the amount of money received if x items are sold. The cost function $C(x)$ gives the total cost of producing x items. The profit is $P(x) = R(x) - C(x)$.

              34 What is the maximum profit if $D(x) = -3x + 18$ and $C(x) = 9x + 4$? (x may represent 1 unit or 1000 units or even 100,000 units.)
              35 What is the maximum profit if $D(x) = -5x + 22$ and $C(x) = 7x + 6$?
              36 What is the maximum profit if $D(x) = -4x + 33$ and $C(x) = -x^2 + 9x + 7$?

 Inequalities

In Chap. 1 we presented the following four basic rules for inequalities with real numbers a, b, and c:

$$\text{If } a < b \text{ and } b < c, \text{ then } a < c \qquad (1)$$
$$\text{If } a < b, \text{ then } a + c < b + c \qquad (2)$$
$$\text{If } a < b \text{ and } c > 0, \text{ then } ac < bc \qquad (3)$$
$$\text{If } a < b \text{ and } c < 0, \text{ then } ac > bc \qquad (4)$$

To prove (3) for instance, we use the fact that the product of two positive numbers is positive.

$$b - a > 0 \qquad \text{since } a < b$$
$$(b - a)c > 0 \qquad \text{since } c > 0 \text{ and } b - a > 0$$
$$bc - ac > 0 \qquad \text{distributive law}$$
$$bc > ac \qquad \text{adding } ac \text{ to both members}$$

and this is equivalent to (3).

● **Example 1**   By (1): $-5 < -1$ and $-1 < 17$, so $-5 < 17$.
              By (2): since $-6 < -3$, then $-6 + 19 < -3 + 19$.
              By (2): since $-5 < 5$, then $-5 + (-20) < 5 + (-20)$.
              By (3): since $8 < 18$ and $3 > 0$, then $24 < 54$.
              By (4): since $4 < 22$ and $-3 < 0$, then $-12 > -66$.   ●

Solution   If the inequality involves an unknown, then a **solution** is a value of the unknown which makes the inequality true. The set of all solutions, is the solution set. If two inequalities have the same solution set, they are called **equivalent.**

● **Example 2**   Solve $12x + 1 < 5x + 3$.

Solution

$$12x - 5x < 3 - 1 \qquad \text{adding } -5x - 1$$
$$7x < 2$$
$$x < \tfrac{2}{7} \qquad \text{multiplying by } \tfrac{1}{7} \text{ since } \tfrac{1}{7} > 0 \qquad ●$$

The solutions of an inequality are often **intervals.** The notations used are

Open interval
Closed interval
$$(a, b) = \{x \mid a < x < b\}$$
$$[a, b] = \{x \mid a \le x \le b\}$$

Infinite intervals
$$(a, \infty) = \{x \mid x > a\} \qquad (-\infty, a) = \{x \mid x < a\}$$
$$[a, \infty) = \{x \mid x \ge a\} \qquad (-\infty, a] = \{x \mid x \le a\}$$

Thus the solution to Example 2 is $(-\infty, \tfrac{2}{7})$,

● **Example 3**   Find the solution of $\tfrac{1}{6}x - \tfrac{3}{4} < \tfrac{3}{8}x + \tfrac{1}{2}$.

Solution

$$\tfrac{1}{6}x - \tfrac{3}{4} < \tfrac{3}{8}x + \tfrac{1}{2} \qquad \text{given}$$
$$24(\tfrac{1}{6}x - \tfrac{3}{4}) < 24(\tfrac{3}{8}x + \tfrac{1}{2}) \qquad 24 > 0$$
$$4x - 18 < 9x + 12 \qquad \text{by the distributive axiom}$$
$$4x - 9x < 12 + 18 \qquad \text{adding } -9x + 18$$
$$-5x < 30 \qquad \text{combining terms}$$
$$x > -6 \qquad \text{multiplying by } -6: -6 < 0 \qquad ●$$

In intervals, the solution is $(-6, \infty)$.

Nonlinear inequalities

The solution of many nonlinear inequalities involves two or more linear inequalities. We must find the *common solution* to them all. One way to do this is by factoring if necessary, then using sets on the number line and a
Sign chart   **sign chart.**

● **Example 4**   Solve $x^2 - 2x - 3 > 0$.

Solution

$$(x - 3)(x + 1) > 0 \qquad \text{factoring}$$

Setting each factor equal to zero gives $x = 3$ and $x = -1$, which determine three sets when put on a number line.

Sign of $(x - 3)$:	$-$	$-$	$+$
Sign of $(x + 1)$:	$-$	$+$	$+$
Sign of product:	$+$	$-$	$+$

$$-1 \qquad 3$$

The sign chart shows that $(x - 3)(x + 1) > 0$ if $x < -1$ or $x > 3$. This solution can be written in interval notation as

$$(-\infty, -1) \cup (3, \infty)$$

Example 5 Solve $\dfrac{2x + 1}{x - 3} > 3$.

Solution   We change this from a fractional inequality to a quadratic inequality by multiplying both members by the *positive* expression $(x - 3)^2$.

$$\left(\frac{2x + 1}{x - 3}\right)(x - 3)^2 > 3(x - 3)^2 \qquad \text{multiplying by } (x - 3)^2$$
$$(2x + 1)(x - 3) > 3(x^2 - 6x + 9)$$
$$2x^2 - 5x - 3 > 3x^2 - 18x + 27$$
$$0 > x^2 - 13x + 30 \qquad \text{combining terms}$$
$$0 > (x - 3)(x - 10) \qquad \text{factoring}$$

Sign of $(x - 10)$:	$-$	$-$	$+$
Sign of $(x - 3)$:	$-$	$+$	$+$
Sign of product:	$+$	$-$	$+$

$$3 \qquad\qquad 10$$

The sign chart shows that $(x - 10)(x - 3) < 0$ if $3 < x < 10$, or in the interval $(3, 10)$.

Note   It is tempting to multiply both members of the inequality by $x - 3$, giving $2x + 1 > 3(x - 3)$. This is incorrect, since $x - 3$ may be positive or negative, and where it is negative the sense of the inequality is changed by multiplying by $x - 3$.

Example 6 Solve $(2x - 1)(x - 3)(3x - 16) > 0$.

Solution   We put $\frac{1}{2}$, 3, and $\frac{16}{3}$ on the sign chart.

Sign of $(2x - 1)$:	$-$	$+$	$+$	$+$
Sign of $(x - 3)$:	$-$	$-$	$+$	$+$
Sign of $(3x - 16)$:	$-$	$-$	$-$	$+$
Sign of product:	$-$	$+$	$-$	$+$

$$\tfrac{1}{2} \qquad\quad 3 \qquad\quad \tfrac{16}{3}$$

The solution is $(\frac{1}{2}, 3) \cup (\frac{16}{3}, \infty)$.

Exercise 5.6   Find the solution of each of the following inequalities.

1 $4x > 8$ **2** $3x < 12$ **3** $-5x < 10$ **4** $-9x > 36$

5 $4x - 1 > 3x + 1$

6 $5x + 2 > 3x + 8$

7 $5x + 7 > x - 1$

8 $7x + 12 > 3 - 2x$

9 $4x - 5 < 2x + 1$

10 $5x - 13 < 3x + 11$

11 $7x + 19 < 3x + 7$

12 $5x - 3 < 3x + 11$

13 $2x + 3 > 3x + 1$

14 $3x + 13 < 7x + 5$

15 $3x - 12 < 5x - 20$

16 $4x + 9 > 7x + 15$

17 $\frac{3}{4}x + \frac{5}{6} < \frac{1}{2}x - \frac{2}{3}$

18 $\frac{3}{10}x + \frac{7}{20} < \frac{2}{5}x + \frac{1}{4}$

19 $\frac{1}{8}x + \frac{5}{6} < \frac{7}{8}x + \frac{1}{12}$

20 $\frac{5}{8}x + \frac{3}{4} < \frac{1}{4}x - \frac{2}{3}$

21 $\frac{1}{2}x - \frac{1}{3} > \frac{3}{4}x + \frac{7}{6}$

22 $\frac{1}{6}x + \frac{1}{3} > \frac{1}{2}x + \frac{2}{3}$

23 $\frac{4}{7}x + \frac{3}{4} > \frac{1}{2}x + \frac{1}{4}$

24 $\frac{5}{9}x + \frac{5}{6} > \frac{1}{2}x + \frac{1}{2}$

25 $(x - 2)(x + 1) > 0$

26 $(x - 4)(x - 2) > 0$

27 $(x + 2)(x + 3) > 0$

28 $(x - 2)(x - 5) > 0$

29 $(2x + 3)(5x - 2) < 0$

30 $(2x - 7)(7x + 2) < 0$

31 $(5x - 3)(2x + 7) < 0$

32 $(3x + 5)(2x - 3) < 0$

33 $-x^2 + 4x - 3 < 0$

34 $-x^2 - x + 6 > 0$

35 $-x^2 + 6x - 9 > 0$

36 $-x^2 + 4x - 5 < 0$

37 $3x^2 > 2x - 5$

38 $-11x > 2x^2 + 12$

39 $6x^2 - 6 < 5x$

40 $12 < 6x^2 + x$

41 $\dfrac{x - 1}{x + 3} > 0$

42 $\dfrac{x + 4}{x - 2} < 0$

43 $\dfrac{2x + 1}{x + 3} < 0$

44 $\dfrac{4x - 3}{2x + 5} > 0$

45 $\dfrac{x + 5}{x - 1} > 2$

46 $\dfrac{1 - x}{3 + x} < 4$

47 $\dfrac{2x + 7}{8x - 5} > -3$

48 $\dfrac{2 - 3x}{2 + 3x} < -4$

49 $\dfrac{x^2 - x + 2}{x - 1} < x + 3$

50 $\dfrac{x^2 + 2x + 3}{x + 1} < x + 2$

51 $\dfrac{2x^2 + 5x - 1}{2x - 1} < x + 2$

52 $\dfrac{3x^2 + 8x - 2}{3x - 4} < x - 2$

53 $(x + 1)(2x - 1)(3x + 4) > 0$ **54** $(x - 3)(3x + 1)(2x + 3) > 0$

55 $(3x - 7)(7x + 3)(x + 2) > 0$ **56** $(2x + 5)(x - 3)(2x - 3) > 0$

57 $(x + 2)(2x + 1)(3x - 5) < 0$ **58** $(3x - 8)(x + 2)(5x - 1) < 0$

59 $(2x - 7)(7x + 2)(x - 4) < 0$ **60** $(4x - 7)(x + 1)(2x - 7) < 0$

61 $(x - 1)(2x + 1) < (3x + 1)(x + 2)$

62 $(x + 3)(x - 5) > (2x + 1)(x - 4)$

63 $(2x + 3)(3x - 2) > (4x - 1)(x - 4)$

64 $(x - 5)(x + 2) < (2x - 5)(x - 2)$

65 Find k so that $(k + 1)x^2 - 2kx + 3 = 0$ has two real solutions.

66 Find k so that $(k + 2)x^2 + 5x + 2k - 1 = 0$ has two real solutions.

67 Find k so that $x^2 + (k - 1)x + 3k = 0$ has two imaginary solutions.

68 Find k so that $kx^2 + (k + 1)x + k + 2 = 0$ has two imaginary solutions.

Physics **69** If a gun is fired upward with a velocity 100 m/s, its height after t s is $-5t^2 + 100t$ m. During what time is it at least 250 m high?

Area **70** If the length of a rectangle is 4 m more than its width, find all possible lengths so that the area of the rectangle will be at least 25 m².

Economics **71** If a company sells x items, its total cost is $145x + 245$ and its total income is $2x^2 - 144x + 100$. How many items will it take to show a profit?

Agriculture **72** A farmer has 15,000 potatoes in the ground which average $\frac{1}{3}$ kg. For every day he waits to harvest, he gains 175 potatoes but loses an average of 1 percent of the weight because of shrinkage. When should he harvest so as to get at least 5050 kg?

Management **73** To determine when to change from double declining balance to straight-line depreciation, solve the following inequality for t:

$$\frac{1}{N - t + 1} \geq \frac{2}{N}$$

5.7 Equations and inequalities that involve absolute values

To solve an equation that involves absolute values, we make use of the definition of $|a|$, which states that $|a| = a$ for $a \geq 0$ and $|a| = -a$ for $a < 0$. We will illustrate the procedure for solving equations that involve absolute values with three examples.

● **Example 1** Solve $|x| = 4$.

Solution If we make use of the absolute value of x, the equation becomes $x = 4$ for $x > 0$ and $-x = 4$ for $x < 0$. Consequently, the solutions are 4 and -4.

●

● **Example 2** Solve $|2x - 1| = 5$.   (1)

Solution By use of the definition of absolute value, the given equation becomes

$$2x - 1 = 5 \quad \text{for} \quad 2x - 1 \geq 0 \tag{2}$$

and $-2x + 1 = 5 \quad \text{for} \quad 2x - 1 < 0$ (3)

From (2), we have $x = 3$ for $x \geq \frac{1}{2}$, and from (3) we get $x = -2$ for $x < \frac{1}{2}$. Consequently, the solutions are -2 and 3.

FIGURE 5.4

Alternate solution   Another way to solve $|2x - 1| = 5$ is to interpret $|a - b|$ as the distance between a and b. Dividing each member by 2 gives

$$\left|\frac{1}{2}(2x - 1)\right| = \frac{1}{2}(5)$$

$$\left|x - \frac{1}{2}\right| = \frac{5}{2}$$

This equation states that the distance from x to $\frac{1}{2}$ is $\frac{5}{2}$. From Fig. 5.4, we see that $x = \frac{1}{2} + \frac{5}{2} = \frac{6}{2} = 3$ or $x = \frac{1}{2} - \frac{5}{2} = -\frac{4}{2} = -2$.   ●

● **Example 3**   Solve $|3x - 2| = |x + 2|$.

Solution   We know that the two absolute values are equal if the two numbers are equal and also if they are numerically equal but of opposite sign. We shall solve accordingly.

$$3x - 2 = x + 2 \qquad \text{or} \qquad 3x - 2 = -x - 2$$
$$2x = 4 \qquad\qquad\qquad\qquad 4x = 0$$
$$x = 2 \qquad\qquad\qquad\qquad x = 0$$

Alternate solution   Since $|t|^2 = t^2$, we may square both sides and get

$$(3x - 2)^2 = (x + 2)^2$$
$$9x^2 - 12x + 4 = x^2 + 4x + 4$$
$$8x^2 - 16x = 0$$
$$8x(x - 2) = 0$$

$$x = 0 \text{ or } x = 2$$ ●

The absolute value $|ax + b|$ is either $ax + b$ or $-(ax + b)$. Thus $|ax + b| < c$ is equivalent to the two inequalities

$$ax + b < c \qquad \text{and} \qquad -(ax + b) < c$$
$$ax + b > -c \qquad \text{multiplying by } -1$$

These two may be combined by writing

$|ax + b| < c$ $-c < ax + b < c$ (4)

● **Example 4**   Solve $|3x - 4| < 5$.

Solution

$$-5 < 3x - 4 < 5 \qquad \text{by (4)}$$

$$-1 < 3x < 9 \qquad \text{adding 4}$$

$$-\tfrac{1}{3} < x < 3 \qquad \text{dividing by positive 3}$$

● **Example 5** Solve $|-2x + 7| < 9$.

Solution

$$-9 < -2x + 7 < 9 \qquad \text{by (4)}$$

$$-16 < -2x < 2 \qquad \text{subtracting 7}$$

$$\frac{-16}{-2} > x > \frac{2}{-2} \qquad \text{dividing by } -2$$

$$8 > x > -1$$

The result of Example 5 shows clearly that the solution of $|-2x + 7| \leq 9$ is $-1 \leq x \leq 8$. We have just changed $<$ to $\leq$. The inequalities

$$|-2x + 7| \leq 9 \qquad \text{and} \qquad |-2x + 7| > 9$$

have solutions which are complements of each other. Thus the solution to $|-2x + 7| > 9$ is the complement of $[-1, 8]$, which is

$$x < -1 \qquad \text{or} \qquad x > 8$$

that is, $(-\infty, -1) \cup (8, \infty)$. Therefore instead of introducing a new technique to solve

Note

$$|ax + b| > c$$

we just solve the related problem $|ax + b| \leq c$ and then take the complement of this solution.

● **Example 6** Solve $|3x + 2| > 4$.

Solution We solve first $|3x + 2| \leq 4$.

$$-4 \leq 3x + 2 \leq 4$$

$$-6 \leq 3x \leq 2 \qquad \text{subtracting 2}$$

$$-2 \leq x \leq \tfrac{2}{3} \qquad \text{dividing by } +3$$

This solution is $[-2, \tfrac{2}{3}]$, and the solution to the given problem is therefore $(-\infty, -2) \cup (\tfrac{2}{3}, \infty)$.

● **Example 7** Solve $|3x - 2| > |x + 2|$. \hfill (5)

Solution For positive numbers a and b, $a > b$ if and only if $a^2 > b^2$. Further, $|x|^2 = x^2$. Hence (5) is equivalent to

$$(3x - 2)^2 > (x + 2)^2$$

$$9x^2 - 12x + 4 > x^2 + 4x + 4 \qquad \text{squaring}$$

$$8x^2 - 16x > 0 \qquad \text{combining terms}$$

$$x^2 - 2x > 0 \qquad \text{dividing by 8}$$

$$x(x - 2) > 0 \qquad \text{factoring}$$

This inequality may be solved with a sign chart, and the solution is $(-\infty, 0) \cup (2, \infty)$. See Example 3 for the solution of the *equality*.   ●

Exercise 5.7
equations and
inequalities
that involve
absolute values

Solve each of the following equations and inequalities.

1 $|x| = 2$ **2** $|x| = 0$ **3** $|-x| = 3$

4 $|-x| = 5$ **5** $|x - 1| = 1$ **6** $|x + 3| = 7$

7 $|x + 4| = 6$ **8** $|x - 2| = 3$ **9** $|2x - 1| = 5$

10 $|3x + 4| = 10$ **11** $|5x + 3| = 12$ **12** $|2x - 3| = 3$

13 $|-2x + 7| = 3$ **14** $|-3x + 1| = 7$ **15** $|-4x - 5| = 5$

16 $|-6x + 7| = 13$ **17** $|x + 2| = |x - 1|$

18 $|x - 3| = |-x + 1|$ **19** $|x - 3| = |-x + 3|$

20 $|x + 4| = |x - 6|$ **21** $|2x - 5| = |x + 4|$

22 $|3x + 7| = |2x - 3|$ **23** $|4x + 3| = |3x + 4|$

24 $|5x - 2| = |3x + 4|$ **25** $|x| < 2$

26 $|x| < 5$ **27** $|x| > 3$

28 $|x| > 7$ **29** $|x + 3| < 4$

30 $|x - 2| < 6$ **31** $|x - 5| < 3$

32 $|x + 7| < 5$ **33** $|2x - 1| > 5$

34 $|3x - 5| > 1$ **35** $|3x + 4| > 2$

36 $|5x - 7| > -3$ **37** $|-x + 2| < 3$

38 $|-2x - 1| < 5$ **39** $|-3x + 5| > 2$

40 $|-5x + 3| > 2$ **41** $|x + 2| < |x - 4|$

42 $|x - 1| < |x - 3|$ **43** $|2x - 3| < |x + 1|$

44 $|3x + 2| < |2x + 3|$ **45** $|3x - 4| > |x + 6|$

46 $|2x - 1| > |-2x + 1|$ **47** $|4x + 1| > |2x - 3|$

48 $|5x - 2| > |3x + 4|$

5.8 Key concepts

Make sure you understand each of the following important words and ideas.

Equation	Fractional equation
Root	Inequality
Solution	Equivalent inequalities
Solution set	How to solve stated problems
Conditional equation	Quadratic equation
Identity	Solution by completing the square
Equivalent equations	Maximum or minimum of $ax^2 + bx + c$
Transposing	Solution by the quadratic formula

Linear equation in one variable
Use of the discriminant to determine the nature of the roots
Sum and product of the roots
Writing an equation given its roots
Equations and problems leading to quadratic equations
Quadratic inequalities
Sign chart

$a = b$ and $a + c = b + c$ are equivalent
$a = b$ and $ac = bc$, $c \neq 0$, are equivalent

$$(5.1) \quad x = \frac{-b}{a} \qquad\qquad (5.2) \quad x = \frac{-b \pm \sqrt{b^2 - 4ac}}{2a}$$

$$(5.3) \quad r + s = \frac{-b}{a} \qquad \text{sum of the roots}$$

$$(5.4) \quad rs = \frac{c}{a} \qquad \text{product of the roots}$$

$a < b$ is equivalent to:

$$a + c < b + c \qquad \text{for any real } c$$

$$ac < bc \qquad \text{for } c > 0$$

$$ac > bc \qquad \text{for } c < 0$$

Exercise 5.8   Are the following pairs of equations or inequalities equivalent?

1 $3x = 2x + 7$
    $1.5x = 7$

2 $2x + 5 = 7x + 10$
   $x + 11 = 3x + 15$

3 $|2x + 1| = 7$
    $2x + 1 = 7$

4 $-5x < 15$
     $x < -3$

5 $\dfrac{3x - 1}{8 - x} > 0$

   $(3x - 1)(8 - x) > 0$

6 $7x + 3 < 4x + 6$
   $2x + 9 > 5x + 6$

Solve the equations in Probs. 7 to 44.

7 $3x - 2 = x + 4$

8 $5x + 3 = 2x - 3$

9 $5(x + 3) = 4(2x + 1) + 5$

10 $3(x - 1) = 4(x - 3) + 4$

11 $\frac{2}{3}(x - 3) = \frac{1}{6}(x + 6)$

12 $\frac{5}{8}(3x - 8) = \frac{1}{2}(x + 12)$

13 $4(\frac{2}{3}x - \frac{9}{4}) = 3(\frac{2}{5}x + \frac{4}{3})$

14 $2(\frac{7}{9}x - \frac{3}{2}) = 3(\frac{1}{3}x + \frac{7}{3})$

15 $\dfrac{23x + 7}{3} = \dfrac{4x + 5}{5} + x + 2$

16 $\dfrac{2x - 6}{3} = \dfrac{3x - 6}{2} - x + 5$

17 $\dfrac{x + 2}{x - 2} = \dfrac{x + 8}{x}$

18 $\dfrac{6x + 2}{2x - 1} = \dfrac{3x + 7}{x + 1}$

19 $\dfrac{2}{x + 2} + \dfrac{1}{x - 2} = \dfrac{16}{x^2 - 4}$

20 $\dfrac{12}{x - 8} - \dfrac{7}{x - 5} = \dfrac{56}{x^2 - 13x + 40}$

21 $16x^2 = 1$

22 $16x^2 = x$

23 $6x^2 - 5x - 4 = 0$

24 $9x^2 + 18x + 2 = 0$

25 $x^2 - 4x + 1 = 0$

26 $4x^2 + 12x + 3 = 0$

27 $x^2 - 4x + 7 = 0$

28 $4x^2 - 20x + 34 = 0$

29 $16x^2 + 8x + 5 = 0$

30 $24x^2 - 2x - 15 = 0$

31 $7(x - 5)^2 + 189 = 0$

32 $x^4 - 10x^2 + 9 = 0$

33 $2(x^2 - 4x)^2 - 5(x^2 - 4x) - 3 = 0$

34 $x - 3\sqrt{x} - 4 = 0$

35 $\sqrt{3x + 1} = x - 1$

36 $\sqrt{4x - 3} + \sqrt{2x - 2} = 5$

37 $\dfrac{2x + 2}{2x - 3} = \dfrac{3x + 2}{x + 3}$

38 $\dfrac{4}{x - 1} + \dfrac{6}{x - 2} = 3$

39 $|x + 3| = 4$

40 $|-2x + 1| = 7$

41 $|x + 2| = |2x + 1|$

42 $|5x - 4| = |2x + 8|$

43 Solve $F = 32 + 9C/5$ for C.

44 Solve $S = \dfrac{n}{2}(a + l)$ for l.

Solve the inequalities in Probs. 45 to 58.

45 $3x - 5 > x + 1$

46 $5x + 2 > 8x - 4$

47 $4x + 3 < x + 6$

48 $7x - 9 < 2x + 11$

49 $(2x - 1)(3x + 7) < 0$

50 $(5x - 8)(3x + 1) > 0$

51 $\dfrac{5x + 9}{6x - 5} > 0$

52 $\dfrac{3x - 7}{7x + 3} < 0$

53 $(x + 2)(x - 4)(2x + 9) < 0$ **54** $(3x + 8)(2x - 5)(5x + 3) > 0$

55 $|x + 3| < 4$ **56** $|-2x + 1| > 7$

57 $|x + 2| > |2x + 1|$ **58** $|5x - 4| < |2x + 11|$

59 The correct formula for converting from Celsius to Fahrenheit temperature is $F = 1.8C + 32$, but a much easier mental calculation is obtained by use of $f = 2C + 30$. Show that $|F - f| < 5°$ if $5° < F < 95°$.

60 Divide 48 into two parts so that their product is 287.

61 Divide 48 into two parts so that their product is a maximum.

62 Find the maximum value of $6 - 3x - 2x^2$.

63 What is the maximum area which can be enclosed in a rectangular field with 400 m of fencing if one side of the field needs no fencing because it is next to a river?

64 Find the sum and product of the roots of the equation $5x^2 + 6x - 9 = 0$.

65 Find k so that there are two real roots of $(k - 2)x^2 + 4x - 2k + 1 = 0$.

66 Solve the inequality $10x^2 - 29x + 10 < 0$.

67 Solve the inequality $16x^2 - 24x + 11 > 0$.

68 Find the quadratic equation whose roots are $\frac{4}{3}$ and $-\frac{3}{5}$.

69 Find the quadratic equation whose roots are $2 \pm 3i$.

70 Is the quadratic $24x^2 + 106x - 105$ factorable with rational coefficients in the factors?

Physics   **71** A projectile is $S = -5t^2 + 18t + 20$ m high after t s.
(a)   When is it first 25 m high?
(b)   When is it at its maximum height, and what is the maximum height?

72 Show that the roots of $ax^2 + bx + c = 0$ are the reciprocals of the roots of $cx^2 + bx + a = 0$, if $a \neq 0$ and $c \neq 0$.

73 Show that the two roots of $ax^2 + bx + a = 0$ are reciprocals of each other.

74 Show that the roots of $ax^2 + bx + c = 0$ are the negatives of the roots of $ax^2 - bx + c = 0$.

75 Show that the average of the roots of $ax^2 + bx + c = 0$ is the x coordinate of the maximum or minimum point of $y = ax^2 + bx + c$.

76 Show that the sum of the roots equals the product of the roots in the equation $ax^2 + bx - b = 0$.

77 Suppose that a, b, and c are rational. Show that $ax^2 + bx + c = 0$ cannot have one rational root and one irrational root. HINT: Suppose p is rational, y is irrational, and $ap^2 + bp + c = 0 = ay^2 + by + c$. Then $ap^2 - ay^2 = by - bp$. Solve this for y, and arrive at a contradiction (which proves the statement).

78 If $f(x) = ax^2 + bx + c$, show that the graph is symmetric about the vertical line $x = -(b/2a)$ by showing that for any h we have

$$f\left(-\frac{b}{2a} + h\right) = f\left(-\frac{b}{2a} - h\right)$$

79 Assume that $ax^2 + bx + c$ has a, b and c rational. Let $D = b^2 - 4ac$. Why is it true that

(i) D is a perfect square if and only if $ax^2 + bx + c$ can be factored with rational coefficients in the factors.

(ii) $D = 0$ if and only if $ax^2 + bx + c$ is the perfect square $(px + q)^2$ *with p and q rational.*

80 Show that the roots of $ax^2 + bx + c = 0$, given by the quadratic formula (5.2) may also be written

$$x = \frac{2c}{-b \pm \sqrt{b^2 - 4ac}}$$

81 Show that, for $a < 0$, the maximum value of $ax^2 + bx + c$ is

$$\frac{-1}{4a}(b^2 - 4ac)$$

Chemistry **82** In the equation below, n is the refraction index of a liquid of density d for a light of a given wavelength, and the refractivity is defined as

$$r = \frac{n^2 - 1}{d(n^2 + 2)}$$

(a) Solve for n.

(b) Find r for $n = \sqrt{2}$ and $d = \frac{1}{4}$.

Chemistry **83** If we start with a moles of C_5H_{10} and b moles of acetic acid, and if there are x moles of the ester at equilibrium, then the equilibrium constant is

$$K_x = \frac{x(a + b - x)}{(a - x)(b - x)}$$

(a) Find K_x for $a = 3$, $b = 4$, and $x = 2$.

(b) Find x if $K_x = 1$, $a = 3$, and $b = 4$.

CHAPTER 6 · GRAPHS AND FUNCTIONS

6.1 The cartesian coordinate system

Many relationships between two variables are represented by graphs. Some of the many examples are blood pressure and time, profit and production, ACT score and z score (adjusted raw score), area and radius, and distance and time. These graphs show a lot of information in an easy-to-understand format. To draw graphs, we will use a two-dimensional coordinate system. This is a straightforward extension of the one-dimensional number line we introduced in Chap. 1.

The method we present is only one of several possibilities for the plane. Another standard method is polar coordinates. Our present method was invented by the French mathematician and philosopher Rene Descartes (1596–1650), and it is called the cartesian or rectangular coordinate system for the plane.

To set up this system, we construct two perpendicular number lines in the plane and choose a suitable scale on each. For convenience these lines are horizontal and vertical, and the unit length on each is the same (see Fig. 6.1), although neither restriction is necessary. The two lines are called *Coordinate axes* — the *coordinate axes,* the horizontal line being the x axis and the vertical *x axis, y axis, origin* — line the y axis. The intersection of the two lines is the *origin,* designated by the letter O. The coordinate axes divide the plane into four sections called *quadrants*. These quadrants are numbered I, II, III, and IV counterclockwise, as indicated in Fig. 6.1*a.*

Next it is agreed that horizontal distances measured to the right from the y axis are positive, and horizontal distances measured to the left are negative. Similarly, vertical distances measured upward from the x axis are positive, and vertical distances measured downward are negative. These *Directed distance* — distances, because of their signs, are called *directed distances*. Finally, we agree that the first number in an ordered pair of numbers represents the directed distance from the y axis to a point and the second number in the

163

FIGURE 6.1

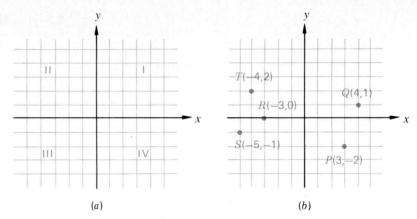

(a)   (b)

pair represents the directed distance from the x axis to the point. It follows then that an ordered pair of numbers uniquely determines the position of a point in the plane. For example, (4, 1) determines the point that is 4 units to the right of the y axis and 1 unit above the x axis. This point is designated by Q in Fig. 6.1b. Similarly, the ordered pair $(-5, -1)$ determines the point S in Fig. 6.1b that is 5 units to the left of the Y axis and 1 unit below the x axis. Conversely, each point in the plane determines a unique ordered pair of numbers. For example, the point P in Fig. 6.1b is 3 units to the right of the y axis and 2 units below the x axis, and so P determines the ordered pair $(3, -2)$. A plane in which the coordinate axes have been constructed

Cartesian plane is called a *cartesian plane*.

The two numbers in an ordered pair that is associated with a point in the

Coordinates cartesian plane are called the *coordinates* of the point. The first number is
Abscissa called the *abscissa* of the point, and it is the directed distance from the y
Ordinate axis to the point. The second number in the pair is the *ordinate* of the point, and it represents the directed distance from the x axis to the point. The procedure for locating a point in the plane by means of its coordinates is
Plotting a point called *plotting* the point. The notation $P(a, b)$ means that P is the point whose coordinates are (a, b). To plot the point $T(-4, 2)$, we count 4 units to the left of the origin on the x axis and then upward 2 units and thus arrive at the point. Similarly, the point $R(-3, 0)$ is 3 units to the left of the origin and on the x axis. The general point and its coordinates are written $P(x, y)$. See Fig. 6.1b.

Ordered pair The pair of numbers (a, b) is called an **ordered pair** because the order in which the numbers appear makes a difference. For instance $(5, 8) \neq (8, 5)$.

Two ordered pairs (a, b) and (c, d) are **equal** means, by definition,

Equality $$(a, b) = (c, d) \qquad \textbf{if and only if} \qquad a = c \text{ and } b = d \qquad (6.1)$$

● **Example 1** $(3a - 1, a - 2b) = (5, -8)$ requires that

$$3a - 1 = 5 \qquad \text{and} \qquad a - 2b = -8$$

The first equation has the solution $a = 2$, and the second then says $2 - 2b = -8$, that is, $-2b = -10$ or $b = 5$.

Note   The standard notation leads to possible confusion since here (a, b) means an ordered pair, while in Chap. 5 we also used (a, b) for the open interval from a to b. This will cause you no problems since the situation will always make it clear whether an ordered pair of numbers, or an interval, is intended.

The distance formula

We will now use the pythagorean theorem, and the distance between two points that are equidistant from a coordinate axis, in developing a formula for the distance between any two points $P_1(x_1, y_1)$ and $P_2(x_2, y_2)$ in the plane. If two points are the same distance from the y axis as shown in Fig. 6.2, then the distance between them is $|x_2 - x_1|$ since x_2 is the distance and direction from the y axis to P_2, and x_1 is the distance and direction from the y axis to P_1. Similarly, the distance between two points $P_1(x_1, y_1)$ and $P_4(x_1, y_2)$ that are equidistant from the x axis is $|y_2 - y_1|$. For example, the distance between $(2, 6)$ and $(7, 6)$ is $|7 - 2| = 5$ and the distance between $(5, 2)$ and $(-3, 2)$ is $|-3 - 5| = 8$.

We will now consider any two points $P_1(x_1, y_1)$ and $P_2(x_2, y_2)$ as shown in Fig. 6.3 and find the distance between them. The point $P_3(x_2, y_1)$ is also shown in the figure and is determined by the intersection of a line parallel

FIGURE 6.2

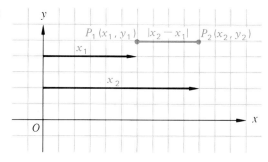

FIGURE 6.3

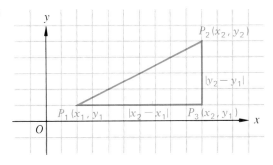

to the x axis through P_1 and a line parallel to the y axis through P_2. Furthermore, the lengths of P_1P_3 and P_3P_2 are $|x_2 - x_1|$ and $|y_2 - y_1|$, as shown in the figure, since they are parallel to the x and y axes, respectively.

Since $|t|^2 = t^2$, we find, by use of the pythagorean theorem, that

$$(P_1P_2)^2 = (P_1P_3)^2 + (P_3P_2)^2$$
$$= (x_2 - x_1)^2 + (y_2 - y_1)^2$$

Therefore, *the distance d between $P_1(x_1, y_1)$ and $P_2(x_2, y_2)$ is*

Distance formula
$$d = \sqrt{(x_2 - x_1)^2 + (y_2 - y_1)^2} \qquad (6.2)$$

● **Example 2**   Find the distance between $(3, -8)$ and $(-2, 4)$.

Solution   The distance is

$$\sqrt{[3 - (-2)]^2 + (-8 - 4)^2} = \sqrt{25 + 144} = \sqrt{169} = 13 \qquad ●$$

Note   Whether we write $(x_2 - x_1)^2$ or $(x_1 - x_2)^2$, and similarly for y values, is immaterial since $(x_2 - x_1)^2 = (x_1 - x_2)^2$.

● **Example 3**   Find the distance between $(-7, 14)$ and $(-1, 12)$.

Solution   The distance is

$$\sqrt{(-7 + 1)^2 + (14 - 12)^2} = \sqrt{36 + 4} = \sqrt{40} = \sqrt{4(10)} = 2\sqrt{10} \qquad ●$$

This is the *exact* distance. A decimal *approximation* by calculator is 6.32456.

● **Example 4**   Show that the triangle with vertices $A(17, 4)$, $B(12, 7)$, and $C(23, 14)$ is a right triangle.

Solution   By the distance formula,

$$[d(A, B)]^2 = (17 - 12)^2 + (4 - 7)^2 = 25 + 9 = 34$$

$$[d(B, C)]^2 = (12 - 23)^2 + (7 - 14)^2 = 121 + 49 = 170$$

$$[d(A, C)]^2 = (17 - 23)^2 + (4 - 14)^2 = 36 + 100 = 136$$

Since $34 + 136 = 170$, the converse of the pythagorean theorem shows that it is a right triangle with the $90°$ angle at $A(17, 4)$.   ●

MIDPOINT   We will now use the distance formula to verify that the **midpoint of the line segment** joining the two points (a, b) and (c, d) is

Midpoint formula
$$\left(\frac{a + c}{2}, \frac{b + d}{2} \right) \qquad (6.3)$$

A proof using perpendicular lines can be given; however, there is clearly only one midpoint of a line segment. The fact that (6.3) is equidistant from (a, b) and (c, d) is shown by letting

$$M = \left(\frac{a + c}{2}, \frac{b + d}{2}\right) \qquad P = (a, b) \qquad \text{and} \qquad Q = (c, d)$$

and then calculating $[d(M, P)]^2$ and $[d(M, Q)]^2$:

$$[d(M, P)]^2 = \left(\frac{a + c}{2} - a\right)^2 + \left(\frac{b + d}{2} - b\right)^2 = \left(\frac{c - a}{2}\right)^2 + \left(\frac{d - b}{2}\right)^2$$

$$[d(M, Q)]^2 = \left(\frac{a + c}{2} - c\right)^2 + \left(\frac{b + d}{2} - d\right)^2 = \left(\frac{a - c}{2}\right)^2 + \left(\frac{b - d}{2}\right)^2$$

$$= \left(\frac{c - a}{2}\right)^2 + \left(\frac{d - b}{2}\right)^2$$

● **Example 5**   Find the midpoint of the line segment with ends $(12, 7)$ and $(-4, 21)$.

Solution   By the midpoint formula we have

$$\left(\frac{12 - 4}{2}, \frac{7 + 21}{2}\right) = \left(\frac{8}{2}, \frac{28}{2}\right) = (4, 14) \qquad\qquad ●$$

● **Example 6**   The points $A(5, 1)$, $B(11, 2)$, $C(8, 7)$, and $D(2, 6)$ taken in the given order are the vertices of a parallelogram. Show that the diagonals AC and BD bisect each other.

Solution   It is enough to show that the midpoints of AC and BD are the same point (*midpoint* implies bisect).

$$\text{Midpoint of } AC = \left(\frac{5 + 8}{2}, \frac{1 + 7}{2}\right) = \left(\frac{13}{2}, \frac{8}{2}\right) = \left(\frac{13}{2}, 4\right)$$

$$\text{Midpoint of } BD = \left(\frac{11 + 2}{2}, \frac{2 + 6}{2}\right) = \left(\frac{13}{2}, \frac{8}{2}\right) = \left(\frac{13}{2}, 4\right) \qquad ●$$

Exercise 6.1   In which quadrant is the given point?

1 $(18, -5)$ **2** $(6, 111)$ **3** $(-43, 61)$ **4** $(-99, -55)$
5 $(-10, -66)$ **6** $(14, -92)$ **7** $(-17, -76)$ **8** $(-19, 45)$

Plot the points in Probs. 9 to 16.

9 $(-6, 4)$ **10** $(-4, -3)$ **11** $(0, 2)$ **12** $(6, -1)$ **13** $(3, 3)$
14 $(-2, -3)$ **15** $(5, -4)$ **16** $(-4, 0)$

Describe in words the given set of points in the plane.

17 All $(x, 0)$ with x real   **18** All $(0, y)$ with y negative

19 All (x, y) with $x > 0$, $y < 0$ **20** All $(x, 4)$ with x real
21 $y > 5$ **22** $xy > 0$ **23** $x = -2$ **24** $xy = 0$

In Probs. 25 to 36, find the distance between the two points, and also the midpoint of the line segment joining them.

25 $(6, 4)$ and $(-8, 2)$ **26** $(8, -5)$ and $(4, 3)$
27 $(9, 4)$ and $(-3, 8)$ **28** $(5, 7)$ and $(7, -5)$
29 $(3, 7)$ and $(8, -5)$ **30** $(-4, -8)$ and $(-12, 7)$
31 $(6, -2)$ and $(-1, 22)$ **32** $(10, 0)$ and $(-10, -21)$
33 $(-11, -3)$ and $(-5, 2)$ **34** $(5, 5)$ and $(8, -2)$
35 $(-9, 4)$ and $(4, 6)$ **36** $(2, -5)$ and $(-7, -1)$

In Probs. 37 to 40, find the other endpoint of the line segment having the given midpoint and endpoint.

37 Midpoint $(8, 2)$, endpoint $(5, 8)$
38 Midpoint $(\frac{3}{2}, -6)$, endpoint $(-3, -7)$
39 Midpoint $(\frac{3}{2}, 2)$, endpoint $(-2, 9)$
40 Midpoint $(\frac{13}{2}, -\frac{1}{2})$, endpoint $(6, -8)$

In Probs. 41 to 44, use the distance formula to show that the given three points are the vertices of a right triangle.

41 $(5, 8)$, $(7, 5)$, $(10, 7)$ **42** $(3, -2)$, $(8, 1)$, $(-3, 8)$
43 $(7, 1)$, $(4, 7)$, $(1, -2)$ **44** $(11, 4)$, $(-4, -1)$, $(-3, -4)$
45 If $A = (2, 3)$ and $B = (10, 9)$, find the point which is $\frac{3}{4}$ of the way from A to B.
46 Show that $(9, 5)$ is equidistant from $(8, -2)$ and $(2, 6)$.
47 Show that the midpoint of the hypotenuse of a right triangle is equidistant from all three vertices.
48 Write an equation satisfied by all points (x, y) which are 3 units from $(5, -10)$.

In Probs. 49 to 52, use the distance formula to show that the given points are all on the same line.

49 $(1, 5)$, $(2, 7)$, $(4, 11)$ **50** $(-2, -3)$, $(2, -1)$, $(8, 2)$
51 $(0, 1)$, $(3, 3)$, $(9, 7)$ **52** $(2, 2)$, $(-8, -2)$, $(7, 4)$

6.2   Graphs

In our visually oriented culture, pictures are a most important method of communication. By use of the rectangular coordinate system, we can obtain a geometric representation, or a geometric "picture," of an equation. In order to do this, we require that each ordered pair of numbers (x, y) be the coordinates of a point in the cartesian plane, with x as the abscissa and y as the ordinate. Now we define the graph of an equation as follows:

Graph of an equation   The graph of an equation is the set of all points $P(x, y)$ whose coordinates x and y satisfy the equation.

Plotting points   In this section we will find the graph of an equation by *plotting points*, lots of points. In later sections and chapters in this book, we will learn how to graph an equation more quickly and accurately by recognizing certain standard forms and some modifications of them. There are other, more advanced, methods which require calculus.

After plotting enough points, connect them in some reasonable order. A pattern will usually emerge if enough points are included and nearby points are joined together. It helps to make a table of values with increasing x values or y values.

● **Example 1**   Draw the graph of the equation

$$y = x^2 - 2x - 2 \tag{1}$$

The first step is to assign several values to x and then calculate each corresponding value of y. Before doing this, however, it is advisable to make a table like the one below in which to record the corresponding values of x and y.

x	
y	

The values selected for x, in most cases in this section, should be small, usually integers less than 10 in absolute value. In this case, we start with the integer 0 and then successively assign the integers 1, 2, 3, 4, -1, and -2 to x. These integers were selected arbitrarily and, as we will see, will be used to determine a portion of the graph. We next calculate the value of y corresponding to each of the selected integers by replacing x in (1) by each of them. Thus we get the corresponding number pairs

$$x = 0 \qquad y = 0^2 - 2(0) - 2 = -2$$
$$x = 1 \qquad y = 1^2 - 2(1) - 2 = -3$$
$$x = 2 \qquad y = 2^2 - 2(2) - 2 = -2$$

and so on until all the integers listed above have been used. When a number is assigned to x, it should be recorded in the table and the corresponding value of y entered below it. The numbers assigned to x should be entered in order of magnitude from left to right. When the value of y for each of the values of x has been calculated and the results entered in the table, we have

x	-2	-1	0	1	2	3	4
y	6	1	-2	-3	-2	1	6

FIGURE 6.4

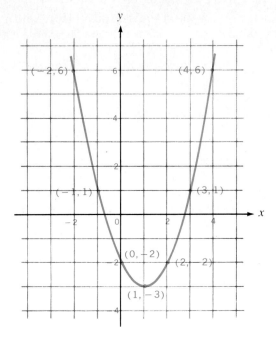

Graphing by
plotting points

Now we *plot the points* (x, y) thus determined, as shown in Fig. 6.4, and connect them with a smooth curve. This curve is a portion of the complete graph. ●

● **Example 2** To graph $y = x^3 - 2x^2 - 5x + 6$, we make the following table of values:

x	-3	-2	-1	0	1	2	3	4
y	-24	0	8	6	0	-4	0	18

For example, if $x = -1$, then $y = (-1)^3 - 2(-1)^2 - 5(-1) + 6 = -1 - 2 + 5 + 6 = 8$ and if $x = 2$, then $y = 2^3 - 2(2^2) - 5(2) + 6 = 8 - 8 - 10 + 6 = -4$. The graph is shown in Fig. 6.5. The exact maximum point occurs near $(-0.8, 8.2)$, a fact for which we need calculus. Also, the minimum is near $(2.1, -4.06)$. The exact x values are

$$\frac{2 \pm \sqrt{19}}{3}$$

●

● **Example 3** The table of values for $y = x^4 - 10x^2 + 9$ is given here.

x	-3	-2	-1	0	1	2	3
y	0	-15	0	9	0	-15	0

The graph is shown in Fig. 6.6.   ●

FIGURE 6.5

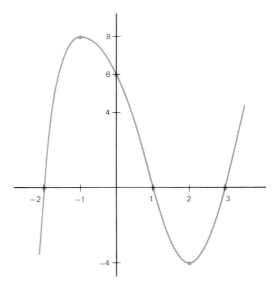

FIGURE 6.6

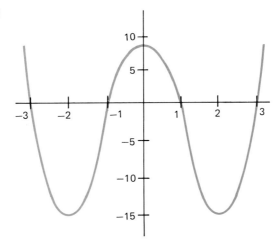

● **Example 4** An important function in the theory of numbers is the bracket function $[x]$. By the notation $[x]$, we mean the *greatest integer* that is less than or equal to x. Hence in the equation

$$y = [x]$$

if $x = \frac{1}{2}$, then $y = [\frac{1}{2}] = 0$ since 0 is the greatest integer less than $\frac{1}{2}$. Similarly, if $x = 2\frac{1}{2}$, $y = [2\frac{1}{2}] = 2$ since 2 is the greatest integer less than $2\frac{1}{2}$. Also, $[-\frac{1}{2}] = -1$, and $[-\frac{4}{3}] = -2$. Since $[x]$ is the greatest integer less than or *equal* to x, then $[n] = n$ if n is an integer. In general, if x is in the interval $0 \le x < 1$, then $y = 0$; and for x such that $1 \le x < 2$, $y = 1$. Hence, by the above and by similar arguments, we have the following corresponding values of x and y:

$$-1 \leq x < 0 \qquad y = -1$$
$$0 \leq x < 1 \qquad y = 0$$
$$1 \leq x < 2 \qquad y = 1$$
$$2 \leq x < 3 \qquad y = 2$$
$$\cdots\cdots\cdots\cdots\cdots$$
$$n \leq x < n + 1 \quad y = n$$

Consequently, the graph of $y = [x]$ is a set of horizontal line segments. Notice that $[x] \leq x$. The graph is in Fig. 6.7. ●

● **Example 5**   The graph of $y = 2x - 1$ is shown in Fig. 6.8. In Sec. 6.4 we will see how to graph this line quickly without plotting a lot of points.

x	-1	0	1	2	3
y	-3	-1	1	3	5

●

FIGURE 6.7

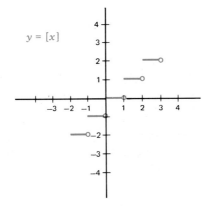

$y = [x]$

FIGURE 6.8

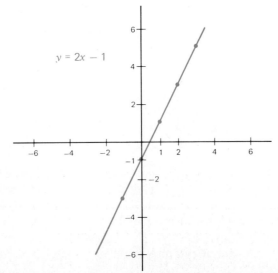

$y = 2x - 1$

FIGURE 6.9

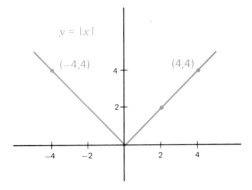

● **Example 6**   If $y = |x|$, then $y = 3$ for both $x = 3$ and $x = -3$, and similarly $y = 1$ if $x = 1$ or $x = -1$. If $x = 0$, $y = 0$. The graph is shown in Fig. 6.9. We also give the graph of $y = |x + 1| + |x - 2|$, in Fig. 6.10, and its table of values below.

x	-3	-2	-1	0	1	2	3	4
$y = |x + 1| + |x - 2|$	7	5	3	3	3	3	5	7

Graph of an equation

So far we have drawn the graphs only of equations which were solved for y, but we may do the same for any equation. The *graph* is still the set of all points $P(x, y)$ whose coordinates satisfy the equation.

● **Example 7**   Interchanging x and y in Example 2 gives the equation $x = y^3 - 2y^2 - 5y + 6$. Hence the table of values is as given below, and the graph is similar to Fig. 6.5. It is shown in Fig. 6.11.

y	-3	-2	-1	0	1	2	3	4
x	-24	0	8	6	0	-4	0	18

FIGURE 6.10

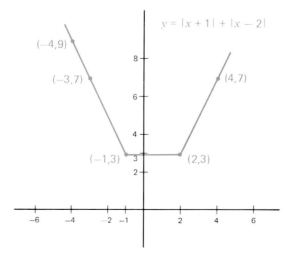

FIGURE 6.11

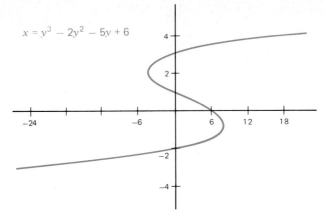

$$x = y^3 - 2y^2 - 5y + 6$$

CIRCLES

● **Example 8**   The graph of $x^2 + y^2 = 25$ is a circle and has the graph shown in Fig. 6.12. We will learn how to graph any circle quickly after this example. A table of values follows.

x	-5	-3	0	4	5
y	0	± 4	± 5	± 3	0

●

The result of Example 8 can be generalized to a circle with any center and any radius. If the center is the point $C(h, k)$ and the radius is the positive number r, then by definition of a circle, the point $P(x, y)$ will be

FIGURE 6.12

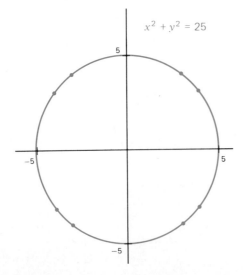

$$x^2 + y^2 = 25$$

on the circle if $d(C, P) = r$, that is,

$$\sqrt{(x - h)^2 + (y - k)^2} = r$$

After squaring we get the *standard form*

Circle standard form $\qquad\qquad (x - h)^2 + (y - k)^2 = r^2$ $\qquad\qquad$ (6.4)

● **Example 9** The graph of the equation $(x - 3)^2 + (y + 5)^2 = 7$ is a circle with center $(3, -5)$ and radius $\sqrt{7}$. ●

If the equation is not given in standard form, we can put it in standard form by completing the square for the x terms and y terms separately. The most general form of the equation of a circle is

General form $\qquad\qquad ax^2 + ay^2 + cx + dy + e = 0$

in which x^2 and y^2 have the *same* coefficient.

● **Example 10** Write $2x^2 + 2y^2 + 8x - 10y - 3 = 0$ in standard form.

Solution To prepare for completing the square, we *divide by the common coefficient* of x^2 and y^2, thereby making their new coefficients 1.

$$x^2 + y^2 + 4x - 5y - \tfrac{3}{2} = 0 \qquad \text{dividing by 2}$$
$$x^2 + 4x + y^2 - 5y = \tfrac{3}{2}$$

Completing the square Remember that we complete the square by adding the square of half the coefficient of x (or of y), and that this must be added to both sides of the equation to produce an equation equivalent to the given one.

$$x^2 + 4x + \left(\frac{4}{2}\right)^2 + y^2 - 5y + \left(\frac{-5}{2}\right)^2 = \frac{3}{2} + \left(\frac{4}{2}\right)^2 + \left(\frac{-5}{2}\right)^2$$

$$x^2 + 4x + 4 + y^2 - 5y + \frac{25}{4} = \frac{3}{2} + 4 + \frac{25}{4}$$

$$(x + 2)^2 + \left(y - \frac{5}{2}\right)^2 = \frac{6 + 16 + 25}{4} = \frac{47}{4}$$

Thus the circle has its center at $(-2, \tfrac{5}{2})$ and a radius of $\sqrt{47/4} = \sqrt{47}/2$. ●

If completing the squares yields an equation $(x - h)^2 + (y - k)^2 = n$ where $n < 0$, there is no graph. If $n = 0$, the graph is the single point (h, k).

Exercise 6.2 Graph the equations by plotting points.

1 $3x + 5y = 15$	**2** $x - 4y = 8$
3 $-4x + 3y = 6$	**4** $6x + y = 6$
5 $y = 3x - 5$	**6** $y = -2x + 4$

7 $x = 2y + 5$

8 $x = -4y - 1$

9 $y = x^2 + 4$

10 $y = -x^2 + 5x$

11 $y = 2x^2 + 12x + 15$

12 $y = -4x^2 + 8x + 3$

13 $x = y^2$

14 $x = -y^2 + 3$

15 $x = 2y^2 - 6y$

16 $x = 3y^2 + 6$

17 $y = x^3 - 2x^2 - x + 2$

18 $y = x^3 + x^2 - 4x - 4$

19 $y - 7 = 2x^3 - 7x^2 - 2x$

20 $y + 3 = 4x^3 + 12x^2 - x$

21 $4y = x^3 + 3x^2 + 12x + 8$

22 $3y = x^3 - 3x^2 + 4x + 1$

23 $5y - 5 = x^3 + 6x^2 + 13x$

24 $3y + 6 = x^3 - 6x^2 + 14x$

25 $2y = x^4 - 13x^2 + 36$

26 $4y = 4x^4 - 25x^2 + 36$

27 $y = x^4 - 15x^2 - 10x + 24$

28 $y = x^4 + 2x^3 - 13x^2 - 14x + 24$

29 $x = y^3 - 3y^2 - 4y + 12$

30 $5x = y^3 - y^2 + y + 4$

31 $20x = 4y^4 - 41y^2 + 100$

32 $6x = y^4 - 2y^3 - y^2 + 2y$

33 $x^2 + 4y^2 = 16$

34 $4x^2 + 9y^2 = 36$

35 $5x^2 + y^2 = 25$

36 $25x^2 + 4y^2 = 100$

37 $xy = 4$

38 $xy = 7$

39 $xy = -3$

40 $xy = -1$

41 $y = |x - 1| + |x - 4|$

42 $y = |x + 2| + |x - 2|$

43 $y = |x + 2| - |x - 1|$

44 $y = |2x - 1| - |x - 2|$

45 $y = \sqrt{x}$

46 $y = \sqrt{-x}$

47 $y = -x^5$

48 $y = x^6$

49 $y = x + |x|$

50 $y = x|x|$

51 $|y| = |x|$

52 $|x| + |y| = 1$

Graph the circles in Probs. 53 to 68.

53 $(x - 4)^2 + (y - 5)^2 = 4$

54 $(x - 2)^2 + (y + 3)^2 = 9$

55 $(x + 5)^2 + (y - 1)^2 = 1$

56 $(x + 2)^2 + (y + 2)^2 = 16$

57 $x^2 + (y + 3)^2 = 4$

58 $(x - 4)^2 + (y - 3)^2 = 25$

59 $(x - 6)^2 + y^2 = 36$

60 $(x - 5)^2 + (y + 12)^2 = 169$

61 $x^2 + y^2 + 6x + 4y = 5$

62 $x^2 + y^2 + 8x - 12y = -27$

63 $x^2 + y^2 - 4x + 8y = 10$

64 $x^2 + y^2 - 2x - 10y = -1$

65 $2x^2 + 2y^2 + 6x - 8y = 1$

66 $3x^2 + 3y^2 + 12x + 21y = 5$

67 $5x^2 + 5y^2 - 10x - 20y = 9$

68 $6x^2 + 6y^2 - 12x + 18y = 13$

69 In Example 4 we graphed $y = [x]$. Graph $y = x - [x]$. Notice that this graph is *periodic;* that is, its y values repeat. Other examples of periodic graphs come from the trigonometric functions, which are studied in another course.

6.3   Symmetry and translation of graphs

In this section we will learn how to start with a graph of an equation and get several other related graphs easily. We begin by considering symmetry with respect to certain lines and the origin.

Symmetric to x axis   The graph of an equation is **symmetric with respect to the x axis** if replacing y by $-y$ yields an equivalent equation. This corresponds to the fact that $(x, -y)$ is on the graph whenever (x, y) is; see P and Q in Fig. 6.13. Notice that these two points are each $|y|$ units from the x axis. The circle $x^2 + y^2 = 25$ is symmetric with respect to the x axis; see Example 8, Sec. 6.2.

● **Example 1**   The graph of $y^4 = x^3 + x^2 + x + 1$ will not be drawn, but the graph is symmetric with respect to the x axis since $(-y)^4 = y^4$.   ●

Symmetric to y axis   The graph of an equation is **symmetric with respect to the y axis** if replacing x by $-x$ yields an equivalent equation. This means that $(-x, y)$ is on the graph whenever (x, y) is. These two points are the same distance, namely, $|x|$, from the y axis. See points P and R in Fig. 6.13.

● **Example 2**   Each of the following is symmetric with respect to the y axis:
(a) $y = |x|$; see Example 6, Sec. 6.2.
(b) $y = x^4 - 10x^2 + 9$; see Example 3, Sec. 6.2.
(c)   The circle $x^2 + y^2 = 25$; see Example 8, Sec. 6.2.
(d) $x^2 + y^6x^{14} + y^{13} = 3$ since $(-x)^2 = x^2$ and $(-x)^{14} = x^{14}$.   ●

Symmetric with respect   The graph of an equation is **symmetric with respect to the origin** if replacing
to the origin x and y by $-x$ and $-y$, respectively, yields an equivalent equation.Thus $(-x, -y)$ is on the graph whenever (x, y) is. See points P and S in Fig. 6.13.

● **Example 3**   (a)   The circle $x^2 + y^2 = 25$ is symmetric wih respect to the origin. See Example 8, Sec. 6.2.
(b) $y = x^3$ is symmetric with respect to the origin since replacing (x, y) by $(-x, -y)$ gives $-y = (-x)^3$, which is $-y = -(x^3)$ or $y = x^3$.   ●

Symmetric to $y = x$ The graph of an equation is **symmetric with respect to the line $y = x$** if interchanging x and y yields an equivalent equation. Thus (y, x) is on the graph whenever (x, y) is. See points P and T in Fig. 6.13.

● **Example 4**   Each of the following is symmetric with respect to the line $y = x$:
(a)   The circle $x^2 + y^2 = 25$; see Example 8, sec. 6.2.

FIGURE 6.13

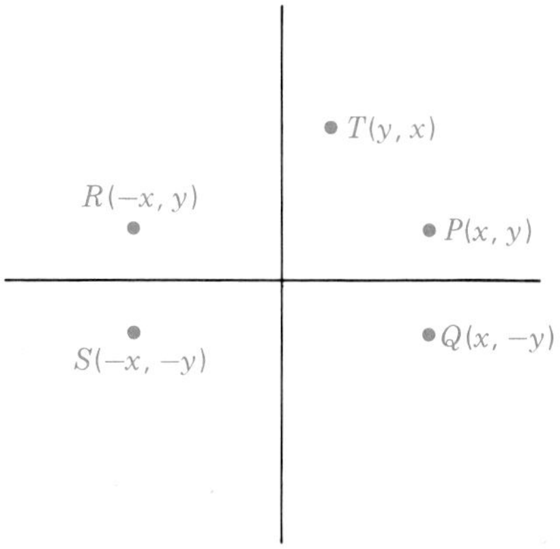

$\bullet\, T(y, x)$

$R(-x, y)$
$\bullet$

$\bullet\, P(x, y)$

$\bullet$
$S(-x, -y)$

$\bullet\, Q(x, -y)$

(b) The line $y = -x$ since interchanging x and y gives $x = -y$ which is clearly equivalent to $y = -x$.

(c) The hyperbola $xy = 4$; see Prob. 37, Exercise 6.2 (the graph is in the Answer section). ●

The above discussion of symmetry is summarized in the table below:

The graph of an equation is symmetric with respect to the	If an equivalent equation results when (x, y) is replaced by
x axis	$(x, -y)$
y axis	$(-x, y)$
Origin	$(-x, -y)$
Line $y = x$	(y, x)

We can also use the facts in the table above to get related graphs from a given one. For instance if we have the graph of an equation and if we replace (x, y) by $(x, -y)$ in the equation, we get a new equation whose graph is the original graph reflected through the x axis. That is, the new graph is the mirror image of the given graph through the x axis. Another way to say this is that if the plane is folded along the x axis, then the two graphs match exactly.

● **Example 5** The graph of $2x + 3y = 6$ is the line shown in Fig. 6.14. Replacing (x, y) by $(x, -y)$ gives $2x - 3y = 6$, whose graph is shown dashed in the figure. Each graph is the reflection of the other one through the x axis. ●

In a similar way, we can get a new graph which is the reflection of a given graph through the y axis, the origin, or the line $y = x$. Again we summarize the information in a table.

FIGURE 6.14

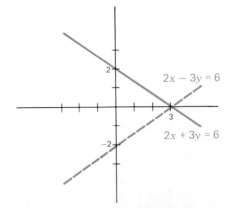

$2x - 3y = 6$

$2x + 3y = 6$

To reflect a graph through the	Replace (x, y) in the given equation by
x axis	$(x, -y)$
y axis	$(-x, y)$
Origin	$(-x, -y)$
Line $y = x$	(y, x)

● **Example 6**   In $y = |x - 1|$, if we replace (x, y) with $(-x, y)$, we get $y = |-x - 1|$. The graph of each is the reflection of the other in the y axis. See Fig. 6.15. Notice that $|-x - 1| = |x + 1|$.   ●

● **Example 7**   In the circle $(x - 2)^2 + (y - 1)^2 = 1$, if we replace (x, y) with $(-x, -y)$, we get $(-x - 2)^2 + (-y - 1)^2 = 1$, or $(x + 2)^2 + (y + 1)^2 = 1$. The graph of each is the reflection of the other through the origin. See Fig. 6.16.   ●

We will now discuss translating a graph, that is, shifting horizontally or vertically. To begin, we consider $y = x^2$, $y - 3 = x^2$, and $y = (x - 4)^2$. Their graphs are shown in Figs. 6.17 and 6.18.

FIGURE 6.15

$y = |x - 1|$ is solid
$y = |-x - 1|$ is dotted

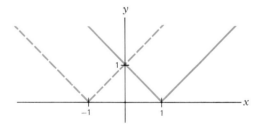

FIGURE 6.16

$(x - 2)^2 + (y - 1)^2 = 1$ is solid
$(x + 2)^2 + (y + 1)^2 = 1$ is dotted

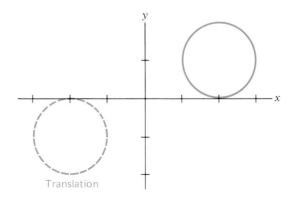

Translation

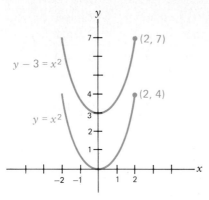

FIGURE 6.17   FIGURE 6.18

As we can see, the graph of $y = x^2$ is translated or shifted 3 units up in the first case, and it is moved to the right 4 units in the second. Although we could graph each by plotting points, it is clearly easier to merely shift the graph of $y = x^2$ the correct amount.

In the following summary, we take $b > 0$:

Replacing	Translates the graph b units
x by $x - b$	To the right
x by $x + b$	To the left
y by $y - b$	Up
y by $y + b$	Down

● Example 8   To translate the graph of $x^2 + y^2 = 1$ by 3 units to the left, we replace x by $x + 3$ and have $(x + 3)^2 + y^2 = 1$. See Fig. 6.19.   ●

● Example 9   In the equation $x = |y|$, we will replace x by $x + 4$ and y by $y - 1$, getting $x + 4 = |y - 1|$. This simultaneously shifts the graph 4 units to the left and 1 unit up. See Fig. 6.20.   ●

FIGURE 6.19

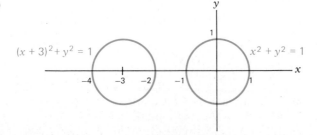

FIGURE 6.20

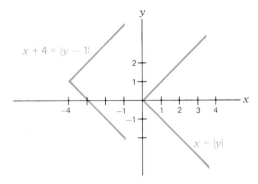

Exercise 6.3   In Probs. 1 to 4, write the point which is symmetric to the given point with respect to (a) the x axis and (b) the origin.

1 $(3, -5)$ **2** $(-2, 1)$ **3** $(-3, -8)$ **4** $(5, 16)$

In Probs. 5 to 8, write the point which is symmetric to the given point with respect to (a) the y axis and (b) the line $y = x$.

5 $(-2, -7)$ **6** $(6, -1)$ **7** $(11, 4)$ **8** $(-7, 3)$

In Probs. 9 to 16 an equation is given. State which symmetry properties, of the four studied, its graph has.

9 $x^2 + 6y^2 = 12$ **10** $x^2 + 6y = 12$
11 $(6x + 1)(3y^4 + 1) = 1$ **12** $|x + y| = 10$
13 $xy = 8$ **14** $x^2 - y^2 = 8$
15 $x^{12}y^{13} + x^{22}y^{33} = 80$ **16** $x^6y^9 + x^9y^6 = 96$

In Probs. 17 to 20, write an equation whose graph is the graph of

$$xy(x + y) = 5$$

reflected as stated.

17 Reflected through the x axis   **18** Reflected through the y axis
19 Reflected through the origin   **20** Reflected through the line $y = x$

In Probs. 21 to 28, write the equation for the stated translation.

21 $x^2 + (y - 5)^2 = 4$ shifted 3 units to the right
22 $(x - 2)^2 - (y + 3)^2 = 11$ shifted 4 units to the left
23 $xy + x = 5$ shifted 1 unit up
24 $6x + 3y = 10$ shifted 7 units down
25 $7x - 2y = 3$ shifted right 3 units, down 2 units
26 $xy = 15$ shifted right 3 units, up 5 units
27 $x^2 + 3x + 2y = 4$ shifted left 2 units, up 3 units
28 $(x - 1)^2 = 4y$ shifted left 1 unit, down 2 units

In Probs. 29 to 36, draw the graphs of both equations on the same set of axes. The second is a translation of the first.

29 $y = x^3$ and $y = (x - 2)^3$
30 $2y = x^2$ and $2y = (x + 3)^2$
31 $3x + 5y = 15$ and $3x + 5(y + 2) = 15$
32 $y = x^4$ and $y - 3 = x^4$
33 $y = |x|$ and $y + 4 = |x - 2|$
34 $y = |x|$ and $y - 2 = |x - 1|$
35 $x = y^2$ and $x + 1 = (y - 3)^2$
36 $x = y^3$ and $x + 2 = (y + 1)^3$

In Probs. 37 to 40, use the same set of axes to draw the graph of $(x - 3)^2 + (y - 1)^2 = 1$ and the given equation (which is a reflection of the previously written circle).

37 $(-x - 3)^2 + (y - 1)^2 = 1$ **38** $(x - 3)^2 + (-y - 1)^2 = 1$
39 $(-x - 3)^2 + (-y - 1)^2 = 1$ **40** $(y - 3)^2 + (x - 1)^2 = 1$

Draw the graphs in Probs. 41 to 52.

41 $y = (x + 3)^2$ **42** $y = (x + 2)^3$
43 $x = (y - 1)^4$ **44** $x = (y + 2)^3$
45 $y + 1 = |x - 4|$ **46** $y - 2 = -|x + 3|$
47 $x - 3 = -|y + 1|$ **48** $x + 4 = |y - 2|$
49 $-y = (x - 3)^2$ **50** $-x = y^3$
51 $-x = |y|$ **52** $-y = |x - 2|$

53 Translation and reflection do not commute with each other. The following specific case verifies this.
   (a) Find the equation obtained by first translating $y = x^3$ by 2 units to the right, then reflecting it in the y axis.
   (b) Find the equation obtained by first reflecting $y = x^3$ in the y axis, then translating to the right by 2 units.

6.4 Functions

In graphing equations earlier in this chapter, we dealt with equations such as

$$3x + 4y = 36$$
$$y = x^2 + 2x + 3$$
$$x^2 + y^2 = 4$$

The first two are fundamentally different from the third. In each of the first two, if a specific number is used in place of x, then the equation can be solved for *exactly one* value of y. For instance, if we take $x = 4$ in the first equation, then

$$3(4) + 4y = 36 \qquad \text{is satisfied if } y = 6$$

In the second equation, if we let $x = 5$ then

$$y = 25 + 10 + 3 = 38$$

However, in the third equation if we take $x = 1$, then

$$1 + y^2 = 4 \qquad \text{and so} \qquad y^2 = 3$$

which has the *two* solutions $y = +\sqrt{3}$ and $y = -\sqrt{3}$.

Equations which can be solved for exactly one value of y if a specific x is given define *functions*. In other words, given an x, the equation provides a way to find a unique y.

Function   More generally we say that we have a **function** from a set A to a set B if *there is a rule, or correspondence, which assigns to each member of A* Domain   *exactly one element of B.* The set A is called the **domain** of the function. We will usually take A and B to be sets of real numbers in this book.

Note   Not every element in B needs to be used, and notice especially that *an element of B may be used more than once.* The set of values in B which Range   are actually used is called the **range** of the function.

For instance, with the function defined above by $y = x^2 + 2x + 3$, we choose the domain to be the set of all real numbers. If $x = -3$, then $y = 9 - 6 + 3 = 6$, while if $x = 1$, then $y = 1 + 2 + 3 = 6$ again. However 1 is not in the range since $1 = x^2 + 2x + 3$ has only complex roots (which are not in the domain).

The correspondence or rule defining the function can be given by an equation, a formula, a rule, a chart, a table, or simply a listing of the ordered pairs (x, y) in the function. For a given x, the corresponding y is called the Value   **value** of the function at x, or also the *image* of x. Thus the range is merely the set of all images.

If no domain is specified, then it is understood to be as large as possible; that is, it includes all real numbers for which all expressions are defined. Specifically, we exclude any number which would lead to division by 0, to square roots of negative numbers, etc.

● **Example 1**   The equation $y = x^2$ defines a function. The domain is the set of all real numbers because x^2 is a real number for all real numbers x. It is a function because the number x is paired with the unique number x^2. It contains the ordered pair $(7, 49)$.   ●

● **Example 2**   The equation $y = \sqrt{x}$ defines a function. The domain is $\{x \mid x \geq 0\}$ because $\sqrt{x}$ is a real number only for $x \geq 0$. It contains the ordered pair $(49, 7)$.   ●

● **Example 3**   The equation $y^2 = x$ is *not* a function since it contains both $(49, 7)$ and $(49, -7)$. We may use any x which is positive or 0.   ●

● **Example 4**   The equation $x^2 + y^2 = 169$ does not define a function, since $(5, 12)$ and $(5, -12)$ both satisfy the equation.   ●

● **Example 5**   The equation $y = -\sqrt{169 - x^2}$ does define a function. Notice that $(5, -12)$ satisfies the equation but $(5, 12)$ does not. The domain is $-13 \le x \le 13$.

●

Functional notation

Independent variable   If a function is defined by an equation, we often use the letters f or g or h to name the function. Thus we write $y = f(x)$, for instance. With this notation, we call x the *independent variable* and y the *dependent variable*.

Note   We must distinguish between f and $f(x)$. The letter f is used to denote the function itself, i.e., the rule or correspondence.

The symbol $f(x)$ is used to denote the function value of the dependent variable—it is the second member of the ordered pair (x, y). If the function is given by an equation, then $f(x)$ is a number. To find this number, we simply replace x everywhere in the equation by its value, then do the indicated calculation.

Instead of saying, for instance, that "f is the function defined by the equation $y = x^2 + 2x + 3$," we often use a verbal shorthand and say "the function $f(x) = x^2 + 2x + 3$."

● **Example 6**   Let $f(x) = 4x - 3$. Then

For $x = 3$, $f(3) = 4(3) - 3 = 12 - 3 = 9$
For $x = -2$, $f(-2) = 4(-2) - 3 = -8 - 3 = -11$
For $x = 2t$, $f(2t) = 4(2t) - 3 = 8t - 3$

●

● **Example 7**   Let $f(x) = 2x^2 - 3x + 6$. Then

For $x = 4$, $f(4) = 2(16) - 3(4) + 6 = 32 - 12 + 6 = 26$
For $x = -4$, $f(-4) = 2(16) - 3(-4) + 6 = 32 + 12 + 6 = 50$
For $x = a + b$, $f(a + b) = 2(a + b)^2 - 3(a + b) + 6$
$$= 2(a^2 + 2ab + b^2) - 3a - 3b + 6$$
$$= 2a^2 + 4ab + 2b^2 - 3a - 3b + 6$$

●

Substitute   Remember, we merely have to *substitute* in order to find function values. Everywhere we see x (or the independent variable) in the formula, we substitute.

● **Example 8**   If $f(x) = (x - 2)/(x + 1)$, then

$$f(2) = \frac{2 - 2}{2 + 1} = \frac{0}{3} = 0 \qquad f(\tfrac{1}{2}) = \frac{\tfrac{1}{2} - 2}{\tfrac{1}{2} + 1} = \frac{-\tfrac{3}{2}}{\tfrac{3}{2}} = -1$$

$$f\left(\frac{1}{w}\right) = \frac{\dfrac{1}{w} - 2}{\dfrac{1}{w} + 1} \cdot \frac{w}{w} = \frac{1 - 2w}{1 + w}$$

●

If we have functions f and g, we may form new functions in a natural way by using the four fundamental operations. We define

$h = f + g$	by	$h(x) = f(x) + g(x)$
$h = f - g$	by	$h(x) = f(x) - g(x)$
$h = fg$	by	$h(x) = f(x) \cdot g(x)$
$h = f/g$	by	$h(x) = f(x)/g(x)$ whenever $g(x) \neq 0$

● **Example 9** If $f(x) = 2x + 3$ and $g(x) = x^2 - 1$, then

$$(f + g)(4) = f(4) + g(4) = (8 + 3) + (16 - 1) = 26$$
$$(f - g)(5) = f(5) - g(5) = (10 + 3) - (25 - 1) = -11$$
$$(fg)(-2) = f(-2) \cdot g(-2) = (-4 + 3) \cdot (4 - 1) = -3$$
$$(f/g)(0) = f(0)/g(0) = (0 + 3)/(0 - 1) = -3$$

●

Composite function

If f and g are functions such that the range of g is a subset of the domain of f, we define the *composite function* $f \circ g$ by

$$(f \circ g)(x) = f(g(x))$$

Note   Notice that we evaluate f at $g(x)$—we do *not* multiply $f(x)$ and $g(x)$.

● **Example 10** If $f(x) = x^2 + 1$ and $g(x) = 3x - 1$, then $g(2) = 6 - 1 = 5$ and

$$(f \circ g)(2) = f(g(2)) = f(5) = 25 + 1 = 26$$

Also

$$(f \circ g)(0) = f(g(0)) = f(-1) = 1 + 1 = 2$$

In general $g \circ f$ need not be defined if $f \circ g$ is. Even if both are defined, they are not always equal. For instance,

$$(f \circ g)(4) = f(g(4)) = f(11) = 121 + 1 = 122$$
$$(g \circ f)(4) = g(f(4)) = g(17) = 51 - 1 = 50$$

●

So far we have considered only functions of one variable. Later in the book, we will work with functions of two variables (for instance, in linear programming and in permutations and combinations) and functions of a set (in probability).

If $f(x, y) = 4x + 9y - 3$, then

$$f(2, 5) = 4(2) + 9(5) - 3 = 8 + 45 - 3 = 50$$
$$f(5, 2) = 4(5) + 9(2) - 3 = 20 + 18 - 3 = 35$$
$$f(3t, t) = 4(3t) + 9(t) - 3 = 12t + 9t - 3 = 21t - 3$$

Graph of a function

If a function is defined by the equation $y = f(x)$, the **graph of the function** f is the graph of the equation $y = f(x)$. That is, the graph of the function f is

the set of all points $P(x, y)$ such that $y = f(x)$

● **Example 11** The graph of $f(x) = x^2$ is the graph of the equation $y = x^2$.   ●

We normally consider x as the independent variable and speak of a function of x. The domain then consists of all the x values. If the function

Largest possible domain is given by an equation, we will assume that the domain is the largest possible set for which the expression is defined.

● **Example 12**   The domain of $f(x) = \sqrt{3x - 4}$ is all real numbers satisfying $3x - 4 \geq 0$, which gives $x \geq \frac{4}{3}$ or interval $[\frac{4}{3}, \infty)$.   ●

Vertical line test   When a graph is given, there is an easy way to see if it is the graph of a function. This method is called the **vertical line test** for a function:

In the graph of a function, each vertical line intersects the graph at most once.

● **Example 13**

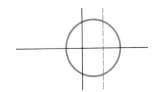

This *is not* the graph of a function

FIGURE 6.21

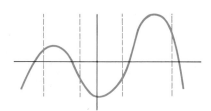

This *is* the graph of a function

FIGURE 6.22   ●

Alternate definition of function   An **alternate definition** of a function is a set of ordered pairs (x, y) such that if the first entries are the same, so are the second entries. In other words, if $x_1 = x_2$, then $y_1 = y_2$. This may be rephrased by saying if $y_1 \neq y_2$, then $x_1 \neq x_2$.
   We will use this formulation, and a horizontal line test, in Sec. 6.6 on inverse functions.

Exercise 6.4   In Probs. 1 to 8, find the largest possible domain of the function.

1 $f(x) = \sqrt{5x + 1}$ **2** $f(x) = \sqrt[3]{2x + 3}$
3 $f(x) = \sqrt{x^2 + 1}$ **4** $f(x) = \sqrt{-6x + 5}$

5 $f(x) = \dfrac{4}{x^2 - 4}$ **6** $f(x) = \dfrac{|x|}{x}$

7 $f(x) = \dfrac{9}{x^2 + 9}$ **8** $f(x) = \dfrac{x + 3}{2x^2 + 11x + 15}$

In Probs. 9 to 12, find a value of x for which $f(x) = 5$.

9 $f(x) = 6x + 1$ **10** $f(x) = x^2$
11 $f(x) = \sqrt{x}$ **12** $f(x) = x^2 - 4x + 8$

In Probs. 13 to 20, does the equation define a function (with x in the domain)?

13 $5y = x^2 + x - 5$ **14** $15y + 4x^3 = 3x - 1$ **15** $x = y^4$

16 $x = y^3$ **17** $x^2 + y^4 = 10$ **18** $y = \sqrt[4]{x}$

19 $3y + 4/(x + 1) = 6$ **20** $y = \sqrt{x} + \sqrt[3]{x}$

Find the function values in Probs. 21 to 36.

21 If $f(x) = 2x + 3$, find $f(0)$ and $f(3)$.
22 If $f(x) = x + 5$, find $f(-4)$ and $f(5)$.
23 If $f(x) = 3x + 2$, find $f(-2)$ and $f(4)$.
24 If $f(x) = 5x - 4$, find $f(5)$ and $f(-4)$.
25 If $f(x) = x^2 - 5x + 2$, find $f(3)$ and $f(-1)$.
26 If $f(x) = x^2 + 6x - 4$, find $f(0)$ and $f(3)$.
27 If $f(x) = 2x^2 + 3x - 8$, find $f(7)$ and $f(-2)$.
28 If $f(x) = -3x^2 + 4x - 1$, find $f(3)$ and $f(-3)$.
29 If $f(x) = 3x - 7$, find $f(3) + f(h)$ and $f(3 + h)$.
30 If $f(x) = 2x + 9$, find $f(2) + f(h)$ and $f(2 + h)$.
31 If $f(x) = 2x^2 - 3x + 4$, find $f(1) + f(h)$ and $f(1 + h)$.
32 If $f(x) = 3x^2 + 4x - 5$, find $f(4) + f(h)$ and $f(4 + h)$.
33 If $f(x) = x^2 + x - 2$, find $f(x + h) - f(x)$.
34 If $f(x) = 2x^2 - 5x + 1$, find $f(x + h) - f(x)$.
35 If $f(x) = 5x^2 - 5x + 7$, find $f(x + h) - f(x)$.
36 If $f(x) = 4x^2 + 3x - 2$, find $f(x + h) - f(x)$.

In Probs. 37 to 56, let $f(x) = 5x - 1$, $g(x) = 2x^2 - 7$, $h(x) = 1/(x + 1)$, and find the following function values.

37 $(f + g)(1)$ **38** $(g - h)(0)$ **39** $(fh)(3)$
40 $(f/g)(2)$ **41** $f(1) - g(-1)$ **42** $g(3) + h(-0.9)$
43 $f(4)/g(3)$ **44** $f(-1)g(-2)h(-3)$ **45** $f(g(1))$
46 $g(f(1))$ **47** $g(h(0))$ **48** $h(g(0))$
49 $f(g(2))$ **50** $f(g(-2))$ **51** $h(f(2))$
52 $h(g(1))$ **53** $f(f(0))$ **54** $g(g(-2))$
55 $f(h(1)) + f(1) + h(1)$ **56** $f(g(3) + h(3))$

A function is called even if $f(-x) = f(x)$ and odd if $f(-x) = -f(x)$.

Even   **57** Show that $f(x) = x^3$ and $g(x) = x^5$ are odd.
and   **58** Show that $f(x) = x^4$ and $g(x) = x^{14}$ are even.
odd   **59** Let

$$h(x) = \frac{f(x) + f(-x)}{2}$$

Show that h is even no matter what function f is.

60 Show that for any function f, if $g(x) = \dfrac{f(x) - f(-x)}{2}$, then g is odd.

61 If $f(x, y) = 2x + 7y + 8$, find $f(2, 3)$ and $f(3, 2)$.

62 If $g(x, w) = 5x - 4w - 3$, find $g(0, 6)$ and $g(6, 0)$.

63 Find $f(1, 2, 3)$ if $f(x, y, z) = x + y^2 + 6/z$.

64 Find $h(0, 1, 0)$ if $h(x, t, u) = xt + tu + ux$.

Find $\dfrac{f(x + h) - f(x)}{h}$ in Probs. 65 to 72.

Calculus

65 $f(x) = 2x - 5$ **66** $f(x) = 2x/3$ **67** $f(x) = 6 - (x/5)$

68 $f(x) = 4/3 - 3x/8$ **69** $f(x) = x^2 + 3x - 1$ **70** $f(x) = 1/x$

71 $f(x) = -2x + 5x^2$ **72** $f(x) = 1/x^2$

In Probs. 73 to 76, graph the function.

73 $f(x) = 2x - 5$ **74** $f(x) = x^2 - 3x$

75 $f(x) = 1/x$ **76** $f(x) = |x - 4|$

Education

77 For a distribution of numbers, the z score is the number of standard deviations away from the mean. The z score and T score are related by

$$T = 10z + 50$$

Find (a) T if $z = 2.4$; (b) z if $T = 58$.

Education

78 If C is the CEEB score (College Board exam) and T is the T score on the test, then $C = 10T$. Find (a) C if $T = 42$; (b) T if $C = 540$.

Education

79 If A is the ACT score, C is the CEEB score, and T is the T score on a college entrance test, then $C = 10T$ and $C = 5A$.

(a) Express A as a function of T.

(b) Find A if $T = 61$.

Education

80 If Q is the Stanford-Binet intelligence quotient of a person, then

$$Q = 100 + 16z$$

and

$$T = 10z + 50$$

(a) Express Q as a function of T.

(b) Find the value of Q (the IQ score) if $z = 1.5$ (1.5 standard deviations above the average on the normal "bell-shaped" curve).

Management

81 The cost of a report is

$$T(x, y) = 4.16x + 0.43y + 0.012xy$$

where x is the number of pages in the report and y is the number of copies of the report. Find $T(x, y)$ if (a) $x = 48$ and $y = 155$, and (b) $x = 21$ and $y = 4618$.

Management

82 If cost $= C(x, y) = 8x + (2y/3) + (xy/15)$, find y as a function of x if the cost must be 1400.

Management

83 If C = total cost = production cost + holding cost, then

$$C(x, y, z) = x^2 + y^2 + z^2 + 2(x - 4) + 2(x + y - 16)$$

Find C if $x = 8$, $y = 9$, $z = 11$.

Business   **84** A store rents a rotary tiller for a fee of $5 plus $2 per hour. Express the cost as a linear function of the number of hours the tiller is used, and evaluate for 3.5 h of use.

Anthropology   **85** In the study of social stratification, we have the formula

$$S = \frac{mn(m + n + 1)}{12}$$

Find S for $m = 3n = 18$.

Agriculture   **86** The number of bees in a hive at the end of w weeks is $H(w) = 2700 + 1000w$ for $w = 0$ to $w = 8$, at which time the bees swarm. Find the number of bees in the hive at the end of 1, 2, 3, and 8 weeks.

Management   **87** In $f(x, y) = 2x^2 + \sqrt{3}xy + y^2 + 4$, let

$$x = \frac{\sqrt{3}}{2}t - \frac{u}{2} \quad \text{and} \quad y = \frac{t}{2} + \frac{\sqrt{3}}{2}u$$

and find f as a function of t and u.

Management   **88** The earnings of a drugstore, in units of $100,000, are

$$E(x, y) = 3x + 5y + 2xy + 5 - x^2 - y^2$$

where x is investment in inventory in millions of dollars and y is floor space in 10,000 ft². Find E if $x = \$7,000,000$ and $y = 40,000$ ft².

Education   **89** The Harris-Jacobsen readability formula for material below fourth grade level is

$$S = 0.094x + 0.168y + 0.502$$

where x is the percent of unique unfamiliar words and y is the average sentence length in words. Find S for $x = 7$ percent and $y = 8.3$.

Anthropology   **90** In the study of social stratification, the expected number of runs or clusters according to the Wald-Wolfowitz run test is

$$d = \frac{2mn}{m + n} + 1$$

(a) Calculate d if $m = 7$ and $n = 10$.
(b) Find m if $n = 12$ and $d = 11.3$.
(c) What is d when $n = 2m$?

Education   **91** The Fog index is $0.4(S + P)$, where S is the average sentence length in words and P is the percentage of words of 3 or more syllables. What is the Fog index if $S = 9.8$ and $P = 32$?

Economics   **92** Under certain assumptions, the utility u is related to income y by

$$u = \begin{cases} y/p - Ap & \text{if } y \leq np \\ n - Ay/n & \text{if } y \geq np \end{cases}$$

What is the maximum value of the utility u (all constants are positive)?

93 Let $f(x, y) = 2x^5 + 17x^2y^3 - 6xy^4$. Show that $f(tx, ty) = t^5 f(x, y)$.

94 Let $g(x, y) = (x^2 + 5y^2)^{1/3}$. Show that $g(tx, ty) = t^{2/3}g(x, y)$.

95 Let $h(x, y) = \dfrac{xy}{x^2 + y^2}$. Show that $h(x, y) = h(1, y/x)$.

96 Let $k(x, y) = (x^3 - y^3)(x^2 + y^2)^{-1}$. Show that $k(12, 18) = 6k(2, 3)$.

 Linear functions

The straight line is important because it describes many actual situations and can be used to approximate others. It occurs in the relationship between distance and speed, between simple interest and rate, and between lift on an airplane wing and density of air, to give just a few examples. Moreover, lines are the easiest type of functional relationship to deal with. The slope of a line (discussed later in this section) is simply the measure of one quantity relative to another.

With the help of analytic geometry, it can be proved that every line is the graph of an equation with the form

$$Ax + By + C = 0 \qquad (1)$$

where A, B, and C are real numbers. It is also true that every equation of the form (1) represents a line. We assume A and B are not both zero. It is important to distinguish between the cases $B = 0$ and $B \neq 0$.

$B = 0$ If $B = 0$, then we have $Ax + C = 0$ and

$$x = -\frac{C}{A} \qquad (2)$$

Hence x is a constant, and the graph is a vertical line. Some examples are given in Fig. 6.23.

The graph of $x = 0$ is the y axis. We see from the graphs that (2) is not a function because it fails the vertical line test.

$B \neq 0$ If $B \neq 0$ in (1), then we may solve (1) for y.

$$By = -Ax - C \qquad \text{transposing}$$

$$y = \left(\frac{-A}{B}\right)x - \frac{C}{B} \qquad \text{dividing by } B \qquad (3)$$

FIGURE 6.23

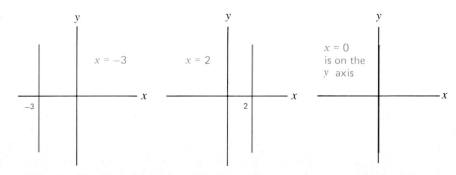

This equation defines a function. More generally

Linear function   A function f is **linear** if $f(x) = ax + b$ where a and b are real numbers.

Now in Eq. (3) we take any two distinct points (x_1, y_1) and (x_2, y_2). Then we may calculate the change in y, $y_2 - y_1$, as follows:

$$y_2 - y_1 = -\frac{A}{B}x_2 - \frac{C}{B} - \left(-\frac{A}{B}x_1 - \frac{C}{B}\right)$$

$$= -\frac{A}{B}x_2 + \frac{A}{B}x_1$$

$$= -\frac{A}{B}(x_2 - x_1)$$

Thus we have, for $x_1 \neq x_2$,

$$\frac{y_2 - y_1}{x_2 - x_1} = -\frac{A}{B} = \text{a constant}$$

Slope   This constant is usually written as m, and it is called the **slope** of the line. Notice that we may use any two distinct points on the line, and we may call either point (x_1, y_1) since

$$m = \frac{y_2 - y_1}{x_2 - x_1} = \frac{y_1 - y_2}{x_1 - x_2} \qquad (6.5)$$

Note   The value of the slope shows how much y changes in proportion to the change in x. Hence for a slope of $\frac{1}{2}$, if x changes by 6 then y changes by 3, and if x changes by 2 then y changes by 1. For a slope of -3, if x changes by 6 then y changes by -18, and if x changes by -1 then y changes by 3.

Slope 0   The slope of a horizontal line is zero since $y_1 = y_2$.

● **Example 1**   Find the slope of the line through the points (a) (2, 5) and (4, -1); (b) $(-1, 3)$ and (3, 5).

Solution   Using (6.5) gives

$$m = \begin{cases} \dfrac{5 - (-1)}{2 - 4} = \dfrac{6}{-2} = -3 & \text{for (a)} \\[2ex] \dfrac{3 - 5}{-1 - 3} = \dfrac{-2}{-4} = \dfrac{1}{2} & \text{for (b)} \end{cases}$$ ●

In Fig. 6.24, we show lines with different slopes. If the slope is positive, the line rises as we move to the right, whereas for a negative slope the line

No slope   falls. Vertical lines have no slope since $x_1 = x_2$ would mean division by 0. If (x_1, y_1) and (x_2, y_2) are specific points on a line and (x, y) is any other

FIGURE 6.24

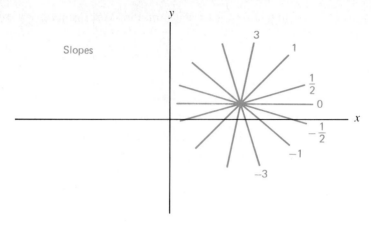

point on the line, then by (6.5) both

$$\frac{y_2 - y_1}{x_2 - x_1} \quad \text{and} \quad \frac{y - y_1}{x - x_1}$$

are expressions for the slope; hence they are equal. Thus we have

Two-point form
$$y - y_1 = \frac{y_2 - y_1}{x_2 - x_1}(x - x_1) \qquad (6.6)$$

which is called the *two-point form* of the equation of a line.

● **Example 2** Find the equation of the line through $(6, -4)$ and $(1, -1)$.

Solution
$$y - (-4) = \frac{-1 - (-4)}{1 - 6}(x - 6) \qquad \text{by (6.6) with } (x_1, y_1) = (6, -4)$$

$$y + 4 = \frac{3}{-5}(x - 6)$$

$$-5(y + 4) = 3(x - 6) \qquad\qquad \text{multiplying by } -5$$

$$-5y - 20 = 3x - 18$$

$$0 = 3x + 5y + 2 \qquad\qquad \text{combining terms} \qquad ●$$

By (6.5), we have $m = (y_2 - y_1)/(x_2 - x_1)$. Combining this with (6.6) gives us a new standard form for the equation of a straight line.

Point-slope form The *point-slope form* of the equation of a line is

$$y - y_1 = m(x - x_1) \qquad (6.7)$$

● **Example 3** Find the equation of the line through $(\frac{4}{3}, 4)$ with slope 6.

Solution Using $(x_1, y_1) = (\frac{4}{3}, 4)$ and $m = 6$ in (6.7) gives

$$y - 4 = 6(x - \tfrac{4}{3}) \qquad \text{by (6.7)}$$
$$y - 4 = 6x - 8 \qquad \text{multiplying}$$
$$0 = 6x - y - 4 \qquad \text{combining terms} \qquad ●$$

The y intercept is the y coordinate of the point where a line crosses the y axis. If we use $(x_1, y_1) = (0, b)$ in (6.7), we have

$$y - b = m(x - 0) = mx$$

$$y = mx + b \qquad (6.8)$$

where b is the y intercept.

Slope-intercept form   The *slope-intercept form* given in (6.8) is very useful in graphing. All we need to do is solve the equation of a line for y. Then the coefficient of x is the slope and the constant term is the y intercept.

● **Example 4**   Find the slope and y intercept of $2x + 3y + 6 = 0$.

Solution
$$
\begin{aligned}
2x + 3y + 6 &= 0 && \text{given} \\
3y &= -2x - 6 && \text{transposing} \\
y &= (-\tfrac{2}{3})x - 2 && \text{dividing by 3}
\end{aligned}
$$

Hence $m = -\tfrac{2}{3}$ is the slope and $b = -2$ is the y intercept.
The graph of $2x + 3y + 6 = 0$ is shown in Fig. 6.25.   ●

Parallel lines   Two lines $y = m_1x + b_1$ and $y = m_2x + b_2$ are *parallel* if and only if

$$m_1 = m_2 \qquad (6.9)$$

Although we will not prove this, it is evident geometrically when we recall from the definition of slope that if x changes by 1, then y changes by m_1 and m_2, respectively, for the two lines. Thus corresponding changes are equal when $m_1 = m_2$. See Fig. 6.26.

Perpendicular lines   We will now prove that $y = m_1x + b_1$ and $y = m_2x + b_2$ are *perpendicular* if and only if

$$m_1m_2 = -1 \qquad (6.10)$$

We assume here, of course, that $m_1 \neq 0$ and $m_2 \neq 0$.

FIGURE 6.25

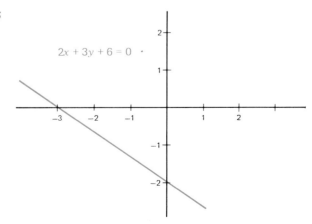

FIGURE 6.26

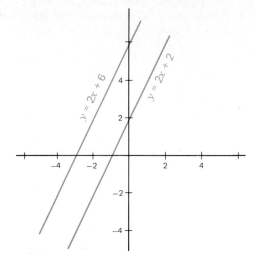

We may, in Fig. 6.27, assume without any loss of generality that the lines intersect at the origin—this only makes the notation easier to follow. In that case $b_1 = 0 = b_2$, and so (r, s) on the line $y = m_1x + b_1$ implies that $s = m_1r$. Similarly, $d = m_2c$, for (c, d) on the second line; hence

$$\frac{s}{r} = m_1 \quad \text{and} \quad \frac{d}{c} = m_2 \tag{4}$$

Suppose that the lines intersect at a right angle, and that (r, s) is on the first line and (c, d) is on the second line. Connecting the two points makes a right triangle, and using the pythagorean theorem gives

$$[(r - 0)^2 + (s - 0)^2] + [(c - 0)^2 + (d - 0)^2] = (r - c)^2 + (s - d)^2$$

$$r^2 + s^2 + c^2 + d^2 = r^2 - 2rc + c^2 + s^2 - 2sd + d^2$$

$$0 = -2rc - 2sd \qquad \text{combining terms}$$

$$2sd = -2rc \qquad \text{transposing}$$

$$\frac{s}{r} \cdot \frac{d}{c} = -1 \qquad \text{dividing by } 2rc$$

$$m_1m_2 = -1 \qquad \text{by (4)}$$

These steps may be reversed to show that if (6.10) holds, then the lines are perpendicular.

● **Example 5**   Classify the following pairs of lines as parallel, perpendicular, or neither:
  (a)   Through $(2, 5)$ and $(4, 9)$, and through $(3, -1)$ and $(6, 5)$
  (b)   Through $(4, 0)$ and $(2, -1)$, and through $(2, 5)$ and $(5, 11)$
  (c)   Through $(12, 5)$ and $(10, 4)$, and through $(-1, 0)$ and $(0, -2)$.

FIGURE 6.27

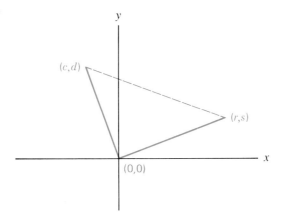

Solution   (a) $m_1 = \dfrac{9 - 5}{4 - 2} = \dfrac{4}{2} = 2,$ $m_2 = \dfrac{5 - (-1)}{6 - 3} = \dfrac{6}{3} = 2$ hence parallel

(b) $m_1 = \dfrac{0 - (-1)}{4 - 2} = \dfrac{1}{2},$ $m_2 = \dfrac{5 - 11}{2 - 5} = \dfrac{-6}{-3} = 2$ neither

(c) $m_1 = \dfrac{5 - 4}{12 - 10} = \dfrac{1}{2},$ $m_2 = \dfrac{-2 - 0}{0 - (-1)} = -2$

$m_1 m_2 = -1$ so perpendicular ●

Exercise 6.5   Find the slope of the line through the two given points.

1 (3, 7) and (2,1)   **2** (5, 4) and (2, 3)
3 (4, −2) and (0, 2)   **4** (−2, 6) and (3, 1)

Find the equation of the line through the two given points.

5 (5, 3) and (1, −1)   **6** (3, 0) and (1, −2)
7 (−6, 4) and (−2, 1)   **8** (5, 0) and (1, −3)
9 (4, 1) and (7, 3)   **10** (−5, −9) and (−3, −4)
11 (−3, 6) and (0, 1)   **12** (−3, −4) and (1, −2)

Find the equations of the lines in Probs. 13 to 20.

13 Through (4, 1) with slope 2   **14** Through (−2, 2) with slope −3
15 Through (0, 3) with slope $\frac{3}{4}$ **16** Through (−3, −5) with slope −$\frac{2}{3}$
17 Slope 5 and *y* intercept 3   **18** Slope −4 and *y* intercept −1
19 Slope −3 and *y* intercept $\frac{2}{3}$ **20** Slope −$\frac{3}{4}$ and *y* intercept −2

Find the slope and *y* intercept of the following lines.

21 $3x - y = 7$ **22** $5x + y = 1$ **23** $5x + 2y = 4$ **24** $2x + 3y = 6$

The equation of the line with a and b as x and y intercepts is

$$\frac{x}{a} + \frac{y}{b} = 1$$

This is called the *intercept form* of the equation of a straight line. Find the x and y intercepts of the following lines:

25 $\dfrac{x}{3} + \dfrac{y}{5} = 1$ **26** $\dfrac{x}{4} - \dfrac{y}{7} = 1$ **27** $\dfrac{2x}{3} + \dfrac{y}{8} = 1$ **28** $\dfrac{x}{4} + \dfrac{y}{3} = 2$

Are the following pairs of lines parallel, perpendicular, or neither?

29 $6x + 3y = 4$ **30** $4x - 11y = 8$ **31** $8x - 2y = 5$
   $2x + y = -5$ $11x + 4y = 3$ $x + 4y = 15$
32 $5x + 15y = 8$ **33** $-7x + 21y = 6$ **34** $8x + 5y = 3$
   $6x + 2y = 1$ $6x + 2y = 7$ $-8x + 5y = 6$
35 $-x + 6y = 18$ **36** $6x - 9y = 14$
   $2x - 12y = -9$ $-4x + 6y = -15$

Show that the three points in each problem are on the same line.

37 $(-2, -10), (1, -1), (2, 2)$ **38** $(-1, 8), (0, 3), (2, -7)$
39 $(-3, -7), (1, 3), (7, 18)$ **40** $(-3, 12), (-1, 5), (3, -9)$
41 What is the line through $(3, 2)$ with equal, nonzero intercepts?
42 Find the equation of the line through $(4, 5)$ which is perpendicular to the line $7x + 6y = -3$.
43 Find the equation of the line with the same y intercept as $2x + 5y = -25$ and twice the slope of $9x - 3y = 4$.
44 Find the equation of the line whose y intercept is twice its x intercept and is three times its slope.

The distance between the point (x_1, y_1) and the line $Ax + By + C = 0$ is

$$\frac{|Ax_1 + By_1 + C|}{\sqrt{A^2 + B^2}}$$

Use this to find the distance between the following points and lines:

45 $(3, 5); 2x + 4y - 1 = 0$ **46** $(4, -1); 3x + 7y - 12 = 0$
47 $(-3, 2); 5x + y = 7$ **48** $(5, 1); x - 3y = -8$

Graph the following lines by using the slope-intercept form:

49 $y = 3x + 1$ **50** $y = -2x - 3$ **51** $2y = x - 5$
52 $3y = -2x + 2$ **53** $6x + 5y + 14 = 0$ **54** $5x + 6y - 23 = 0$
55 $-4x + y + 6 = 0$ **56** $3x - 7y + 15 = 0$
57 If f is a linear function such that $f(4) = 7$ and $f(7) = 16$, find the equation for $f(x)$.

58 Find an equation for the perpendicular bisector of the line segment joining $(4, -3)$ and $(8, 5)$.

59 Find an equation for the altitude through A in the triangle with vertices $A(1, 5)$, $B(2, -1)$, and $C(5, 0)$.

60 Show that the equation of the line through $(a, 0)$ and $(0, b)$ can be put in the form

$$\frac{x}{a} + \frac{y}{b} = 1$$

In Probs. 61 to 64, use the distance formula to find an equation satisfied by all points $P(x, y)$ which are equidistant from the two given points. Notice that each equation is a straight line.

61 $(3, 8)$ and $(-1, 4)$ **62** $(11, -3)$ and $(4, 7)$
63 $(-2, 5)$ and $(-4, -14)$ **64** $(1, -9)$ and $(12, -4)$

6.6 Inverse functions

For the function $f(x) = 3x + 1$ or $y = 3x + 1$, we know how to find $f(4)$ by substitution: $f(4) = 3(4) + 1 = 13$. That is, given a value for x, we can find a value for y. However, suppose we have a value for y given, can we find x? For this function the answer is yes. In fact if we are given $y = -5$, we just solve the equation

$$-5 = 3x + 1$$
$$-6 = 3x$$
$$-2 = x$$

What we want to do in this section is investigate the general problem of solving $y = f(x)$ for x. Moreover, we want the solution to be a single, well-defined real number for each x. This solution will be called the **inverse function.**

● **Example 1**   Find the inverse of $y = 3x + 1$.

Solution   We solve for x by writing

$$y - 1 = 3x$$

$$\frac{y - 1}{3} = x$$ ●

We use the notation

$$f^{-1}(y) = \frac{y - 1}{3}$$

for the inverse function. Note that $f^{-1}(y)$ and $1/f(y)$ are different things entirely.

Since the normal functional notation has x as the independent variable, we write

$$f^{-1}(x) = \frac{x - 1}{3}$$

This leads to the following rule for finding the inverse of a function *if it exists*:

Inverse function (a) $y = f(x)$ is given.
(b) Interchange x and y, getting $x = f(y)$.
(c) Solve for y, and write the solution as $y = f^{-1}(x)$.

● **Example 2** Find the inverse function for

(a) $f(x) = \dfrac{2x + 1}{x - 3}$ (b) $f(x) = x^2 + 2x + 5$

Solution (a)

$$y = \frac{2x + 1}{x - 3} \qquad \text{given}$$

$$x = \frac{2y + 1}{y - 3} \qquad \text{interchange } x \text{ and } y$$

$$xy - 3x = 2y + 1$$

$$xy - 2y = 1 + 3x \qquad \text{transposing}$$

$$y = \frac{1 + 3x}{x - 2} \qquad \text{solving for } y$$

Thus the inverse function is

$$f^{-1}(x) = \frac{3x + 1}{x - 2}$$

(b) $y = x^2 + 2x + 5$ given

$x = y^2 + 2y + 5$ interchange x and y

$0 = y^2 + 2y + (5 - x)$

$$y = \frac{-2 \pm \sqrt{4 - 4(5 - x)}}{2} \qquad \begin{array}{l} \text{quadratic formula} \\ \text{with } a = 1, b = 2, c = 5 - x \end{array}$$

This gives two values, not just one value as we were seeking. Hence an inverse *function* does not exist, even though we were able to solve for y.   ●

The question thus arises: When can we find an inverse function to a given function $y = f(x)$? The answer is that an inverse function exists if and only if f is one-to-one. This means by definition that

One-to-one function   If $x_1 \neq x_2$, then $y_1 \neq y_2$ (1)

For a function, we require that each x correspond to exactly one y. For a function to be one-to-one, we require further that different x values correspond to different y values.

The condition in (1) is equivalent to

$$\text{If } y_1 = y_2, \text{ then } x_1 = x_2$$

But this is precisely what we need in order for the inverse to be a function; that is, each given y must correspond to exactly one x.

The most common way for a function to be one-to-one is for it to be increasing (or decreasing). Graphically this means that as we move from left to right, the graph rises or falls. Algebraically it means

Increasing $$\text{If } x_1 < x_2, \text{ then } y_1 < y_2$$

Decreasing $$\text{If } x_1 < x_2, \text{ then } y_1 > y_2$$

The function $y = f(x) = x^2$ is not one-to-one because different x values can produce the same y value:

$$\begin{aligned} x = -12 &\qquad \text{gives } y = (-12)^2 = 144 \\ x = 12 &\qquad \text{gives } y = 12^2 = 144 \end{aligned}$$

However we can restrict the domain to the interval $[0, \infty)$, thereby getting a new function, say, F. By definition $F(x) = x^2$ is defined only for $x \geq 0$. For these values of x,

$$x_1 < x_2 \qquad \text{implies} \qquad x_1^2 < x_2^2$$

since the latter is equivalent to $0 < x_2^2 - x_1^2 = (x_2 - x_1)(x_2 + x_1)$. But $x_2 - x_1 > 0$ since $x_1 < x_2$ is assumed, and $x_2 + x_1 > 0$ since the domain requires that $x_1 \geq 0$ and $x_2 > 0$. Hence $(x_2 - x_1)(x_2 + x_1)$ is the product of positive numbers. Thus F is increasing, hence one-to-one, and so has an inverse function F^{-1}. We find F^{-1} in the usual way:

$$\begin{aligned} y &= x^2 &\qquad (x \geq 0) \text{ definition of } F \\ x &= y^2 &\qquad \text{interchange } x \text{ and } y \text{ (now } y \geq 0) \\ \sqrt{x} &= y &\qquad \text{solve for } y \end{aligned}$$

Hence $F^{-1}(x) = \sqrt{x}$. Note that y is $+\sqrt{x}$, not $\pm\sqrt{x}$, since $y \geq 0$. It is true that $F^{-1}(F(x)) = x$ and $F(F^{-1}(x)) = x$ since

Note

$$\begin{aligned} F^{-1}(F(x)) &= F^{-1}(x^2) = \sqrt{x^2} = x &\qquad (\text{recall } x \geq 0) \\ F(F^{-1}(x)) &= F(\sqrt{x}) = (\sqrt{x})^2 = x \end{aligned}$$

This is always true for any one-to-one function f and its inverse function f^{-1}, and it could have been taken as the *definition* of the inverse function:

$$f(f^{-1}(x)) = x \qquad \text{and} \qquad f^{-1}(f(x)) = x \qquad (6.11)$$

f and f^{-1} are inverses   Thus f^{-1} undoes what f does. We may use (6.11) to check whether our solution for f^{-1} really is the inverse function of the function f.

● **Example 3**   For $f(x) = (x - 1)^3$:
(a) Show f is one-to-one.   (b)   Find $f^{-1}(x)$.
(c) Show (6.11) holds.   (d)   Graph $y = f(x)$ and $y = f^{-1}(x)$.

Solution   (a)   If $x_1 < x_2$ then $x_1 - 1 < x_2 - 1$, and also $(x_1 - 1)^3 < (x_2 - 1)^3$. Hence f is increasing, and thus one-to-one.

(b)
$$y = (x - 1)^3 \qquad \text{given}$$
$$x = (y - 1)^3 \qquad \text{interchanging } x \text{ and } y$$
$$\sqrt[3]{x} = y - 1$$
$$\sqrt[3]{x} + 1 = y = f^{-1}(x) \qquad \text{inverse function}$$

(c) $f(f^{-1}(x)) = f(\sqrt[3]{x} + 1) = [(\sqrt[3]{x} + 1) - 1]^3 = (\sqrt[3]{x})^3 = x$

$f^{-1}(f(x)) = f^{-1}[(x - 1)^3] = \sqrt[3]{(x - 1)^3} + 1 = x - 1 + 1 = x$

(d)   Here is a table of values for f.

x	-1	0	1	2	3
$f(x)$	-8	-1	0	1	8

A table of values for f^{-1} can be made from this by interchanging the x and $f(x)$ values. See Fig. 6.28.

x	-8	-1	0	1	8
$f^{-1}(x)$	-1	0	1	2	3

●

As the graphs show, the graph of $y = f^{-1}(x)$ is the reflection of the graph

FIGURE 6.28

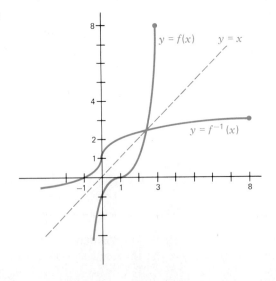

of $y = f(x)$ through the line $y = x$. This comes about because to find $f^{-1}(x)$, we interchanged x and y—recall symmetry from Sec. 6.3.

There is an easily applied graphical method to see whether a function is one-to-one, that is, to determine if an inverse function exists. A function is **Horizontal line test** one-to-one if **every horizontal line intersects the graph in at most one point.**

The vertical line test tells whether a graph is the graph of a function. The horizontal line test tells whether a function is one-to-one.

Domain and The domain of f^{-1} is the range of f. The range of f^{-1} is the domain of f. **range of f^{-1}** This is true because of the interchange of the x and y values.

In this section we have considered functions which are one-to-one, increasing, or decreasing in order to study inverse functions. We will now look at some other properties functions can have.

Even The function f is **even** if $f(-x) = f(x)$. Hence the graph is symmetric with respect to the y axis. Some examples are $f(x) = |x|$, $f(x) = x^2$, $f(x) = x^4$, $f(x) = 5x^6 - 18x^4 + 111x^2 - 200$, etc.

Odd The function f is **odd** if $f(-x) = -f(x)$. Such functions have graphs which are symmetric with respect to the origin. Some examples are $f(x) = x$, $f(x) = x^3$, $f(x) = x^5$, $f(x) = x^7 + 28x^5 - 14x^3 + 1111x$, and other polynomials involving only odd powers of x. Also $f(x) = 1/x$ is odd.

One reason for looking at even and odd functions is that every function can be written as the sum of an even function and an odd function. In fact

$$f(x) = \frac{f(x) + f(-x)}{2} + \frac{f(x) - f(-x)}{2} \qquad (2)$$
$$= \quad \text{even} \quad + \quad \text{odd}$$

See Probs. 59 and 60 in Exercise 6.4.

Continuous Another special type of function is a **continuous** function. A complete description of a continuous function comes in a calculus class. However, the graph of a continuous function can be drawn without lifting the pencil from the paper—there are no holes or gaps in the graph. Every polynomial is continuous. So is $f(x) = |x|$, even though it has a corner at the origin.

The last problem in Exercise 6.2 is to graph $f(x) = x - [x]$, where $[x]$ is **Periodic** the greatest integer $\leq x$. The graph is shown in Fig. 6.29, and the function is called periodic because its values repeat. It is true that $f(x + 1) = f(x)$ for all x. Note that this function is not continuous.

FIGURE 6.29

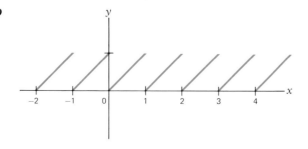

Exercise 6.6   Is the function in Probs. 1 to 8 one-to-one?

1 $y = \sqrt{x}$ 2 $y = x^2$ 3 $y = x^3$ 4 $y = \sqrt[3]{x}$
5 $y = |x|$ 6 $y = [x]$ 7 $y = x$ 8 $y = 1/x$

In Probs. 9 to 24, find the inverse of $y = f(x)$.

9 $y = 3x + 2$ 10 $y = 5x - 1$ 11 $y = -2x + 7$

12 $y = -4x - 3$ 13 $y = \dfrac{x + 4}{3}$ 14 $y = \dfrac{2x - 1}{7}$

15 $y = 3x - \frac{1}{4}$ 16 $y = \dfrac{x}{8} + 2$ 17 $y = (x + 1)^3$

18 $y = x^3 - 3$ 19 $y = x^5 + 4$ 20 $y = 2x^5 - 5$

21 $y = \dfrac{1}{x}$ 22 $y = \dfrac{4x}{3}$ 23 $y = \dfrac{1}{2x - 3}$ 24 $y = \dfrac{5}{3x + 1}$

25 If $f(x) = 3x - 1$, find $f(2)$ and $f^{-1}(2)$.
26 If $f(x) = 2x + 7$, find $f(-3.5)$ and $f^{-1}(-3.5)$.
27 If $f(x) = 4x - 3$, find $f(1)$ and $f^{-1}(1)$.
28 If $f(x) = -3x + 5$, find $f(3)$ and $f^{-1}(3)$.

In Probs. 29 to 32, let $f(x) = 3^x$. Find $f^{-1}(t)$ for the given value of t. Use the fact that if $f^{-1}(t) = w$, then $t = f(w)$.

29 $t = 3$ 30 $t = 81$ 31 $t = \frac{1}{9}$ 32 $t = \frac{1}{243}$

In Probs. 33 to 36, graph $y = f(x)$ and $y = f^{-1}(x)$. Verify the fact that the domain and range of f^{-1} are the range and domain, respectively, of f.

33 $y = 2x - 4$ for $0 \le x \le 6$ 34 $y = 1/x$ for $\frac{1}{8} \le x \le 8$
35 $y - 1 = (x + 1)^3$ for $-3 \le x \le 1$ 36 $y = \sqrt{x}$ for $0 \le x \le 9$

In Probs. 37 to 40, show that $f[f^{-1}(x)] = x$ and $f^{-1}[f(x)] = x$.

37 $f(x) = 3x + 5$ 38 $f(x) = \dfrac{3}{x}$

39 $f(x) = \dfrac{3}{x + 5}$ 40 $f(x) = \dfrac{1}{3x + 5}$

Show that the functions in Probs. 41 to 44 are increasing.

41 $f(x) = x^2 + 2x + 5$ on $(-1, \infty)$ 42 $f(x) = 3x + 5$
43 $f(x) = x^3 + 2x$ 44 $f(x) = -1/x$ for $x > 0$

45 Let $f(x) = \dfrac{x + 1}{x - 1}$. Show that f is decreasing on $(1, \infty)$.

46 Let $g(x) = -x + b$. Show that $g^{-1}(x) = g(x)$.
47 Let $f(x)$ be the remainder when x is divided by 3. What is the smallest positive value of k such that $f(x + k) = f(x)$ for all x?
48 Show that (a, b) and (b, a) are symmetric with respect to the line $y =$

x by showing that $y = x$ is the perpendicular bisector of the line segment joining (a, b) and (b, a).

The number p is called a **fixed point** of the function f if $f(p) = p$. Find at least one fixed point for each function in Probs. 49 to 52.

49 $f(x) = 4x - 9$ **50** $f(x) = \sqrt{x}$

51 $f(x) = 2x^2 - 2x - 20$ **52** $f(x) = \dfrac{4x + 8}{15x + 2}$

53 Let $h(x) = \dfrac{6x + 7}{5x - 6}$. Show that $h(h(4)) = 4$ and $h(h(-9)) = -9$.

54 Let $M(x) = \dfrac{x + 9}{x - 1}$. Show that $M^{-1}(6) = M(6)$.

55 Let $f(x) = \dfrac{ax + b}{cx - a}$. Show that $f(f(x)) = x$.

56 Let $g(x) = (1 - x^{2/3})^{3/2}$. Show that $g(g(x)) = x$, and that $g^{-1}(x) = g(x)$.

Variation

In many applications of mathematics to science, one of several basic relationships described by functions is the key. There is a special name for each of these types of functions.

Direct variation   **1**   The phrase y **varies directly** as x means

$$y = kx$$

The word "directly" is often omitted, hence "y varies as x" means "y varies directly as x." We also say y is directly proportional to x. If the weight w of a piece of pipe varies directly as its length L, then $w = kL$.

Inverse variation   **2**   The phrase y **varies inversely** as x means

$$y = \frac{k}{x}$$

This is also written y is inversely proportional to x. Boyle's law states that the volume V of a confined mass of gas at a constant temperature varies inversely as the pressure P. Hence $V = k/P$, or $PV = k$.

Joint variation   **3**   The phrase y **varies jointly** as x and w means

$$y = kxw$$

This can also be used for more than two quantities. Thus if the volume V of a box varies jointly as its length L, width W, and height H, then $V = kLWH$.

    4   If one number varies jointly as several others and inversely as still Combined variation   others, then the variation is referred to as a *combined variation*. Thus, if y varies jointly as x and z and inversely as w, then $y = kxz/w$.

Constant of variation   In each of the four types of variation defined above, k is called the *constant of variation*. The consant can be determined if a set of values for the variables is known. Thus if in the example given for direct variation, $w = 90$ for $L = 15$, then $90 = k(15)$ and $k = 6$; hence, $w = 6L$.

In variation problems, the actual constant of variation does not *need* to be calculated. In inverse variation, for instance, we can write $xy = k$. For two sets of values for x and y, we then have

$$x_1 y_1 = x_2 y_2$$

and information will be given so that only one unknown remains in this equation. However, actually finding the constant allows the function to be written.

A typical problem in variation involves a set of values for all the variables and a second set for all but one of the variables. After the variation has been expressed as an equation, the value of the constant of variation can be found by making use of the complete set of values of the variables as in the previous example. Finally, the value of the variable not included in the incomplete set of values can be determined by use of the incomplete set. If we want to find the value of w for $L = 17$, we need only substitute 17 for L in $w = 6L$ and thereby get $w = 6(17) = 102$.

Frequently situations involve relations that are combinations of the above types of variation. For example, Newton's law of gravitation states that the gravitational attraction between two bodies varies directly as the product of their masses and inversely as the square of the distance between their centers of gravity. If we let G, M, m, and d, respectively, represent the gravitational attraction, the two masses, and the distance, then the law states that

$$G = k\left(\frac{Mm}{d^2}\right)$$

● **Example 1**   The pressure on the bottom of a swimming pool varies directly as the depth. If the pressure is 624,000 lb/ft^2 when the water is 2 ft deep, find the pressure when it is $4\frac{1}{2}$ ft deep.

Solution   If we let P represent the pressure and d the depth, then P varies directly as d. Hence,

$$P = kd$$

Now, when $P = 624,000$ lb, $d = 2$ ft, then

$$624,000 = 2k$$

Hence,

$$k = 624,000/2$$
$$= 312,000$$

Thus,

$$P = (312,000)d$$

Consequently, if $d = 4\frac{1}{2}$ ft, then

$$P = (312,000)4\frac{1}{2}$$
$$= 1,404,000 \text{ lb}$$

● **Example 2**   The horsepower required to propel a ship varies as the cube of the speed. If the horsepower required for a speed of 15 knots is 10,125, find the horsepower required for a speed of 20 knots.

Solution   If we let

$$P = \text{required horsepower}$$
$$s = \text{speed, knots}$$

then, since P varies as s^3, we have

$$P = ks^3 \tag{1}$$

We are given that $P = 10,125$ for $s = 15$ knots. By substituting these values in (1), we get

$$10,125 = k(15^3)$$

and $$k = \frac{10,125}{15^3} = \frac{10,125}{3375} = 3$$

Now we substitute $k = 3$ and $s = 20$ in (1) and have

$$P = 3(20^3) = 3(8000)$$
$$= 24,000 \text{ hp}$$

● **Example 3**   If the volume of a mass of gas at a given temperature is 56 in³ when the pressure is 18 lb, use Boyle's law to find the volume when the pressure is 16 lb.

Solution   Boyle's law states that the volume varies inversely as the pressure. Hence, if we let

$$V = \text{the volume} \qquad \text{and} \qquad p = \text{the pressure}$$

we have

$$V = k\left(\frac{1}{p}\right) \qquad \text{by definition of inverse variation} \tag{2}$$

Thus, if $V = 56$ when p is 18, we have

$$56 = k(\tfrac{1}{18})$$

Therefore,

$$k = (56)(18)$$
$$= 1008$$

Now we substitute this value for k in (2) and get

$$V = 1008\left(\frac{1}{p}\right)$$

Hence, when $p = 16$, we have

$$V = 1008(\tfrac{1}{16}) = 63 \text{ in}^3$$

● **Example 4**   The weight of rectangular block of metal varies jointly as the length, the width, and the thickness. If the weight of a 12- by 8- by 6-in block of aluminum is 18.7 lb, find the weight of a 16- by 10- by 4-in block.

Solution   We let

$$W = \text{weight, lb}$$
$$l = \text{length, in}$$
$$w = \text{width, in}$$
$$t = \text{thickness, in}$$

Then, since the weight varies jointly as the length, width, and thickness, we have

$$W = klwt$$

When $l = 12$ in, $w = 8$ in, and $t = 6$ in, $W = 18.7$ lb. Therefore,

$$18.7 = k(12)(8)(6)$$
$$= 576k$$

and

$$k = \frac{18.7}{576}$$

On substituting $k = 18.7/576$, $l = 16$, $w = 10$, and $t = 4$ in the equation $W = klwt$, we obtain

$$W = \frac{18.7}{576}(16)(10)(4)$$

$$= 20.8 \text{ lb}$$

as the weight of the 16- by 10- by 4-in block. The reader should note that in the example k is the weight of 1 in³ of aluminum. ●

● **Example 5**   The amount of coal used by a steamship traveling at a uniform speed varies jointly as the distance traveled and the square of the speed. If a steamship uses 45 tons of coal traveling 80 miles at 15 knots, how many tons will it use if it travels 120 miles at 20 knots?

Solution   We let

$$T = \text{number of tons used}$$
$$s = \text{distance in miles}$$
$$v = \text{the speed in knots}$$

then

$$T = k(sv^2) \qquad \text{definition of joint variation} \qquad (3)$$

Hence, when $T = 45$, $s = 80$, and $v = 15$, we have

$$45 = k(80)(15^2)$$

Therefore,
$$k = \frac{45}{(80)(225)}$$
$$= 1/400$$

If we substitute this value for k in (3), we have

$$T = (1/400)(sv^2)$$

Now, when $s = 120$ and $v = 20$, it follows that

$$T = (1/400)(120)(20^2)$$
$$= 48{,}000/400 = 120 \text{ tons}$$

● **Example 6** The safe load of a beam with a rectangular cross section that is supported at each end varies jointly as the product of the width and the square of the depth and inversely as the length of the beam between supports. If the safe load of a beam 3 in wide and 6 in deep with supports 8 ft apart is 2700 lb, find the safe load of a beam of the same material that is 4 in wide and 10 in deep with supports 12 ft apart.

Solution We let

$$w = \text{width of beam, in}$$
$$d = \text{depth of beam, in}$$
$$l = \text{length between supports, ft}$$
$$L = \text{safe load, lb}$$

Then
$$L = \frac{kwd^2}{l}$$

According to the first set of data, when $w = 3$, $d = 6$, and $l = 8$, then $L = 2700$. Therefore,

$$2700 = \frac{k(3)(6^2)}{8}$$

$$21{,}600 = 108k$$

and
$$k = 200$$

Consequently, if $w = 4$, $d = 10$, $l = 12$, and $k = 200$, we have

$$L = \frac{200(4)(10^2)}{12}$$

$$= 6666\tfrac{2}{3}$$

Direct variation problems have equations such as $y = kx$, which can also be written as $y/x = k$ and stated as y is *directly proportional* to x. If one set of values for x and y is a and b, and another set is c and d, then $b/a = k = d/c$. This is equivalent to

$$\frac{a}{b} = \frac{c}{d} \qquad\qquad (4)$$

Proportion An equation such as this is called a **proportion.** The proportion (4) is equivalent to several others

$$\frac{a + b}{b} = \frac{c + d}{d} \qquad \frac{a - b}{b} = \frac{c - d}{d}$$

$$\frac{a + b}{a - b} = \frac{c + d}{c - d} \qquad \frac{a}{b} = \frac{a + c}{b + d}$$

The first proportion of these four follows from (4) merely by adding 1 to both sides since

$$\frac{a}{b} + 1 = \frac{a}{b} + \frac{b}{b} = \frac{a + b}{b}$$

and similarly

$$\frac{c}{d} + 1 = \frac{c + d}{d}$$

Exercise 6.7 Express the statements in Probs. 1 to 4 as equations.

1 m varies directly as n^2
2 s varies inversely as t
3 p varies jointly as q and r^3
4 w varies directly as x and inversely as the square of y
5 If y varies directly as x and is 10 when $x = 5$, find the value of y if $x = 7$.
6 If w varies directly as x and is 10 when $x = 5$, find the value of w if $x = 3$.
7 Given that y varies inversely as x. If $y = 3$ when $x = 4$, find the value of y when $x = 6$.
8 If w varies inversely as y and is equal to 2 when $y = 3$, find the value of w if $y = 6$.
9 If y varies jointly as x and w and is 30 when $x = 2$ and $w = 3$, find the value of y if $x = 4$ and $w = 5$.
10 Given that x varies jointly as w and y, and also that $x = 24$ when $w = 3$ and $y = 4$, find the value of x if $w = 4$ and $y = 5$.
11 Given that w varies directly as the product of x and y and inversely as the square of z. If $w = 9$ when $x = 6$, $y = 27$, and $z = 3$, find the value of w when $x = 4$, $y = 7$, and $z = 2$.

12 Suppose p varies directly as b^2 and inversely as s^3. If p is $\frac{3}{4}$ when $b = 6$ and $s = 2$, find b when p is 6 and s is 4.

Geometry **13** The volume of a right circular cylinder varies jointly as the height and the square of the radius. If the volume of a right circular cylinder of radius 4 in and height 7 in is 352 in³, find the volume of another of radius 8 in and height 14 in.

Physics **14** In comparing Celsius and Fahrenheit thermometers, it has been found that the Celsius reading varies directly as the difference between the Fahrenheit reading and 32°F. If a Celsius thermometer reads 100° when a Fahrenheit reads 212°, what will the Celsius read when the Fahrenheit reads 100°?

Engineering **15** The intensity of light varies inversely as the square of the distance from the source. Compare the intensity on a screen that is 5 ft from a given source with that on a screen 7 ft from the source.

Photography **16** The time necessary to make an enlargement from a photographic negative varies directly as the area. If 6 s are required for an enlargement that is 4 by 5 in, how many seconds are required to make an enlargement that is 8 by 10 in from the same negative?

Engineering **17** The amount of oil used by a ship traveling at a uniform speed varies jointly as the distance traveled and the square of the velocity. If a certain ship used 600 bbl of oil on a 300-mi trip at 20 knots, how much oil would it use on a trip of 1200 mi at 10 knots?

Engineering **18** How much oil would be used by the ship described in Prob. 17 on a trip of 200 mi at 30 knots?

Physics **19** The wind force on a flat vertical surface varies jointly as the area of the surface and the square of the wind velocity. If the pressure on 1 ft² is 1 lb when the wind velocity is 15 mi/h, find the force on an 8- by 10-ft sign in a storm with a wind velocity at 60 mi/h.

Navigation **20** On the ocean, the square of the distance in miles to the horizon varies as the height in feet that the observer is above the surface of the water. If a 6-ft man on a surfboard can see 3 mi, how far can he see if he is in a plane that is 1000 ft above the water?

Strength of a beam **21** The strength of a rectangular horizontal beam that is supported at the ends varies jointly as the width w and the square of the depth d and inversely as the length L. Compare the strengths of two beams if one of them is 20 ft long, 4 in wide, and 6 in deep and the other is 10 ft long, 2 in wide, and 4 in deep.

Engineering **22** The horsepower required to propel a ship varies as the cube of its speed. Find the ratio of the power required at 14 knots to that required at 7 knots.

Physics **23** The kinetic energy of a body varies as the square of its velocity. Find the ratio of the kinetic energy of a car at 20 mi/h to that of the same car at 50 mi/h.

Physics **24** As a body falls from rest, its velocity varies as the time in flight. If the velocity of a body at the end of 2 s is 64.4 ft/s, find the velocity at the end of 5 s.

Physics   **25** If a body is above the surface of the earth, its weight varies inversely as the square of the distance of the body from the center of the earth. If a man weighs 160 lb on the surface of the earth, how much will he weigh 200 mi above the surface? Assume the radius of the earth is 4000 mi.

Engineering **26** The current I varies as the electromotive force E and inversely as the resistance R. If in a system a current of 20 A flows through a resistance of 20 Ω with an electromotive force of 100 V, find the current that 150 V will send through the system.

Physics   **27** The illumination produced on a surface by a source of light varies directly as the candlepower of the source and inversely as the square of the distance between the source and the surface. Compare the illumination produced by a 512-candela lamp that is 8 ft from a surface with that of a 72-candela lamp 2 ft from the surface.

Physics   **28** If other factors are equal, the centrifugal force on a circular curve varies inversely as the radius of the turn. If the centrifugal force is 18,000 lb for a curve of radius 50 ft, find the centrifugal force on a curve of radius 150 ft.

            29 If y varies jointly as x and w and is 72 for $x = 9$ and $w = 4$, find the value of y for $x = 18$ and $w = 1.5$.

Economics   **30** The simple interest earned in a given time varies jointly as the principal and the rate. If $300 earned $45 at a 6 percent rate, how much would be earned by $500 at 5 percent in the same time?

Volume   **31** The volume of a regular pyramid varies jointly as the altitude and the area of the base. If the volume of a regular pyramid of altitude 5 in and base 9 in^2 is 15 in^3, find the volume of another of altitude 4 in and base area 6 in^2.

Engineering **32** The crushing load of a circular pillar varies as the fourth power of the diameter and inversely as the square of the height of the pillar. If 256 tons is needed to crush a pillar 8 in in diameter and 20 ft high, find the load needed to crush a pillar 10 in in diameter and 15 ft high.

Engineering **33** The mechanical advantage of a jackscrew varies directly as the length of the lever arm. If the mechanical advantage of a jackscrew is 192 when a 3-ft lever arm is used, what is the mechanical advantage for a 2-ft arm?

Aerodynamics **34** If the other factors are fixed, the lift on a wing of a plane varies with the density of the air. If the density of air at sea level is 0.08 lb/ft^3 and the lift on a wing is 2500 lb, find the lift at such an altitude that the density is 0.06 lb/ft^3.

Physics   **35** For a given load, the amount a wire stretches varies inversely as the square of the diameter. If a wire with a diameter of 0.6 in is stretched 0.006 in by a given load, how much will a wire of the same material with a diameter of 0.2 in be stretched by the same load?

Engineering **36** Under the same load, the sag of beams of the same material, length, and width varies inversely as the cube of the thickness. If a beam 4 in thick sags $\frac{1}{64}$ in when a load is placed on it, find the sag of a beam 2 in thick under the same load.

Education   **37** The Wechsler score W is $W = 10 + 3z$.
              (a) Find z in terms of W.
              (b) Find z if $W = 8$.

Earth science **38** The integrated intensity reflected by a perfect crystal is

$$I = \frac{L^2F\sqrt{0.69/\pi}}{0.64V}$$

for an angle of $20°$. What is the effect on I if L is doubled and V is halved?

Biology   **39** The volume V of the body of any animal is given by $V = kL^3$, where L is one of its linear dimensions. Find the relative body size of two animals if their lengths are 12 and 21 in.

Nutrition   **40** If a child is 110 cm tall on her thirteenth birthday and 120 cm tall on her fourteenth, what is the constant monthly growth rate?

 Key concepts

Be certain that you understand and can use the following important words and ideas.

Ordered pair   Parabola
Relation   Circle
Domain   Ellipse
Range   Hyperbola
Function   One-to-one function
Function value   Inverse function
Composite function   Increasing function
Coordinate axes   Decreasing function
Origin   Vertical line test for a function
Graph   Horizontal line test for one-to-one function
Symmetry   Graph of inverse function
Translation   Direct variation
Straight line   Inverse variation
Slope   Joint variation
Equations of a line

(6.1) $(a, b) = (c, d)$ if and only if $a = c$ and $b = d$

(6.2)   Distance $= \sqrt{(x_2 - x_1)^2 + (y_2 - y_1)^2}$

(6.3)   Midpoint $= \left(\dfrac{a + c}{2}, \dfrac{b + d}{2}\right)$

(6.4) $(x - h)^2 + (y - k)^2 = r^2$

(6.5) Slope $= m = \dfrac{y_2 - y_1}{x_2 - x_1}$

(6.6) $y - y_1 = \dfrac{y_2 - y_1}{x_2 - x_1}(x - x_1)$

(6.7) $y - y_1 = m(x - x_1)$

(6.8) $y = mx + b$

(6.9) $m_1 = m_2$ for parallel lines

(6.10) $m_1 m_2 = -1$ for perpendicular lines

(6.11) $f(f^{-1}(x)) = x = f^{-1}(f(x))$ for inverse functions

Exercise 6.8 review

1 Does the equation define a function of x? (a) $x^2 + y^2 = 6$; (b) $x + y^2 = 6$; (c) $x^2 + y = 6$; (d) $x + y = 6$.

2 If $f(x) = 3x - 4$, find $f(2)$, $f(-2)$, and $f(2t)$.

3 If $g(x) = x^2 - 4x + 1$, find $g(3)$, $g(1)$, and $g(g(0))$.

4 If $f(x) = 4x - 1$ and $g(x) = 1/(x^2 + 1)$, find $f(g(0))$ and $g(f(0))$.

5 Find the distance between the points $(8, 13)$ and $(-12, -8)$.

6 Are the points $(2, 1)$, $(-1, 5)$, and $(5, -2)$ all on the same line?

7 Find the equation of the line through $(1, -4)$ and $(3, 1)$.

8 Find the equation of the line through $(3, -7)$ and parallel to $6x + 2y = 1$.

9 Find the equation of the line with x intercept 5 which is perpendicular to $7x - 2y = 13$.

10 Find the distance from the point $(4, 1)$ to the line $3x - 4y + 7 = 0$.

11 Is $y = \sqrt{25 - x^2}$ one-to-one?

12 If $f(x) = 4x - 3$, find $f(-1)$ and $f^{-1}(-1)$.

13 Find $f^{-1}(x)$ if $f(x) = \dfrac{2x - 1}{x + 3}$.

14 Show that if $f(x) = -x + 5$, then $f^{-1}(x) = f(x)$.

15 If $f(x) = 6x - 5$, show that $(f^{-1})^{-1} = f$ by finding $f^{-1}(x)$ and then finding the inverse of $f^{-1}(x)$.

16 Find the midpoint of the line segment which joins $(\frac{4}{5}, \frac{2}{3})$ and $(-\frac{1}{3}, \frac{2}{5})$.

17 Show that in any triangle, the line joining the midpoints of two sides is parallel to the third side.

Graph the following equations and functions.

18 $y = x + |x| + 1$

19 $x = y^2 - 4y + 3$

20 $y = -5x + 1$ and $y = x/5 + 1$

21 $x = 3y$ and its inverse

22 $y = x^3 + 1$ and its inverse

23 $y + 1 = -4(x - 2)^2$

24 $y = (x - 1)^4$ and $y - 1 = x^4$

25 $x = y^2 + 1$ and $x = (y + 1)^2$

In Probs. 26 to 37, tell what letter of the alphabet the graph most looks like.

26 $(x - 4)^2 + (y - 5)^2 = 9$ **27** $x = 1$ **28** $y = 2(x - 4)^2$

29 $y = 4x^4 - 65x^2 + 16$ **30** $y = |x + 5|$ **31** $|y + 2| = |x - 3|$

32 $-y = 4x^4 - 65x^2 + 16$ **33** $x = y^4 - 26y^2 + 25$

34 $x + 3 = (y + 5)^2$ **35** $y = x^3 - 16x$

36 $x = y^3 - 9y$ **37** $-x = y^3 - 25y$

38 Show that the graph of $y = ax^2 + bx + c = f(x)$ is symmetric about the vertical line $x = -b/2a$ by proving that for any $h > 0$

$$f\left(-\frac{b}{2a} + h\right) = f\left(-\frac{b}{2a} - h\right)$$

39 Show that the graph of $y = ax^3 + bx^2 + cx + d = f(x)$ is symmetric about the point $(-b/3a, f(-b/3a))$ by showing that for any $h > 0$

$$\frac{f\left(-\frac{b}{3a} + h\right) + f\left(-\frac{b}{3a} - h\right)}{2} = f\left(-\frac{b}{3a}\right) \qquad (*)$$

Another way to express (*) is to write

$$g(h) = -g(-h)$$

where $g(h) = f\left(-\frac{b}{3a} + h\right) - f\left(-\frac{b}{3a}\right)$.

Education **40** The Dale-Chall reading score is

$$R = 0.1579x + 0.0496y + 3.6365$$

where x is the percentage of words outside the Dale list of 3000 easy words and y is the average sentence length in words. Find R if $x = 28$ and $y = 11.3$.

Education **41** The Flesch readability formula is

$$r = 206.8 - 0.85W - 1.015S$$

where W is the average number of syllables per 100 words and S is the average number of words per sentence. Calculate r if $W = 178$ and $S = 8.3$.

Psychology **42** Weber's law in psychology states that $D/S = c$, where D is the minimum amount of stimulus change required to produce a sensation difference, S is the magnitude of the stimulus, and c is a constant. Calculate D if $S = 4000$ and $c = 0.20$.

Agriculture **43** The amount y of potassium absorbed by *Zea mays* (corn) leaf is a linear function of time t up to 4 h. Thus $y = at$, where a is called the rate of absorption and is 18 μmoles per unit of leaf in darkness and 4.0 μmoles in a light intensity of 2×10^4 lumens/m². How long in darkness is required for the same absorption by a leaf as in 1 h of 2×10^4 lumens/m² of light?

Photography   **44** The exposure time necessary to obtain a good negative varies directly as the square of the f numbers of the camera shutter. If $\frac{1}{25}$ s is required when the shutter is set at $f/16$, what exposure is required under the same conditions if the shutter is set at $f/8$?

Physics   **45** The force of attraction between two spheres varies directly as the product of their masses and inversely as the square of the distance between them. Compare the attraction between two bodies with masses m_1 and m_2 that are separated by the distance d with that between two other bodies of masses $2m_1$ and $8m_2$ that are separated by the distance $4d$.

Energy   **46** If the kinetic energy of a body varies as the square of its velocity, compare the kinetic energy of a car traveling at 10 mi/h with that of the same car traveling at 50 mi/h.

Motion of the planets   **47** One of Kepler's laws states that the square of the time required by a planet to make one revolution about the sun varies directly as the cube of the average distance of the planet from the sun. If Mars is $1\frac{1}{2}$ times as far from the sun, on the average, as the earth, find the approximate length of time required for it to make a revolution about the sun. Use 1 year = 365 days.

Satellites   **48** The gravitational attraction of the earth for an object varies inversely as the square of the distance of the object from the center of the earth. If a space satellite weighs 100 lb on the earth's surface, how much does it weigh when it is 1000 mi from the earth's surface? (Assume the radius of the earth to be 4000 mi.)

CHAPTER 7·POLYNOMIAL FUNCTIONS AND CONICS

7.1 Quadratic functions

Polynomial functions are functions of the form

$$f(x) = a_nx^n + a_{n-1}x^{n-1} + \cdots + a_2x^2 + a_1x + a_0$$

Degree   where $a_n, \ldots, a_1, a_0$ are real numbers. If $a_n \neq 0$, then the **degree** of the polynomial is n.

We studied linear functions $f(x) = ax + b$, the case $n = 1$, in Sec. 6.5. We will consider polynomials with $n > 2$ in Sec. 7.2 and Chap. 11.

Quadratic functions   In this section we will consider the case $n = 2$, that is, **quadratic functions**

$$f(x) = ax^2 + bx + c$$

Parabola   with $a \neq 0$. The graph of this polynomial function is the same as the graph of the equation $y = ax^2 + bx + c$, which is called a **parabola**. Parabolas are polynomials as well as conics; see Sec. 7.4.

The most basic aid in graphing a parabola is knowing whether $a > 0$ (graph opens up) or $a < 0$ (graph opens down). The two prototypes are $y = x^2$ and $y = -x^2$. Their graphs, obtained by plotting points from a table of values, are shown in Figs. 7.1 and 7.2. Notice that either is the reflection of the other through the x axis.

x	-3	-2	-1	0	1	2	3
$y = x^2$	9	4	1	0	1	4	9

Each graph is symmetric about the y axis since $(-x)^2 = x^2$. The extreme point is $(0, 0)$; it is the lowest or highest point for the graphs.

The graphs of $y = 3x^2$ and $y = \frac{1}{4}x^2$ are shown in Figs. 7.3 and 7.4, along with $y = x^2$. These show the effect of changing the coefficient a in $y = ax^2$.

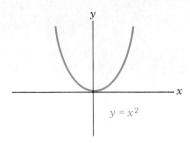

FIGURE 7.1   **FIGURE 7.2**

The graph of $y = ax^2$ can be translated by replacing x by $x - h$ and y by $y - k$, as in Sec. 6.3. The graph of

Standard form of a parabola

$$y - k = a(x - h)^2 \qquad\qquad (7.1)$$

Vertex

is shown in Fig. 7.5 in the case $a > 0$. The lowest or minimum point is (h, k), and it is called the **vertex.** If $a < 0$, the vertex is the highest point because the graph opens down.

If $f(x) = ax^2 + bx + c$, then we may put it in standard form by completing the square. The graph will be a parabola.

● **Example 1**   For the equation $y = 2x^2 + 4x + 5$, put it in standard form and draw the graph.

Solution

$$y - 5 = 2x^2 + 4x \qquad\qquad \text{transposing } 5$$
$$y - 5 = 2(x^2 + 2x) \qquad\qquad \text{factoring coefficient of } x^2$$
$$y - 5 = 2[(x^2 + 2x + 1) - 1] \qquad \text{adding and subtracting the square}$$
$$\qquad\qquad\qquad\qquad\qquad\qquad\qquad \text{of } \tfrac{1}{2} \text{ the coefficient of } x$$
$$y - 5 = 2[(x + 1)^2 - 1] \qquad\qquad \text{factoring as a square}$$
$$y - 5 = 2(x + 1)^2 - 2 \qquad\qquad \text{multiplying}$$
$$y - 3 = 2(x + 1)^2 \qquad\qquad\qquad \text{transposing } -2$$

The graph is thus a parabola with vertex at $(-1, 3)$, and its graph is in Fig. 7.6. When $x = 0$, $y = 5$. Since $a = 2 > 0$, the graph opens up.   ●

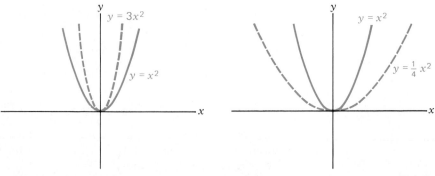

FIGURE 7.3   **FIGURE 7.4**

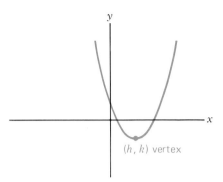

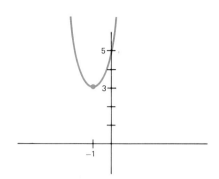

FIGURE 7.5   **FIGURE 7.6**

The graph in Example 1 is symmetric about the vertical line $x = -1$. One way to see this is to first write $y = f(x) = 2(x + 1)^2 + 3$, then show that $f(-1 + t) = f(-1 - t)$ for any $t > 0$. In fact

$$f(-1 - t) = 2(-1 - t + 1)^2 + 3 = 2(-t)^2 + 3 = 2t^2 + 3$$
$$f(-1 + t) = 2(-1 + t + 1)^2 + 3 = 2(t)^2 + 3 = 2t^2 + 3$$

The symmetry about the line $x = -1$ follows since $-1 - t$ and $-1 + t$ are both $|t|$ units from the line $x = -1$.

Of course, we can find the points where the graph crosses the x axis by solving the quadratic equation $ax^2 + bx + c = 0$ by using the quadratic formula.

$$x = \frac{-b \pm \sqrt{b^2 - 4ac}}{2a}$$

Note Since we are interested only in real values, we see that if $b^2 - 4ac < 0$, then the graph does not cross the x axis.

We saw in Sec. 5.3 that the maximum or minimum point of the graph of $f(x) = ax^2 + bx + c$ occurs for

$$x = \frac{-b}{2a}$$

This value is the average of the x intercepts. The maximum or minimum value is $f(-b/2a)$.

We now summarize the previous information about quadratic functions $f(x) = ax^2 + bx + c$ and their graphs:

Parabola Opens up if $a > 0$ and opens down if $a < 0$
Vertex $(-b/2a, f(-b/2a))$
Axis of symmetry The vertical line $x = -b/2a$
x intercepts Solve $ax^2 + bx + c = 0$ by quadratic formula:
    Two intercepts if $b^2 - 4ac > 0$
    One intercept if $b^2 - 4ac = 0$
    No intercepts if $b^2 - 4ac < 0$
y intercept $(0, c)$

FIGURE 7.7

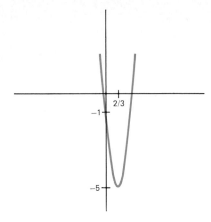

● **Example 2**   Graph $y = 9x^2 - 12x - 1$.

Solution   This is a parabola which opens up since $a = 9 > 0$. The axis of symmetry is the vertical line $x = -b/2a = -(-12)/2(9) = \frac{2}{3}$. The minimum value is

$$f\left(\tfrac{2}{3}\right) = 9\left(\tfrac{2}{3}\right)^2 - 12\left(\tfrac{2}{3}\right) - 1$$
$$= 9\left(\tfrac{4}{9}\right) - 8 - 1 = 4 - 9 = -5$$

The solutions of $9x^2 - 12x - 1 = 0$ are

$$x = \frac{-(-12) \pm \sqrt{144 - 4(9)(-1)}}{18} = \frac{12 \pm \sqrt{180}}{18} = \frac{2 \pm \sqrt{5}}{3}$$

These values are aproximately 1.4 and -0.1. The y intercept is $(0, -1)$. The graph is given in Fig. 7.7.   ●

● **Example 3**   Graph $y = -4x^2 + 8x - 5$.

Solution   The axis of symmetry is $x = -b/2a = -8/2(-4) = 1$. The parabola opens down since $a = -4 < 0$. The maximum value is

$$f(1) = -4 + 8 - 5 = -1$$

Since $b^2 - 4ac = 64 - 4(-4)(-5) = 64 - 80 < 0$, there are no x intercepts. The y intercept is $(0, -5)$, and the graph is in Fig. 7.8.   ●

For equations $x = ay^2 + by + c$, x and y have been interchanged and the graph opens to the right if $a > 0$ and to the left if $a < 0$.

● **Example 4**   Graph $x = y^2$ and $y = \sqrt{x}$.

Solution   The graph of $x = y^2$ is just the graph of $y = x^2$ reflected through the line $y = x$. The graph of $y = \sqrt{x}$ is the *top half* of the graph of $x = y^2$ since $\sqrt{x} \geq 0$ means y cannot be negative. See Fig. 7.9.   ●

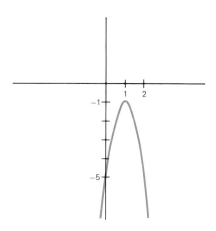

FIGURE 7.8   **FIGURE 7.9**

● **Example 5**   Graph $x - 3 = -(y - 2)^2$.

Solution   This is a parabola which opens to the left, and the vertex is (3, 2). The axis of symmetry is the *horizontal* line $y = 2$. When $y = 0$, we have $x = -1$. If $x = 0$, then $(y - 2)^2 = 3$ and so $y = 2 \pm \sqrt{3}$. These values are approximately 3.7 and 0.3. The graph is in Fig. 7.10.   ●

Uses of parabolas   Parabolas are used in making flashlights and headlights for cars and "dishes" that receive signals from satellites. A cross section through the extreme point is a parabola. The reason parabolas are used is that rays which come in parallel to the axis of symmetry are all reflected off the surface and then pass through a common point (called the focus). This same property makes them useful in picking up sound from a distance—say, sideline microphones at a football game being televised.

         Another use for parabolas is in describing the path of a projectile if we ignore air resistance.

FIGURE 7.10

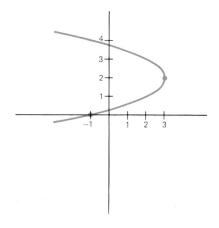

Exercise 7.1   Draw the graph of the parabola in Probs. 1 to 20.

1 $y = 2x^2$ **2** $y = -3x^2$ **3** $y = 6x^2$ **4** $y = -4x^2$

5 $y = \dfrac{x^2}{2}$ **6** $y = \dfrac{x^2}{5}$ **7** $y = \dfrac{-x^2}{3}$ **8** $y = \dfrac{-x^2}{7}$

9 $y = (x + 3)^2$ **10** $y = -(x - 4)^2$ **11** $y = -(x + 1)^2$

12 $y = (x - 2)^2$ **13** $y - 2 = -(x + 1)^2$ **14** $y + 1 = -(x - 1)^2$

15 $y - 1 = (x + 3)^2$ **16** $y + 2 = (x - 3)^2$ **17** $y - 2 = 2(x - 3)^2$

18 $y + 3 = 3(x - 1)^2$ **19** $y - 4 = -3(x + 1)^2$ **20** $y + 1 = -2(x - 2)^2$

In Probs. 21 to 28, find the axis of symmetry and the vertex.

21 $y = 3x^2 + 6x - 5$ **22** $y = -2x^2 + 8x - 3$

23 $y = -x^2 + 4x + 1$ **24** $y = 5x^2 - 10x + 6$

25 $x = y^2 + 2y + 3$ **26** $x = 2y^2 + 12y + 7$

27 $x = -2y^2 - 8y - 5$ **28** $x = 3y^2 - 6y - 8$

In Probs. 29 to 32, write the equation in the standard form $y - k = a(x - h)^2$.

29 $y = 2x^2 + 8x + 11$ **30** $y = 3x^2 + 30x + 71$

31 $y = 5x^2 - 10x + 2$ **32** $y = 4x^2 - 16x + 22$

Draw the graphs in Probs. 33 to 40.

33 $y = 2x^2 - 4x + 6$ **34** $y = -6x^2 - 12x - 11$

35 $y = 3x^2 + 18x + 30$ **36** $y = 2x^2 - 12x + 16$

37 $x = -3y^2 - 6y + 2$ **38** $x = 2y^2 - 8y + 4$

39 $x = 4y^2 + 16y + 19$ **40** $x = -y^2 + 6y - 10$

If a projectile is thrown or shot up with a velocity v m/s from a height H m above ground, its height after t s is

$$f(t) = -10t^2 + vt + H$$

In Probs. 41 to 44, find the maximum height of the projectile and the number of seconds before it hits the ground.

41 $f(t) = -10t^2 + 40t + 25$ **42** $f(t) = -10t^2 + 52t + 17$

43 $f(t) = -10t^2 + 36t + 42$ **44** $f(t) = -10t^2 + 64t + 14$

45 For which values of x is $f(x) = ax^2 + bx + c$ increasing if $a > 0$?

7.2   Graphs of polynomial functions

We will graph polynomial functions

$$y = a_n x^n + a_{n-1} x^{n-1} + a_{n-2} x^{n-2} + \cdots + a_1 x + a_0 \qquad (1)$$

in this section, where the coefficients a_i are real numbers. For $n = 1$, we

FIGURE 7.11

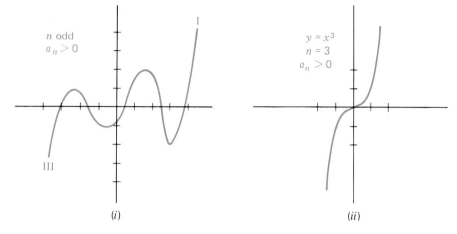

(i) (ii)

Plot points   know that the graphs are straight lines, and for $n = 2$ they are parabolas. In general we will be forced to *plot points*. All polynomial graphs have as their domain the set of real numbers.

Note   In addition to plotting points, the two most helpful things to know are *whether n is odd or even, and whether the leading coefficient is positive or negative*. The basic reason for this is that, for large values of $|x|$ (x either positive or negative), the main term in the polynomial of Eq. (1) is $a_n x^n$. This allows us to tell whether y is positive or negative for large $|x|$, and thus which quadrant that part of the graph is in. See Figs. 7.11 to 7.14 for typical graphs in these four basic cases. The part (ii) graphs give the prototypes $y = x^3$, $y = -x^3$, $y = x^2$, and $y = -x^2$.

If n is odd, then

n odd   (a)   The range (set of all y values) is all real numbers.

FIGURE 7.12

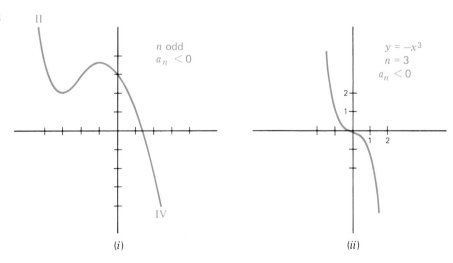

(i) (ii)

FIGURE 7.13

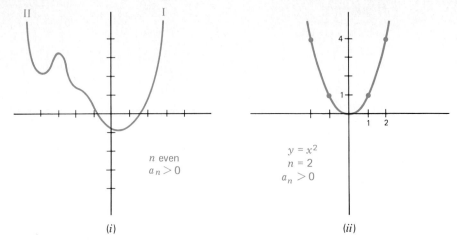

(b) For very large values of $|x|$, y has opposite signs when x is positive and when x is negative.

On the other hand, if n is even, then

n even (c) For $a_n > 0$, the range is $\{y \mid y \geq m\}$ for the minimum value m. For $a_n < 0$, the range is $\{y \mid y \leq M\}$ for the maximum value M.

(d) For very large values of $|x|$, y has the same sign when x is positive and when x is negative.

All polynomial graphs are continuous—they have no breaks or abrupt changes. If the polynomial has degree n, its graph crosses the x axis at most n times (in fact, it crosses *any* horizontal line at most n times). And if the degree is n, it has at most $n - 1$ extreme points (maximum or minimum points).

To graph properly and to fully use all possible information, calculus must

FIGURE 7.14

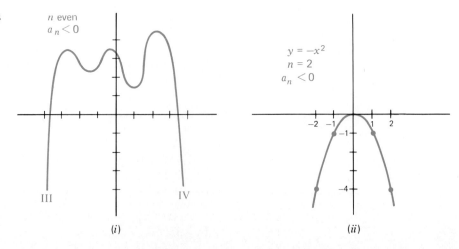

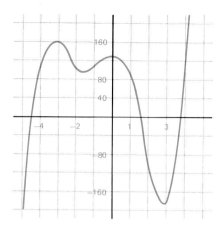

FIGURE 7.15   **FIGURE 7.16**

be used. We will content ourselves with a quick, accurate sketch of the graph.

● **Example 1**   Graph (a) $y = x^5 + 2x^4 - 17x^3 - 34x^2 + 15x + 132$
          (b) $y = (x - 2)^5 + 2(x - 2)^4 - 17(x - 2)^3 - 34(x - 2)^2$
$$+ 15(x - 2) + 132$$

Solution   (a)   Here n is odd and a_n is positive; so the graph resembles Fig. 7.11. See Fig. 7.15.

x	-5	-4	-3	-2	-1	0	1	2	3	4	5
y	-543	104	159	102	101	132	99	-46	-183	96	1607

(b)   If we translate the graph of (a) two units to the right, we get the graph of (b). See Fig. 7.16.   ●

● **Example 2**   Graph (a) $y = -8x^3 + 36x^2 - 54x + 11$
          (b) $x = -8y^3 + 36y^2 - 54y + 11$

Solution   (a)   In this case, n is odd and a_n is negative; so the graph resembles Fig. 7.12. See Fig. 7.17.

x	-1	0	1	2	3	4
y	109	11	-15	-17	-43	-141

(b)   If we reflect the graph of (a) through the line $y = x$, we get the graph of (b). See Fig. 7.18.   ●

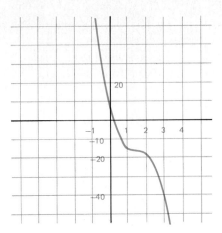

FIGURE 7.17   **FIGURE 7.18**

● **Example 3**   Graph (a) $y = x^6 - 10x^4 + 9x^2$
            (b) $y = -x^6 + 10x^4 - 9x^2$

Solution   (a)   For this function, n is even and $a_n > 0$; so we refer to Fig. 7.13. The graph is in Fig. 7.19.

x	-4	-3	-2	-1	0	1	2	3	4
y	1680	0	-60	0	0	0	-60	0	1680

(b)   Here n is even and $a_n < 0$, so we refer to Fig. 7.14. The graph of (a) reflected through the x axis is the graph of (b). See Fig. 7.20.   ●

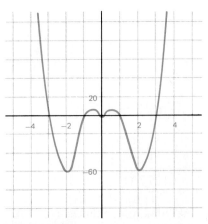

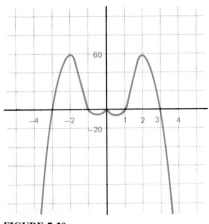

FIGURE 7.19   **FIGURE 7.20**

Factored polynomials   Any polynomial with real coefficients can be expressed as the product of linear and irreducible quadratic factors, each factor having real coefficients. We will assume in the next two examples that such a factorization has been found. Chapter 11 shows how to find these factors in certain cases.

● **Example 4**   Graph $y = (x - 1)(x - 2)^2(x - 3)^3$.

Solution   If $x = 1$, 2, or 3, then $y = 0$. However if x is close to 1, then y is approximately

$$(x - 1)(1 - 2)^2(1 - 3)^3 = (x - 1)(1)(-8) = -8(x - 1)$$

Similarly if x is near 2, then y is nearly

$$(2 - 1)(x - 2)^2(2 - 3)^3 = -(x - 2)^2$$

and if x is near 3, then y is nearly

$$(3 - 1)(3 - 2)^2(x - 3)^3 = 2(x - 3)^3$$

We summarize this information graphically in Fig. 7.21, where we have translated the graphs of $y = -8x$, $y = -x^2$, and $y = 2x^3$ to the right by 1, 2, and 3, respectively. In Fig. 7.22, we complete our sketch of the graph by joining the parts of Fig. 7.21. Refer to Fig. 7.13 since $n = 6$ is even and $a_n > 0$.   ●

● **Example 5**   Graph (a) $y = (x + 2)^3(x + 1)^2(2x - 3)(2x - 5)$
                        (b) $y = (x + 2)^4(x + 1)^2(2x - 3)(2x - 5)$

Solution   The only difference between (a) and (b) is that (b) has one more factor of $x + 2$. Both have $a_n > 0$, but in (a) $n = 7$ is odd while in (b) $n = 8$ is even.

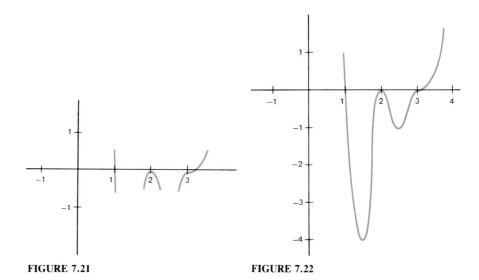

FIGURE 7.21   FIGURE 7.22

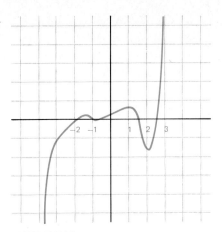

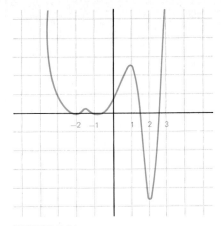

FIGURE 7.23   **FIGURE 7.24**

The relevant pictures then are Figs. 7.11 and 7.13. In both curves $y = 0$ if $x = -2, -1, \frac{3}{2}$, or $\frac{5}{2}$. See Figs. 7.23 and 7.24. Notice that for values of x near -2, the graph of (a) is approximately

$$(x + 2)^3(-2 + 1)^2(-4 - 3)(-4 - 5) = 63(x + 2)^3$$

while the graph of (b) is approximately $63(x + 2)^4$.   ●

Exercise 7.2   Sketch the graphs of the following polynomial functions.

1 $y = x^3 - 5x^2 + 5x + 3$ **2** $y = x^3 - 9x^2 + 21x - 5$
3 $y = x^3 - x^2 - 8x - 8$ **4** $y = x^3 - 10x^2 + 25x - 6$
5 $y = -x^3 - 2x^2 + 5x + 6$ **6** $y = -x^3 + 7x^2 - 5x - 13$
7 $y = -x^3 + 3x^2 + 14x - 12$ **8** $y = -x^3 + 9x^2 - 11x - 21$
9 $y = x^4 - x^3 - 7x^2 + x + 6$ **10** $y = x^4 - x^3 - 5x^2 - x - 6$
11 $y = x^4 - 22x^2 + 49$ **12** $y = x^4 - 10x^2 + 4$
13 $y = -x^4 - 4x^3 + 12x^2 + 32x + 17$
14 $y = -x^4 + 4x^3 + 12x^2 - 32x + 17$
15 $y = -x^4 - x^3 + 18x^2$ **16** $y = -x^4 - 9x^3 - 12x^2 + 28x + 48$
17 $y = 4x^5 - 17x^3 + 4x$ **18** $y = x^5 - 3x^4 - 5x^3 + 15x^2 + 4x - 12$
19 $y = -x^5 + x^4 + 4x - 4$ **20** $y = x^5 - x^3 - 12x + 1$
21 $y = -x^6 + 14x^4 - 49x^2 + 36$ **22** $y = 2x^6 - 9x^4 + 7x^2 + 3$
23 $y = -x^6 + x^5 + 8x^4 - 8x^3 + 9x^2 - 9x + 4$
24 $y = x^6 - 4x^4 + 2x^2 - 3$ **25** $y = x^3$ and $y = -x^5$
26 $y = x^4$ and $y = -x^6$ **27** $y = 2x^3$ and $y = 2(x - 3)^3$
28 $y = -x^4$ and $x = -y^4$ **29** $y = (x - 1)^2(x - 2)^3(x - 3)^2$
30 $y = (x - 1)^2(x - 2)^2(x - 3)^2$ **31** $y = (x - 1)^3(x - 2)^2(x - 3)^3$
32 $y = (x - 1)^3(x - 2)^3(x - 3)^2$
33 $y = (x + 3)(x + 1)^2(x - 3)^3(x - 1)$

34 $y = (x + 3)(x + 1)^3(x - 1)^3(x - 3)$
35 $y = (x + 2)(x^2)(x - 1)^4(x - 2)^2$
36 $y = (x + 2)(x^2)(x - 1)^3(x - 3)^3$

7.3 Graphs of rational functions

A function $y = f(x)$ defined by

Rational function

$$f(x) = \frac{P(x)}{Q(x)} \qquad (1)$$

No common factor

Note

where $P(x)$ and $Q(x)$ are polynomials, is called a **rational function.** We assume that $P(x)$ and $Q(x)$ have *no common factor.* The most important facts to know about a rational function are the degree of $P(x)$ and of $Q(x)$, and the values for which $Q(x) = 0$. The sign of $f(x)$ in different regions is also helpful.

● **Example 1** Graph (a) $y = \dfrac{1}{2x + 3}$ (b) $y = \dfrac{1}{(2x + 3)^2}$

Solution (a) First, we emphasize that the function is *not defined* when $2x + 3 = 0$, hence for $x = -\frac{3}{2}$. We cannot divide by zero. Second, $y > 0$ if $2x + 3 > 0$, hence for $x > -\frac{3}{2}$; also $y < 0$ if $x < -\frac{3}{2}$.

x	-999	-2	-1.51	-1.499	-1.495	-1.49	-1.4	-1	0	5	999
y	$-\frac{1}{1995}$	-1	-50	500	100	50	5	1	$\frac{1}{3}$	$\frac{1}{13}$	$\frac{1}{2001}$

Vertical asymptote

Horizontal asymptote

As the table of values shows the values of y will become as large as desired by making x get closer and closer to $-\frac{3}{2}$. The vertical line $x = -\frac{3}{2}$ is called a *vertical asymptote*, since the graph of the function approaches this vertical line if x is near $-\frac{3}{2}$. Furthermore, the table of values shows that as $|x|$ becomes arbitrarily large, y approaches 0. For this reason the x axis is called a *horizontal asymptote*. The graph is shown in Fig. 7.25.

(b) This graph is very similar to (a), except that all function values are positive. The same vertical and horizontal asymptotes are present. See Fig. 7.26. Some of the function values are $f(-2) = 1 = f(-1)$ and $f(-4) = \frac{1}{25} = f(1)$.   ●

Note In the previous example and the next example, the degree of the denominator is greater than the degree of the numerator. In such cases, the x axis is always a *horizontal asymptote.* The *vertical asymptotes* $x = r$ occur whenever r is a zero of the denominator, hence when $Q(r) = 0$.

● **Example 2** Graph (a) $y = \dfrac{x - 2}{x^2 - 2x - 3}$ (b) $y = \dfrac{x - 2}{x^2 - 2x + 3}$

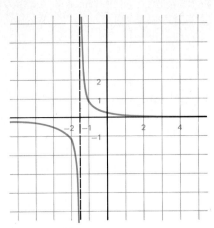

FIGURE 7.25   **FIGURE 7.26**

Solution   Both graphs have the x axis as a horizontal asymptote since the degree of the denominator is greater than that of the numerator. Both cross the x axis at $x = 2$. For (a) we have

$$y = \frac{x - 2}{x^2 - 2x - 3} = \frac{x - 2}{(x - 3)(x + 1)}$$

while for (b) we have, upon completing the square,

$$y = \frac{x - 2}{x^2 - 2x + 3} = \frac{x - 2}{x^2 - 2x + 1 + 2} = \frac{x - 2}{(x - 1)^2 + 2}$$

Thus (a) has two vertical asymptotes, $x = 3$, and $x = -1$, while (b) has no vertical asymptotes since its denominator is never zero. See Figs. 7.27 and 7.28.

FIGURE 7.27

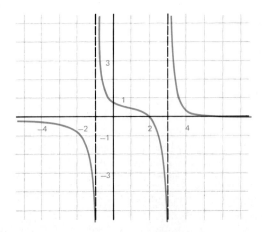

FIGURE 7.28

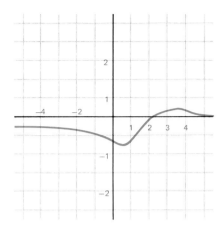

Values for (a)

x	-10	-2	-1	0	1	2	3	4	10
y	$-\frac{4}{39}$	$-\frac{4}{5}$	not defined	$\frac{2}{3}$	$\frac{1}{4}$	0	not defined	$\frac{2}{5}$	$\frac{8}{77}$

Values for (b)

x	-10	-5	0	1	2	3	4	8
y	$-\frac{12}{123}$	$-\frac{7}{38}$	$-\frac{2}{3}$	$-\frac{1}{2}$	0	$\frac{1}{6}$	$\frac{2}{11}$	$\frac{2}{17}$

Note   If the degree of the numerator is at least as large as the degree of the denominator, we *perform the division* and get a quotient $g(x)$ and remainder $r(x)$ for which

$$\frac{P(x)}{Q(x)} = g(x) + \frac{r(x)}{Q(x)} \tag{2}$$

where the degree of $r(x)$ is less than the degree of $Q(x)$. Thus as $|x|$ becomes large, $r(x)/Q(x)$ approaches 0, and hence, by Eq. (2), $P(x)/Q(x)$ approaches $g(x)$. If the graph of $y = g(x)$ is a line, then it is an *asymptote* to the graph of $y = P(x)/Q(x)$.

● **Example 3**   Graph (a) $y = \dfrac{8x + 13}{2x + 3}$ (b) $y = \dfrac{2x^2 - x - 5}{2x + 3}$

Solution   (a)   Division gives $y = 4 + 1/(2x + 3)$. Since $1/(2x + 3)$ approaches 0 as $|x|$ gets large, we have the horizontal asymptote $y = 4$. The graph is the same as the graph in Fig. 7.25 shifted up four units. See Fig. 7.29.

(b)   Division gives $y = x - 2 + 1/(2x + 3)$. As in (a), $1/(2x + 3)$ approaches 0 as $|x|$ gets large; so $y = x - 2$ is an asymptote. It is neither horizontal nor vertical, and is just called an asymptote (or sometimes an oblique asymptote). The denominator is zero if $2x + 3 = 0$; hence $x = -\frac{3}{2}$ is a vertical asymptote. See Fig. 7.30.

FIGURE 7.29

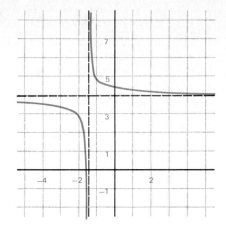

Values for (b)

x	-10	-2	$-\frac{3}{2}$	-1.4	-1	0	2	5	10
y	$-12\frac{1}{17}$	-5	not defined	1.6	-2	$-\frac{5}{3}$	$\frac{1}{7}$	$3 + \frac{1}{13}$	$8 + \frac{1}{13}$

Exercise 7.3   Sketch the graph of each function given in Probs. 1 to 24.

1 $y = \dfrac{2}{x - 3}$ **2** $y = \dfrac{3}{x + 4}$ **3** $y = \dfrac{5}{(x - 1)^2}$ **4** $y = \dfrac{4}{(x + 2)^2}$

5 $y = \dfrac{4}{(x - 1)((x + 2)}$ **6** $y = \dfrac{3}{(x - 2)(x - 4)}$

7 $y = \dfrac{x - 1}{(x + 1)(x - 2)}$ **8** $y = \dfrac{x + 1}{(x - 1)((x + 2)}$

FIGURE 7.30

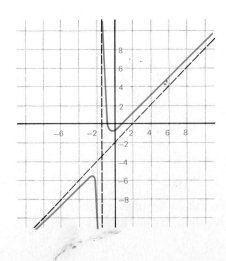

9 $y = \dfrac{12}{(x + 1)(x - 2)(x + 3)}$ **10** $y = \dfrac{-8}{(x + 2)(x - 1)(x - 3)}$

11 $y = \dfrac{3}{x^2 + x - 2}$ **12** $y = \dfrac{x + 1}{x^2 + x - 6}$ **13** $y = \dfrac{12}{x^2 + x + 3}$

14 $y = \dfrac{9}{x^2 + 4x + 6}$ **15** $y = \dfrac{1 - x}{x^2 - 2x + 3}$ **16** $y = \dfrac{x - 2}{x^2 - 6x + 12}$

17 $y = \dfrac{4x + 1}{2x - 3}$ **18** $y = \dfrac{12x - 15}{3x - 5}$ **19** $y = \dfrac{3x^2 - 3x - 2}{x^2 - x - 2}$

20 $y = \dfrac{2x^2 - x - 14}{x^2 - x - 6}$ **21** $y = \dfrac{2x^2 + x + 2}{x + 1}$ **22** $y = \dfrac{2x^2 - x - 12}{2x - 5}$

23 $y = \dfrac{3x^2 - x - 1}{3x - 4}$ **24** $y = \dfrac{3x^2 + 4x - 3}{3x + 7}$

7.4 Graphs of conics

Conic If a plane and a right circular cone intersect, the curve of intersection is called a *conic*. If the plane does not go through the vertex of the cone, then the conic is one of the following curves (see Fig. 7.31):

Parabola (a) A *parabola* if the plane is parallel to a linear element of the cone
Ellipse (b) An *ellipse* if the plane cuts across one end of the cone
Hyperbola (c) A *hyperbola* if the plane cuts both halves of the cone

A circle is a special case of an ellipse.

Ellipse   The definition of an **ellipse** requires that we begin with two fixed points
Focus F_1 and F_2 which are called the *foci* (each is a focus). A point P is on the ellipse if

$$d(P, F_1) + d(P, F_2) \text{ is constant} \tag{1}$$

FIGURE 7.31

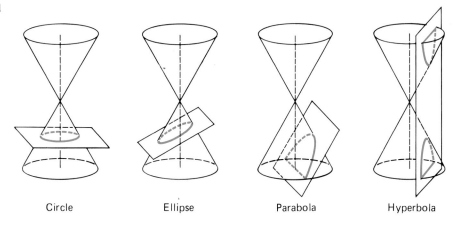

Circle   Ellipse   Parabola   Hyperbola

That is, the sum of the distances from a point on the ellipse to the foci is the same, regardless of which point on the ellipse is chosen.

Taking the foci on the x axis at $F_1(-c, 0)$ and $F_2(c, 0)$ makes the equation simple. Choosing the sum of the distances to be $2a$ also makes the form of the equation simple. We use the distance formula for the distance from $P(x, y)$ to $F_1(-c, 0)$, and from $P(x, y)$ to $F_2(c, 0)$. Then (1) becomes

$$\sqrt{(x + c)^2 + y^2} + \sqrt{(x - c)^2 + y^2} = 2a$$

Transposing the second radical, then squaring both sides gives

$$x^2 + 2cx + c^2 + y^2 = 4a^2 + x^2 - 2cx + c^2 + y^2 - 4a\sqrt{(x - c)^2 + y^2}$$

Combining terms and dividing by 4, we have

$$a\sqrt{(x - c)^2 + y^2} = a^2 - cx$$

$$a^2(x^2 - 2cx + c^2 + y^2) = a^4 - 2a^2cx + c^2x^2 \qquad \text{squaring both sides}$$

$$a^2x^2 - 2a^2cx + a^2c^2 + a^2y^2 = a^4 - 2a^2cx + c^2x^2 \qquad \text{distributive axiom}$$

$$(a^2 - c^2)x^2 + a^2y^2 = a^2(a^2 - c^2) \qquad \text{combining terms and factoring}$$

If we now divide by $a^2(a^2 - c^2)$, we have

$$\frac{x^2}{a^2} + \frac{y^2}{a^2 - c^2} = 1$$

We must have $a^2 - c^2 > 0$ (see Fig. 7.32), and we may thus replace $a^2 - c^2$ by b^2. Hence the standard form for the equation of the ellipse with both foci on the x axis at $(\pm c, 0)$ is

$$\frac{x^2}{a^2} + \frac{y^2}{b^2} = 1 \qquad (7.2a)$$

By taking $x = 0$ we have $y = \pm b$, while $y = 0$ gives $x = \pm a$. See Fig. 7.32.

By taking the foci $F_1(0, -c)$ and $F_2(0, c)$ on the y axis, the equation $d(P, F_1) + d(P, F_2) = 2a$ becomes

$$\frac{y^2}{a^2} + \frac{x^2}{b^2} = 1 \qquad (7.2b)$$

FIGURE 7.32

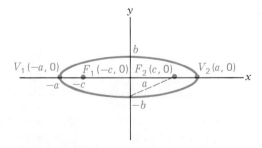

FIGURE 7.33

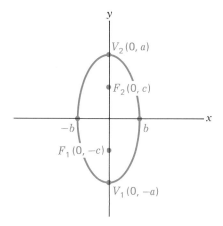

Major axis
Minor axis

Center

$a^2 = b^2 + c^2$

The line segment from $(-a, 0)$ to $(a, 0)$ is called the **major axis** of the ellipse (7.2a), and it is longer than the **minor axis,** which goes from $(0, -b)$ to $(0, b)$. The foci are always on the major axis. The length of the major axis is $2a$, which is the same constant as in the definition of the ellipse. The **center** of the ellipse is the point $(0, 0)$—it is *not* on the ellipse itself. The vertices are the endpoints of the major axis.

Since $a^2 = b^2 + c^2$, the number a is larger than either b or c. In graphing an ellipse in standard form:

1 The larger number (under x^2 or y^2) is a^2.
2 If a^2 is under x^2, then the major axis, the foci, and the vertices are on the x axis.
3 If a^2 is under y^2, then the major axis, the foci, and the vertices are on the y axis.
4 The distance from the center to each vertex is a.
5 The distance from the center to each focus is c.
6 The distance from the center to each intercept on the minor axis is b.
7 $a^2 = b^2 + c^2$ and $a > c$ and $a > b$.
8 The graph is symmetric about its center.

● **Example 1** Graph $\dfrac{x^2}{9} + \dfrac{y^2}{4} = 1$.

Solution Since $9 > 4$, we have $a^2 = 9$ and $b^2 = 4$. Thus $c^2 = a^2 - b^2 = 9 - 4 = 5$, and so

$$a = 3 \qquad b = 2 \qquad c = \sqrt{5}$$

The major axis, the foci, and the verticles, are on the x axis. See Fig. 7.34. ●

FIGURE 7.34

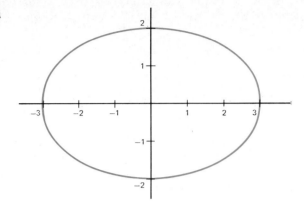

● **Example 2**   Graph $25x^2 + 4y^2 = 100$.

    Solution   Since we need the constant term to be 1, we divide by 100.

$$\frac{x^2}{4} + \frac{y^2}{25} = 1$$

Now $25 > 4$, hence $a^2 = 25$, $b^2 = 4$, and $c^2 = a^2 - b^2 = 25 - 4 = 21$. This gives

$$a = 5 \qquad b = 2 \qquad c = \sqrt{21}$$

The major axis, the foci, and the vertices are on the y axis. See Fig. 7.35.

                                                            ●

    The definition of a hyperbola is the set of all points $P(x, y)$ whose distances from two fixed points F_1 and F_2 (the foci) differ by a constant. Thus

$$d(P, F_1) - d(P, F_2) = \pm 2a$$

FIGURE 7.35

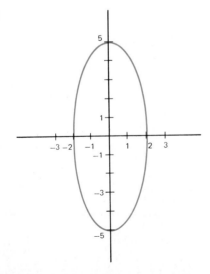

Taking the foci on the x axis at $(\pm c, 0)$ gives the equation

$$\frac{x^2}{a^2} - \frac{y^2}{b^2} = 1 \qquad \text{with } b^2 = c^2 - a^2 \qquad\qquad (7.3a)$$

while if the foci are taken on the y axis at $(0, \pm c)$, the equation becomes

$$\frac{y^2}{a^2} - \frac{x^2}{b^2} = 1 \qquad \text{with } b^2 = c^2 - a^2 \qquad\qquad (7.3b)$$

See Figs. 7.36 and 7.37. In each case, we begin with the distance formula and simplify as in the derivation of the equation of the ellipse. There are two symmetric halves to a hyperbola. The hyperbola crosses one axis at two points called the vertices but does not cross the other axis.

1 a^2 is in the term with the positive sign.
2 If a^2 is under x^2, then the foci and vertices are on the x axis.
3 If a^2 is under y^2, then the foci and vertices are on the y axis.
4 The distance from the center $(0, 0)$ to each vertex is a.
5 The distance from the center to each focus is c.

Asymptotes 6 The *asymptotes* are the two lines found by replacing 1 by 0 in (7.3).
7 $c^2 = b^2 + a^2$ and $c > a$ and $c > b$.
8 The graph is symmetric about its center.

Only fact 6 is radically different from the facts about ellipses. Neither of the two halves of a hyperbola is a parabola—in fact, they approach straight lines called asymptotes. To find the asymptotes of

$$\frac{y^2}{a^2} - \frac{x^2}{b^2} = 1$$

for instance, we solve

$$\frac{y^2}{a^2} - \frac{x^2}{b^2} = 0$$

and get the two lines

$$y = \frac{a}{b}x \qquad \text{and} \qquad y = -\frac{a}{b}x$$

FIGURE 7.37

FIGURE 7.36

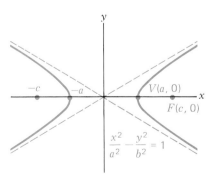

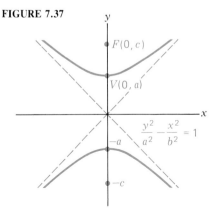

● **Example 3**   Graph $\dfrac{x^2}{16} - \dfrac{y^2}{1} = 1$.

Solution   The x^2 term is positive, hence $a^2 = 16$, $b^2 = 1$, and $c^2 = b^2 + a^2 = 1 + 16 = 17$. Hence

$$a = 4 \qquad b = 1 \qquad c = \sqrt{17}$$

The foci and vertices are on the x axis. The asymptotes are $y = \pm\frac{1}{4}x$. See Fig. 7.38.   ●

● **Example 4**   Graph $9x^2 - 25y^2 = -225$.

Solution   Dividing by -225 gives

$$\frac{y^2}{9} - \frac{x^2}{25} = 1$$

Here the y^2 term is positive, giving $a^2 = 9$, $b^2 = 25$, and $c^2 = 25 + 9 = 34$. Hence

$$a = 3 \qquad b = 5 \qquad c = \sqrt{34}$$

The foci and vertices are on the y axis. The asymptotes are $y = \pm\frac{3}{5}x$. See Fig. 7.39.   ●

In Eqs. (7.2) and (7.3) for ellipses and hyperbolas, a and b are positive numbers. The special cases when $a = b$ deserve particular attention.

In Eq. (7.2), if $a = b$, then $x^2/a^2 + y^2/a^2 = 1$. Multiplying by a^2 gives

$$x^2 + y^2 = a^2 \tag{2}$$

Circle   which is the equation of a *circle* with center at the origin and radius a. See Sec. 6.2. Hence a circle is a special case of an ellipse.

In Eq. (7.3), if $a = b$, then the two equations become

$$\frac{x^2}{a^2} - \frac{y^2}{a^2} = 1 \qquad \text{and} \qquad \frac{y^2}{a^2} - \frac{x^2}{a^2} = 1 \tag{3}$$

FIGURE 7.38

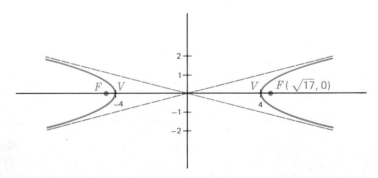

FIGURE 7.39

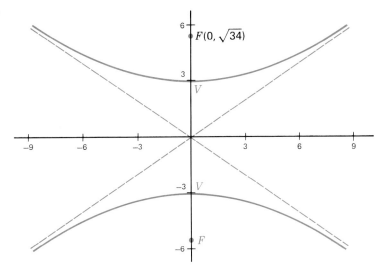

Equilateral hyperbola   Each of these hyperbolas is called an *equilateral hyperbola*. The graphs of

$$\frac{x^2}{4} - \frac{y^2}{4} = 1 \quad\text{and}\quad y^2 - x^2 = 1$$

are shown in Fig. 7.40 in a solid line and dashed line, respectively. The dashed straight lines are $y = x$ and $y = -x$, which are asymptotes for both hyperbolas.

It may be shown that the graph of

$$xy = k \qquad\text{for } k \neq 0 \tag{4}$$

is also a *hyperbola*. The graph of $xy = 3$ is shown in Fig. 7.41.

FIGURE 7.40

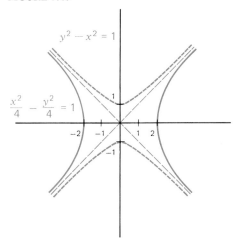

FIGURE 7.41

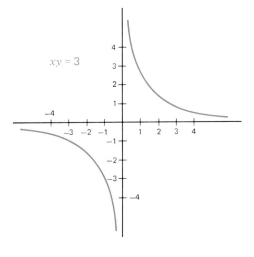

We have now studied standard forms for the equations of straight lines, parabolas, ellipses, and hyperbolas. These are all special forms of one equation. In fact, if

Lines and conics

$$Ax^2 + Cy^2 + Dx + Ey + F = 0 \qquad (7.4)$$

then the graph is in general:

(a)   A straight line if $A = C = 0$ (and $DE \neq 0$)
(b)   A parabola if either A or C (but not both) is 0
(c)   A circle if $A = C \neq 0$
(d)   An ellipse if A and C have the same sign (hence $AC > 0$)
(e)   A hyperbola if A and C have opposite signs (hence $AC < 0$)

In some cases, (b) may be one line or two parallel lines, (c) may be a point, (d) may be a point, and (e) may be two lines which intersect.

Complete the square   Recall that to *complete the square* of $x^2 + bx$, we write

$$x^2 + bx = x^2 + bx + \left(\frac{b}{2}\right)^2 - \left(\frac{b}{2}\right)^2$$

$$= \left(x + \frac{b}{2}\right)^2 - \left(\frac{b}{2}\right)^2$$

● **Example 5**   Graph $9x^2 + 4y^2 - 54x + 16y + 61 = 0$.

Solution   From (7.4) we have $A = 9$ and $C = 4$, and so the graph will be an ellipse by (d). We now complete the square of the terms in x and those in y.

$$9(x^2 - 6x) + 4(y^2 + 4y) = -61$$

$$9(x^2 - 6x + 9) + 4(y^2 + 4y + 4) = -61 + 81 + 16$$

$$9(x - 3)^2 + 4(y + 2)^2 = 36$$

$$\frac{(x - 3)^2}{4} + \frac{(y + 2)^2}{9} = 1 \qquad (5)$$

This is the ellipse $x^2/4 + y^2/9 = 1$ translated 3 units to the right and 2 units down. See Fig. 7.42. The center is the point $(3, -2)$. ●

● **Example 6**   Graph $4y^2 - 25x^2 - 24y - 50x = 89$.

Solution   By (7.4) we have a hyperbola since A and C have opposite signs. Completing the square gives

$$4(y^2 - 6y) - 25(x^2 + 2x) = 89$$

$$4(y^2 - 6y + 9) - 25(x^2 + 2x + 1) = 89 + 36 - 25$$

$$4(y - 3)^2 - 25(x + 1)^2 = 100$$

$$\frac{(y - 3)^2}{5^2} - \frac{(x + 1)^2}{2^2} = 1 \qquad (6)$$

FIGURE 7.42

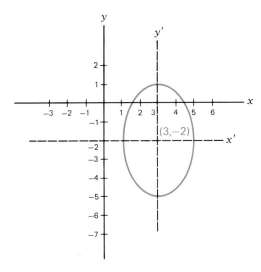

This hyperbola crosses the y axis but not the x axis. The asymptotes are found by replacing 1 by 0 in (6) and solving for the two lines. They are

$$y - 3 = \tfrac{5}{2}(x + 1) \qquad \text{and} \qquad y - 3 = -\tfrac{5}{2}(x + 1)$$

and each asymptote passes through the point $(-1, 3)$. Taking $x = -1$ in (6) gives $(y - 3)^2 = 5^2$ or $y - 3 = \pm 5$ or $y = 8$, $y = -2$, and so $(-1, 8)$ and $(-1, -2)$ are the vertices of the hyperbola. See Fig. 7.43.   ●

Some uses of parabolas were mentioned at the end of Sec. 7.1. Ellipses are important since the paths of the planets and some comets are ellipses with the sun at one focus. Ellipses also have an important reflecting property—rays from one focus bounce off the ellipse and go through the other focus. They are also used in the study of atomic particles. Some

FIGURE 7.43

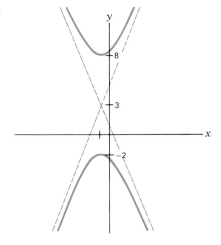

comets travel in hyperbolic paths. When a jet plane breaks the sound barrier, the front wave of the sonic boom on the ground is in the shape of a hyperbola. Hyperbolas are approximated by straight lines for large values of the variables.

Exercise 7.4 Graph the following equations:

1 $x^2/9 + y^2 = 1$ 2 $x^2/9 + y^2/16 = 1$ 3 $x^2/4 + y^2/64 = 1$
4 $x^2 + y^2/81 = 1$ 5 $9x^2 + y^2 = 9$ 6 $25x^2 + 4y^2 = 100$
7 $x^2 + 9y^2 = 81$ 8 $x^2 + 64y^2 = 64$ 9 $x^2/9 - y^2/25 = 1$
10 $x^2/25 - y^2/4 = 1$ 11 $y^2/49 - x^2 = 1$ 12 $y^2/9 - x^2/16 = 1$
13 $25x^2 - 4y^2 = 100$ 14 $x^2 - 36y^2 = 36$ 15 $25y^2 - x^2 = 25$
16 $25y^2 - 4x^2 = 100$ 17 $x^2 - y^2 = 4$ 18 $y^2 - x^2 = 9$

19 $\dfrac{y^2}{20} - \dfrac{x^2}{20} = 1$ 20 $\dfrac{x^2}{25} - \dfrac{y^2}{25} = 1$ 21 $xy = 1$

22 $y = 5/x$ 23 $xy = -2$ 24 $y = -8/x$

25 $\dfrac{(x-1)^2}{4} + \dfrac{(y-3)^2}{16} = 1$ 26 $\dfrac{(x-1)^2}{4} - \dfrac{(y-3)^2}{16} = 1$

27 $(y-2)^2 - \dfrac{(x+2)^2}{4} = 1$ 28 $(y-2)^2 + \dfrac{(x+2)^2}{4} = 1$

29 $\dfrac{(x+3)^2}{9} - \dfrac{(y-5)^2}{49} = 1$ 30 $\dfrac{(x-4)^2}{16} + \dfrac{(y+1)^2}{49} = 1$

31 $\dfrac{(x+2)^2}{36} + \dfrac{(y+2)^2}{25} = 1$ 32 $\dfrac{(x+4)^2}{9} - \dfrac{(y-3)^2}{9} = 1$

33 $4x^2 + 20y^2 - 12x + 40y + 29 = 0$
34 $4x^2 - 20y^2 + 12x - 40y - 11 = 0$
35 $4y^2 + 8 = x^2 + 4x + 16y$
36 $x^2 + 4x + 4y^2 + 16 = 16y$
37 $3x^2 + y^2 + 6x - 8y + 15 = 0$
38 $4x^2 + 2y^2 + 16x + 6y + \frac{9}{2} = 0$
39 $2x^2 - 3y^2 - 4x + 12y - 28 = 0$
40 $-x^2 + 2y^2 + 4x - 4y + 7 = 0$

An *ellipse* was defined as the set of all points the sum of whose distances from two fixed points is a constant. Each of the fixed points was called a *focus*. See Fig. 7.44, where

$$\overline{PF}_1 + \overline{PF}_2 = \text{a constant}$$

Verify this property in Probs. 41 to 44 for the fixed points F_1 and F_2 and the given points P.

41 $x^2/25 + y^2/16 = 1$; $F_1(-3, 0)$, $F_2(3, 0)$; $P(0, 4)$ and also $P(5, 0)$.
42 Same as Prob. 41 except $P(0, -4)$ and $P(4, 12/5)$.
43 $x^2/25 + y^2/169 = 1$; $F_1(0, -12)$, $F_2(0, 12)$; $P(5, 0)$ and also $P(0, 13)$.
44 Same as Prob. 43 except $P(-5, 0)$ and $P(60/13, 5)$.

FIGURE 7.44

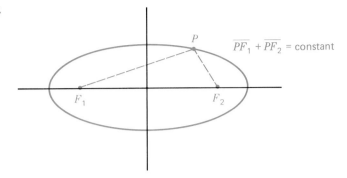

A *hyperbola* was defined as the set of all points the difference of whose distances from two fixed points is a constant. Each of the fixed points was called a *focus*. See Fig. 7.45, where

$$|\overline{PF_1} - \overline{PF_2}| = \text{a constant}$$

Verify this property in Probs. 45 to 48 for the fixed points F_1 and F_2 and the given points P.

45 $x^2/16 - y^2/9 = 1$; $F_1(-5, 0)$, $F_2(5, 0)$; $P(4, 0)$ and also $P(5, 9/4)$.

46 Same as Prob. 45 except $P(-4, 0)$ and $P(20/3, 4)$.

47 $y^2/144 - x^2/25 = 1$; $F_1(0, -13)$, $F_2(0, 13)$; $P(0, 12)$ and also $P(12, 156/5)$.

48 Same as Prob. 47 except $P(0, -12)$ and $P(25/12, 13)$.

49 The previous eight problems treat sums and differences of distances. This one deals with quotients of distances. Suppose that $Q(a, b)$ and $R(c, d)$ are two fixed points and that k is a positive constant. Let P have coordinates (x, y). Show that if

$$\frac{d(P, Q)}{d(P, R)} = k$$

then P lies on a circle if $k \neq 1$, and on a line if $k = 1$.

FIGURE 7.45

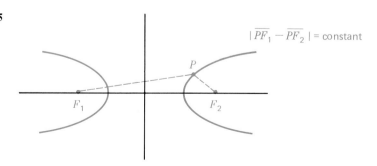

50 The *eccentricity* of the ellipse $x^2/a^2 + y^2/b^2 = 1$, with $a > b$, is by definition

$$e = \frac{\sqrt{a^2 - b^2}}{a}$$

Show that $0 < e < 1$. What shape does the ellipse approach if e is close to 0?

51 The graphs of the hyperbolas

$$\frac{x^2}{a^2} - \frac{y^2}{b^2} = 1 \quad \text{and} \quad \frac{y^2}{a^2} - \frac{x^2}{b^2} = 1$$

are called *conjugate hyperbolas*. Draw their graphs on different sets of axes if $a = 1$ and $b = 3$.

Area of ellipse **52** The **area** inside the ellipse $x^2/a^2 + y^2/b^2 = 1$ is πab. Show that this formula also holds in the special case when the ellipse reduces to a circle.

7.5 Inequalities

In Chap. 5, we studied inequalities in one variable, inequalities such as $3x + 5 < 0$ or $x^2 + 6x + 7 > 0$. In this section we will look at inequalities in two variables, such as $3x - 7y > 9$ or $x^2 + y^2 < 4$.

In addition to learning about inequalities, we will get a good review of graphing. This is because our main technique in solving an inequality will

Graph the equality be to graph the corresponding equality. We will then complete the solution by considering one or more "checkpoints" (see Example 1).

● **Example 1** Solve the inequality $x^2 + y^2 < 4$.

Solution We first graph $x^2 + y^2 = 4$. This is a circle whose center is $(0, 0)$ and radius is 2. The graph is in Fig. 7.46. The graph divides the plane into two regions, and we will choose points which we know definitely to be in each of the regions. We call such a point a **checkpoint** because:

Checkpoint

If a checkpoint satisfies the given inequality, then every point in the same region also satisfies it.

FIGURE 7.46

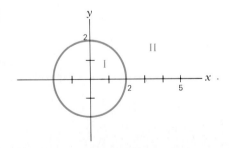

In this case the origin is clearly in region I, and $(5, 0)$ is plainly in region II. We now check each point in the given inequality and see whether the resulting statement is true or false.

Region I $\qquad\qquad\qquad\qquad 0^2 + 0^2 < 4?\qquad$ true since $0 < 4$
Region II $\qquad\qquad\qquad\qquad 5^2 + 0^2 < 4?\qquad$ false since $25 > 4$

Thus the solution of $x^2 + y^2 < 4$ is all points in region I. $\qquad\bullet$

Choosing checkpoints $\qquad$ In choosing checkpoints, choose points which are *clearly* in the various regions. This normally means either using our previous knowledge of standard forms of graphs or else choosing points far away from the graph of the equality.

Note $\qquad$ When determining the regions, *ignore* the x axis and y axis.

$\bullet$ **Example 2** $\quad$ Solve $4x - 3y > 12$.

Solution $\quad$ The graph of the line $4x - 3y = 12$ is in Fig. 7.47. Our checkpoint in region I will be $(0, 0)$, and the checkpoint in region II will be $(2, -6)$. Substituting these points in the given inequality gives

Region I $\qquad\qquad\qquad 4(0) - 3(0) > 12?\qquad$ false since $0 < 12$
Region II $\qquad\qquad\qquad 4(2) - 3(-6) > 12?\qquad$ true since $26 > 12$

The solution is region II. $\qquad\bullet$

$\bullet$ **Example 3** $\quad$ Solve the inequality $y - 4 < -3(x + 1)^2$.

Solution $\quad$ The graph of the parabola $y - 4 = -3(x + 1)^2$ is shown in Fig. 7.48. We will use the checkpoints $(-1, 0)$ in region I and $(0, 5)$ in region II.

Region I $\qquad\qquad\qquad 0 - 4 < -3(-1 + 1)^2?\qquad$ true since $-4 < 0$
Region II $\qquad\qquad\qquad 5 - 4 < -3(0 + 1)^2?\qquad$ false since $1 > -3$

The solution is region I. $\qquad\bullet$

FIGURE 7.47

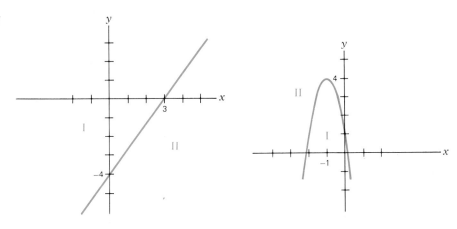

FIGURE 7.49

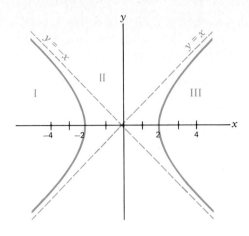

● **Example 4**   Solve the inequality $x^2 - y^2 > 4$.

Solution   The graph of the hyperbola $x^2 - y^2 = 4$ is given in Fig. 7.49. We choose the checkpoints $(-4, 0)$ in region I, $(0, 0)$ in region II, and $(4, 0)$ in region III.

Region I	$(-4)^2 - 0^2 > 4$?	true since $16 > 4$
Region II	$0^2 - 0^2 > 4$?	false since $0 < 4$
Region III	$4^2 - 0^2 > 4$?	true since $16 > 4$

The solution is regions I and III.   ●

● **Example 5**   Solve the inequality $y < \dfrac{1}{2x + 3}$.

Solution   The rational function $y = \dfrac{1}{2x + 3}$ was graphed in Example 1, Sec. 7.3, and we reproduce the graph in Fig. 7.50. Because $1/(2x + 3)$ is not defined for $x = -\frac{3}{2}$, there will be four regions to consider—two to the left of $x = -\frac{3}{2}$ (I above and II below the curve) and two to the right of $x = -\frac{3}{2}$ (III above and IV below).

Note   Remember to *disregard* the x axis and y axis in visualizing the regions.

Alternate method   Although we could again use checkpoints in each of the four regions, there is a quicker way since the inequality is of the form $y < f(x)$, that is, it is "solved" for y. The graph shows where y *equals* $f(x)$ for a fixed x, so we will have y *less than* $f(x)$ for points below the graph. The solution is regions II and IV.   ●

Note $y < \dfrac{1}{2x + 3}$ and $y(2x + 3) < 1$ have different solutions, because multiplying by $2x + 3$ will change the sense of the first inequality whenever

$$2x + 3 < 0$$

FIGURE 7.50

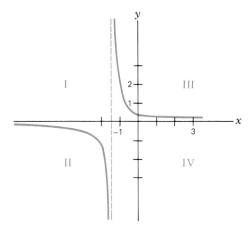

Exercise 7.5

Solve these inequalities graphically.

1 $x^2 + (y - 1)^2 < 4$

2 $(x - 2)^2 + (y + 3)^2 < 9$

3 $(x + 3)^2 + y^2 > 16$

4 $(x - 1)^2 + (y + 1)^2 > 1$

5 $\dfrac{(x - 1)^2}{4} + \dfrac{(y - 2)^2}{9} < 1$

6 $\dfrac{(x + 2)^2}{9} + \dfrac{(y - 2)^2}{16} > 1$

7 $\dfrac{(x - 3)^2}{9} + (y + 1)^2 > 1$

8 $\dfrac{(x + 1)^2}{25} + \dfrac{(y - 1)^2}{4} < 1$

9 $6x + 5y > 30$

10 $y - 4 > 3(x - 1)$

11 $y < 2x - 3$

12 $\dfrac{x}{3} + \dfrac{y}{-2} < 1$

13 $y > \dfrac{4}{x}$

14 $y + 2 < \dfrac{4}{x - 1}$

15 $y - 3 < \dfrac{4}{x + 2}$

16 $y + 2 > \dfrac{4}{x + 1}$

17 $y - 1 < 2(x - 3)^2$

18 $-y - 1 < 2(x - 3)^2$

19 $y - 1 > 2(-x - 3)^2$

20 $-y - 1 > 2(-x - 3)^2$

21 $y > (x + 1)(x - 1)(x - 3)$

22 $y > -(x + 3)(x + 1)(x - 2)$

23 $y < (x + 2)(x + 1)(x - 1)(x - 2)$

24 $y < -(x + 3)(x + 1)(x - 1)(x - 2)$

25 $y < \dfrac{1}{x^2 + 1}$

26 $y > \dfrac{x}{x^2 + 1}$

27 $y < \dfrac{1}{x^2 - 1}$

28 $y > \dfrac{x}{x^2 - 1}$

29 $x > 3y + 6$

30 $x - 3 < 2(y - 2)^2$

31 $x > (y + 1)(y - 1)(y - 3)$
32 $x < (y + 4)(y + 2)(y - 2)(y - 4)$

7.6 Key concepts

Be sure you understand and can use each of the following words and ideas.

Quadratic function
Parabola
Vertex
Axis of symmetry is $x = -b/2a$
Maximum or minimum value is $f(-b/2a)$
Intercepts
Polynomial function
Degree of a polynomial
   n even and odd
     Leading coefficient positive or
       negative
     $y = \pm x^2, \qquad y = \pm x^3$

Factored polynomial
Rational function
Vertical asymptote
Horizontal asymptote
Conic
Ellipse
Hyperbola
Asymptotes for hyperbola
Complete the square
Inequalities in two variables
Checkpoint

(7.1) $y - k = a(x - h)^2$

(7.2)   Ellipse $\dfrac{x^2}{a^2} + \dfrac{y^2}{b^2} = 1$ or $\dfrac{y^2}{a^2} + \dfrac{x^2}{b^2} = 1$ with $a > b$

(7.3)   Hyperbola $\dfrac{x^2}{a^2} - \dfrac{y^2}{b^2} = 1$ or $\dfrac{y^2}{a^2} - \dfrac{x^2}{b^2} = 1$
      with a^2 in the positive term

(7.4) $Ax^2 + Cy^2 + Dx + Ey + F = 0$:
      Line if $A = C = 0$
      Parabola if $A = 0$ or $C = 0$, but not both
      Circle if $A = C \neq 0$
      Ellipse if A and C have same sign
      Hyperbola if A and C have opposite sign

**Exercise 7.6
Review**

Draw the parabolas in Probs. 1 to 8.

1 $y = 2(x - 3)^2$
3 $y - 3 = 2x^2$
5 $x - 2 = -5(y + 3)^2$
7 $x^2 + 4x + 8y = 20$

2 $y = 2(x + 3)^2$
4 $y + 2 = 5(x + 3)^2$
6 $y^2 = 4x + 2y + 19$
8 $-y = 4x^2$

In Probs. 9 to 11, give the coordinates of the vertex, and state whether it is a maximum or a minimum point for the parabola.

9 $y = 4x^2 + 12x + 17$
11 $3x + 4y = 3x^2 + 4$

10 $y = -3x^2 + 12x + 5$

In Probs. 12 to 17, give the name of the conic (parabola, ellipse, hyperbola).

12 $x^2 + 15x + 4y^2 = 21y + 53$ **13** $3x^2 + 7x = 19 - 4y$
14 $121x + 144y^2 = 169$ **15** $4y^2 + 13y = 22x^2 + 31x$
16 $3y^2 - 5x^2 = 3y - 5x + 35$ **17** $x^2 + 2x + 1 = 3 - 4y - 5y^2$

Graph the functions in Probs. 18 to 26.

18 $y = x^3 - 4x^2 + x + 6$ **19** $-y = x^3 - 2x^2 - 5x + 6$
20 $y = x^4 + 2x^3 - 7x^2 - 8x + 12$ **21** $y = (x + 1)^2(x - 1)^2(x - 3)^3$

22 $y = (x + 2)^3(x - 1)^2(x - 4)^3$ **23** $y = \dfrac{4}{x - 2}$

24 $y = \dfrac{2}{(x + 1)(x - 2)}$ **25** $y = \dfrac{6x + 1}{3x - 2}$

26 $y = \dfrac{(x + 1)(x - 1)}{(x + 2)(x - 2)}$

Graph the conics in Probs. 27 to 33.

27 $(x - 2)^2 + 4(y + 2)^2 = 4$ **28** $9(x - 1)^2 + (y + 1)^2 = 9$

29 $\dfrac{(x - 1)^2}{16} + \dfrac{(y - 3)^2}{4} = 1$ **30** $\dfrac{(x - 1)^2}{16} - \dfrac{(y - 3)^2}{4} = 1$

31 $\dfrac{(y - 2)^2}{9} - \dfrac{(x + 1)^2}{4} = 1$ **32** $4x^2 + 2y = y^2 + 16x + 1$

33 $9x^2 + y^2 + 36 = 36x + 2y$

Solve these inequalities graphically.

34 $4x + 9y < 36$ **35** $4x^2 + 9y^2 < 36$
36 $9x^2 - 4y^2 > 36$ **37** $|x| + |y - 2| < 1$

38 $\dfrac{(x - 2)^2}{4} + \dfrac{(y + 1)^2}{25} > 1$ **39** $9y^2 - x^2 > 9$

40 $y - 2 < \dfrac{3}{x - 1}$

CHAPTER 8 · EXPONENTIAL AND LOGARITHMIC FUNCTIONS

The linear, quadratic, polynomial, and rational functions we have dealt with so far in this book are formed with the four fundamental operations and the extraction of roots. These are called *algebraic* functions. If a function is not algebraic, we refer to it as *transcendental*. In this chapter we will introduce two types of transcendental functions, namely, exponential and logarithmic functions.

8.1 Exponential functions

In earlier chapters, we dealt with the following definitions and facts concerning exponential expressions with the base $b > 0$ and m and n positive integers.

$$b^m = b \cdot b \cdot \cdots \cdot b \qquad m \text{ factors of } b$$

$$b^{-m} = \frac{1}{b^m} \qquad b^0 = 1$$

$$b^m \cdot b^n = b^{m+n} \qquad \frac{b^m}{b^n} = b^{m-n} \qquad (b^m)^n = b^{mn}$$

We also saw that for $b > 0$ the above rules apply with negative exponents and rational exponents m/n where

$$b^{m/n} = (b^{1/n})^m = (\sqrt[n]{b})^m = \sqrt[n]{b^m} = (b^m)^{1/n}$$

The definitions above allow us to define $f(x) = 3^x$ for any rational value of x. The decimal approximations in the table below came from a calculator, but the exact values merely use the rules for rational exponents.

x	-4	$-\frac{1}{2}$	$\frac{1}{5}$	$\frac{4}{7}$	$1.4 = \frac{7}{5}$	$1.41 = \frac{141}{100}$
Exact value of 3^x	$3^{-4} = \dfrac{1}{81}$	$3^{-1/2} = \dfrac{1}{\sqrt{3}}$	$3^{1/5} = \sqrt[5]{3}$	$3^{4/7} = (\sqrt[7]{3})^4$	$3^{1.4} = (\sqrt[5]{3})^7$	$3^{1.41} = (\sqrt[100]{3})^{141}$
Decimal approximation of 3^x	0.012	0.577	1.25	1.87	4.66	4.71

To define an exponential function $f(x) = 3^x$ for all real numbers x, we need to consider not only rational values of x as above, but also irrational values as well. We will start with $3^{\sqrt{2}}$.

When we write $\sqrt{2} = 1.41421 \cdots$, we mean that $\sqrt{2}$ is approximated better and better by the rational numbers

$$1.4 \quad 1.41 \quad 1.414 \quad 1.4142 \quad 1.41421 \quad \text{etc.}$$

We can actually calculate each of the numbers

$$3^{1.4} \quad 3^{1.41} \quad 3^{1.414} \quad 3^{1.4142} \quad 3^{1.41421} \quad \text{etc.}$$

since the exponents are rational. It seems reasonable to assume that these numbers approximate some real number better and better, and we call this number $3^{\sqrt{2}}$.

In fact, $3^{1.414} \approx 4.728$ and $3^{1.4142} \approx 4.729$ to four significant digits.

A similar thing may be done for any positive base b and any irrational exponent x, although higher mathematics is required for a proof. Hence we will simply assume that

Definition of b^x, $b > 0$　　　　For $b > 0$　　$f(x) = b^x$ is defined

for each real number x.

If $b = 1$, then $f(x) = 1^x = 1$, which is a constant function whose graph is a horizontal line. If $b \neq 1$, then the graph of $y = b^x$ is one of two basic types, depending upon whether $b > 1$ or $0 < b < 1$. The following table of values allows us to graph $y = 2^x$ and $y = (\frac{1}{2})^x = 2^{-x}$ in Fig. 8.1, and $y = 2^x$

Asymptote　　and $y = 3^x$ in Fig. 8.2. The x axis is a horizontal asymptote for both.

x	-3	-2	-1	0	1	2	3
2^x	$\frac{1}{8}$	$\frac{1}{4}$	$\frac{1}{2}$	1	2	4	8
$(\frac{1}{2})^x$	8	4	2	1	$\frac{1}{2}$	$\frac{1}{4}$	$\frac{1}{8}$
3^x	$\frac{1}{27}$	$\frac{1}{9}$	$\frac{1}{3}$	1	3	9	27

FIGURE 8.1

Graph of
$y = b^x$

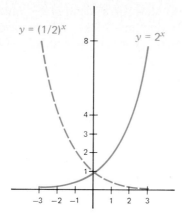

FIGURE 8.2

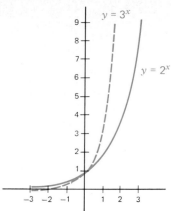

The graphs in Figs. 8.1 and 8.2, along with the table of values for 2^x, $(\frac{1}{2})^x$, and 3^x, illustrate the following statements. They are true for every exponential function $f(x) = b^x$ where $b > 0$ and $b \neq 1$.

Properties of $f(x) = b^x$

1 The domain is the set of all *real* numbers.
2 The range is the set of all *positive real* numbers.
3 If $x = 0$, then $y = 1$.
4 If $b > 1$, then $f(x) = b^x$ is increasing. If $0 < b < 1$, then $f(x) = b^x$ is decreasing.
5 $b^x = b^w$ if and only if $x = w$.

Property 1 means that whether x is positive or negative, rational or irrational, b^x is a well-defined real number. Property 2 implies that $b^x > 0$; hence the graph of $y = b^x$ is always above the x axis. Property 3 tells us that, just as all roads lead to Rome, all of the graphs of $y = b^x$, $b > 0$, go through the point $(0, 1)$.

The meaning of property 4 is that if r and s are real numbers with $r < s$, then the following situation holds:

$$\text{If } b > 1, \text{ then } b^r < b^s$$
$$\text{If } 0 < b < 1, \text{ then } b^r > b^s$$

For example since $2 > 1$,

$$2^4 < 2^5 \qquad \text{and} \qquad 2^{1.4} < 2^{1.41} < 2^{1.414}$$

and since $\frac{1}{2} < 1$ we have

$$(\tfrac{1}{2})^6 > (\tfrac{1}{2})^8 \qquad \text{and} \qquad (\tfrac{1}{2})^{0.3} > (\tfrac{1}{2})^{0.4} > (\tfrac{1}{2})^{0.5}$$

Property 5 follows directly from property 4. As a result of this, the only way that b^x can equal b^w is for x to equal w. For example, if $3^x = 1/9$, then since $1/9 = 1/3^2 = 3^{-2}$, we may write $3^x = 3^{-2}$ and we must conclude that $x = -2$.

The function $f(x) = b^x$ is a one-to-one function.

FIGURE 8.3

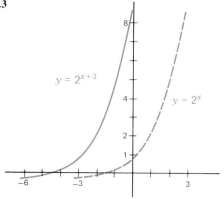

FIGURE 8.4

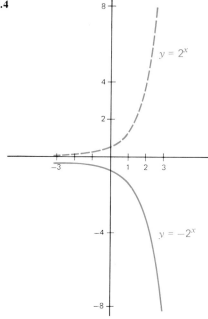

● **Example 1**   Graph $y = 2^{x+3}$ and $y = -2^x$.

Solution   We may graph $y = 2^{x+3}$ by translating the graph of $y = 2^x$ three units to the left, as we have seen in Chap. 6. An alternate method is to write $y = 2^{x+3} = 2^x \cdot 2^3 = 8(2^x)$, which shows that we just multiply each y value in $y = 2^x$ by 8. For $y = -2^x$, we multiply each y value in $y = 2^x$ by -1. This is equivalent to reflecting the graph of $y = 2^x$ in the x axis. See Figs. 8.3 and 8.4.   ●

It is interesting and important to note that for sufficiently large values of x, exponential functions (with $b > 1$) increase more rapidly than polynomial functions. See Example 2.

● **Example 2** 2^x eventually becomes and remains larger than x^{1000}. This is true because x^{1000} has 1000 factors, while 2^x has more and more factors of 2 as x increases. For instance, if $x = 14,000$ we have

$$2^x = 2^{14,000} = (2^{14})^{1000} = (16,384)^{1000} > (14,000)^{1000} = x^{1000}$$ ●

The number e Of particular interest is the number e, an irrational number which is approximately 2.718281828. The actual value of e is the limiting value of $(1 + 1/n)^n$ as n gets larger and larger. See also Prob. 65.

n	1	2	10	100	1000	100,000
$(1 + 1/n)^n$	2	2.25	2.59	2.705	2.717	2.7182818

The graph of $y = e^x$, although not shown, is between the two graphs in Fig. 8.2. The number e occurs often in both practical applications of actual physical situations and in developing the mathematical theory of certain ideas.

● **Example 3**   Solve   (a) $10^x = 1/10,000$ (b) $5^{x(x-1)} = 25$ (c) $3^x = 9^{2x-1} \cdot 27^{4-2x}$
(d) $3(9^x) - 28(3^x) + 9 = 0$

Solution   (a) $10^x = 1/10,000 = \dfrac{1}{10^4} = 10^{-4}$; thus $x = -4$

(b) $5^{x(x-1)} = 25 = 5^2$; thus $x(x - 1) = 2$, $x^2 - x - 2 = 0$, and $(x - 2)(x + 1) = 0$. Hence $x = 2$ or $x = -1$.

(c) $3^x = (3^2)^{2x-1} \cdot (3^3)^{4-2x} = 3^{4x-2} \cdot 3^{12-6x} = 3^{-2x+10}$

Hence $x = -2x + 10$, $x = \frac{10}{3}$.

(d)   The given equation may be written as $3(3^x)^2 - 28(3^x) + 9 = 0$ since $9^x = (3^2)^x = 3^{2x} = (3^x)^2$.
Letting $y = 3^x$ gives $3y^2 - 28y + 9 = 0$, $(3y - 1)(y - 9) = 0$, $y = 9$ or $y = \frac{1}{3}$.
Hence $3^x = 9 = 3^2$ and $x = 2$, or $3^x = \frac{1}{3} = 3^{-1}$ and $x = -1$.   ●

Many situations are governed by exponential functions of the form

$$f(t) = A \cdot b^{kt}$$

Exponential growth and decay

where A, b, and k are constants and t represents time. Taking $t = 0$ gives $f(0) = A \cdot b^0 = A$, and so A is the initial amount (the amount at the beginning). If $k > 0$ we have **exponential growth,** while if $k < 0$ we have **exponential decay.** The base b is positive, but not 1. In different situations, different bases are desirable. Often the base is chosen to be 2, e, or 10. We will see in Sec. 8.4 that we can change from one base to another by suitably changing the constant k.

In Example 4 below, we are given the exponential function to work with, while in Example 5 part of the problem is to find the function.

● **Example 4**   With stable market conditions and no advertising, the sales per week after t weeks is given by

$$f(t) = (86,400)e^{-t/10}$$

What are the sales at the beginning? After 1 week? After 2 weeks? After 7 weeks?

Solution   The sales at the beginning are

$$f(0) = (86,400)e^0 = 86,400$$

After 1 week, 2 weeks, and 7 weeks, the sales are

$$f(1) = (86,400)e^{-1/10} = 86,400e^{-0.1}$$
$$f(2) = (86,400)e^{-0.2}$$
$$f(7) = (86,400)e^{-0.7}$$

Decimal approximations to these values, using a calculator, are

$$f(1) \approx 86,400(0.905) \approx 78,178$$
$$f(2) \approx 86,400(0.819) \approx 70,738$$
$$f(7) \approx 86,400(0.497) \approx 42,905$$

Thus after 7 weeks, with no advertising, sales are about 49.7 percent of what they were initially. ●

● **Example 5** If the birthrate and other factors remain constant over a period of time, then the population after t years is $P(t) = A \cdot 2^{kt}$ where A is the initial population and k is a constant which depends on the situation. After 4 years, the population is $\sqrt[8]{2}$ as much as at the beginning (about a 9 percent increase).

(a) What is the value of k?
(b) What is $P(t)$?
(c) How long will it take for the population to double?

Solution (a) Since $P(4) = A \cdot \sqrt[8]{2}$, we have

$$A\sqrt[8]{2} = A \cdot 2^{k(4)} \qquad P(t) = A \cdot 2^{kt}$$

Hence $\sqrt[8]{2} = 2^{4k} \qquad 2^{1/8} = 2^{4k} \qquad 4k = \tfrac{1}{8} \qquad k = \tfrac{1}{32}$

(b) Since $k = \tfrac{1}{32}$, we have $P(t) = A \cdot 2^{t/32}$.
(c) We want to find t so that $P(t) = 2A$. Thus

$$A \cdot 2^{t/32} = 2A$$
$$2^{t/32} = 2 = 2^1$$
$$\frac{t}{32} = 1$$
$$t = 32 \text{ years}$$ ●

Exercise 8.1 Graph each of the following:

1 $y = 4^x$ **2** $y = 5^x$ **3** $y = e^x$ **4** $y = 10^x$
5 $y = (\tfrac{1}{3})^x$ **6** $y = (\tfrac{1}{4})^x$ **7** $y = (\tfrac{5}{8})^x$ **8** $y = (\tfrac{3}{5})^x$
9 $y = 2^{-x}$ **10** $y = -(\tfrac{1}{5})^{-x}$ **11** $y = -1/7^x$ **12** $y = 1/8^{-x}$
13 $y = 3^{x+3}$ **14** $y = 3^{x-3}$ **15** $y = -(\tfrac{2}{3})^{x-5}$
16 $y = -(\tfrac{1}{5})^{x+4}$ **17** $y = 2^{|x|}$ **18** $y = 2^{-(x^2)}$
19 $y = 2^x + 2^{-x}$ **20** $y = 2^x - 2^{-x}$

Solve the following equations.

21 $16^x = \frac{1}{8}$ **22** $81^{x/4} = \frac{1}{27}$ **23** $(7^{-x})^2 = 343$ **24** $(9^{3x})^3 = 243$

25 $5^{-(2x-1)x} = \frac{1}{5}$ **26** $3^{3(x+2)} \cdot 27 = 3^{2x+1}$

27 $2^{3(x+2)} \cdot 8 = 2^{2x+1} \cdot 8^{2/3}$ **28** $7^{3(x-1)} \cdot 343 = 7^{x+1} \cdot 49^{3/2}$

29 $(2^x)^2 - 20(2^x) + 64 = 0$ **30** $(3^x)^2 - 10(3^x) + 9 = 0$

31 $25^x - 6(5^x) + 5 = 0$ **32** $49^x - 50(7^x) + 49 = 0$

Problems 33 to 40 deal with equations of the form $f(t) = A \cdot b^{kt}$. If $k > 0$, we have exponential growth, while if $k < 0$ we have exponential decay.

Biology **33** The number of bacteria of type T_2 in a certain person's body doubles every 40 min. If 6 are present at 2 A.M., how many will there be at 12 noon? *Hint:* Take $t = 0$ at 2 A.M. Then $6 = f(0) = A \cdot b^{k(0)} = A$, and so $f(t) = 6 \cdot b^{kt}$. We may use $b = 2$ since we are given information about doubling; thus $f(t) = 6 \cdot 2^{kt}$. Determine k from the doubling every 40 min. Measure t in hours.

Physics **34** Under certain conditions, $I = 10^{D/10}$, where D is the number of decibels of a sound and I is its intensity. If he shouts with 30 more decibels than she talks with, what is the ratio of his intensity to hers?

Economics **35** $A = D(1 + i/n)^n$, where D dollars are invested at i compounded n times per year, and A is the value of the investment at the end of the year. Is it better to invest $1000 at $6\frac{1}{4}$ percent simple interest for 1 year ($n = 1$), or to invest it at 6 percent compounded monthly? *Hint:* $x^{12} = [(x^2)^2]^3$.

Anthropology **36** The population of a country is $P = 10^8 \cdot (1.5)^{t/20}$, where t is measured in years. How long will it take for the population to increase 125 percent? *Hint:* If the population increases by 100 percent, it doubles.

Physics **37** Atmospheric pressure P in pounds per square inch and altitude h in feet are related by $P = (14.7)2^{-0.0000555h}$. What is the approximate pressure at 18,000 ft?

Economics **38** If we want P dollars n years from now, we must invest $P(1 + i)^{-n}$ dollars now at i per year compounded annually. On a bond that pays 9 percent per year, about how much should be invested now in order to have $10,000 in 12 years?

Engineering **39** The temperature of a rod is 100°C initially, and it cools in air kept at 20°C. After t min, the temperature is $20 + 80(3^{-t})$. How long will it take to get below 21°C?

Physics **40** The half-life of radioactive material is the time in which half of it will disintegrate. If the amount after t years is $3(4^{-t/120})$ g, what is its half-life?

True or false

41 $2^{1/5} < 3^{1/8}$ *Hint:* Raise each number to the power $5(8) = 40$

42 $3^{1/4} < 5^{1/6}$ **43** $(\frac{4}{5})^{1/2} > \frac{4}{5}$ **44** $(\frac{3}{4})^{3/4} > \frac{3}{4}$

Simplify

45 $(7^{\sqrt{2}})^{\sqrt{8}}$ **46** $(4^{\sqrt{5}})^{-\sqrt{5}}$ **47** $(\sqrt{6}^{\sqrt{2}})^{\sqrt{2}}$ **48** $(\sqrt{5}^{\sqrt{2}/2})^{\sqrt{18}}$

Let $f(x) = 2^x + 2^{-x}$ and $g(x) = 2^x - 2^{-x}$. Find

49 $f(3) + g(3)$ **50** $f(3) - g(3)$ **51** $f(4)g(4)$ **52** $f^2(5) - g^2(5)$

In Probs. 53 to 56, show that $a^b = b^a$. You need not compute the actual values—just use the properties of exponents carefully.

53 $a = 2, b = 4$ **54** $a = (\frac{3}{2})^2, b = (\frac{3}{2})^3$

55 $a = (\frac{5}{3})^{3/2}, b = (\frac{5}{3})^{5/2}$ **56** $a = \left(\dfrac{1 + \sqrt{2}}{\sqrt{2}}\right)^{\sqrt{2}}, b = \left(\dfrac{1 + \sqrt{2}}{\sqrt{2}}\right)^{1+\sqrt{2}}$

Calculator   A calculator is useful for Probs. 57 to 60.

57 Find $5^{1.7}$, $5^{1.73}$, and $5^{1.732}$ **58** Find $5^{\sqrt{3}}$

59 Find $\sqrt{2 + \sqrt{2}}$ and $\sqrt{2 + \sqrt{2 + \sqrt{2 + \sqrt{2}}}}$ **60** $(\sqrt{1.8})^{\sqrt[3]{6}}$

Engineering   **61** If

$$\tanh x = \frac{e^x - e^{-x}}{e^x + e^{-x}}$$

show that $0.5(1 + \tanh x)$ can be written in the form $1/(1 + e^{-2x})$. This is called the *logistic function*.

Anthropology   **62** The equation

$$y = \frac{b \cdot 2^{ax}}{1 + c \cdot 2^{ax}}$$

occurs in studying population growth, where a, b, and c are positive numbers. What happens to y as x becomes larger and larger? *Hint:* Divide numerator and denominator by 2^{ax}.

Management   **63** The proportion of people informed by advertising after t weeks is

$$P = \frac{a - b}{a + c \cdot e^{(b-a)t}}$$

for appropriate constants a, b, and c. Find the value which P approaches if $1 > b/a$, as t becomes larger and larger.

Cooling   **64** According to Newton's law of cooling, if an object at temperature B is surrounded by a medium (air or water) at temperature A with $A < B$, then the temperature of the object after t min is

$$f(t) = A + (B - A)10^{-kt}$$

where k is a positive constant. What value does $f(t)$ approach as t becomes larger and larger?

Calculus and calculator   **65** It is proved in calculus that

$$f(n) = 1 + \frac{1}{1!} + \frac{1}{2!} + \frac{1}{3!} + \frac{1}{4!} + \cdots + \frac{1}{n!}$$

gets closer and closer to the number e as n gets larger and larger. Calculate $f(4)$, $f(6)$, and $f(8)$ to five decimal places.

66 Show that $4^{y+2} - 4^y = 15(4^y)$.

67 Show that $5^{t+2} + 3(5^{t+1}) - 40(5^t) = 0$.

68 Show $2^x + 2^{2x} = 2^x(1 + 2^x)$.

8.2 Logarithmic functions

We may solve the equation $2^x = 8$ since $8 = 2^3$. The solution is $x = 3$. Similarly the solution of $2^x = \frac{1}{4} = 2^{-2}$ is $x = -2$. However, $2^x = 5$ may not be treated in the same manner since 5 is not a "nice" power of 2.

Nonetheless, there is a solution to the equation $2^x = 5$. In fact, there is a solution x for the equation $2^x = N$ for any positive N. See Fig. 8.5. This solution is called $\log_2 N$, the logarithm of N to the base 2.

The same thing may be done with any base $b > 0$, $b \neq 1$. We recall from Sec. 8.1 that if $b > 1$, then $f(x) = b^x$ is increasing. We learned in Chap. 6 that an increasing function has an inverse function, and this inverse is also increasing. Similar comments apply for decreasing functions if $0 < b < 1$. Hence $y = b^x$ may be solved, uniquely, for x. This solution is called the logarithm of y to the base b and is written $x = \log_b y$.

We thus have the following definition of a logarithm for $b > 0$, $b \neq 1$, and $N > 0$.

Definition of $\log_b N$ $\qquad\qquad\qquad \log_b N = L \qquad$ if and only if $\qquad b^L = N \qquad\qquad$ (8.1)

A logarithm is an exponent. In fact, $\log_b N$ is the exponent which b must have to produce N. This follows directly from (8.1). We also have

$$\log_b 1 = 0 \qquad \text{since } b^0 = 1$$
$$\log_b b = 1 \qquad \text{since } b^1 = b$$
$$L = \log_b N = \log_b(b^L) \qquad \text{by (8.1)}$$
$$N = b^L = b^{(\log_b N)} \qquad \text{by (8.1)}$$

FIGURE 8.5

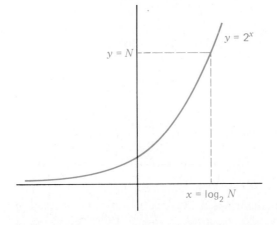

From the above definition and the table of values in Sec. 8.1, we may construct the following table of values for logarithms to the base 2:

N	$\frac{1}{8} = 2^{-3}$	$\frac{1}{4} = 2^{-2}$	$\frac{1}{2} = 2^{-1}$	$1 = 2^0$	$2 = 2^1$	$4 = 2^2$	$8 = 2^3$
$\log_2 N$	-3	-2	-1	0	1	2	3

● **Example 1**

$$\log_2 8 = 3 \qquad \text{since } 2^3 = 8$$
$$\log_4 8 = \tfrac{3}{2} \qquad \text{since } 4^{3/2} = (4^{1/2})^3 = 2^3 = 8$$
$$\log_{1/9} 3 = -\tfrac{1}{2} \qquad \text{since } (\tfrac{1}{9})^{-1/2} = (3^{-2})^{-1/2} = 3^1 = 3$$
$$\log_{16} \tfrac{1}{8} = -\tfrac{3}{4} \qquad \text{since } 16^{-3/4} = (2^4)^{-3/4} = 2^{-3} = \tfrac{1}{8}$$
$$\log_6(6^4) = 4$$
$$3^{\log_3 5} = 5$$

It is important to be able to change an equation from logarithmic form to exponential form, and vice versa.

● **Example 2**

Logarithmic equation	Exponential equation
$\log_4 3 = 5x$	$4^{5x} = 3$
$\log_5 4x = -\tfrac{1}{3}$	$5^{-1/3} = 4x$
$\log_x 2 = 6$	$x^6 = 2$
$\log_7(x - 3) = \tfrac{13}{3}$	$x - 3 = 7^{13/3}$

The following properties of $f(x) = \log_b x$, $b > 0$, $b \neq 1$, parallel the properties of exponential functions given in Sec. 8.1. Remember that $g(x) = b^x$ and $f(x) = \log_b x$ are inverse functions.

Properties of $f(x) = \log_b x$

1   The domain is the set of *positive real* numbers.
2   The range is the set of all *real* numbers.
3   $\log_b 1 = 0$.
4   If $b > 1$, then $f(x) = \log_b x$ is increasing. If $0 < b < 1$, then $f(x) = \log_b x$ is decreasing.
5   $\log_b x = \log_b w$ if and only if $x = w$.

● **Example 3**   Solve the following equations:
(a) $\log_b 32 = \tfrac{5}{2}$ (b) $\log_9 \tfrac{1}{27} = L$ (c) $6^L = 7$

Solution   (a) $\log_b 32 = \tfrac{5}{2}$ is equivalent to $b^{5/2} = 32$; hence

$$b = 32^{2/5} = (\sqrt[5]{32})^2 = 2^2 = 4$$

(b)   This is equivalent to $9^L = \tfrac{1}{27}$; hence

$$(3^2)^L = 3^{-3}$$
$$3^{2L} = 3^{-3}$$
$$2L = -3$$
$$L = -\tfrac{3}{2}$$

(c) $6^L = 7$ is the same as $L = \log_6 7$, by (8.1).   ●

We know that $y = b^x$ is equivalent to $x = \log_b y$. If we exchange x and y in the last equation to get the inverse, we have $y = \log_b x$. Since the inverse of $f(x) = b^x$ is $f^{-1}(x) = \log_b x$ the definition (8.1) shows that

$$f^{-1}(f(x)) = f^{-1}(b^x) = \log_b (b^x) = x$$
$$f(f^{-1}(x)) = f(\log_b x) = b^{\log_b x} = x$$

Inverse functions   By Chap. 6, we know that the graphs of f and f^{-1} are reflections of each other through the line $y = x$. See Fig. 8.6, in the case $b > 1$.

Notice that $(3, 8)$ and $(8, 3)$ are, respectively, on the graphs of $y = 2^x$ and $y = \log_2 x$. Similarly for $(-4, \frac{1}{16})$ and $(\frac{1}{16}, -4)$.

In graphing $y = \log_2(3x - 5)$, we must remember that $3x - 5$ must be positive; hence $3x > 5$, or $x > \frac{5}{3}$. The domain is thus $\{x \mid x > \frac{5}{3}\}$.

In both calculations and in simplifying expressions, the following rules are basic.

Laws of logarithms

$$\log_b(MN) = \log_b M + \log_b N \tag{8.2}$$

$$\log_b\left(\frac{M}{N}\right) = \log_b M - \log_b N \tag{8.3}$$

$$\log_b(N^p) = p \cdot \log_b N \tag{8.4}$$

We will prove only (8.2) here. Let $x = \log_b M$ and $y = \log_b N$. Then

$$
\begin{array}{ll}
M = b^x \quad \text{and} \quad N = b^y & \text{definition of log} \\
MN = b^x \cdot b^y = b^{x+y} & \text{multiplying} \\
\log_b(MN) = x + y & \text{using logarithmic form} \\
\qquad\quad = \log_b M + \log_b N & \text{definition of } x \text{ and } y
\end{array}
$$

FIGURE 8.6

Graph of
$y = \log_b x$

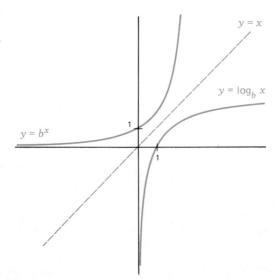

It is important to remember (8.2), (8.3), and (8.4) in words.

(8.2)   The logarithm of a product is the sum of the logarithms.
(8.3)   The logarithm of a quotient is the difference of the logarithms.
(8.4)   The logarithm of a power of a number is the product of the exponent and the logarithm of the number.

Note   In general, $\log_b(M + N) \neq \log_b M + \log_b N$

and $\log_b(MN) \neq (\log_b M)(\log_b N)$

Also $(\log_b x)^2 = (\log_b x)(\log_b x)$

but $\log_b(x^2) = 2 \log_b x$

● **Example 4** $\log_b \dfrac{xy^3}{w^2} = \log_b(xy^3) - \log_b(w^2)$ by (8.3)

$$= \log_b x + \log_b(y^3) - \log_b(w^2)   \quad \text{by (8.2)}$$

$$= \log_b x + 3 \log_b y - 2 \log_b w   \quad \text{(by (8.4)}$$

These steps in this example may be reversed to write the last expression as a single logarithm.   ●

● **Example 5**   If $\log_{10} 2 = 0.30$ and $\log_{10} 3 = 0.48$, then

$$\log_{10} 180 = \log_{10}(10)(18) = \log_{10}(10)(2)(3^2)   \quad \text{factoring}$$
$$= \log_{10} 10 + \log_{10} 2 + 2 \log_{10} 3   \quad \text{by (8.2) and (8.4)}$$
$$= 1 + 0.30 + 2(0.48) = 2.26$$ ●

One case of (8.4) deserves special mention. Since $\sqrt[r]{N} = N^{1/r}$, we have

$$\log_b \sqrt[r]{N} = \frac{1}{r} \log_b N \tag{8.5}$$

● **Example 6** $\log_{10} \sqrt[3]{6} = (\tfrac{1}{3})\log_{10} 6$

$$= (\tfrac{1}{3})(\log_{10} 2 + \log_{10} 3) = (\tfrac{1}{3})(0.30 + 0.48)$$
$$= (\tfrac{1}{3})(0.78) = 0.26$$ ●

● **Example 7**   Suppose $\log_7 2 = a$, $\log_7 3 = b$, and $\log_7 5 = c$. Then

(i) $\log_7 28 = \log_7(2 \cdot 2 \cdot 7) = \log_7 2 + \log_7 2 + \log_7 7 = 2a + 1$

(ii) $\log_7(\tfrac{27}{25}) = \log_7 27 - \log_7 25 = \log_7(3^3) - \log_7(5^2)$
$$= 3 \log_7 3 - 2 \log_7 5 = 3b - 2c$$

(iii) $\log_7 \sqrt{48} = \log_7(48)^{1/2} = \tfrac{1}{2}\log_7 48 = \tfrac{1}{2}\log_7(2^4 \cdot 3)$

$$= \tfrac{1}{2}\log_7 2^4 + \tfrac{1}{2}\log_7 3 = 4(\tfrac{1}{2})(\log_7 2) + \tfrac{1}{2}\log_7 3 = 2a + \frac{b}{2}$$

(iv) $\dfrac{\log_7 3}{\log_7 5} = \dfrac{b}{c}$

and no further simplification by the present laws is possible. However, later, in Eq. (8.7), we will find that $\log_7 3/\log_7 5 = \log_5 3 \approx 0.6826$. In particular, notice that the values of $\log_7 3/\log_7 5$ and $\log_7 \dfrac{3}{5}$ are different. ●

As we stated in the definition of a logarithm, the base b must be positive and not 1. The bases most often used are 10 and e. We will discuss logarithms with base 10 in the next section on common logarithms. For logarithms with base e, see Eq. (8.7).

Note   Remember:

A logarithm is an exponent.
$\log_b N$ is the exponent which b must have to give N.
$\log_b N$ is only defined for $N > 0$ and $b > 0$ ($b \neq 1$).
The value of $\log_b N$ may be positive, 0, or negative.

Exercise 8.2

Change Probs. 1 to 8 to logarithmic form and Probs. 9 to 16 to exponential form.

1 $2^3 = 8$ **2** $3^5 = 243$ **3** $5^{-2} = \frac{1}{25}$ **4** $7^{-1} = \frac{1}{7}$
5 $32^{3/5} = 8$ **6** $27^{2/3} = 9$ **7** $(\frac{1}{2})^{-4} = 16$ **8** $(\frac{1}{3})^{-5} = 243$
9 $\log_3 81 = 4$ **10** $\log_7 49 = 2$ **11** $\log_5 \frac{1}{125} = -3$
12 $\log_4 \frac{1}{64} = -3$ **13** $\log_8 32 = \frac{5}{3}$ **14** $\log_{32} 8 = \frac{3}{5}$
15 $\log_{64} 16 = \frac{2}{3}$ **16** $\log_{27} 81 = \frac{4}{3}$

Solve each of the following equations by using (8.1).

17 $\log_2 32 = L$ **18** $\log_7 343 = L$ **19** $\log_5 625 = L$ **20** $\log_6 216 = L$
21 $\log_3 N = 2$ **22** $\log_2 N = 5$ **23** $\log_3 N = 4$ **24** $\log_3 N = 5$
25 $\log_b 8 = 3$ **26** $\log_b 81 = 4$ **27** $\log_b 125 = \frac{3}{2}$ **28** $\log_b 64 = \frac{6}{5}$

Sketch the graph of the following equations:

29 $y = \log_3 x$ and $y = 3^x$ **30** $y = \log_5 x$ and $y = 5^x$
31 $y = \log_2 x$ and $y = \log_3 x$, $x > 1$
32 $y = \log_5 x$ and $y = \log_{10} x$, $x > 1$
33 $y = \log_5(x^2)$ **34** $y = \log_5(3x)$ **35** $y = \log_2(x - 3)$
36 $y = \log_3(2x + 5)$ **37** $x = \log_2 y$ **38** $y = \log_{1/2} x$
39 $2x = 3^y$ **40** $3y = 2^{(\log_2 x^2)}$

Calculus   In calculus, it is important to simplify certain functions before working with them further (finding a derivative, for distance). In Probs. 41 to 44, use (8.2), (8.3), and (8.4), to change as much as possible the expression to sums and differences of logarithms.

41 $\log_e \left[\dfrac{x(x^3 - 1)}{5x + 3} \right]$ **42** $\log_e \left[\dfrac{x^4(x + 1)^{2/3}}{x + 23} \right]$

43 $\log_e \left[\dfrac{(x - 4)(2x + 7)^5}{x(3x + 8)^3} \right]$ **44** $\log_e \left[\dfrac{(x^2 + 5)(7x - 96)}{3x \sqrt{5x + 37}} \right]$

Express as a single logarithm.

45 $\log_b 3 + 2 \log_b x + 3 \log_b y^3$ **46** $2 \log_b x - \frac{1}{2} \log_b y + \frac{1}{3} \log_b z$
47 $\log_b xy^2 + 2 \log_b x/y$ **48** $\log_b(x + y) + 2 \log_b y - 3 \log_b x$

Use $\log_{10} 2 = 0.30$, $\log_{10} 3 = 0.48$, and $\log_{10} 7 = 0.85$ to find the following logarithms.

49 $\log_{10} 28$ **50** $\log_{10} 630$ **51** $\log_{10} \left(\dfrac{27}{\sqrt[4]{3}} \right)$ **52** $\log_{10} \left(\dfrac{49}{36} \right)$

Is the statement in each of Probs. 53 to 60 true or false?

True or
false

53 $(\log_2 1)(\log_4 5) = \log_2 8 - \log_3 27$ **54** $2 + \log_{16} 16 = 5^{\log_5 3}$
55 $3^{2 \log_3 4} = (\log_7 7^8)(\log_7 14)$ **56** $\log_6 6^6 = 6 \log_6 6$
57 $\log_3 17 < \log_4 17$ **58** $\log_9 2 + \log_9 3 > \log_9 5$
59 $\log_3(3 + 9) < \log_6 6 + \log_3 4$ **60** $(\log_{10} 4)(\log_{10} 3) < \log_{10} 12$

Proof

61 Show that $\log_b N = \log_{(b^3)}(N^3)$. *Hint:* If $x = \log_b N$, then $b^x = N$ and $(b^x)^3 = N^3$.
62 Show that $x^{\log_2(\log_2 x)} = (\log_2 x)^{\log_2 x}$. *Hint:* Take the logarithm of both sides.
63 Show that $x^5 = 10^{5 \log_{10} x}$. *Hint:* $5 \log_{10} x = \log_{10}(x^5)$.
64 Show that $(\log_b a)(\log_a b) = 1$. *Hint:* If $x = \log_b a$, then $b^x = a$ and $b = a^{1/x}$.

Use a calculator, if one is available, to verify each of the following equations. Change the equation from logarithmic to exponential form first.

Calculator

65 $\log_3 46.77 = \frac{7}{2}$ **66** $\log_5 11.18 = \frac{3}{2}$ **67** $\log_4 13.93 = 1.9$
68 $\log_6 442.3 = 3.4$

Anthropology

69 Under random genetic drift, the kinship coefficient f between populations is related to the time of separation t by

$$f = 1 - e^{-t/2N}$$

where N is the population size. Solve for the time t.

Anthropology

70 The formula for measuring information, developed by Shannon, is

$$H = -p_1 \log_2 p_1 - p_2 \log_2 p_2 - \cdots - p_n \log_2 p_n$$

(a) Show $H = p_1 \log_2(1/p_1) + p_2 \log_2(1/p_2) + \cdots + p_n \log_2(1/p_n)$.
(b) Calculate H for $n = 4$, $p_1 = \frac{1}{8}$, $p_2 = \frac{1}{8}$, $p_3 = \frac{1}{4}$, $p_4 = \frac{1}{2}$.

Anthropology   **71** The equation

$$F = \frac{\log_e\left(\dfrac{N-s}{N}\right)}{\log_e\left(\dfrac{N-1}{N}\right)}$$

arises in the statistical theory of marriage structures. Solve it for s.

Management   **72** In a department store with sales of S boxes per month, n boxes of each brand, and v the number of brands, we have $nv = c = $ constant and want to maximize $S = k \cdot c^v \cdot v^{c/v - v}$ as a function of v.

(a)   Show that S has a maximum, as a function of v, if

$$\log k + v \log c + \left(\frac{c}{v} - v\right)\log v$$

has a maximum.

(b)   If $\log_e c < 4$, then $v = \sqrt{c}$ makes S a maximum. Find $S(\sqrt{c})$.

Psychology   **73** Fechner's law in psychology about a stimulus S is

$$S = S_0(1 + c)^r$$

where S_0 is a standard stimulus, c is a constant, and r is the number of thresholds. Solve the equation for r.

Management   **74** In a short, intense advertising campaign, the response R and dollars received A are related by

$$R = c \log_{10} A + d$$

where $c > 0$ and $d < 0$.

(a)   Find the threshold amount of advertising by solving for A the equation

$$0 = c \log_{10} A + d$$

(b)   If there are x campaigns with each one getting an equal amount of the total of A dollars, what is the total response from all campaigns?

Anthropology   **75** In the study of agricultural terraces in Ecuador, the following equation arises:

$$A = \frac{WH}{2} \log_e\left(\frac{x^2}{W^2} + 1\right)$$

Solve the equation for x.

8.3   Common logarithms

We have defined $\log_b N$ for any base $b > 0$, $b \neq 1$. Since our decimal system of numbers is based on 10, the most convenient base for logarithms

Common logarithms   in calculation is base 10. Logarithms with base 10 are called *common logarithms* or *Briggs logarithms*.

The other base which is used often is e, the number defined in Sec. 8.1;
Natural | e is approximately 2.72. Logarithms will base e are called *natural or naperian*

Natural
logarithms

The other base which is used often is e, the number defined in Sec. 8.1; e is approximately 2.72. Logarithms will base e are called *natural or naperian logarithms,* and they are the most convenient for theoretical work. See also Eq. (8.8) for a relationship between common and natural logarithms. The notation used is $\ln x = \log_e x$.

$\log N$ means $\log_{10} N$

For the remainder of this chapter, $\log N$ will mean $\log_{10} N$.

If N is a power of 10, its common logarithm is the integer exponent of 10. For example,

$$\log 0.01 = \log_{10} 10^{-2} = -2 \log_{10} 10 = -2(1) = -2$$
$$\log 0.1 = \log 10^{-1} = -1$$
$$\log 1 = \log 10^0 = 0$$
$$\log 10 = \log 10^1 = 1$$
$$\log 100 = \log 10^2 = 2$$
$$\log 1000 = \log 10^3 = 3$$

Scientific
notation

Regardless of whether N is a power of 10, we may write N in scientific notation as $N = m \cdot 10^c$, where $1 \le m < 10$ and c is an integer. Then

$$\log N = \log(m \cdot 10^c)$$
$$= \log m + \log 10^c$$
$$= c + \log m$$

Since $1 \le m < 10$, then $\log 1 \le \log m < \log 10$, i.e., $0 \le \log m < 1$. We call c the *characteristic* of $\log N$, and we call $\log m$ the *mantissa*. Thus

$$\log N = c + \log m$$
$$= \text{characteristic} + \text{mantissa} \qquad (8.6)$$

Characteristic
Mantissa
Position of the
decimal point

where the *characteristic* is an integer $(\ldots, -3, -2, -1, 0, 1, 2, 3, \ldots)$
        the *mantissa* is 0 or between 0 and 1 (it is *positive* or 0)

The characteristic of $\log N$ is determined completely by the position of the decimal point in N. For example, if N is any number between 10 and 100, then $\log N$ will have the characteristic 1.

Sequence of digits

The mantissa of $\log N$ is determined completely by the sequence of digits in N. The position of the decimal point is immaterial. For example, the mantissa of $\log N$ is the same whether N is 483 or 48.3 or 0.000483.

There are several ways to find the value of $\log N$ when N is given. All the methods must use some form of approximation to give a decimal value since $\log 7$, for example, is irrational while the approximations 0.8451 and 0.84509804 are rational.

They are good approximations, accurate to four or eight decimal places, but nonetheless approximations. As such, slight differences may appear in the last decimal place depending on round-off error, order of operations, and other factors. Whole books have been written on error analysis.

One method for finding $\log N$ is to use a calculator. With so many models available, the best advice is simply to read *and understand* the owner's manual. This is especially true in a sequence of calculations, such as

$$\frac{3 + \log 2.8}{(6 \log 4.1)^{1/2}} \approx 1.797761$$

Calculator   Generally, to find log N on a calculator, the number N must be entered on the display and then the log key is pushed. For instance, log 5.84 is found by pushing the keys.

$$\boxed{5}\ \boxed{\cdot}\ \boxed{8}\ \boxed{4}\ \boxed{\log} \qquad \text{giving } 0.76641285$$

to eight decimal places. Thus the mantissa is 0.76641285 and the characteristic is 0.

We may also use Table A.1 to find log N for any positive N. Actually the values in the table are the values of the mantissa of log N. They are printed without decimal points. We must figure out the characteristic for ourselves. We have included a portion of the table below. See also Fig. 8.7.

Part of a table of logarithms

N	0	1	2	3	4	5	6	7	8	9
55	7404	7412	7419	7427	7435	7443	7451	7459	7466	7474
56	7482	7490	7497	7505	7513	7520	7528	7536	7543	7551
57	7559	7566	7574	7582	7589	7597	7604	7612	7619	7627
58	7634	7642	7649	7657	7664	7672	7679	7686	7694	7701
59	7709	7716	7723	7731	7738	7745	7752	7760	7767	7774
60	7782	7789	7796	7803	7810	7818	7825	7832	7839	7846
61	7853	7860	7868	7875	7882	7889	7896	7903	7910	7917
62	7924	7931	7938	7945	7952	7959	7966	7973	7980	7987
63	7993	8000	8007	8014	8021	8028	8035	8041	8048	8055
64	8062	8069	8075	8082	8089	8096	8102	8109	8116	8122

● **Example 1**
Given N, to find log N

To find log 5.84, we first find 58 in the left column, then find 4 in the top row. The number in the intersection of this column and row is 7664. Hence the mantissa is 0.7664. Since $5.84 = 5.84(10^0)$, the characteristic is 0. Hence

$$\log 5.84 = 0 + 0.7664 = 0.7664$$

Further, since $58.4 = 5.84(10^1)$ and $58,400 = 5.84(10^4)$ we have

$$\log 58.4 = 1 + 0.7664 = 1.7664$$
$$\log 58,400 = 4 + 0.7664 = 4.7664$$

●

FIGURE 8.7

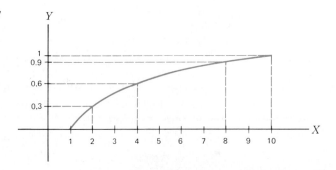

● **Example 2**

N	$\log N$ by table	$\log N$ by calculator
6.40	0.8062	0.80617997
59.5	1.7745	1.77451697
571	2.7566	2.75663611
6090	3.7846	3.78461729

All the examples above have $N > 1$, for which the characteristic is positive or zero. If $0 < N < 1$, then $\log N < 0$ and the characteristic is negative. To find $\log 0.0579$, we see from the table that the mantissa is 0.7627. Since $0.0579 = 5.79(10^{-2})$, then

$$\log 0.0579 = -2 + 0.7627$$

Note   We may write this as $-2 + 0.7627 = -1.2373$, which is the correct value for $\log 0.0579$. However, it is not the standard form since $-1.2373 = -1 - 0.2373$; there is no mantissa here since -0.2373 is negative, whereas a mantissa is by definition positive or 0. The standard method is to write -2 as $8 - 10$, although $18 - 20$ or $58 - 60$ are sometimes advantageous. Thus

$$\log 0.0579 = -2 + 0.7627 = 8.7627 - 10$$

By calculator, $\log 0.0579 = -1.23732144 = 8.76267856 - 10$.

● **Example 3**

N	$\log N$ by table	$\log N$ by calculator
0.576	$9.7604 - 10$	-0.23957752
0.0551	$8.7412 - 10$	-1.25884840
0.00622	$7.7938 - 10$	-2.20620962
0.000646	$6.8102 - 10$	-3.18976748

Given $\log N$, to find N   If we know $\log N$, we may reverse the steps above to find N. We first locate the mantissa in the body of the table, and then get the sequence of digits in N from the left and top of the table. The characteristic will then enable us to place the decimal point properly.

● **Example 4**   Suppose $\log N = 1.7875$. Since 7875 occurs at the intersection of 61 and 3, we have 613 as the digits in N. Since the characteristic is 1, $N = 61.3$. Similarly if $\log N = 8.7875 - 10$, then $N = 0.0613$.

Antilog   The number N is sometimes called the *antilog* of $\log N$. By the definition of logarithms and antilogs

$$\text{antilog}(\log N) = N = 10^{\log N}$$

Using a calculator to find N given log N

The equation $N = 10^{\log N}$ can be used to find N on a calculator if we are given log N. This requires that the calculator have a key labeled $\boxed{x^y}$ or $\boxed{y^x}$ or $\boxed{10^x}$. As in Example 4, if log $N = 1.7875$, then

$$N = 10^{1.7875} = 61.30557921$$

to eight decimal places by using the $\boxed{10^x}$ key if 1.7875 is in the display.

● **Example 5**

log N	N by table	N by calculator
4.8000	63,100	63,095.734
2.7903	617	617.02108
9.7774 − 10	0.599	0.59896301
7.7574 − 10	0.00572	0.0057200523

●

In all our examples so far we have used number N with three significant digits. If we are dealing with approximate numbers with more than three significant digits, we may either:

(a) Round off to three significant digits, then proceed as earlier.
(b) Round off to four significant digits, if necessary, then *interpolate*—we will do that shortly.

Calculator

If a calculator is used, interpolation is not necessary. The value of log 59.56, for example, is 1.77495469 to eight decimal places.

Interpolation

If we are asked to find log 59.56 directly from the table, we cannot do it since there are four significant digits in 59.56. However, since $y = \log x$ is increasing, we know that

$$\log 59.50 < \log 59.56 < \log 59.60$$

Furthermore, since 59.56 is $\frac{6}{10}$ of the way from 59.50 to 59.60, we will

FIGURE 8.8

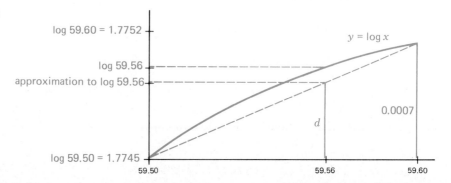

assume that log 59.56 is $\frac{6}{10}$ of the way from log 59.50 to log 59.60; that is, $\frac{6}{10}$ of the way from 1.7745 to 1.7752. Now

$$(\tfrac{6}{10})(1.7752 - 1.7745) = (\tfrac{6}{10})(0.0007) = 0.0004 \qquad (1)$$

to four decimal places; so we write

$$\log 59.56 = 1.7745 + 0.0004 = 1.7749 \qquad (2)$$

This work may be summarized as follows:

$$
\begin{array}{ccc}
 & N & \log N \\
 & 59.50 & 1.7745 \\
10\begin{bmatrix}6\begin{bmatrix}\\ 59.56\end{bmatrix} & & \end{bmatrix}d \Big] 0.0007 \\
 & 59.60 & 1.7752
\end{array}
$$

$$\frac{6}{10} = \frac{d}{0.0007} \qquad d = \frac{6}{10}(0.0007) = 0.0004 \qquad (3)$$

This takes us through Eq. (1), and the interpolation is finished as above.

Interpolation   This method of *interpolation* is also called linear interpolation because it is based on replacing a small part of the graph of $y = \log x$ by a straight line as in Fig. 8.8.

Equation (3) comes from equating the ratios of corresponding sides of similar triangles.

● **Example 6**   Find log 1.703.

Solution

$$
\begin{array}{ccc}
 & N & \log N \\
 & 1.700 & 0.2304 \\
10\begin{bmatrix}3\begin{bmatrix}\\ 1.703\end{bmatrix} & & \end{bmatrix}d \Big] 0.0026 \\
 & 1.710 & 0.2330
\end{array}
$$

Finding
log N by
interpolation

$$\frac{3}{10} = \frac{d}{0.0026} \qquad d = \frac{3}{10}(0.0026) = 0.0008$$

$$\log 1.703 = 0.2304 + 0.0008 = 0.2312$$

By calculator, log 1.703 = 0.23121465.   ●

We may use the same principle to find antilogs; that is, to find N if log N is given to four decimal places and the four-place mantissa is not in the logarithm tables.

● **Example 7**   Find N if log $N = 2.9256$.

Solution

$$
\begin{array}{cc}
N & \log N
\end{array}
$$

$$
1.0\left[d\left[\begin{array}{cc} 842.0 & 2.9253 \\ & 2.9256 \end{array}\right] 0.0003 \right] 0.00005
$$

$$
\begin{array}{cc} 843.0 & 2.9258 \end{array}
$$

Finding N by interpolation

$$
\frac{d}{1.0} = \frac{0.0003}{0.0005} \qquad d = \frac{3}{5} = 0.6
$$

$$
N = 842.0 + 0.6 = 842.6
$$

By calculator, $N = 10^{\log N} = 10^{2.9256} = 842.55838$.   ●

The examples above show that the differences used in the ratios depend on the mantissas, not on the characteristics. Thus we may treat numbers with negative logarithms in a similar manner.

● **Example 8**   Find N if $\log N = 9.4835 - 10$.

Solution

$$
\begin{array}{cc}
N & \log N
\end{array}
$$

$$
0.001\left[d\left[\begin{array}{cc} 0.304 & 9.4829 - 10 \\ & 9.4835 - 10 \end{array}\right] 0.0006 \right] 0.0014
$$

$$
\begin{array}{cc} 0.305 & 9.4843 - 10 \end{array}
$$

$$
\frac{d}{0.001} = \frac{0.0006}{0.0014} \qquad d = 0.001\left(\frac{6}{14}\right) = 0.0004
$$

$$
N = 0.304 + 0.0004 = 0.3044
$$

By calculator, $N = 10^{\log N} = 10^{9.4835 - 10} = 10^{-0.5165} = 0.30443880$.   ●

Exercise 8.3 use of tables, interpolation

Verify that the logarithms of the numbers in each problem have the same characteristic.

1 10.3; 93.5; 68.44; 31.002; 85
2 2024; 8100; 4343.43; $6.3(10^3)$; 2500
3 0.122; 0.3589; 0.18162; $4.3(10^{-1})$; 0.9934
4 0.0022; 0.00103; 0.008765; $5.23(10^{-3})$; 0.00680

Verify that the logarithms of the numbers in each problem have the same mantissa.

5 642; 64.2; 0.00642; $6.42(10^7)$
6 530; 0.000530; 0.0530; $5.30(10^{-8})$
7 9173; 0.009173; 0.9173; $9.173(10^4)$
8 666; 0.0666; 6.66; $6.66(10^{-5})$

Find the common logarithms of the following numbers. Interpolate in Probs. 25 to 32. Using a calculator is optional.

9 3.76   **10** 9.83   **11** 2.05   **12** 8.64   **13** 38.9   **14** 50.2
15 673   **16** 4680   **17** 0.323   **18** 0.711   **19** 0.0307   **20** 0.0276
21 0.0052   **22** 0.000218   **23** 0.00923   **24** 0.0000606
25 2762   **26** 5839   **27** 77.82   **28** 187.7
29 0.5723   **30** 0.8702   **31** 0.05006   **32** 0.003008

If log N is the number given in each of Probs. 33 to 48, find N. Using a calculator is optional.

33 0.3284   **34** 1.4871   **35** 2.7627   **36** 10.9375
37 3.6646   **38** 4.5694   **39** 6.9106   **40** 5.7767
41 9.9661 − 10   **42** 8.2504 − 10   **43** 7.4683 − 10   **44** 8.5866 − 10
45 7.9576 − 10   **46** 6.0719 − 10   **47** 9.8215 − 10   **48** 8.7059 − 10

In each of Probs. 49 to 52, log N is given, and the value of N to three significant digits is to be found by using the entry in the table that is at least as near the given value of log N as any other entry in the table. Using a calculator is optional.

49 1.3427   **50** 2.4973   **51** 8.7186 − 10   **52** 7.8999 − 10

By use of interpolation, find the value of N to four digits if log N is the number given in each of Probs. 53 to 60. Using a calculator is optional.

53 2.2933   **54** 1.4892   **55** 0.0940   **56** 4.7764
57 9.7013 − 10   **58** 7.7770 − 10   **59** 8.7654 − 10   **60** 6.2882 − 10

Use the characteristic to determine the number of digits in each of the following numbers:

61 8^{17} **62** 6^{73} **63** 92^{61} **64** 117^{54}

Anthropology and farming   **65** The equation $y = a + bx + cx^2 + d \log_{10} x$ is used for population growth changes, the change of size of eggs with successive layings, and change of milk production with age. By fitting data from the census reports of 1790–1860 to an equation, we get

$$y = f(x) = 8,600,000 - 5,700,000x + 82,000x^2 + 96,000,000 \log_{10} x$$

for population. Find $f(50)$, thereby "predicting" the 1910 population of the United States 50 years in advance.

Genealogy   **66** The maximum likelihood estimate of the mean in the study of evolutionary tree structures in genealogy is

$$m = \frac{p - 1}{p \log_{10} p} \qquad \text{where } 0 < p < 1$$

Calculate m for $p = 0.5$ and 0.1.

Management **67** Storing wine increases its value, but also costs more. One of the equations which must be solved for t, in years, is

$$0.81 = \log_e(1 + t) - \frac{t}{t + 1} = 2.3 \log_{10}(1 + t) - \frac{t}{1 + t}$$

Which positive integer comes closest to satisfying this equation: 4, 5, or 6?

Geology **68** The temperature of lava after a volcanic eruption is $y = 15 + (800)10^{-0.06t}$, where t is in hours and y is the Celsius reading in degrees. Show that

$$t = -\frac{1}{0.06} [\log(y - 15) - \log 800]$$

and find the time when the temperature is 100°C.

  Applications

Approximate numbers Recall that in a number which is an approximation the digits known to be correct are called significant digits. The digits 9, 8, 7, 6, 5, 4, 3, 2, 1 are always significant, whereas 0 sometimes is and sometimes is not.

Do not confuse *significant digits* with *decimal place accuracy*. The number 63.492 has five significant digits but only three decimal places of accuracy.

Our rules for calculating with approximate numbers were given in Sec. 4.1. They are:

Products If x and y each has n significant digits, then round off xy and x/y to n significant digits.

Sums If x and y each is accurate to n decimal places, then round off $x + y$ and $x - y$ to n decimal places.

Since powers and roots are defined ultimately by means of products, we have the following rule:

Powers and roots If x has n significant digits, then round off x^r to n significant digits for any real number r.

If the number N is accurate to three significant digits, then the mantissa of $\log N$ is usually accurate to only three or four decimal places. For instance, if N is the approximate number 9.97, then we have N between 9.965 and 9.975, and the logarithms are

$$\log 9.965 = 0.99847730$$
$$\log 9.97 = 0.99869516$$
$$\log 9.975 = 0.99891290$$

For this reason, if N is given as an approximation to three or four digits, then the last few digits of $\log N$ on a calculator display are inconsequential—they represent only apparent accuracy.

Note Furthermore, since values in a table are rounded off, there will inevitably be some inconsistencies in Table A.1. For instance, log 36 = 1.5563, and 36 = 6², but 2 log 6 = 2(0.7782) = 1.5564. In short, the numbers in any log table, from a calculator or the table in this book, are decimal approximations; so do not be surprised or upset if answers obtained in different ways are slightly different.

Calculators are *very* useful for their intended purposes of computing and serving as tables. However, the person using a calculator or a table must still think and understand the situation. The equation must be set up, then solved in a manner ready for calculation.

● **Example 1**   Use logarithms to find $P = (2.36)(0.358)(719)$.

Solution   By (8.2), the logarithm of a product is the sum of the logarithms, and so

$$\begin{aligned}\log P &= \log 2.36 + \log 0.358 + \log 719 \\ &= 0.3729 + [9.5539 - 10] + 2.8567 \\ &= 12.7835 - 10 = 2.7835\end{aligned}$$

Using the nearest entry to the mantissa 0.7835 in the table gives

$$P = \text{antilog } 2.7835 = 607$$

By calculator, the value is 607.46872, which is 607 rounded off to three digits.   ●

● **Example 2**   Find $Q = \dfrac{473}{(28.1)(75.9)}$.

Solution   By (8.3) and (8.2),

$$\begin{aligned}\log Q &= \log 473 - [\log 28.1 + \log 75.9] \\ &= 2.6749 - 1.4487 - 1.8802 \\ &= -0.6540 = (-0.6540 + 10) - 10 = 9.3460 - 10\end{aligned}$$

Using the nearest entry to mantissa 0.3460 in the table gives $Q = 0.222$. By calculator, the value is 0.22177523, which is 0.222 rounded off to three digits.   ●

● **Example 3**   Find $R = \sqrt[4]{\dfrac{(0.318)^3}{16.5}}$.

Solution   Using (8.3), (8.4), and (8.5) gives

$$\begin{aligned}\log R &= \tfrac{1}{4}[\log(0.318^3) - \log 16.5] \\ &= \tfrac{1}{4}[3(9.5024 - 10) - 1.2175] \\ &= \tfrac{1}{4}[28.5072 - 30 - 1.2175] \\ &= \tfrac{1}{4}[-2.7103] = \tfrac{1}{4}[37.2897 - 40] \qquad \text{adding } 40 - 40 \text{ so} \\ &\qquad\qquad\qquad\qquad\qquad\qquad\qquad\qquad\quad \text{we may divide by 4}\end{aligned}$$

$$\begin{aligned}&= 9.3224 - 10 \\ R &= 0.210\end{aligned}$$

On a calculator with algebraic logic, the sequence 0.318, x^y, 3, =, $\div$, 16.5, =, x^y, 0.25, = gives 0.21011127, which is 0.210 rounded off to three digits.   ●

Although Table A.1 gives logarithms to the base 10, it could be used to find logarithms to the base 2 by dividing each entry by 0.3010. See Example 5 below. Any base other than 2 could be treated similarly, as a consequence of the next theorem.

Change of base
$$\log_a N = \frac{\log_b N}{\log_b a} \tag{8.7}$$

To prove (8.7), we let $\log_a N = p$. Then

$$
\begin{array}{ll}
N = a^p & \text{exponential form} \\
\log_b N = \log_b(a^p) & \text{taking logs to base } b \\
\log_b N = p \log_b a & \text{by (8.4)} \\
\dfrac{\log_b N}{\log_b a} = p &
\end{array}
$$

which proves (8.7), since $p = \log_a N$.

Taking $N = b$ in (8.7) gives another proof of the result stated in Prob. 64, Exercise 8.2, namely,

$$(\log_b a)(\log_a b) = 1$$

For instance $(\log_3 5)(\log_5 3) = 1$, and also

$$\log_7 10 = \frac{1}{\log 7} \approx \frac{1}{0.8451} \approx 1.18$$

● **Example 4** $\log_5 103 = \dfrac{\log_{10} 103}{\log_{10} 5} = \dfrac{2.0128}{0.6990} = 2.8795 = 2.88 \qquad$ to three digits   ●

Note that $\log_{10} 103$ and $\log_{10} 5$ are *divided* above, not subtracted.

● **Example 5** $\log_2 N = \dfrac{\log_{10} N}{\log_{10} 2} = \dfrac{\log_{10} N}{0.3010} = 3.322 \log_{10} N$ ●

Logarithms to the base e (see Sec. 8.1) are called natural logarithms and are denoted by $\ln x$. Thus by (8.7)

$$\ln x = \log_e x = \frac{\log x}{\log_{10} e} = \frac{\log x}{0.4343} = 2.303 \log x \tag{8.8}$$

Many laws of exponential growth or decay are expressed as powers of e or of 2. These forms are essentially the same since

$$e = 2^{(\log_2 e)}$$ definition of
                                                logarithm
$$= 2^{(\log e)/(\log 2)}$$ by (8.7)
$$= 2^{0.4343/0.3010} = 2^{1.443}$$

Thus $$(A)e^{kt} = (A)2^{1.443kt}$$

In fact

$$e^x = 2^{1.443x} = 10^{0.4343x}$$
$$2^x = e^{0.6931x} = 10^{0.3010x}$$
$$10^x = e^{2.303x} = 2^{3.322x}$$

and furthermore $e^x = [17.43^{(\log_{17.43} e)}]^x = 17.43^{0.350x}$, which serves to emphasize that *any* base may be used (as long as it is positive).

● **Example 6**   The Richter scale, which is used in measuring the severity of an earthquake, is defined by

$$R = \log \frac{I}{I_0} \tag{1}$$

where I is the intensity of an earthquake and I_0 is a standard intensity. (a) What is the magnitude on the Richter scale of an earthquake which is 1,230,000 times as intense as the standard one? (b) If two earthquakes have Richter numbers of 7.3 and 5.7, what is their relative intensity?

Solution   (a)   Since $I = 1,230,000I_0$, we have

$$R = \log 1,230,000 = \log(1.23)10^6 = 6.0899 \approx 6.1$$

(b)   Solving (1) for I gives

$$\frac{I}{I_0} = 10^R \quad \text{or} \quad I = I_0(10^R)$$

Thus we see that the ratio of the intensities is

$$\frac{I_0(10^{7.3})}{I_0(10^{5.7})} = 10^{(7.3-5.7)} = 10^{1.6} = 10^{1.6000} = 39.8$$

by Table A.1 since 39.8 is the antilog of 1.6000. By calculator,

$$10^{1.6} = 39.810717 \approx 39.8$$ ●

● **Example 7**   If P dollars are invested for n years at an interest rate j compounded m times per year, then the amount at the end of the n years is

$$A = P\left(1 + \frac{j}{m}\right)^{nm}$$

If \$20,000 is invested at 6 percent compounded semiannually, how long will it take for it to accumulate to \$36,120?

Solution   We have $P = 20,000$, $j = 0.06$, $m = 2$, $A = 36,120$, and want n.

$$36,120 = 20,000\left(1 + \frac{0.06}{2}\right)^{2n}$$

$$1.806 = (1.03)^{2n}$$

$$\log 1.806 = 2n \log 1.03$$

The exact value of n is

$$n = \frac{\log 1.806}{2 \log 1.03}$$

By Table A.1 and interpolation,

$$n = \frac{0.2567}{2(0.0128)} \approx 10.03$$

By a calculator with algebraic logic, the key sequence 1.806, log, $\div$, 2, $\div$, 1.03, log, $=$ gives 9.9989582. It takes 10 years.   ●

Exercise 8.4 applications   Use the laws of logarithms to find each of the following numbers to three significant digits. Use the closest entry in the table. Do not interpolate. Using a calculator is optional.

1 (3.47)(23.6)   2 (58.6)(13.8)   3 (80.3)(0.762)   4 (47.1)(0.174)
5 976/3.42   6 67.9/34.9   7 6.07/87.4   8 593/9530
9 (34.6)(0.503)/7.97   10 (9050)(3.17)/476
11 0.784/[(4.36)(0.0347)]   12 0.0975/[(3.97)(0.0564)]
13 5.97^3 14 47.1^4 15 0.319^4 16 0.0977^3

17 $\sqrt{76.3}$ 18 $\sqrt[3]{4970}$ 19 $\sqrt[5]{2.38}$ 20 $\sqrt[7]{9850}$

21 $\sqrt[3]{0.737}$ 22 $\sqrt[4]{0.0126}$ 23 $\sqrt[7]{0.843}$ 24 $\sqrt[6]{0.714}$

25 $\sqrt{6.71}\,\sqrt[3]{0.0953}/2.54$ 26 $\sqrt[5]{939}\,\sqrt[3]{0.744}/1.80$

27 $\sqrt[3]{763}(4.49)/\sqrt[7]{0.0567}$ 28 $\sqrt[4]{94.9}(3.35)/\sqrt[3]{75400}$

If $0 < x < y$, then

$$\frac{2xy}{x + y} < \sqrt{xy} < \frac{y - x}{\ln y - \ln x} < \frac{x + y}{2}$$

Verify this by calculation in Probs. 29 and 30.

29 $x = 4.67$, $y = 8.03$ 30 $x = 2.38$, $y = 3.28$

31 Show that $\dfrac{6.1 + 8.2 + 11.4 + 14.3}{4} > [(6.1)(8.2)(11.4)(14.3)]^{1/4}$.

32 Show that $\dfrac{8.92 + 5.77 + 6.17}{3} > \sqrt[3]{(8.92)(5.77)(6.17)}$.

Find the following numbers to four significant digits. Use interpolation. Using a calculator is optional.

33 (34.72)(5.039)   **34** (0.8077)(1,153)
35 (0.3427)(0.01863)   **36** (71.18)(60.33)
37 2843/1596   **38** 9437/7562   **39** 3491/8743   **40** 2399/8276

Use (8.7) with $b = 10$ to find the following logarithms:

41 $\log_5 292$ **42** $\log_7 292$ **43** $\log_8 47.3$ **44** $\log_8 69.1$

Use (8.7) to verify the following equations.

45 $\log_{16} 32 = \frac{5}{4}$ **46** $\log_{27} \frac{1}{9} = -\frac{2}{3}$
47 $\log_{0.25} \frac{1}{16} = 2$ **48** $\log_{4/9}(\frac{8}{27}) = \frac{3}{2}$

Use (8.8) to find the following natural logarithms:

49 ln 20   **50** ln 65   **51** ln 111   **52** ln 1000

The following problems deal with equations of the type

$$f(t) = A \cdot b^{kt}$$

Sociology **53** If the population of a country increases 2 percent each year, then $P(t) = A(1.02)^t$, t in years. How long will it take the population to double? to multiply by 4?

54 Express the population in Prob. 53 in terms of a power of 2; i.e., find k so that $P(t) = A \cdot 2^{kt}$.

Physics **55** If the half-life of a substance is 4.34 min, then half of it disintegrates in that time. Express the amount after t min as an exponential function with base 2 if there are 6.72 g to begin with. Express the amount with base e.

56 In Prob. 55, how long will it take for there to be only 1.53 g? *Hint:* Use the answer to Prob. 55.

Economics **57** If $1000 is invested at 7 percent compounded quarterly, how much is there after n years? (See Example 7.) How long will it take for the amount to double?

58 If $1000 is invested for 12 years compounded quarterly, what interest rate is necessary for the money to double?

Earth science **59** Compare the intensities of two earthquakes whose Richter numbers are 7.5 and 6.3. See Example 6.

60 If the Los Angeles earthquake of 1971 had a Richter number of 6.7, what would the Richter number be for an earthquake which is three times as intense?

Sociology   **61** The score on an industrial safety test after t months is $67 - 15 \log(1 + t)$. What is the average monthly score from 1 month to 6 months?

62 In Prob. 61, how long will it take for the score to decrease to 44?

Engineering   **63** The temperature of an ingot upon being immersed in water kept at 20°C is $20 + 670e^{-5t}$ after t min. How long will it take for the temperature to reach 25°C?

64 Express the temperature in Prob. 63 with base 10.

Chemistry   **65** The pH of a chemical substance is $pH = -\log[H^+]$, where $[H^+]$ measures the hydrogen concentration. Find the pH of milk if its hydrogen concentration is $4(10^{-7})$.

66 If the pH of a solution is 11.4, what is its hydrogen concentration? See Prob. 65.

Meteorology   **67** If atmospheric pressure at h ft is $P = (14.7)2^{-0.0000555h}$, express P at m mi with base 10.

68 In Prob. 67, what height (in feet) is required for the pressure to be 4.71?

Ecology   **69** In the Poisson distribution, if x is the average occurrence of an event, then the probability of n occurrences is $e^{-x}x^n/n!$. In an 8-mi stretch of beach, there were 20 beached whales. What is the probability of finding, in a 1-mi stretch of this beach, (a) two whales? (b) three whales? Use $x = 2.5$.

Sociology   **70** The population of a city t years after it is P is approximately $P \cdot e^{0.05t}$. Find the population of a city of 100,000 after 5 years, 10 years, and 20 years.

Immunology   **71** If one person in an urban area of 50,000 gets the flu, the number who will have it after t days is approximately

$$N(t) = \frac{10,000e^t}{e^t + 10,000}$$

Approximately how many will have the disease after 1 day, 4 days, 10 days?

Economics   **72** It can be shown that if the interest is a nominal rate j compounded continuously for n years, a principal of P will accumulate to $A = Pe^{nj}$, where e is approximately 2.718. Make use of this formula and a calculator or logarithms to find the accumulated value if $1000 is invested for a year at 6 percent compounded continuously.

73 Show that for $a > 0, b > 0, b \neq 1$,

$$\frac{\log a}{\log b} = \frac{\ln a}{\ln b}$$

8.5 Exponential and logarithmic equations and inequalities

Exponential and logarithmic equations

An equation in which the unknown occurs in one or more exponents is called an *exponential equation*. A *logarithmic equation* is an equation in which the logarithm of the unknown or a function of it occurs. For example, $3^{x+2} = 7^{x-1}$ is an exponential equation, and $\log_2(x - 2) + \log_2(x + 1) = 2$ is a logarithmic equation.

● **Example 1**   Solve $5^{2x} = 7^{x+1}$.

Solution   If we take the common logarithm of each member of the given equation, we have

$$\log 5^{2x} = \log 7^{x+1}$$
$$2x \log 5 = (x + 1)\log 7 \qquad \text{by (8.4)}$$
$$x(2 \log 5 - \log 7) = \log 7 \qquad \text{collecting terms}$$
$$x(\log 25 - \log 7) = \log 7 \qquad \text{by (8.4)}$$
$$x = \frac{\log 7}{\log 25 - \log 7} \qquad \text{exact value of } x$$

By Table A.1

$$x = \frac{0.8451}{1.3979 - 0.8451} = \frac{0.8451}{0.5528} = 1.529$$

By a calculator with algebraic logic, the key sequence 7, log, ÷, (, 25, log, −, 7, log,), = gives 1.52864306, which rounds off to 1.529.   ●

● **Example 2**   Solve $\log_2(x - 1) + \log_2(x + 1) = 3$.

Solution   If we apply (8.2) to the given equation, we get

$$\log_2(x - 1)(x + 1) = 3$$

Hence, changing to exponential form, we have

$$(x - 1)(x + 1) = 2^3 = 8$$
$$x^2 - 9 = 0 \qquad \text{adding } -8 \text{ to each side}$$

Therefore, 3 and -3 are the possible solutions, but we must check each in the given equation.
  If $x = 3$, the left member of the given equation becomes

$$\log_2 2 + \log_2 4 = 1 + 2 = 3$$

as stated in the given equation. If $x = -3$, the left member of the given equation becomes $\log_2(-4) + \log_2(-2)$; hence, $x = -3$ cannot be admitted as a root since the logarithm of a negative number has not been defined.   ●

● **Example 3**   Solve the inequality $(0.3)^x < 5$.

Solution   Since the logarithm function is increasing, the given inequality is equivalent to

$$\log(0.3)^x < \log 5$$
$$x \log 0.3 < \log 5$$

Now $0 < 0.3 < 1$; hence $\log 0.3 < 0$. Thus if we divide by the negative number $\log 0.3$, we must change the sense of the inequality, giving

$$x > \frac{\log 5}{\log 0.3}$$

By Table A.1,

$$\frac{\log 5}{\log 0.3} = \frac{0.6990}{-0.5229} = -1.34$$

By calculator,

$$\frac{\log 5}{\log 0.3} = -1.33677264 \approx -1.34$$ ●

● **Example 4**   Solve the inequality $2^x \leq 3x + 4$.

Solution   Taking logarithms will not help (try it). However, we may use graphs and a table of values to approximate the solutions. We graph $y = 2^x$ and $y = 3x + 4$ on the same set of axes in Fig. 8.9.

x	$-\frac{4}{3}$	-1.2	0	4
2^x	$2^{-4/3}$	0.44	1	16
$3x + 4$	0	0.40	4	16

It is apparent from the graph that 2^x is less than $3x + 4$ for all x between the two points of intersection. From the table of values, one point has $x = 4$ exactly; the other has $x = -1.2$ approximately. The solution is thus $-1.2 \leq x \leq 4$, to one decimal place.   ●

FIGURE 8.9

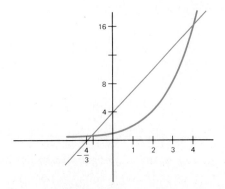

● **Example 5**   Solve the following equations simultaneously:

$$5^{x+y} = 100 \tag{1}$$

$$2^{2x-y} = 10 \tag{2}$$

Solution   If we equate the common logarithms of the members of each of (1) and of (2), we get

$$(x + y)\log 5 = 2 \tag{3}$$

$$(2x - y)\log 2 = 1 \tag{4}$$

If we solve these equations for $x + y$ and $2x - y$ and then solve the resulting equations simultaneously, we obtain

$$\left. \begin{aligned} x + y &= \frac{2}{\log 5} = 2.86 \\[2mm] 2x - y &= \frac{1}{\log 2} = 3.32 \end{aligned} \right\} \begin{aligned} &\text{to 3 digits by} \\ &\quad \text{Table A.1 or calculator} \end{aligned} \tag{5} \tag{6}$$

$$\overline{\quad\quad\quad\quad\quad\quad}$$

$$3x \qquad\qquad = 6.18 \qquad \text{adding (5) and (6)}$$

$$x = 2.06$$

Substituting 2.06 for x in (5) and solving for y gives $y = 0.80$.   ●

Exercise 8.5 exponential and logarithmic equations and inequalities

Solve each equation in Probs. 1 to 8 for x in terms of y.

1 $y = 10^x$ **2** $y = 10^{-x}$ **3** $y = 5(10^{-2x})$

4 $y = 3(10^{2x})$ **5** $y = \dfrac{e^x + e^{-x}}{2}$ **6** $y = \dfrac{e^x - e^{-x}}{2}$

7 $y = \dfrac{e^x + e^{-x}}{e^x - e^{-x}}$ **8** $y = \dfrac{e^x - e^{-x}}{e^x + e^{-x}}$

Solve Probs. 9 to 28 for x.

9 $5^x = 625$ **10** $2^x = 64$ **11** $3^{x^2-1} = 27$ **12** $5^{x^2-4x+3} = 1$
13 $2^x = 5^{2x-1}$ **14** $5^{x+1} = 3^{2x+1}$ **15** $19.5^x = 8.6^{3x-1}$
16 $3.02^x = 2.87^{x+2}$ **17** $\log_5(x - 1) + \log_5(x + 3) = 1$
18 $\log_5(2x + 1) + \log_5(3x - 1) = 2$ **19** $\log_6 3 + \log_6(x + 6) = 2$
20 $\log_3(x + 2) + \log_3(x + 4) = 1$ **21** $\log_5(2x + 4) - \log_5(x - 1) = 1$
22 $\log_2(3x + 1) - \log_2(x - 3) = 3$ **23** $\log_3(x + 11) - \log_3(x + 3) = 2$
24 $\log_5(3x + 7) - \log_5(x - 5) = 2$
25 $\log_2(x^2 - 1) - \log_2(x - 1) = \log_2(3x - 5)$
26 $\log_5(2x^2 - 7) - \log_5(x + 1) = \log_5(2x - 3)$
27 $\log_6(x^2 + x - 6) - \log_6 x = 1 + \log_6(x - 5)$
28 $\log_7(2x^2 - 1) - \log_7(x + 2) = 1 + \log_7(x - 4)$

Solve Probs. 29 to 36 for x and y.

29 $5^{x+y} = 120$
$\quad 3^{2x-y} = 25$
30 $4^{x+y} = 250$
$\quad 7^{y-2x} = 8$
31 $4^{2x-y} = 15$
$\quad 3^{x+y} = 80$

32 $8^{2x-7y} = 9$
$\quad 5^{x-y} = 120$
33 $5^{x+2y} = 28$
$\quad 4x + 2y = 3$
34 $2^{2x+3y} = 48$
$\quad x + y = 2$

35 $10^{x+2y} = 2$
$\quad \log 3x - \log 2y = 1$
36 $10^{x-3y} = 3$
$\quad \log 2x - \log y = 1$

Solve the following inequalities:

37 $3.1^x < 5.5$
38 $6.7^{2x} < 4.4$
39 $(0.7)^{3x} < 3.5$

40 $(0.84)^x < 12$
41 $x^{4.3} < 6.6$
42 $(3x)^{3.8} < 88$

43 $(5x)^{0.55} < 14$
44 $(4x)^{0.18} < 2.1$
45 $3^x \le 8x + 3$

46 $5^x \le 4x + 17$
47 $2^x \le 3x - 1$
48 $3^x \le 5x - 1$

Use a calculator, if available, to solve the following equations:

49 $4.31^x = 166.215$
50 $3.035^x = 7.0176$

51 $6.35^{2x} = 3.78^{2x+1.07}$
52 $4.63^{x-1} = 11.9^{2x-3.74}$

8.6 Key concepts

Make sure you understand each of the following important words and ideas.

b^x for x rational and x irrational
Graphing exponential and logarithmic functions
Definition of $\log_b N$ for $N > 0$, $b > 0$, $b \ne 1$
Exponential and logarithmic functions, as inverses
Common logarithm = characteristic + mantissa
Natural logarithms
Logs and antilogs by table and calculator
Significant digits
Rounding off
Interpolation
Change of base
Exponential growth and decay
Exponential and logarithmic equations

(8.1) $\log_b N = L$ means $b^L = N$, and $b^{\log_b N} = N$

(8.2) $\log_b(MN) = \log_b M + \log_b N$

(8.3) $\log_b(M/N) = \log_b M - \log_b N$

(8.4) $\log_b(N^p) = p \log_b N$

(8.5) $\log_b \sqrt[r]{N} = \dfrac{1}{r} \log_b N$

(8.7) $\log_a N = \dfrac{\log_b N}{\log_b a}$

Exercise 8.6 review

Graph the following:

1 $y = \log_7 x$, $y = 7^x$ **2** $y = \log_5(x + 3)$, $y + 3 = \log_5 x$
3 $y = \log |x|$, $y = |\log x|$ **4** $2y = \log x$, $2x = \log y$

Verify the following:

5 $3^{1/5} > 5^{1/8}$ **6** $\sqrt{0.7} > 0.7$ **7** $a^b = b^a$ if $a = (\frac{4}{3})^3$, $b = (\frac{4}{3})^4$
8 $2^{\sqrt{2}} > (\sqrt{2})^2$ **9** $\log_3 3^5 = 5$ **10** $6^{\log_6 8} = 2^3$
11 $\log_{100} 1 = 0$ **12** $\log_{2/5}(\frac{8}{125}) = 3$ **13** $\log_9 243 = \frac{5}{2}$
14 $\log_4 \frac{1}{128} = -\frac{7}{2}$ **15** $\log_2 30 = 1 + \log_2 3 + \log_2 5$

Use $\log_7 2 = x$, $\log_7 3 = y$, and $\log_7 5 = z$ to find the following:

16 $\log_7 8$ **17** $\log_7(\frac{21}{20})$ **18** $\log_7 \sqrt{45}$

Find the following common logarithms:

19 $\log 485$ **20** $\log 67.2$ **21** $\log 0.974$ **22** $\log 10^{0.3347}$

If $\log N$ is the given number, find N.

23 4.7419   **24** 4.9269 – 10   **25** 8.9050 – 10   **26** 14.3345 – 15
27 Use characteristics to decide which is larger, 93^{95} or 95^{93}.

Calculate the following numbers:

28 $(48.3)(6.44)(0.555)$ **29** $\dfrac{64.2}{(9.87)(11.3)}$ **30** $\dfrac{\sqrt{41.8}}{(1.35)^5}$

31 $10^{4.4440}$ **32** $(\log 3)(\log 5)$ **33** $\log_6 12$

34 $\dfrac{8.31 + \sqrt{46.4}}{(\log 74.3)^2}$ **35** $\dfrac{18.3 - \sqrt[3]{500}}{(\log 51)^3}$

36 $\dfrac{5.62}{3.48 + (6.55)(0.392)}$ **37** $\dfrac{2^{0.6} + 3^{0.7}}{4^{0.8}}$

Solve the following equations:

38 $7^x = 343$ **39** $4^{x^2 + 3x} = 256$ **40** $7^x = 243$
41 $\log_5(x - 3) + \log_5(3x + 1) = 3$
42 $\log_5(x^2 - 5x + 1) - \log_5(2x^2 - 3) = -1$
43 $3^{2x+1} = 5^{x-1}$ **44** $\log_4(\log_3 x) = 2$ **45** $\begin{aligned} 4^{2x-y} &= 5 \\ 3^{-4x+3y} &= 4 \end{aligned}$

Solve the following inequalities

46 $4^{3x} < 6.1$ **47** $(3x)^{0.71} < 2$ **48** $5^x < 2 - x$

Biology   **49** The number of bacteria after t h is $4.8(10^4) \cdot 2^{t/3}$. How long will it take for the number to double? how long to triple?

Earth science   **50** A mild earthquake has a Richter number of 4.5 while a severe one has a Richter number of 8.2. What is the ratio of their intensities? See Example 6, Sec. 8.4.

Chemistry   **51** Carbon 14 occurs in organic objects in a fixed percentage Q. When the organism dies, the carbon 14 decays in such a way that after t years the percentage of carbon 14 is $P = (Q)2^{-t/5600}$. If a dead organism is found to have $P = 0.76Q$, approximately how long has it been dead?

Physics   **52** If we take into account air resistance and initial velocity, the velocity in feet per second of a falling body after t s is $v = 120 - 180e^{-t/4}$. For which values of t is $v \geq 115$? What happens to v as t becomes larger and larger (i.e., what is the terminal velocity)?

Economics   **53** With appropriate units and constants, Malthus in 1790 said that population after t years is $P \cdot 2^{t/35}$ while food production is $a \cdot t + b$. Verify graphically what he concluded: that after a long enough period of time, the population will overtake the available food (assuming all other factors do not change).

Theory   **54** Let

$$f(p) = \left(\frac{4^p + 5^p}{2}\right)^{1/p}$$

Show by calculation that $f(2) < f(3) < f(4)$.

Area formula   **55** The area of a triangle with sides a, b, c, and $s = (a + b + c)/2$ is given by Heron's formula $K = \sqrt{s(s - a)(s - b)(s - c)}$. Find the area of a triangle with sides $a = 64.3$, $b = 47.9$, and $c = 38.4$.

Probability   **56** The definition of $n!$, read n factorial, is $n(n - 1)(n - 2) \cdots (4)(3)(2)(1)$, the product of all integers between 1 and n inclusive. Since this involves n factors, a handy approximation for large n is Stirling's formula, $n! \approx n^{(n+\frac{1}{2})}(e^{-n})\sqrt{2\pi}$. Use Stirling's formula to approximate the number

$$\frac{100!}{50!\,50!}$$

Management   **57** New equipment should be installed every x years where

$$e^{-dx} = \frac{1}{1 + \dfrac{da}{mb} + xd}$$

Show that $x = 3.75$ is an approximate solution for the situation where $a = 10,000 = b$, $d = 0.06$, and $m = 2$.

Anthropology   **58** One equation governing population growth in thousands is

$$y = \frac{2.91}{e^{-0.03x} + 0.015}$$

Calculate y for $x = 28$ years.

Physics   **59** The pressure and volume of a certain gas are related by $pv^{1.4} = 14,720$. Find v if $p = 115$.

Theory   **60** Show that if $f(x) = x/\log x$, then

$$f(2x) - f(x) = \frac{x \log (x/2)}{\log x \log 2x}$$

Calculator   **61** For $0 < y < x$, it is true that

$$\frac{2(x-y)}{x+y} < \ln x - \ln y < (\sqrt{x} - \sqrt{y})\left(\frac{1}{\sqrt{x}} + \frac{1}{\sqrt{y}}\right)$$

Verify this by calculation for $x = 8$ and $y = 7$.

CHAPTER 9·SYSTEMS OF EQUATIONS AND OF INEQUALITIES

Most problems that we encounter in life involve more than one variable. If the relations between variables can be put in terms of equations or inequalities, then we may be able to find values for the variables such that the equations or inequalities are true statements. These values are called solutions, and we will consider several ways for determining them in this chapter.

9.1 Linear systems of equations

Two equations in two variables

Solution   The ordered pair (p, q) is a **solution** of the two equations

$$ax + by = c$$
$$Ax + By = C$$

if it satisfies each equation. In this section, we will give three methods of solution: addition and subtraction, substitution, and graphical.

● **Example 1**   Find the simultaneous solution of the pair of equations

$$2x - 3y = 3 \qquad (1)$$

$$3x + 5y = 14 \qquad (2)$$

SOLUTION BY ADDITION AND SUBTRACTION   This method of solution requires that each equation be multiplied by a number such that the coefficients of one of the variables are numerically equal, since we can then eliminate that variable by addition or subtraction of the equations.

If we multiply each member of (1) by 3 and each member of (2) by 2, we get

$$6x - 9y = 9 \tag{3}$$

$$6x + 10y = 28 \tag{4}$$

If we now subtract each member of (3) from the corresponding member of (4), we have

$$19y = 19$$

Threrefore, $y = 1$ and using this in (1) gives

$$2x - 3 = 3$$
$$2x = 6$$
$$x = 3$$

Consequently, $(3, 1)$ is the simultaneous solution of the given pair of equations. We could have eliminated y by multiplying (1) by 5, (2) by 3, and adding the new equations.

SOLUTION BY   In solving a pair of equations by substitution, we solve one of the equations
SUBSTITUTION   for one of the variables and substitute this value into the other equation.
We thereby obtain one equation in one variable and can solve it if it is linear. The solution for this variable can then be put in either original equation, and the resulting equation solved for the remaining variable.
If we solve equation (1) for y, we get

$$y = \frac{2x - 3}{3}$$

Substituting this in (2) gives

$$3x + \frac{5(2x - 3)}{3} = 14$$
$$9x + 10x - 15 = 42 \qquad \text{multiplying by 3}$$
$$19x = 57 \qquad \text{combining terms}$$
$$x = 3$$

If we put $x = 3$ in (2), we have

$$3(3) + 5y = 14$$
$$5y = 5$$
$$y = 1$$

and the solution is $(3, 1)$ as previously found.   ●

Graphs   The graph of one linear equation in two variables is a straight line; consequently, two linear equations in two variables may represent two intersecting lines, two parallel lines, or two coincident lines. If the two lines *intersect,* the simultaneous solution is one ordered pair of numbers and the

FIGURE 9.1

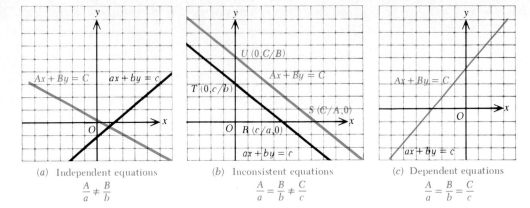

(a) Independent equations

$$\frac{A}{a} \neq \frac{B}{b}$$

(b) Inconsistent equations

$$\frac{A}{a} = \frac{B}{b} \neq \frac{C}{c}$$

(c) Dependent equations

$$\frac{A}{a} = \frac{B}{b} = \frac{C}{c}$$

Independent   equations are said to be **independent.** If the two lines are *parallel and distinct,* there is no point of intersection and therefore no solution. In this
Inconsistent   case the equations are called **inconsistent.** If the two lines *coincide,* then every solution pair of one equation is a solution pair of the other and the
Dependent   equations are said to be **dependent.** Figure 9.1 illustrates the three possibilities.

We now derive a simple criterion that enables us to decide whether two linear equations in two unknowns are independent, inconsistent, or dependent. We consider the equations

$$ax + by = c \tag{5}$$

$$Ax + By = C \tag{6}$$

The graphs of (5) and (6) intersect the x axis at $R(c/a, 0)$ and $S(C/A, 0)$, respectively, and the y axis at the points $T(0, c/b)$ and $U(0, C/B)$, respectively, as illustrated in Fig. 9.1. We can see that the graphs of Eqs. (5) and (6) are parallel if the segments RT and SU are parallel, and these two segments are parallel if and only if the triangles ORT and OSU are similar. The triangles are similar if and only if $OR/OS = OT/OU$.

Now $OR/OS = c/a \div C/A = Ac/aC$, and $OT/OU = c/b \div C/B = Bc/bC$. Consequently, $OR/OS = OT/OU$ if and only if $Ac/aC = Bc/bC$. If we multiply each member of the last equation by C/c, we obtain $A/a = B/b$. Therefore, the graphs of (5) and (6) are parallel if and only if $A/a = B/b$. If in addition to $A/a = B/b$ we have $A/a = B/b = C/c = k$, then $A = ak$ and $C = ck$. Hence, $C/A = ck/ak = c/a$, and the points R and S coincide. Therefore, since the graphs are parallel and have at least one point in common, they coincide. Hence we have the following theorem:

Coefficient   Two linear equations $ax + by = c$ and $Ax + By = C$ are
conditions

Independent if $A/a \neq B/b$
Inconsistent if $A/a = B/b \neq C/c$
Dependent if $A/a = B/b = C/c$

● **Example 2**   The following three examples illustrate the application of the above theorem.
(a)   The equations

Independent

$$2x - 3y = 4$$
$$5x + 2y = 8$$

are independent, since $\frac{2}{5} \neq -\frac{3}{2}$. There is *only one solution*.

(b)   The equations

Inconsistent

$$3x - 9y = 1$$
$$2x - 6y = 2$$

are inconsistent, since $\frac{3}{2} = -9/(-6) \neq \frac{1}{2}$. There are *no solutions*.

(c)   The equations

Dependent

$$2x - 4y = 12$$
$$3x - 6y = 18$$

are dependent, since $\frac{2}{3} = -4/(-6) = \frac{12}{18}$. The *number of solutions* is infinite.   ●

GRAPHICAL   A pair of linear equations in two variables can be solved graphically by
SOLUTION   sketching the graphs of the lines and then estimating the coordinates of the
point of intersection. Even though this gives only approximate solutions,
and we already have methods for exact solutions, it is worthwhile studying
the graphical method in the present simple situation because it can be used
with nonlinear systems in the next section.

● **Example 3**   Sketch the graphs of

$$4x + 3y = 3 \tag{7}$$

$$2x + 5y = -2 \tag{8}$$

and estimate the coordinates of their point of intersection.

Solution   There are of course many ways to graph each straight line. We could use
the $y = mx + b$ form, we could find the intercepts of each line, or we
could use a table of values as given below. We use three points instead of
just two merely for safety. The lines appear to intersect at $(1.5, -1)$, as
shown in Fig. 9.2, and this is in fact the exact solution.

x	2	0	-3
y from (7)	$-\frac{5}{3}$	1	5
y from (8)	$-\frac{6}{5}$	$-\frac{2}{5}$	$\frac{4}{5}$

●

FIGURE 9.2

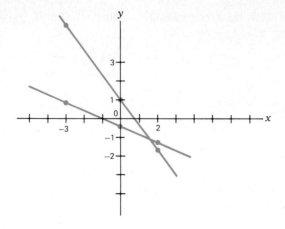

Three equations in three variables

Systems of three linear equations in three variables will be treated with matrices and determinants in Chap. 10, but we will now solve such a system by *addition or subtraction* and by *substitution*. The procedures used are essentially the same as those used in solving a pair of linear equations in two variables. We illustrate with an example.

● **Example 4**   Solve the following set of equations simultaneously.

$$2x + 3y + z = 4 \tag{9}$$

$$3x - 2y - 3z = -1 \tag{10}$$

$$2x + 5y + 2z = 5 \tag{11}$$

SOLUTION BY ADDITION AND SUBTRACTION We begin by eliminating z between equations (9) and (10). This is accomplished by multiplying (9) by 3 and adding the resulting equation term by term to (10). Thus

$$
\begin{array}{ll}
6x + 9y + 3z = 12 & \text{Eq. (9) times 3} \\
\underline{3x - 2y - 3z = -1} & \text{(10) recopied} \\
9x + 7y = 11 & \text{adding}
\end{array}
\tag{12}
$$

We now eliminate z between (9) and (11). Thus

Eliminate z again

$$
\begin{array}{ll}
4x + 6y + 2z = 8 & \text{Eq. (9) times 2} \\
\underline{2x + 5y + 2z = 5} & \text{(11) recopied} \\
2x + y = 3 & \text{subtracting}
\end{array}
\tag{13}
$$

We now have the system (12) and (13) of two equations in two variables, and it may be solved in any manner. We will use substitution since it is easy to solve (13) for y. This gives $y = 3 - 2x$, and substituting in (12) gives the single equation

$$9x + 7(3 - 2x) = 11$$
$$9x + 21 - 14x = 11$$
$$10 = 5x$$
$$2 = x$$

We can now find y by using $x = 2$ in $y = 3 - 2x$. Hence $y = 3 - 4$, and so

$$y = -1$$

We can now find z by substituting $x = 2$ and $y = -1$ in any of the three original equations. Using (9) gives

$$2(2) + 3(-1) + z = 4$$
$$4 - 3 \quad + z = 4$$

Therefore, $z = 3$ and the solution is $(2, -1, 3)$.

SOLUTION BY
SUBSTITUTION

We begin by solving (9) for z and putting its value in (10) and (11).

$z = 4 - 2x - 3y$	solving (9) for z
$3x - 2y - 3(4 - 2x - 3y) = -1$	substituting in (10)
$9x + 7y = 11$	collecting terms (14)
$2x + 5y + 2(4 - 2x - 3y) = 5$	substituting in (11)
$-2x - y = -3$	collecting terms (15)

We now have a system (14) and (15) of two equations in two variables. It happens to be the same system (12) and (13) as in the earlier solution. It can be solved by a variety of methods, and as before we get $x = 2$, $y = -1$. These values in any of the original equations lead to $z = 3$. Consequently, as seen earlier, the solution is $(2, -1, 3)$.   ●

A system of n equations, where $n > 3$, in n variables can be solved similarly by either addition and subtraction or by substitution.

Exercise 9.1

Solve the systems in Probs. 1 to 4 by addition or subtraction and Probs. 5 to 12 by substitution.

1 $3x + y = 7$ **2** $4x + 3y = -2$ **3** $2x + 3y = 3$
$\quad x + 2y = 4$ $\quad 2x - 5y = 12$ $\quad 3x - 7y = 16$

4 $5x - 4y = 2$ **5** $2x + 5y = -4$ **6** $5x + 4y = 1$
$\quad 3x + 2y = 10$ $\quad 3x + 4y = 1$ $\quad 2x - 3y = 5$

7 $3x + 5y = 1$ **8** $7x - 3y = 1$ **9** $2x + 3y = 4$
$\quad 2x - 4y = 8$ $\quad 5x - 2y = 1$ $\quad 5x + 2y = -1$

10 $2x + y = -2$ **11** $3x + 2y = 0$ **12** $3x - 5y = 1$
$\quad 3x + 2y = -1$ $\quad 2x + 3y = 5$ $\quad 2x - 7y = -3$

Solve the systems of equations in Probs. 13 to 24 by any method.

13 $3x - 2y - z = 3$
$\quad 2x - y + z = 4$
$\quad x - 2y + 3z = 3$

14 $3x + 2y + 2z = -1$
$\quad 5x - 3y + 4z = -3$
$\quad 2x + y + 2z = -2$

15 $\quad x + 3y + 3z = -1$
$\quad 2x - y + z = -3$
$\quad 3x + 5y + 7z = -1$

16 $5x + 4y + 7z = 2$
$\quad 3x - 2y + z = 0$
$\quad x + 5y + 8z = -2$

17 $2x - y - z = 0$
$\quad 6x - 7y + 2z = 11$
$\quad 3x + 5y - 3z = 2$

18 $3x + 2y - 5z = 1$
$\quad 2x - 3y - 8z = 1$
$\quad x + 5y + 2z = 1$

19 $\quad x + y + z = 0$
$\quad 2x + 3y - z = 1$
$\quad 3x + 4y - 3z = 4$

20 $\quad x + y - z = 0$
$\quad 3x + 2y - 3z = 3$
$\quad 2x + 3y + 2z = 1$

21 $\quad x + y + 2z + w = 3$
$\quad 2x + 3y - 2w = -6$
$\quad 3x - 2y + z = -2$
$\quad 3y - z + 3w = 5$

22 $\quad x + y + z + w = 1$
$\quad 2x + 3z + 2w = 5$
$\quad 2y + 5z + 3w = 12$
$\quad 3x - y + w = -2$

23 $\quad x + y + z + w + t = 2$
$\quad 2x + 3z - 2w + t = -6$
$\quad 3y - z + 3w - 2t = -5$
$\quad x - y + 2z + w = 0$
$\quad 3x + 5y + 5w + 2t = 4$

24 $\quad x + y + z + w + t = 1$
$\quad x - y - z - t = -4$
$\quad 2x + y - z - w = -7$
$\quad 2x - 2y + 3z + 2t = 5$
$\quad 3x + 4y + 2z + 3w = -5$

Determine whether the system in each of Probs. 25 to 32 is independent, inconsistent, or dependent, and solve each independent system.

25 $3x - 2y = 4$
$\quad 2x - 3y = 1$

26 $x - 2y = 3$
$\quad 3x - 6y = 8$

27 $2x + y = 4$
$\quad 4x + 2y = 8$

28 $3x - 2y = 9$
$\quad 6x - 4y = 16$

29 $2x + 6y = 4$
$\quad 3x + 9y = 6$

30 $3x + y = 1$
$\quad 2x + 4y = -6$

31 $\quad x - 2y = -1$
$\quad 3x - 6y = -3$

32 $\quad x - 2y = -1$
$\quad 3x + 4y = 17$

Solve each of the following systems graphically to the nearest half-unit.

33 $5x + 3y = 11$
$\quad 3x + y = 5$

34 $5x + 2y = 10$
$\quad 2x + y = 4$

35 $3x + 2y = 1$
$\quad x + 4y = 3$

36 $\quad x + y =$
$\quad 4x + 3y =$

9.2 Nonlinear systems

Although linear systems of two equations have only one solution (unless the equations are dependent), in general nonlinear systems have several solutions. The first step in the algebraic procedure for solving two nonlinear equations in two variables is to combine the two equations in such a way as to obtain one equation in one variable each of whose roots is one of the numbers in an ordered pair of the solution set. This process is called **eliminating a variable**. After one number in each solution pair is determined,

Eliminating a variable

the other can be obtained by substitution in either of the original equations. The methods that we will discuss in the subsequent sections of this chapter are elimination by **addition or subtraction,** elimination by **substitution,** and elimination by a **combination** of addition or subtraction and substitution.

Check solutions   Each solution should be checked in each of the given equations.

Elimination by addition and subtraction

TWO EQUATIONS   In Sec. 9.1, we showed the application of this method to a pair of linear
OF THE TYPE   equations. The same procedure is applicable to two equations of the form
$Ax^2 + Cy^2 = F$ $Ax^2 + Cy^2 = F$, and we will go immediately to an example.

● **Example 1**   Find the solution of the system of equations

Ellipse

$$2x^2 + 3y^2 = 21 \tag{1}$$

Hyperbola

$$3x^2 - 4y^2 = 23 \tag{2}$$

Solution   We arbitrarily select y as the variable to be eliminated, and proceed as follows

$$
\begin{array}{ll}
8x^2 + 12y^2 = 84 & \text{multiplying (1) by 4} \tag{3} \\
9x^2 - 12y^2 = 69 & \text{multiplying (2) by 3} \tag{4} \\
\overline{17x^2 \quad\;\; = 153} & \text{equating the sums of the corresponding} \\
& \quad \text{members of (3) and (4)} \\
x^2 = 9 & \text{solving for } x^2 \\
x = \pm 3 &
\end{array}
$$

We obtain the corresponding values of y by replacing x by 3 or -3 in (1) and solving for y. In either case, we get

$$
\begin{array}{ll}
18 + 3y^2 = 21 & \\
3y^2 = 3 & \text{adding } -18 \text{ to each member} \\
y^2 = 1 & \\
y = \pm 1 &
\end{array}
$$

Hence the solutions are $(3, 1)$, $(3, -1)$, $(-3, 1)$, and $(-3, -1)$.

Check   If we replace x and y in the two given equations by 3 and 1, respectively, we get

$$
\begin{array}{ll}
18 + 3 = 21 & \text{from (1)} \\
27 - 4 = 23 & \text{from (2)}
\end{array}
$$

The other solution pairs can be checked in a similar manner. Figure 9.3 shows the graphs of the two equations and the coordinates of their points of intersection.   ●

FIGURE 9.3

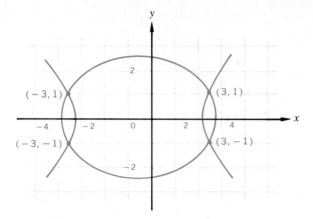

TWO EQUATIONS
OF THE TYPE
$Ax^2 + Cy^2 + Dx = F$

If we have two equations of this form, we begin by eliminating y^2 and thereby obtain a quadratic equation in x. We then solve this quadratic and substitute each root for x in either of the given equations. Finally, solve the resulting quadratic in y and pair the values of x and y.

● **Example 2**

Solve the system of equations

Hyperbola

$$3x^2 - 2y^2 - 6x = -23 \qquad (5)$$

Circle

$$x^2 + y^2 - 4x = 13 \qquad (6)$$

Solution

Since each of the given equations contains one term in y^2 and no other term involving y, we eliminate y^2 and then complete the process of solving as outlined above.

$$
\begin{array}{ll}
3x^2 - 2y^2 - 6x = -23 & \text{(5) recopied} \\
\underline{2x^2 + 2y^2 - 8x = 26} & \text{multiplying (6) by 2} \qquad (7) \\
5x^2 \qquad\quad - 14x = 3 & \text{equating the sums of the left and the} \quad (8) \\
 & \text{right members of (5) and (7)}
\end{array}
$$

$$
\begin{array}{ll}
5x^2 - 14x - 3 \;\; = 0 & \text{adding } -3 \text{ to each member of (8)} \quad (9) \\
(x - 3)(5x + 1) = 0 & \text{factoring} \\
\quad x = 3 \text{ or } -\tfrac{1}{5}
\end{array}
$$

Using $x = 3$

$$
\begin{array}{ll}
9 + y^2 - 12 = 13 & \text{replacing } x \text{ by 3 in (6)} \\
y^2 = 16 & \text{adding 3 to each member} \\
y = \pm 4 &
\end{array}
$$

Hence, if $x = 3$, then $y = \pm 4$.

Using $x = -\tfrac{1}{5}$

$$
\begin{array}{ll}
(-\tfrac{1}{5})^2 + y^2 - [4 \times (-\tfrac{1}{5})] = 13 & \text{replacing } x \text{ by } -\tfrac{1}{5} \text{ in (6)} \\
\tfrac{1}{25} + y^2 + \tfrac{4}{5} = 13 & \text{performing the indicated operations} \\
1 + 25y^2 + 20 = 325 & \text{multiplying each member by 25} \\
y^2 = \dfrac{304}{25} & \text{solving for } y^2 \\
y = \pm \sqrt{\dfrac{304}{25}} = \pm \dfrac{4\sqrt{19}}{5} &
\end{array}
$$

FIGURE 9.4

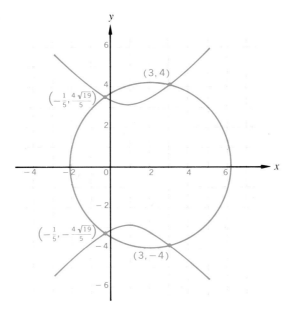

Consequently, if $x = -\frac{1}{5}$, then $y = \pm(4\sqrt{19})/5$. Therefore, the solution set is $\{(3, 4), (3, -4), (-\frac{1}{5}, (4\sqrt{19})/5), (-\frac{1}{5}, (-4\sqrt{19})/5)\}$. Each solution pair can be checked by replacing x and y in the given equations by the appropriate number from the solution pair. The graphs of (1) and (2) are shown in Fig. 9.4 together with the coordinates of their points of intersection. ●

TWO EQUATIONS OF THE FORM $Ax^2 + Bxy + Dx = F$ We solve two equations of the type $Ax^2 + Bxy + Dx = F$ by first eliminating xy and then solving the resulting equation for x. The corresponding value of y can then be found by substitution. The method is illustrated in Example 3.

● **Example 3**
Hyperbola

Hyperbola

Solve the system of equations
$$x^2 + 4xy - 7x = 12 \tag{10}$$
$$3x^2 - 4xy + 4x = 15 \tag{11}$$

Solution

Eliminating the xy term

$x^2 + 4xy - 7x = 12$	(10) recopied	
$3x^2 - 4xy + 4x = 15$	(11) recopied	
$4x^2 \qquad\quad - 3x = 27$	adding (10) and (11)	(12)
$4x^2 - 3x - 27 = 0$		(13)
$(x - 3)(4x + 9) = 0$	factoring	
$x = 3$ and $-\frac{9}{4}$		

Finally, we replace x in (10) successively by 3 and $-\frac{9}{4}$, solve for y, and get $y = 2$ and $\frac{47}{48}$. Consequently, the solution set is $\{(3, 2), (-\frac{9}{4}, \frac{47}{48})\}$. The graphs of (10) and (11), together with the coordinates of their points of intersection, are shown in Fig. 9.5. ●

FIGURE 9.5

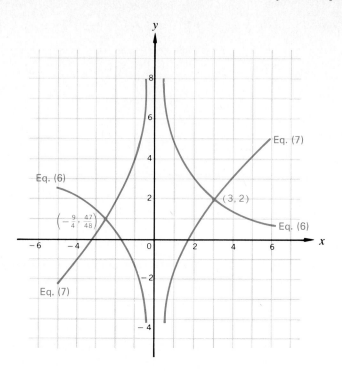

The procedure for solving two equations of the type $Ax^2 + Cy^2 + Ey = F$ is the same as that used in Example 2, except that we eliminate x^2 as the first step and solve the resulting equation for y. To solve two equations of the type $Bxy + Cy^2 + Ey = F$, we proceed as in Example 3. In this case, however, the equation obtained after eliminating xy involves the variable y instead of x.

Elimination by substitution

If, in a system of two equations in two variables, one equation can be solved for one variable in terms of the other, this variable can be eliminated by substitution. We will assume that the variables are x and y, that one equation can be solved for y in terms of x, and that the solution is $y = f(x)$. If we get a linear or a quadratic equation by substituting $f(x)$ for y, we can solve it. We then have the values of x and can find the corresponding values of y by substituting in $y = f(x)$.

● **Example 4**   Obtain the solution of the system of equations

Ellipse
$$x^2 + 2y^2 = 54 \tag{14}$$

Line
$$2x - y = -9 \tag{15}$$

by the method of substitution.

Solution   We first solve (15) for y and get

$$y = 2x + 9 \qquad (16)$$

Next, we replace y by $2x + 9$ in (14) and obtain

Substitute $x^2 + 2(2x + 9)^2 = 54$

which we solve as follows:

$x^2 + 2(4x^2 + 36x + 81) = 54$	squaring $2x + 9$
$x^2 + 8x^2 + 72x + 162 = 54$	by the distributive axiom
$9x^2 + 72x + 108 = 0$	combining similar terms
$x^2 + 8x + 12 = 0$	dividing by 9
$(x + 6)(x + 2) = 0$	factoring
$x = -6$	setting $x + 6 = 0$ and solving
$x = -2$	setting $x + 2 = 0$ and solving

We now replace x in (16) by -6 and then by -2, solve for y, and get

$$y = [2 \times (-6)] + 9 = -12 + 9 = -3$$
$$y = [2 \times (-2)] + 9 = -4 + 9 = 5$$

Therefore, the solution of the system is $(-6, -3)$, $(-2, 5)$. See Fig. 9.6. These solutions can be checked in the original equations.   ●

● **Example 5**   Solve the system of equations

Hyperbola $4x^2 - 2xy - y^2 = -5 \qquad (17)$

Parabola $y + 1 = -x^2 - x \qquad (18)$

by substitution.

FIGURE 9.6

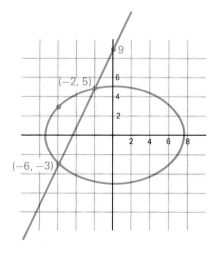

Solution   We first solve (18) for y in terms of x and get

$$y = -x^2 - x - 1 \tag{19}$$

Next we replace y in (17) by the right member of (19) and have

Substitute $$4x^2 - 2x(-x^2 - x - 1) - (-x^2 - x - 1)^2 = -5 \tag{20}$$

Now, if we perform the indicated operations in (20) and combine similar terms, we get

$$-x^4 + 3x^2 + 4 = 0 \tag{21}$$

This is an equation in quadratic form, and we now solve it.

$x^4 - 3x^2 - 4 = 0$	dividing each member of (21) by -1
$(x^2 - 4)(x^2 + 1) = 0$	factoring the left member of (22)
$x^2 = 4$	setting $x^2 - 4 = 0$ and solving for x^2
$x = \pm 2$	equating the square roots
$x^2 = -1$	setting $x^2 + 1 = 0$ and solving for x^2
$x = \pm\sqrt{-1}$	equating the square roots
$x = \pm i$	since $\sqrt{-1} = i$

Now we replace x in (19) successively by 2, -2, i, and $-i$ and obtain each corresponding value of y. This procedure yields the following results:

$y = -2^2 - 2 - 1 = -7$	replacing x by 2
$y = -(-2)^2 - (-2) - 1 = -3$	replacing x by -2
$y = -i^2 - i - 1$	replacing x by i
$\quad = 1 - i - 1 = -i$	since $i^2 = -1$
$y = -(-i)^2 - (-i) - 1$	replacing x by $-i$
$\quad = 1 + i - 1 = i$	

Consequently, the solutions are $(2, -7)$, $(-2, -3)$, $(i, -i)$, $(-i, i)$.   ●

Complex solutions are   Since the coordinates of any point in a cartesian plane are real numbers,
not on the graph $(i, -i)$ and $(-i, i)$ do not represent points on the graph of either equation, although each is a solution pair of both equations. The graphs of (17) and (18) are shown in Fig. 9.7, and they appear to intersect at $(2, -7)$ and $(-2, -3)$.

Elimination by a combination of methods

Frequently, the computation involved in solving the system of equations

$$F(x, y) = K \tag{23}$$

$$f(x, y = k \tag{24}$$

by substitution is very tedious. In such cases, it may be more efficient to use a method that depends upon a theorem in analytic geometry. It states that the graph of

$$mF(x, y) + nf(x, y) = mK + nk \tag{25}$$

FIGURE 9.7

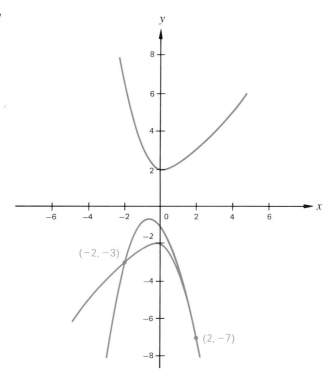

passes through the intersections of the graphs of Eqs. (23) and (24). Hence the solution set of the system (23) and (24) is the same as the solution set of the system composed of either (23) and (25) or of (24) and (25) for all nonzero values of m and n.

Now if we can so determine m and n in (25) that the resulting equation is easily solvable for either unknown in terms of the other, we can obtain the solution set of (23) and (24) by solving either (23) or (24) with (25) by substitution. We will discuss two classes of systems of quadratic equations in two unknowns that can be solved by this method.

TWO EQUATIONS OF THE TYPE $Ax^2 + Bxy + Cy^2 = D$ The first step in solving two equations of the type $Ax^2 + Bxy + Cy^2 = D$ is to *eliminate the constant terms* by addition or subtraction. That is, we combine the two equations in such a way as to obtain an equation in which the constant term is zero. We then solve this equation for one unknown in terms of the other, and complete the process by substitution. For example, if we solve the equation for y in terms of x, we usually obtain two equations of the type $y = rx$ and $y = tx$, where r and t are constants. We then replace y in one of the given equations successively by rx and tx and solve each resulting equation for x. We then find the value of y that corresponds to each root thus obtained by replacing x in $y = rx$ and $y = tx$ by the root. We will illustrate the process with an example.

● **Example 6**   Find the simultaneous solutions of the equations

Ellipse

$$3x^2 - 4xy + 2y^2 = 3 \tag{26}$$

Hyperbola

$$2x^2 - 6xy + y^2 = -6 \tag{27}$$

Solution   We first eliminate the constant terms

$$6x^2 - 8xy + 4y^2 = 6 \qquad \text{multiplying (26) by 2} \tag{28}$$

$$\underline{2x^2 - 6xy + y^2 = -6} \qquad \text{Eq, (27) recopied}$$

$$8x^2 - 14xy + 5y^2 = 0 \qquad \text{equating the corresponding sums} \tag{29}$$

$$(5y - 4x)(y - 2x) = 0 \qquad \text{factoring}$$

$$y = 2x \tag{30}$$

$$y = \frac{4x}{5} \tag{31}$$

We continue the process by replacing y in either given equation (26) or (27), first by $2x$ and then $4x/5$, and solving the resulting equation for x. Substituting $y = 2x$ in (26), we get

Using $y = 2x$

$$3x^2 - 4x(2x) + 2(2x)^2 = 3$$
$$3x^2 - 8x^2 + 8x^2 = 3 \qquad \text{performing the indicated operations}$$
$$3x^2 = 3 \qquad \text{combining terms}$$
$$x^2 = 1$$
$$x = \pm 1$$

We now replace x in (30) by ± 1 and get $y = 2(\pm 1) = \pm 2$. Hence two solutions are $(1, 2)$ and $(-1, -2)$.

Next we replace y by $4x/5$ in (26) and solve the resulting equation for x. Thus we obtain

Using $y = 4x/5$

$$3x^2 - 4x\left(\frac{4x}{5}\right) + 2\left(\frac{4x}{5}\right)^2 = 3$$

$$3x^2 - \frac{16x^2}{5} + \frac{32x^2}{25} = 3$$

$$75x^2 - 80x^2 + 32x^2 = 75 \qquad \text{multiplying each member by 25}$$

$$27x^2 = 75$$

$$x^2 = \tfrac{75}{27} = \tfrac{25}{9}$$

$$x = \pm \tfrac{5}{3}$$

Finally, we replace x by $\pm \tfrac{5}{3}$ in (31) and get $y = \tfrac{4}{5}(\pm \tfrac{5}{3}) = \pm \tfrac{4}{3}$. Therefore, two additional solutions are $(\tfrac{5}{3}, \tfrac{4}{3})$ and $(-\tfrac{5}{3}, -\tfrac{4}{3})$.

Consequently the complete simultaneous solution set is $\{(1, 2), (-1, -2), (\tfrac{5}{3}, \tfrac{4}{3}), (-\tfrac{5}{3}, -\tfrac{4}{3})\}$, which can be checked in the original equations.   ●

Note   *If the constant term in either of the given equations is zero,* as in the system

$$2x^2 - 3xy - 2y^2 = 0$$
$$x^2 + 2xy + 5y^2 = 17$$

it is not necessary to perform the first step in Example 6. We immediately solve the first equation for y in terms of x, or x in terms of y, and then complete the solution by the method of substitution.

THE EQUATIONS OF THE TYPE $Ax^2 + Ay^2 + Dx + Ey = F$ If each of the given equations is of the type $Ax^2 + Ay^2 + Dx + Ey = F$, we can *eliminate the second-degree terms* by addition or subtraction and thereby obtain a linear equation in x and y. We can then solve this equation with one of the given equations by the method illustrated in Example 7 below.

● **Example 7**   Obtain the simultaneous solution set of

Circle
$$3x^2 + 3y^2 + x - 2y = 20 \qquad (32)$$

Circle
$$2x^2 + 2y^2 + 5x + 3y = 9 \qquad (33)$$

Solution

$$
\begin{array}{lll}
6x^2 + 6y^2 + 2x - 4y = 40 & \text{multiplying (32) by 2} & (34)\\
\underline{6x^2 + 6y^2 + 15x + 9y = 27} & \text{multiplying (33) by 3} & (35)\\
\qquad\qquad -13x - 13y = 13 & \text{subtracting} & (36)
\end{array}
$$

Now we solve Eq. (36) simultaneously with Eq. (33) and complete the process of solving as indicated below.

$$
\begin{array}{ll}
y = -x - 1 & \text{solving (36) for } y \qquad (37)\\
2x^2 + 2(-x - 1)^2 + 5x + 3(-x - 1) = 9 & \text{replacing } y \text{ by } -x - 1\\
& \text{in (33)}\\
2x^2 + 2x^2 + 4x + 2 + 5x - 3x - 3 = 9 & \text{performing the indicated}\\
& \text{operations}\\
4x^2 + 6x - 10 = 0 & \text{combining similar terms}\\
2x^2 + 3x - 5 = 0 & \text{dividing by 2}\\
x = 1 \text{ and } -\tfrac{5}{2} & \text{by the quadratic formula}
\end{array}
$$

We find the corresponding values of y from (37) as follows:

$$
\begin{array}{ll}
y = -1 - 1 = -2 & \text{replacing } x \text{ by } 1 \text{ in (37)}\\
y = \tfrac{5}{2} - 1 = \tfrac{3}{2} & \text{replacing } x \text{ by } -\tfrac{5}{2} \text{ in (37)}
\end{array}
$$

Consequently the solution set is $\{(1, -2), (-\tfrac{5}{2}, \tfrac{3}{2})\}$, which can be checked in the original equations.   ●

Exercise 9.2 Solve Probs. 1 to 12 by addition or subtraction.

1 $x^2 + y^2 = 1$
  $2x^2 + 3y^2 = 2$

2 $x^2 + y^2 = 2$
  $3x^2 + 4y^2 = 7$

3 $x^2 + 4y^2 = 5$
  $9x^2 - y^2 = 8$

4 $9x^2 + 4y^2 = 72$
  $x^2 - 9y^2 = -77$

5 $x^2 + y^2 + 2x = 9$
  $x^2 + 4y^2 + 3x = 14$

6 $x^2 + y^2 - 4x = -2$
  $9x^2 + y^2 + 18x = 136$

7 $9x^2 + 4y^2 - 17y = 21$
  $4x^2 - y^2 - 2y = 1$

8 $16x^2 - 9y^2 - 10y = 0$
  $9x^2 + 4y^2 + 12y = 1$

9 $x^2 + xy + 4x = -4$
  $3x^2 - 2xy + x = 4$

10 $x^2 - xy + 5x = 4$
  $2x^2 - 3xy + 10x = -2$

11 $2x^2 + 3xy + x = -2$
  $3x^2 - 2xy + 4x = 1$

12 $x^2 + 4xy - 7x = 10$
  $x^2 + 3xy - 6x = 7$

Use substitution to find the solutions of the following systems of equations.

13 $x - 2y = -5$
  $x^2 - 2y^2 = -23$

14 $9x^2 + 5y^2 = 1$
  $3x - 2y = 1$

15 $y^2 - 2x = -2$
  $x + 2y = 7$

16 $x^2 - 6y = 13$
  $x - 2y = 5$

17 $x^2 - 6xy = -20$
  $x - 2y^2 - 3y = -4$

18 $x^2 - 2xy + y^2 = 9$
  $x^2 + x - y = 4$

19 $x^2 - 12xy + 2y^2 = -72$
  $x = 2y^2 + 6y$

20 $2x^2 - 12xy + y^2 = -36$
  $x = y^2 + 3y$

21 $2x^2 + y^2 = 9$
  $xy = 2$

22 $2x^2 + 3y^2 = 29$
  $xy = 3$

23 $x^2 + 2xy + 2y^2 = 17$
  $xy = -5$

24 $x^2 - 6xy + 4y^2 = 1$
  $2xy = 3$

In Probs. 25 to 32, show that there is no solution. They may be done either algebraically or graphically.

25 $\dfrac{x^2}{4} + y^2 = 1$
  $3x + 5y = 15$

26 $\dfrac{x^2}{4} + \dfrac{y^2}{9} = 1$
  $2y = x - 12$

27 $x^2 - 4y^2 = 1$
  $y = 3x$

28 $4y^2 - 25x^2 = 100$
  $y = x + 1$

29 $(x - 3)^2 + (y - 1)^2 = 4$
  $(x + 1)^2 + (y + 1)^2 = 4$

30 $x^2 + y^2 + x + y + 1 = 0$
  $x^2 + y^2 - 2x - 2y + 3 = 0$

31 $\dfrac{x^2}{16} - \dfrac{y^2}{4} = 1$
  $\dfrac{y^2}{9} + \dfrac{y^2}{25} = 1$

32 $\dfrac{y^2}{4} - \dfrac{x^2}{25} = 1$
  $\dfrac{x^2}{81} + y^2 = 1$

Find the solution set of each of the following pairs of equations.

33 $x^2 + 4xy - 17y^2 = -20$
  $x^2 + xy - 6y^2 = 0$

34 $5x^2 - 2xy - 2y^2 = 1$
  $7x^2 - 3xy - 3y^2 = 1$

35 $x^2 - 2xy - 2y^2 = 2$
$\quad 2x^2 - 4xy - 3y^2 = 3$

36 $x^2 - 6xy + 4y^2 = 1$
$\quad x^2 - 10xy + 8y^2 = -4$

37 $x^2 + 2xy + 2y^2 = 10$
$\quad 2x^2 + xy + 22y^2 = 50$

38 $x^2 + 15xy + 9y^2 = -5$
$\quad 7x^2 + 9xy + 27y^2 = 25$

39 $x^2 + xy + 2y^2 = 8$
$\quad 5x^2 + 8xy - 4y^2 = 32$

40 $2x^2 - 4xy + 3y^2 = 3$
$\quad x^2 + 7xy - 3y^2 = 15$

41 $x^2 + y^2 - 2x = 1$
$\quad x^2 + y^2 - 6y = -1$

42 $x^2 + y^2 - 3x - 2y = 4$
$\quad x^2 + y^2 - 2x - y = 5$

43 $x^2 + y^2 + 2x - y = 2$
$\quad x^2 + y^2 - x + 2y = 5$

44 $x^2 + y^2 - 3x = 1$
$\quad x^2 + y^2 - y = 9$

45 $6x^2 + 6y^2 - 5x + 3y = -1$
$\quad 9x^2 + 9y^2 - 7x + 4y = -1$

46 $x^2 + y^2 - x + 3y = 0$
$\quad 2x^2 + 2y^2 + 3x + 5y = -3$

47 $x^2 + y^2 - 3x = 1$
$\quad x^2 + y^2 + 3y = 7$

48 $x^2 + y^2 - 2x - 3y = 1$
$\quad 3x^2 + 3y^2 - 6x - 4y = 13$

49 $x^2 + y^2 - 5x + y = -4$
$\quad x^2 + y^2 - 3x + 2y = 1$

50 $4x^2 + 4y^2 - 11x = 19$
$\quad 3x^2 + 3y^2 - 11y = 17$

51 $3x^2 + 3y^2 - 13x - y = -2$
$\quad 5x^2 + 5y^2 - 16x + 4y = 25$

52 $x^2 + y^2 + 2x + 2y = 6$
$\quad 2x^2 + 2y^2 + 3x + 3y = 10$

Solve by eliminating the xy term and then substituting from the resulting linear equation.

53 $xy + 2x + 2y = 21$
$\quad 2xy - x - y = 12$

54 $3xy + 4x + 4y = -2$
$\quad xy + x + y = -1$

55 $xy + x + y = 11$
$\quad 2xy - 3x - 3y = -3$

56 $xy - x - y = 0$
$\quad 3xy - 2x - 2y = 4$

57 $2xy + 2x - y = 7$
$\quad 3xy - 2x + 3y = 5$

58 $4xy + 3x - 2y = 1$
$\quad 5xy + 4x - 3y = 2$

59 $3xy + 4x + y = 1$
$\quad 4xy + 5x - 2y = -2$

60 $5xy + 2x + 8y = 4$
$\quad 2xy + x + 3y = 1$

9.3 Applications

A word problem can frequently be best solved if more than one unknown is used; consequently, more than one equation is employed. The method of setting up the equations involves letting the unknowns represent quantities which are called for in the problem. The general rule is that *the number of equations formed must be equal to the number of unknowns introduced.*

● **Example 1**   A real-estate dealer received $4800 in rents on two dwellings in 1981, and one of the dwellings brought $40/month more than the other. How much did she receive per month for each if the more expensive house was vacant for 2 months?

Solution   On inspecting the problem for equations that we might form, we find two basic relations: the connection between the separate rentals and the connection between the monthly rentals and the income per year. Since one house rented for $40 more than the other, we let

$$x = \text{monthly rental on the more expensive house, \$}$$
$$y = \text{monthly rental on the other house, \$}$$

and then $x - y = 40$ (1)

Furthermore, since the first house was rented for 10 months and the other was rented for 12 months, we know that $10x + 12y$ is the total amount received in rentals. Hence,

$$10x + 12y = 4800 \qquad (2)$$

We now have the two equations (1) and (2) in the variables x and y, and we will solve them simultaneously by eliminating y. The solution follows.

$12x - 12y = 480$	(1) × 12	(3)
$10x + 12y = 4800$	(2) recopied	
$22x \quad\quad = 5280$	(3) + (2)	
$x \quad\quad = 240$	dividing both members by 22	

By substituting 240 for x in (1), we get

$$240 - y = 40$$
$$-y = 40 - 240 = -200 \qquad \text{adding } -240 \text{ to each member}$$
$$y = 200 \qquad\qquad\qquad \text{dividing both members by } -1$$

Therefore, the monthly rentals were $240 and $200, respectively.   ●

● **Example 2**   A tobacco dealer mixed 12 lb of one grade of tobacco with 10 lb of another grade to obtain a blend worth $54. He then made a second blend worth $61 by mixing 8 lb of the first grade with 15 lb of the second grade. Find the price per pound of each grade.

Solution   In this problem we find two basic relations that we can use to form two equations. Therefore, we let

$$x = \text{price per pound, \$, of the first grade}$$
$$y = \text{price per pound, \$, of the second grade}$$

and then

$12x + 10y = 54$ by using the numbers of pounds as coefficients and   (4)

$8x + 15y = 61$ the values of the blends as the constant terms   (5)

We will solve these two equations simultaneously by eliminating y and solving for x. The steps in the solution follow.

$$36x + 30y = 162 \qquad (4) \times 3 \qquad\qquad (6)$$

$$16x + 30y = 122 \qquad (5) \times 2 \qquad\qquad (7)$$

$$20x \qquad\quad = 40 \qquad (6) - (7) \qquad\qquad (8)$$

$$x \qquad\quad = 2$$

By substituting 2 for x in (5), we get

$$16 + 15y = 61$$
$$15y = 45$$
$$y = 3$$

Therefore, the prices of the two grades are \$2/lb and \$3/lb. The solution can be checked by substitution in (4) or (5). ●

● **Example 3**   Two airfields A and B are 720 mi apart, and B is due east of A. A plane flew from A to B in 1.8 h and then returned to A in 2 h. If the wind blew with a constant velocity from the west during the entire trip, find the speed of the plane in still air and the speed of the wind.

Solution   The essential point in solving such a problem is that the wind helps the plane in one direction and hinders it in the other. We therefore have the basis for two equations that involve the speed of the plane, the speed of the wind, and the time for the trip. We let

$$x = \text{speed of the plane in still air, mi/h}$$
$$y = \text{speed of the wind, mi/h}$$

Then, since the wind blew constantly from the west,

$$x + y = \text{speed of the plane eastward from } A \text{ to } B \text{ (wind helping)}$$
$$x - y = \text{speed of the plane westward from } B \text{ to } A \text{ (wind hindering)}$$

The distance traveled each way was 720 mi, and we set up equations in time:

$$\frac{720}{x + y} = 1.8 \qquad \text{time required for first half of trip} \qquad (9)$$

$$\frac{720}{x - y} = 2 \qquad \text{time required for second half of trip} \qquad (10)$$

Now we multiply both members of (9) by $5(x + y)$ and of (10) by $x - y$ and get

$$3600 = 9x + 9y \qquad\qquad (11)$$

$$720 = 2x - 2y \qquad\qquad (12)$$

We solve (11) and (12) simultaneously by first eliminating y.

$$7,200 = 18x + 18y \qquad \text{(11)} \times 2 \qquad\qquad (13)$$

$$6,480 = 18x - 18y \qquad \text{(12)} \times 9 \qquad\qquad (14)$$

$$13,680 = 36x \qquad\qquad \text{(13)} + \text{(14)}$$

$$x = 380$$

On substituting 380 for x in (11), we have

$$3600 = (9 \times 380) + 9y$$
$$3600 = 3420 + 9y$$
$$y = 20$$

Hence, the speed of the plane in still air was 380 mi/h, and the speed of the wind was 20 mi/h. The solution can be checked by substitution in (9) and (10). ●

● **Example 4**   The cost for building a rectangular vat with a square base was $128. The base cost $0.30/ft², and the sides cost $0.20/ft². Find the dimensions of the vat if the combined area of the base and the sides was 512 ft².

Solution   We let x = length of one side of the base
                                 y = depth
            Then x^2 = area of the base
                                 $4xy$ = area of the sides

Nonlinear   Hence, $x^2 + 4xy = 512$ (15)

Furthermore, since the costs of the base and sides were $0.30/ft² and $0.20/ft², respectively, then

$$0.30x^2 = 0.3x^2 = \text{cost of base, \$}$$
$$0.80xy = 0.8xy = \text{cost of the sides, \$}$$

Therefore, $0.3x^2 + 0.8xy = 128$ (16)

Hence we have the equations

$$x^2 + 4xy = 512 \qquad \text{(15) recopied}$$
$$0.3x^2 + 0.8xy = 128 \qquad \text{(16) recopied}$$

Since each equation contains a term in xy and no other term involving y, we eliminate xy and complete the solution as follows:

$$0.2x^2 + 0.8xy = 102.4 \qquad \text{multiplying (15) by 0.2} \qquad (17)$$

$$0.3x^2 + 0.8xy = 128 \qquad \text{(16) recopied}$$

$$-0.1x^2 = -25.6 \qquad \text{equating the differences of the}$$
$$\text{members of (16) and (17)}$$

$$x^2 = 256 \qquad \text{multiplying by} -10$$

$$x = 16$$

Note that if $x^2 = 256$, then $x = \pm 16$, but we discard -16 since the dimensions are positive numbers.

Now we replace x by 16 in (15), solve for y, and get

$$256 + 64y = 512$$
$$64y = 256$$
$$y = 4$$

Hence the vat is 16 ft wide by 16 ft long by 4 ft deep.   ●

Exercise 9.3

1 A high school club earned a net profit of $45.80 selling candy apples and suckers, which cost them $0.08 apiece, at a basketball game. If they sold 480 candy apples and 610 suckers and a candy apple and a sucker together sold for $0.25, what was the selling price of each?

2 Karen spent $5.98 for $2\frac{1}{2}$ qt of cream to make ice cream. If whipping cream cost $0.66/$\frac{1}{2}$ pt and half-and-half cost $0.70/pt, how much of each type was used?

3 One term Fred received 22 grade points for making A or B in each of the six subjects he studied. If each A was worth four grade points and each B was worth three grade points, in how many subjects did he make A and in how many did he make B?

4 A music teacher charged $5 for each $\frac{1}{2}$-h organ lesson and $3.50 for each $\frac{1}{2}$-h piano lesson. If in 4 h of teaching she earned $32.50, how many lessons did she teach on each instrument?

5 Tickets for a banquet were $4 for a single ticket or $7.50 for a couple. If 144 people attended the banquet and $549 was collected from ticket sales, how many couples and how many singles attended?

6 A ranger inspecting a forest trail walked at a rate of $3\frac{1}{2}$ mi/h. A second ranger inspecting another portion of the trail walked at a rate of 3 mi/h. If the trail was 42 mi long and the complete inspection required 13 h, how far did each ranger walk?

7 An apartment building contained 20 units, consisting of one-bedroom apartments which rented for $110/month and two-bedroom apartments which rented for $135/month. If the rental from 17 apartments one month was $2045 and three apartments were vacant, how many of each type were rented?

8 Jennifer found that she could drive from the campus to the beach in 5 h by averaging 55 mi/h. However, on one trip, after she had averaged 55 mi/h for awhile, she encountered bad weather and was forced to reduce her speed to 40 mi/h. If this trip required $5\frac{3}{4}$ h, how many miles did she travel at each speed?

9 On a television quiz program, each contestant was given $100 at the start of the program. For each question he answered correctly, he was given a bonus of $100; for each one he missed he was penalized $25. If a contestant attempted 14 questions and ended the game with $875, how many questions were answered correctly and how many were missed?

10 Janice had 6 gal of paint to cover 2380 ft² of fencing. One gallon will cover 470 ft² with one coat and 250 ft² with two coats. If she used all the paint, how many square feet received one coat and how many received two coats?

11 Two different routes between two cities differ by 20 mi. Two people made the trip between the cities in exactly the same time. One traveled the shorter route at 50 mi/h, and the other traveled the longer route at 55 mi/h. Find the length of each route.

12 Mrs. Conner estimated that it cost her $0.55/day to drive to work when she took three passengers, all of whom paid the same daily fee. When two more passengers joined the group, Mrs. Conner cut the fee each paid by $0.10 and found that she earned $0.05 on each trip. Find the total cost of each trip and the fee that the first three passengers paid.

13 On the first day of homecoming weekend, a campus organization earned $650 by selling 450 college pennants and 100 corsages. On the second day they sold the 150 pennants they had left, but the remaining 50 corsages had wilted, so that they lost as much per corsage as they had earned on the previous day. If on the second day they earned $50, how much did they earn on each pennant and on each corsage sold the first day?

14 A biology class of 35 students took a field trip that included a hike of 8 mi. Part of the class also investigated a side trail, which added 3 mi to their hike. If the class walked a total of 331 human-miles, how many students took each hike? (If a group of 20 people walk 10 miles, the group has walked 10 × 20 = 200 human-miles.)

15 One year Bill worked at a part-time job some of the 9 months he was in school. The months he worked he was able to save $65/month, but the months he was idle he had to dip into his savings at the rate of $150/month. If at the end of the school year he had $60 less in his savings account than he had at the beginning, how many months did he work and how many months was he idle?

16 Elizabeth and Catherine signed a lease on an apartment for 9 months. At the end of 6 months, Catherine got married and moved out. She paid the landlord an amount equal to the difference between double-occupancy rental and single-occupancy rental for the remaining 3 months, and Elizabeth paid the single-occupancy rate for the 3 months. If the 9-month rental cost Elizabeth $435 and Catherine $285, what were the single and double monthly rates?

17 The sum of the seven digits in a telephone number is 30. Counting from the left, the first three digits and the last digit are the same. The fourth digit is twice the first, the sum of the fifth digit and the first is 7, and the sixth digit is twice the fifth. Find the number.

18 One summer Elaine earned $12/day and Terry earned $14/day. Together they earned $1142. How long did each work if Terry worked 11 days more than Elaine?

19 Three volunteers assembled 741 newsletters for bulk mailing. The first could assemble 124/h, the second 118/h, and the third 132/h. They worked a total of 6 h. If the first worked 2 h, how long did each of the others work?

20 A class of 32 students was made up of people who were 18, 19, and 20 years of age. The average of their ages was 18.5. How many of each age were in the class if the number of 18-year-olds was six more than the combined number of 19- and 20-year-olds?

21 Toledano sold a square carpet and a rectangular carpet whose length was $\frac{3}{2}$ the width, and the combined area of the two was 375 ft². The price of the first carpet was \$10/yd², and of the second \$12/yd². If \$50 more was received for the square piece than for the other, find the dimensions of each.

22 A tract of land is in the form of a trapezoid with two 90° angles. Two sides which meet at the vertex of one right angle are equal, and the oblique side is 50 rods in length. The area of the tract is 2200 rods², and the perimeter is 200 rods. Find the lengths of the unknown sides.

23 The cost of the material for a rectangular bin with a square base and an open top was \$0.24/ft² for the base and \$0.16/ft² for the sides. The total material cost was \$27.84. Find the dimensions of the bin if the combined area of the base and sides was 156 ft².

24 A rectangle of area 48 ft² is inscribed in a circle of area $78\frac{4}{7}$ ft². Find the dimensions of the rectangle. (Use $\frac{22}{7}$ for pi.)

25 A rectangular pasture with an area of 6400 rods² is divided into three smaller pastures by two fences parallel to the shorter sides. The widths of two of the smaller pastures are the same, and the width of the third is twice that of the others. Find the dimensions of the original pasture if the perimeter of the largest of the subdivisions is 240 rods.

26 A rectangular field has an irrigation well at the midpoint of a longer side. Ditches 5×10^2 rods long run from the well to each opposite corner. If the area of the field is 24×10^4 rods², find the dimensions.

27 A civic club adopted a project that would cost \$960. Before the project was completed, 16 new members joined the club and agreed to pay their share of the cost of the project. The cost per member was thereby reduced by \$2. Find the original number of members and the original cost per member.

28 A block of stock was bought for \$11,000. After 1 year the buyer received a dividend of \$2.20/share and a stock dividend of 20 shares. She then sold the stock for \$2 more per share than it cost and made a profit of \$1980 on the transaction. Find the number of shares bought and the price of each share if no commission is involved.

29 A circle is drawn inside a rectangle and is tangent to each of the longer sides. The difference between the areas of the rectangle and circle is 126 ft², and the sum of their perimeters is 112 ft. Find the dimensions of the rectangle. (Use $\frac{22}{7}$ for pi.)

30 Josephine worked for 90 days in the summer. During the first 60 days she worked 570 h and earned \$1920 in the daytime and \$540 at night.

During the last 30 days she worked 8 h each day and 3 h each night and earned $1500. Find the hourly wage in the daytime and the hourly wage at night, provided that each is an integral number of dollars.

31 A piece of wire 152 in long is cut into two pieces. One piece is bent into a square, and the other into a circle. If the combined area of the square and the circle is 872 in², find the side of the square and the radius of the circle.

32 A swimming pool is in the shape of a rectangle with a semicircle at each end. The area of the pool is $178\frac{2}{7}$ yd², and the perimeter is $57\frac{1}{7}$ yd. Find the width and overall length of the pool.

Graphical solution of a system of inequalities

In this section we will explain the use of the graphical method for finding the solution set of one linear inequality in two variables and also of a system of linear inequalities in two variables.

In most of our discussion, we will be concerned with finding the solution set of an inequality such as $y \geq ax + b$. As in our previous sections, we will use the notation $P(x, y)$ to stand for the point whose coordinates are (x, y).

To solve $y \geq ax + b$, we begin by constructing the graph of $y = ax + b$, as in Fig. 9.8. Now if $P(x, y)$ is on the graph of $y = ax + b$ and if $y' > y$, then $Q(x, y')$ is above the graph. Furthermore, since $y' > y$, it follows that $y' > ax + b$. Therefore, the solution set of $y \geq ax + b$ is

$$\{(x, y) \mid P(x, y) \text{ is on or above the graph of } y = ax + b\}$$

Furthermore, for similar reasons, the solution set of $y \leq ax + b$ is

$$\{(x, y) \mid P(x, y) \text{ is on or below the graph of } y = ax + b\}$$

FIGURE 9.8

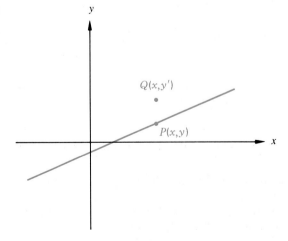

FIGURE 9.9

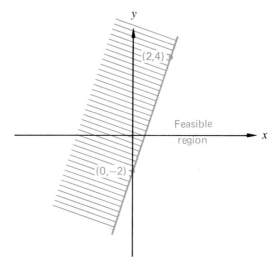

● **Example 1**   The solution set of $y \le 3x - 2$ is $\{(x, y) \mid P(x, y)$ is on or below the graph of $y = 3x - 2\}$. Hence P is in the region not marked in Fig. 9.9.   ●

   In the remainder of this section we will be concerned with determining the region in which $P(x, y)$ lies if (x, y) belongs to the simultaneous solution set of two or more inequalities. The method consists of (1) drawing the graph of each related equation; (2) for each line, *mark out the half plane which we do* **not** *want*; (3) determine the intersection of the regions remaining in the second step. We will refer to this region as the *region determined by*
Feasible region   *the inequalities* or the **feasible region.** Marking out the region **not** wanted makes it easier to see the feasible region when there are several inequalities.

● **Example 2**   Find the region determined by $x \ge 0$, $y \ge 0$, $y \ge 3x - 3$, and $y \le 0.5x + 2$.

Solution   We begin by sketching the graphs of $x = 0$, $y = 0$, $y = 3x - 3$, and $y = 0.5x + 2$. The first one is the y axis, the second is the x axis, the third is the line through $(0, -3)$ and $(1, 0)$, and the last is the line through $(0, 2)$ and $(4, 4)$. The graphs are shown in Fig. 9.10. We now mark out each half plane that we do *not* want. We thus find that the region determined by the given set of inequalities is the boundary and interior of the quadrilateral with vertices at the origin $(0, 0)$, $Q(1, 0)$, $I(2, 3)$, and $R(0, 2)$.   ●

   If **all** points of the line segment PQ are in a set S of the xy plane whenever
Convex region P and Q are in it, we say that the set S is a **convex set** *of points* or a *convex region.*
   According to this definition, the interior of a circle and the quadrilateral $OQIR$ of Fig. 9.10 are convex sets, as is any half plane.
Polygonal set   The intersection of two or more closed half planes is called a *polygonal set of points.* It is also convex.

FIGURE 9.10

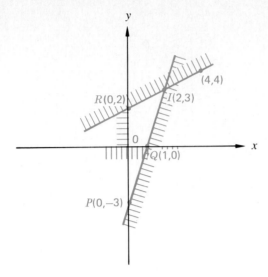

Finite polygonal region   If a polygonal set of points has a finite area, then it is called a *finite*
Convex polygon   *polygonal region* and its boundary is called a *convex polygon*.

The quadrilateral $OQIR$ in Fig. 9.10 is a convex polygon.

Solving a system of nonlinear inequalities is also done by first graphing each corresponding equality. Then we again mark out the parts we do not want.

● **Example 3**   Solve graphically the system of inequalities

$$x^2 + y^2 \leq 25$$
$$4x^2 + y^2 \leq 73$$
$$y \geq 2x - 5$$

Solution   See Fig. 9.11, where we have drawn the corresponding equalities. For each nonlinear inequality, choose a checkpoint which you know definitely to be inside (or outside). For instance, $(0, 0)$ is definitely inside the circle, inside the ellipse, and above the line.   ●

Exercise 9.4   Mark out the half plane ruled out by the inequality in each of Probs. 1 to 4.

1 $x \leq 2$ **2** $y \geq -3$ **3** $2x + y \geq 5$ **4** $3x - y \leq -1$

Show the convex region determined by the set of inequalities in each of Probs. 5 to 24 by marking out the appropriate half planes.

5 $x \geq 0$, $y \geq 0$, $x + 2y \leq 4$ **6** $x \leq 0$, $y \leq 0$, $2x + y \geq -6$
7 $x \geq 1$, $y \leq 2$, $5x - 3y \leq 14$ **8** $x \leq 3$, $y \geq 1$, $x - y \geq 0$
9 $x \geq 0$, $y \leq 0$, $x - y \leq 2$ **10** $x \geq 1$, $y \geq -1$, $x + y \leq 3$

FIGURE 9.11

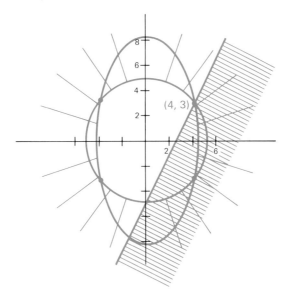

11 $x \le 4$, $y \le 3$, $3x + 4y \ge 12$ 12 $x \ge 1$, $y \ge -1$, $3x + 2y \le 7$

13 $x + y \le 1$, $x - y \le 1$, $x + y \ge -1$, $x - y \ge -1$

14 $2x - y + 2 \ge 0$, $x + y \le 3$, $2x + 3y \ge 6$

15 $x + y \le 2$, $x - y \le 2$, $2x - y \ge 2$, $2x + y \ge 2$

16 $5x - 2y \le 6$, $5x - 2y \ge 4$, $x + y \ge 0$, $5x + 2y \le 10$

17 $y \ge 2x - 1$, $y \ge -x + 5$, $x \ge 0$, $2y + x \le 13$

18 $5x + y \le 25$, $x - 3y \le 5$, $x + y \ge -1$, $3x - 4y \ge -11$

19 $y - x \le 2$, $2x + y \le 2$, $y - 3x \ge -3$, $3x + 2y \ge -6$

20 $4y \ge x$, $2y - 3x + 10 \ge 0$, $4y - x \le 10$, $3x - 2y \ge 0$

21 $x \le 0$, $y \le 0$, $y \le 3x + 2$, $y \ge x - 3$

22 $x \le 1$, $y \ge -1$, $x - 3y + 7 \ge 0$, $4x + y + 6 \ge 0$

23 $x \le 2$, $y \ge 1$, $x - 3y + 2 \ge 0$, $3x + 2y \ge 6$

24 $2x - y + 4 \ge 0$, $x + y + 2 \le 0$, $x \le 1$, $y \ge -3$

Show that the quadrilateral *ABCD* in each of Probs. 25 to 28 is not a convex set.

25 $A(0, 0)$, $B(5, 0)$, $C(2, 1)$, $D(1, 4)$

26 $A(-2, 0)$, $B(3, 0)$, $C(1, 1)$, $D(0, 3)$

27 $A(-2, 1)$, $B(0, -2)$, $C(1, 1)$, $D(3, 2)$

28 $A(4, 0)$, $B(4, 5)$, $C(-1, 4)$, $D(2, 3)$

Show that the inequalities in each of Probs. 29 to 32 do not determine a finite region.

29 $y + x \ge 0$, $y \ge x + 2$, $y \le x + 4$

30 $x \ge 0$, $y \ge 0$, $y \ge x - 2$

31 $4x - y \le 7$, $x + 5y \le 7$, $5x - 4y + 7 \ge 0$

32 $3x + y \ge 4$, $x - 2y + 1 \le 0$, $x + 5y \ge 12$

Find the vertices of the polygonal region determined by the inequalities in each of Probs. 33 to 36.

33 $y + 4x \leq 7$, $2x + 5y \geq -1$, $x - 2y \geq -5$
34 $x + y \leq 3$, $x - 2y \leq 3$, $5x + 2y \geq 3$
35 $x + 5y \leq 11$, $5x + y \leq 7$, $x - 5y \leq 17$, $7x + y \geq -25$
36 $x + 3y \leq 7$, $4x - y \leq 28$, $x + 9y \geq -30$, $4x - 7y \geq 9$

By ruling out the regions not needed, draw the region determined by the inequalities in each of Probs. 37 to 48.

37 $x^2 + y^2 \leq 16$
  $y \geq 2$

38 $3x^2 + y^2 \geq 6$
  $y \leq 3$

39 $y \geq x^2 - 1$
  $y \geq x$

40 $4x^2 - y^2 \leq 4$
  $2x + y \leq 3$

41 $y \geq x^2 - 2$
  $y \leq x + 1$

42 $x^2 - 9y^2 \leq 9$
  $y \leq \dfrac{x}{3} - 1$

43 $x^2 + y^2 \geq 9$
  $y \leq 4$

44 $x^2 + 4y^2 \leq 4$
  $y \leq x$

45 $x^2 + y^2 \leq 13$
  $7x^2 + y^2 \leq 37$
  $y \geq 2x - 1$

46 $x^2 + y^2 \leq 17$
  $18x^2 + 3y^2 \leq 66$
  $2y \geq x + 7$

47 $3x^2 + y^2 \leq 28$
  $y \geq x^2$
  $y \leq x + 2$

48 $10x^2 + 3y^2 \leq 37$
  $y \geq 3x^2$
  $y \leq 2x + 1$

9.5 Linear programming

If there is a largest and a smallest value of a function, they are called the *maximum* and the *minimum*, respectively, and are often referred to as the *extrema*. For example, the maximum value of $y = x^2$, $-1 \leq x \leq 3$, occurs for $x = 3$ and is $3^2 = 9$; the minimum is zero, for $x = 0$.

Maximum
Minimum
Extrema

If a problem involves two variables x and y, if the conditions restrict the domain to a region S determined by a set of linear inequalities, and if a given linear combination of x and y is to be an extremum subject to the restricting inequalities, then the determination of (x, y) and/or the extremum is called *linear programming*.

Linear programming
Constraints
Feasible solutions
Objective function

The inequalities that determine the region S are called *constraints,* the region S is called the set of *feasible solutions,* and the linear function f is called the *objective function.*

We will now state without proof a theorem on linear programming, then give a strictly mathematical application, and follow that by one from business.

○ **Theorem**   If S is a convex polygon, if $f(x, y) = ax + by + c$ has domain S, and if B is the polygonal boundary of S, then f has a maximum and a minimum in S and they occur at vertices of B.   ○

FIGURE 9.12

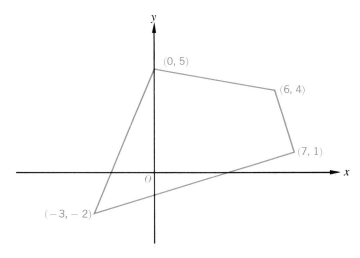

● **Example 1**   Find the maximum and minimum values of $f(x, y) = 2x - 3y + 4$ in the region S with vertices at $(0, 5)$, $(6, 4)$, $(7, 1)$, and $(-3, -2)$.

Solution   Since the region S, shown in Fig. 9.12 with the four given points as vertices, is convex and f is a linear function in two variables, we know by use of the theorem of this section that the extrema of f are attained at vertices of the boundary. Therefore, we will evaluate f at each vertex. The values are

$$f(0, 5) = (2 \times 0) - (3 \times 5) + 4 = -11$$
$$f(6, 4) = (2 \times 6) - (3 \times 4) + 4 = 4$$
$$f(7, 1) = (2 \times 7) - (3 \times 1) + 4 = 15$$
$$f(-3, -2) = [2 \times (-3)] - [3 \times (-2)] + 4 = 4$$

Consequently, the maximum value of f in S is $f(7, 1) = 15$ and the minimum value is $f(0, 5) = -11$.   ●

● **Example 2**   A maker of animal shoes specializes in horseshoes, mule shoes, and oxen shoes and can produce 200 sets of shoes per unit of time. He has standing orders for 60 sets of horseshoes and 20 sets of oxen shoes and can sell at most 150 sets of horseshoes and 50 sets of mule shoes. How many sets of each type should he produce to make a maximum profit provided his profit on a set of shoes is \$0.40 for horseshoes, \$0.50 for mule shoes, and \$0.30 for oxen shoes?

Solution   If we represent the number of sets of horseshoes produced by x and of mule shoes by y, then $200 - x - y$ is the number of oxen shoes. Consequently, his profit in terms of dollars is

$$f(x, y) = 0.4x + 0.5y + 0.3(200 - x - y)$$
$$= 0.1x + 0.2y + 60$$

FIGURE 9.13

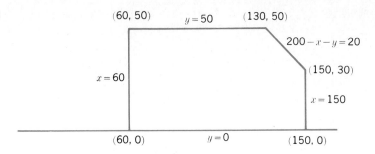

This objective function is subject to the constraints imposed by the problem. They are:

$x \geq 60$	since he has a standing order for 60 sets of horseshoes
$200 - x - y \geq 20$	since he has a standing order for 20 sets of oxen shoes
$x \leq 150$	since he cannot sell more than 150 sets of horseshoes
$y \leq 50$	since he cannot sell more than 50 sets of mule shoes

$$x \geq 0 \qquad y \geq 0$$
$$200 - x - y \geq 0 \qquad \text{since all three are for sale}$$

Figure 9.13 shows these constraints graphically, along with the vertices of the polygonal boundary of the set of feasible solutions as obtained by solving the equations of the pair of bounding lines that meet in a vertex. We now evaluate $f(x, y) = 0.1x + 0.2y + 60$ at each vertex and find that $f(60, 0) = 66$, $f(150, 0) = 75$, $f(150, 30) = 81$, $f(130, 50) = 83$, and $f(60, 50) = 76$. Consequently, the shoemaker has the greatest profit if he makes 130 sets of horseshoes, 50 sets of mule shoes, and 20 sets of oxen shoes. ●

Exercise 9.5   Find the maximum and minimum values of the linear function $f(x, y)$ on the convex polygon determined by the given points as vertices in each of Probs. 1 to 16. Also give the vertex at which the extremes occur.

1 $(2, 0)$, $(4, 1)$, $(3, 5)$, $f(x, y) = 3x - 2y + 1$
2 $(0, 0)$, $(3, 2)$, $(-1, 4)$, $f(x, y) = 2x + y + 3$
3 $(3, -1)$, $(2, 5)$, $(-1, -2)$, $f(x, y) = 5x - y + 4$
4 $(0, -3)$, $(3, 3)$, $(-3, 0)$, $f(x, y) = 3x + 9y - 4$
5 $(-1, -1)$, $(3, 0)$, $(4, 3)$, $(2, 4)$, $f(x, y) = -4x - y + 11$
6 $(0, 5)$, $(8, 2)$, $(2, 7)$, $(-3, -2)$, $f(x, y) = x - 8y + 3$
7 $(0, -2)$, $(5, 0)$, $(0, 6)$, $(-3, 0)$, $f(x, y) = 2x + 3y - 5$
8 $(-1, 4)$, $(4, -1)$, $(1, 4)$, $(-4, 1)$, $f(x, y) = 4x - y + 7$
9 $(3, -1)$, $(4, 3)$, $(0, 2)$, $(-1, -3)$, $f(x, y) = 5x - 3y + 1$
10 $(1, 3)$, $(4, 2)$, $(3, 4)$, $(1, 5)$, $f(x, y) = 6x - 3y + 5$

11 $(2, -2), (4, 3), (-3, 2), (-4, -3), f(x, y) = -4x + 5y + 2$
12 $(3, -2), (5, 3), (-3, 2), (-4, -2), f(x, y) = 5x - y + 8$
13 $(2, -3), (4, 1), (3, 5), (1, 4), (-2, 2), f(x, y) = 3x + 4y - 5$
14 $(1, -3), (2, 2), (1, 4), (-1, 3), (-2, 1), f(x, y) = 6x - 7y + 2$
15 $(3, -1), (6, 2), (4, 3), (2, 3), (0, 1), f(x, y) = 5x + 2y + 7$
16 $(3, -2), (5, 1), (4, 3), (0, 2), (-1, -1), f(x, y) = 4x + 3y - 8$

For each system of constraints in Probs. 17 to 32, find the maximum and minimum values of the objective function f and the coordinates of the vertices at which they are attained.

17 $x \geq 0, y \geq 0, 2x + 3y \leq 6, f(x, y) = 4x + 7y + 1$
18 Same constraints as in 17. $f(x, y) = 5x + y - 2$
19 $x \geq 0, y \leq 2, y \geq x, f(x, y) = 3x + 5y + 2$
20 $x + y \leq 4, 3y \geq x, y \leq 3x, f(x, y) = 6x - y + 5$
21 $-x + y \leq 1, x + y \geq -1, x - y \leq 1, x + y \leq 1, f(x, y) = x + y + 1$
22 $3x - y - 7 \leq 0, x - 4y + 5 \geq 0, 2x + 3y - 1 \geq 0, f(x, y) = 3x + y + 2$
23 $3x - 2y - 5 \leq 0, 4x - y + 2 \geq 0, 3x + y - 2 \leq 0, f(x, y) = 4x - 2y - 1$
24 $3x - y - 7 \leq 0, x - 4y + 5 \geq 0, 2x + 3y - 1 \geq 0, f(x, y) = 7x + 6y - 13$
25 $x \geq 0, y \leq 0, x + y + 1 \geq 0, x - y - 3 \leq 0, f(x, y) = 3x - 4y + 8$
26 $x \leq 0, y \geq 0, 2x + y + 2 \geq 0, x - 2y + 6 \geq 0, f(x, y) = 5x - y - 1$
27 $x \leq 2, y \leq 1, 2x - y + 1 \geq 0, x + 3y + 4 \geq 0, f(x, y) = 3x - 2y + 7$
28 $x \leq 2, y \geq -1, 2x - y + 1 \geq 0, x - 2y + 2 \geq 0, f(x, y) = x + 4y + 9$
29 $x - 4y + 10 \geq 0, 3x + y - 8 \geq 0, x + 5y + 6 \geq 0, 5x + 2y - 16 \leq 0, f(x, y) = 2x + 5y - 3$
30 $x \leq 3, 2x - 3y - 9 \leq 0, 4x + y + 3 \geq 0, x - 4y + 5 \geq 0, f(x, y) = x - y - 2$
31 $2x - 3y + 2 \geq 0, 4x + y - 10 \leq 0, x - 3y - 9 \leq 0, 3x + y + 3 \geq 0, f(x, y) = x + 3y - 4$
32 $x + 3y - 3 \leq 0, 3x - y - 9 \leq 0, x + 4y + 10 \geq 0, 3x - 2y + 2 \geq 0, f(x, y) = 2x - 7y - 5$
33 A manufacturer produces three types of products in 940 h. Product X requires 5h/unit, product Y requires 7h/unit, and product Z takes 3h/unit. The profits per unit are \$60 on X, \$85 on Y, and \$35 on Z. He has orders for 38 X units, 47 Y units, and 100 Z units. If he can sell all he can produce, what combination should he produce to make a maximum profit under the assumption that he can not produce more than $150X$ units or more than $60Y$ units?
34 Repeat Prob. 33 with profit on an X unit changed to \$65.
35 Repeat Prob. 33 with profit on a Z unit changed to \$38.
36 Repeat Prob. 33 with the restriction on X units being that he can not produce more than 140.

37 Feed for animals is to be a mixture of products X and Y. The cost and content of 1 kg of each is:

Product	Cost	g of protein	g of fat	g of carbohydrates
X	$0.80	340	5	520
Y	$0.42	85	15	440

How much of each product should be used to make cost a minimum if each bag must contain a minimum 1530 g of protein and 50 g of fat and not more than 5440 g of carbohydrates? What is the minimum cost of a bag?

38 Repeat Prob. 37 if the cost of 1 kg of product X is reduced to $0.75.

39 Repeat Prob. 37 if the cost of 1 kg of product Y is increased to $0.45.

40 Repeat Prob. 37 if the amount of carbohydrate in Y is increased to 480 g/kg.

41 A shoe manufacturer makes oxfords, high tops, and boots and can produce 500 pairs per unit of time. He has standing orders for 200 pairs of oxfords, and 70 pairs of boots and can sell at most 300 pairs of oxfords and 120 pairs of high tops. How many pairs of each type should be produced to make a maximum profit provided his profit on a pair of footwear is $3.50 for oxfords, $4.25 for high tops, and $6.00 for boots? What is that profit?

42 Repeat Prob. 41 with the maximum production of 400 pairs instead of 500.

43 Repeat Prob. 41 if the number of standing orders for oxfords is increased to 250 pairs.

44 Repeat Prob. 43 if the profit on a pair of oxfords is increased to $4.50 and on a pair of boots decreased to $5.50.

45 A farmer raises corn and wheat. The corn requires 8 units of fertilizer and 11 units of labor, whereas the wheat needs 10 units of fertilizer and 8 units of labor. To maximize his profit, how many acres of each should he plant if the profit per acre on corn is $90 and on wheat $80. Assume he has 7000 units of fertilizer, 7440 units of labor, and 800 acres.

46 A farmer has 600 acres available for planting corn, wheat, and maize. He thinks that he can make a profit of $90/acre on corn, $80/acre on wheat, and $70/acre on maize. He can not take care of more than 200 acres of corn, 400 acres of wheat, or 300 acres of maize. How many acres of each should he plant to make a maximum profit? What is that profit?

47 A floriculturist raises day lilies and amaryllis. For a unit of each he uses labor, fertilizer, and insecticide. To produce 1 unit of day lilies he uses 2 units of labor, 5 of fertilizer, and 1 of insecticide. For 1 unit of amaryllis, he uses 3 units of labor, 4 of fertilizer, and 2 of insecticide. The profit on a unit of each flower is $300 for day lilies and $250 for amaryllis. He has 400 units of labor, 650 units of fertilizer, and 200 units of insecticide. How many units of each flower should be produced to maximize profit? What is that profit?

48 A manufacturer makes two types of wheelbarrows. Type W costs \$18 to produce and sells for \$25, while type B costs \$28 to produce and sells for \$45. The company can manufacture between 400 and 500 of type W and between 150 and 200 of type B, but not more than 650 altogether. How many of each type should the company manufacture to make a maximum profit?

9.6 Partial fractions

We have studied fractions and have seen how to add two fractions, how to divide one fraction by another, and how to reduce a fraction to lowest terms. We will now see how to break a rational function **with a numerator of lower degree than the denominator** into the sum of several fractions with each denominator a linear function, an irreducible quadratic, or an integral power or powers of one or both of these. Furthermore, we will use a constant as the numerator of each linear factor or power thereof and a linear function as the numerator for each irreducible quadratic factor or power thereof. If this is done and the constants evaluated, the rational function is reduced to a form that is readily handled in integral calculus.

In fact, for each power of a linear factor $(ax + b)^m$ in the denominator, we write the sum of m terms

$$\frac{A_1}{ax + b} + \frac{A_2}{(ax + b)^2} + \frac{A_3}{(ax + b)^3} + \cdots + \frac{A_{m-1}}{(ax + b)^{m-1}} + \frac{A_m}{(ax + b)^m}$$

and for each power of an irreducible quadratic factor $(ax^2 + bx + c)^m$, we put

$$\frac{A_1x + B_1}{ax^2 + bx + c} + \frac{A_2x + B_2}{(ax^2 + bx + c)^2} + \cdots$$

$$+ \frac{A_{m-1}x + B_{m-1}}{(ax^2 + bx + c)^{m-1}} + \frac{A_mx + B_m}{(ax^2 + bx + c)^m}$$

Note   The quadratic factor $ax^2 + bx + c$ is **irreducible** (cannot be factored with real coefficients) if $b^2 - 4ac < 0$.

● **Example 1**   To break the fraction

$$\frac{5x^2 - 6x + 10}{(x + 2)(x^2 - 3x + 4)}$$

into partial fractions, we begin by writing

$$\frac{5x^2 - 6x + 10}{(x + 2)(x^2 - 3x + 4)} \equiv \frac{a}{x + 2} + \frac{bx + c}{x^2 - 3x + 4} \tag{1}$$

Later, we will learn how to find a, b, and c. See Example 4.   ●

Degree   If the *numerator is of the same or higher degree than the denominator, we divide* until we get a remainder of lower degree than the denominator, and then apply the above procedure to the fraction that has the remainder as numerator and the original denominator as denominator.

● **Example 2**   Since the numerator of

$$\frac{x^4 - 3x^3 + 5x^2 + 6x - 6}{(x + 2)(x^2 - 3x + 4)}$$

is of higher degree than the denominator, we begin by dividing and thereby get

$$\frac{x^4 - 3x^3 + 5x^2 + 6x - 6}{(x + 2)(x^2 - 3x + 4)} = x - 2 + \frac{5x^2 - 6x + 10}{(x + 2)(x^2 - 3x + 4)}$$

The fraction is the one in Eq. (1).   ●

● **Example 3**   To break

$$\frac{8x^5 + x^4 - 5x^3 + 11x^2 - 17x + 12}{(x - 1)^2(x^2 + 1)(x^2 + x + 2)} = F(x)$$

into partial fractions, we begin by writing

$$F(x) = \frac{a}{x - 1} + \frac{b}{(x - 1)^2} + \frac{cx + d}{x^2 + 1} + \frac{ex + f}{x^2 + x + 2}$$

The constants are not found in this example.   ●

Finding the constants   We must now find how to evaluate the constants that we put in the numerators of the partial fractions. There are **two procedures** that can be used, as well as a combination of them. Both are used in Example 4. One **Equate coefficients** consists of making use of the fact that *two polynomials are identical if and only if the coefficients of each power of the variable in one polynomial are* **Substitute fox** x *equal to the corresponding coefficients in the other polynomial.* For example, $Ax^3 + Bx^2 + Cx + D \equiv 2x^3 - 3x^2 - 5x + 7$ if and only if $A = 2$, $B = -3$, $C = -5$, and $D = 7$. The other method for evaluating the constants in the numerators is to make use of the fact that *if two polynomials are identical, then they are equal for any value assigned to the variable x.* **Note**   For convenience, we select values of x that make each of the linear factors equal to zero.

● **Example 4**   Determine the constants a, b, and c in Eq. (1) of Example 1. The fraction is

$$\frac{5x^2 - 6x + 10}{(x + 2)(x^2 - 3x + 4)} = \frac{a}{x + 2} + \frac{bx + c}{x^2 - 3x + 4}$$

and, after multiplying by the common denominator $(x + 2)(x^2 - 3x + 4)$, it becomes

$$5x^2 - 6x + 10 = a(x^2 - 3x + 4) + (bx + c)(x + 2)$$
$$= x^2(a + b) + x(-3a + 2b + c) + 4a + 2c$$

collecting terms   (2)

We assign -2 to x since then the factor $x + 2$ is zero, and we have

$$5(-2)^2 - 6(-2) + 10 = a[(-2)^2 - 3(-2) + 4] + 0$$
$$20 + 12 + 10 = a(4 + 6 + 4)$$
$$42 = 14a$$

and

$$a = 3$$

Since there is no other real value of x that will make a factor zero, we resort to equating coefficients of like powers; hence, by (2)

$$5 = a + b \qquad \text{from coefficients of } x^2 \qquad (3)$$
$$-6 = -3a + 2b + c \qquad \text{from coefficients of } x \qquad (4)$$
$$10 = 4a + 2c \qquad \text{from constant terms} \qquad (5)$$

Since we know that $a = 3$, we use (3) and get $b = 2$, and use (5) and get $c = -1$. This avoids having to solve from scratch the system made up of (3), (4), and (5). Consequently,

$$\frac{5x^2 - 6x + 10}{(x + 2)(x^2 - 3x + 4)} \equiv \frac{3}{x + 2} + \frac{2x - 1}{x^2 - 3x + 4}$$

This can be verified by adding the fractions on the right.   ●

● **Example 5**   Break

$$\frac{10x^3 + 3x^2 - 7x - 6}{(2x - 1)^2(x^2 + x + 3)}$$

into partial fractions.

Solution   We write

$$\frac{10x^3 + 3x^2 - 7x - 6}{(2x - 1)^2(x^2 + x + 3)} = \frac{a}{2x - 1} + \frac{b}{(2x - 1)^2} + \frac{cx + d}{x^2 + x + 3}$$

and clear of fractions. Accordingly, we get

$$10x^3 + 3x^2 - 7x - 6 = a(2x - 1)(x^2 + x + 3) + b(x^2 + x + 3)$$
$$+ (cx + d)(2x - 1)^2 \qquad (6)$$
$$= x^3(2a + 4c) + x^2(a + b - 4c + 4d)$$
$$+ x(5a + b + c - 4d) - 3a + 3b + d$$

If we use $x = \frac{1}{2}$ in (6), then $2x - 1 = 0$ and we get

$$\tfrac{10}{8} + \tfrac{3}{4} - \tfrac{7}{2} - 6 = a(0) + b(\tfrac{1}{4} + \tfrac{1}{2} + 3) + (cx + d)(0)$$
$$-\tfrac{15}{2} = \tfrac{15}{4} b$$
$$b = -2$$

Note   Since there are no other real values of x that make a coefficient in (6) equal to zero, we resort to equating coefficients of like powers and get

$$10 = 2a + 4c \qquad \text{from coefficients of } x^3$$
$$3 = a + b - 4c + 4d \qquad \text{from coefficients of } x^2$$
$$-7 = 5a + b + c - 4d \qquad \text{from coefficients of } x$$
$$-6 = -3a + 3b + d \qquad \text{from constant terms}$$

Since $b = -2$, these four equations reduce to

$$10 = 2a + 4c \tag{7}$$
$$5 = a - 4c + 4d \tag{8}$$
$$-5 = 5a + c - 4d \tag{9}$$
$$0 = -3a + d \tag{10}$$

We still have four equations but only three unknowns; hence, we can use *any* three of the equations. Using the last three and eliminating c between (8) and (9) gives

$$-15 = 21a - 12d \qquad \text{or} \qquad -5 = 7a - 4d \tag{11}$$

Now in (11) using $d = 3a$ from (10),

$$-5 = 7a - 4(3a)$$
$$-5 = -5a$$
$$1 = a$$

This value of a in (10) leads to $d = 3a = 3(1)$, so

$$3 = d$$

Using $a = 1$ in (7) gives

$$10 = 2 + 4c$$
$$c = 2$$

Consequently,

$$\frac{10x^3 + 3x^2 - 7x - 6}{(2x - 1)^2(x^2 + x + 3)} = \frac{1}{2x - 1} - \frac{2}{(2x - 1)^2} + \frac{2x + 3}{x^2 + x + 3}$$

Exercise 9.6   Break each of the following fractions into partial fractions

1 $\dfrac{1}{(x - 1)(3x - 2)}$ **2** $\dfrac{9}{(2x + 3)(x + 3)}$ **3** $\dfrac{1}{(3x + 7)(2x + 5)}$

4 $\dfrac{-7}{(3x - 2)(4x - 5)}$ **5** $\dfrac{7x}{(2x + 1)(x - 3)}$ **6** $\dfrac{-8x}{(3x + 5)(x + 3)}$

7 $\dfrac{x - 8}{(x - 3)(x - 4)}$ **8** $\dfrac{6x + 9}{(2x - 5)(2x - 1)}$

9 $\dfrac{-4x - 23}{(2x + 1)(x + 2)(x - 3)}$

10 $\dfrac{61x - 1}{(3x - 2)(x + 5)(2x + 1)}$

11 $\dfrac{6x^2 - 4x + 31}{(2x - 3)(x + 4)(3x - 1)}$

12 $\dfrac{-2x^2 - 22x - 24}{(x + 6)(2x + 3)(x + 2)}$

13 $\dfrac{-8x^2 + 35x + 9}{(2x - 1)(x - 4)^2}$

14 $\dfrac{12x^2 - 9x + 20}{(2x + 3)(3x - 1)^2}$

15 $\dfrac{-22x^2 + 32x + 138}{(5x + 2)(2x - 7)^2}$

16 $\dfrac{24x^2 - 94x + 88}{(2x - 3)(3x - 5)^2}$

17 $\dfrac{-2x + 5}{(x - 1)(x^2 + 2)}$

18 $\dfrac{5x^2 + x + 2}{(x + 1)(x^2 + 1)}$

19 $\dfrac{3x^2 - 10x + 16}{(x - 3)(x^2 + x + 1)}$

20 $\dfrac{2x^2 + 4}{(x + 1)(x^2 + 5)}$

21 $\dfrac{5x^3 - 6x^2 + 12x + 13}{(x - 1)(2x + 1)(x^2 + 3)}$

22 $\dfrac{5x^3 + 14x^2 + 71x + 14}{(x + 5)(2x - 1)(x^2 + 3)}$

23 $\dfrac{x^3 + 11x^2 + 13x - 5}{(x - 3)(3x + 1)(x^2 - x + 2)}$

24 $\dfrac{-9x^3 + 19x^2 + 2x + 3}{(2x - 3)(x + 2)(x^2 + 3)}$

25 $\dfrac{3x^3 - 10x^2 + 9x - 6}{(x - 1)^2(x^2 + 1)}$

26 $\dfrac{-x^3 + 8x^2 + 2x - 19}{(x - 2)^2(x^2 + 5)}$

27 $\dfrac{4x^3 - 11x^2 + 12x - 19}{(x - 2)^2(x^2 + x + 1)}$

28 $\dfrac{8x^2 + 19x + 12}{(x + 1)^2(x^2 + 3x + 3)}$

29 $\dfrac{2x^3 - 8x + 4}{(x^2 + 2)(x^2 - 2)}$

30 $\dfrac{3x^3 - 10x^2 + 7x - 3}{(x^2 + 1)(x^2 - 3x - 1)}$

31 $\dfrac{3x^3 - 3x^2 - 4x - 14}{(x^2 - 5)(x^2 - x + 3)}$

32 $\dfrac{-x^3 - 13x^2 + 9}{(x^2 + 3)(x^2 - 3x - 3)}$

33 $\dfrac{x^3 + 2x^2 - 3}{(x^2 - 2)^2}$

34 $\dfrac{2x^3 - 3x^2 + 3x - 10}{(x^2 + 3)^2}$

35 $\dfrac{2x^3 - x^2 - 6x - 8}{(x^2 - x - 3)^2}$

36 $\dfrac{3x^3 + 4x - 5}{(x^2 + 2)^2}$

37 $\dfrac{2x^4 + 13x^3 + 13x^2 - 12x + 2}{x^2(x^2 + 3x - 1)^2}$

38 $\dfrac{x^5 + x^4 - 4x + 4}{x^2(x^2 - x + 2)^2}$

39 $\dfrac{x^4 - x^3 + 1}{x^2(x^2 + 1)^2}$

40 $\dfrac{-2x^5 - 4x^4 + 12x^3 + 9x^2 - 6x + 1}{x^2(x^2 + 3x - 1)^2}$

41 Find A and B so that

$$\frac{cx + d}{(x - r)(x - s)} = \frac{A}{x - r} + \frac{B}{x - s}$$

9.7 Key concepts

Be certain that you understand and can use each of the following words and concepts.

System of equations

Elimination by addition or subtraction

Independent, inconsistent, and dependent equations

Feasible region

Finite polygonal region

Extremes

Maximum

Linear programming

Objective function

System of inequalities

Elimination by substitution

Convex region

Polygonal set

Minimum

Feasible solution

Constraint

Partial fractions

Exercise 9.7 review

State whether the pair of equations in each of Probs. 1, 2, and 3 is independent, inconsistent, or dependent, and solve each independent pair.

1 $3x + 5y = 11$
$6x + 10y = 22$

2 $3x + 5y = 11$
$6x + 10y = 21$

3 $3x + 5y = 11$
$6x + 3y = 15$

Solve the systems in Probs. 4 to 6 for x and y.

4 $3x^2 + 5y^2 = 17$
$2x - y = 3$

5 $x^2 + y^2 = 2$
$5x^2 + 3y^2 = 8$

6 $3x^2 - 2xy + 4y^2 = 23$
$3x^2 + 4y^2 = 19$

7 Solve $y^2 = 3x$ and $2x^2 + y^2 = 11$ simultaneously by graphical methods to the nearest half-unit.

8 Show the simultaneous solution of $5x^2 - y^2 = 6$ and $y = x$ graphically.

9 Show graphically the convex region determined by $x \geq 0$, $3x + 2y \leq 6$, $4x - 3y \leq 12$.

10 Find the vertices of the polygonal region determined by $x - 4y + 3 \geq 0$, $3x + 2y - 5 \geq 0$, $y - 2x \geq -8$.

11 Find the extrema of $f(x, y) = 4x - 5y + 7$ in the convex region with vertices at $(3, -2)$, $(4, 1)$, $(-1, 2)$, and $(-2, -1)$.

12 Find the extrema of f in the region determined by $y \geq 0$, $-2x + y + 6 \geq 0$, $x + 3y \leq 10$, $3x - y \geq 0$, where $f(x, y) = 2x + 3y - 4$.

13 A farmer has 700 acres for growing corn, maize, and wheat. He thinks he can make a profit of $30/acre on corn, $18/acre on maize, and $35/acre on wheat. He must plant 200 acres or less of corn, 250 acres or less of maize, and 400 acres or less of wheat. How many acres of each should he plant to make a maximum profit? What is that profit?

Break the following fractions into partial fractions.

14 $\dfrac{4x + 2}{x(4x + 5)}$

15 $\dfrac{2x - 1}{(x - 3)^2}$

16 $\dfrac{5x^2 + 4x + 5}{(x + 1)(x^2 + x + 3)}$

17 $\dfrac{3x^4 - 9x^3 + 23x^2 - 27x + 11}{(x - 2)(x^2 - x + 3)^2}$

CHAPTER 10 · MATRICES AND DETERMINANTS

The concepts of a matrix and of a determinant are important ones in mathematics. Determinants were invented or devised independently by Kiowa, a Japanese, in 1683 and Leibnitz, a German, in 1693 and were rediscovered in 1750 by Cramer, a Swiss, who used them for solving systems of linear equations. Matrices were invented by Cayley, an Englishman, during the nineteenth century and are used in solving systems of linear equations and in connection with computers.

10.1 Matrices and their basic properties

A rectangular array of numbers of the type

$$
\begin{bmatrix}
a_{11} & a_{12} & \cdots & a_{1j} & \cdots & a_{1n} \\
a_{21} & a_{22} & \cdots & a_{2j} & \cdots & a_{2n} \\
\cdots & \cdots & \cdots & \cdots & \cdots & \cdots \\
a_{i1} & a_{i2} & \cdots & a_{ij} & \cdots & a_{in} \\
\cdots & \cdots & \cdots & \cdots & \cdots & \cdots \\
a_{m1} & a_{m2} & \cdots & a_{mj} & \cdots & a_{mn}
\end{bmatrix}
$$

$m \times n$ matrix

Element

Equality of matrices

is called an $m \times n$ matrix. If $m = n$, it is called a *square matrix*. As seen from the array, an $m \times n$ matrix consists of m rows (horizontal) and n columns (vertical). Each number in the matrix is called an *element* of the matrix. In the above array, a_{ij} is the element in the ith row and jth column. Sometimes $A = (a_{ij})$ is used to designate the matrix. The number i belongs to $\{1, 2, 3, \ldots, m\}$, and j is an element of $\{1, 2, 3, \ldots, n\}$. The symbol $A_{m \times n}$ is often used to indicate that A is an $m \times n$ matrix. The **order** of A is $m \times n$.

Two m by n matrices are equal if and only if each element of one is equal to the corresponding element of the other. Symbolically, if $A = (a_{ij})$ and $B = (b_{ij})$, then $A = B$ means $a_{ij} = b_{ij}$ for all possible subscripts i and j.

● **Example 1**

$$\begin{bmatrix} 3 & 2 & 1 \\ 2 & \sqrt{9} & 7 \end{bmatrix} = \begin{bmatrix} 3 & \sqrt{4} & 1 \\ \sqrt[3]{8} & 3 & 7 \end{bmatrix}$$

but

$$\begin{bmatrix} 5 & -3 \\ 4 & -2 \end{bmatrix} = \begin{bmatrix} 2x+1 & y \\ z+2 & w-1 \end{bmatrix}$$

if and only if $5 = 2x + 1$, $-3 = y$, $4 = z + 2$, and $-2 = w - 1$; hence, if and only if $x = 2$, $y = -3$, $z = 2$, and $w = -1$. Furthermore

$$\begin{bmatrix} 1 & 2 \\ 4 & 5 \\ 8 & 9 \end{bmatrix} \neq \begin{bmatrix} 1 & 2 & 8 \\ 4 & 5 & 9 \end{bmatrix}$$

since the first matrix is a 3×2 matrix while the second one is 2×3.   ●

Sum of two matrices   The sum of two $m \times n$ matrices A and B is the $m \times n$ matrix obtained by adding corresponding elements of A and B. Symbolically, $(a_{ij}) + (b_{ij}) = (a_{ij} + b_{ij})$. The sum is defined if and only if both matrices are $m \times n$. Consequently the sum of

$$A = \begin{bmatrix} 1 & 3 & 5 \\ 2 & 4 & 6 \end{bmatrix} \quad \text{and} \quad B = \begin{bmatrix} -2 & 7 & 8 \\ 9 & -5 & -6 \end{bmatrix}$$

is $$A + B = \begin{bmatrix} 1+(-2) & 3+7 & 5+8 \\ 2+9 & 4+(-5) & 6+(-6) \end{bmatrix} = \begin{bmatrix} -1 & 10 & 13 \\ 11 & -1 & 0 \end{bmatrix}$$

The product kA   The product of a number k and the matrix A is denoted by kA and is the matrix obtained by multiplying each element of A by the constant k. Symbolically $kA = k(a_{ij})$. For example, if

$$A = \begin{bmatrix} 1 & 3 & 6 \\ 0 & -2 & 5 \end{bmatrix} \quad \text{then} \quad 4A = \begin{bmatrix} 4 & 12 & 24 \\ 0 & -8 & 20 \end{bmatrix}$$

Notice that the definitions of $A + A$ and $2A$ both give the same matrix

$$A + A = 2A = \begin{bmatrix} 2 & 6 & 12 \\ 0 & -4 & 10 \end{bmatrix}$$

The product $A_{m \times p} B_{p \times n}$ The product AB of two matrices is defined if and only if the number of columns in A is equal to the numbers of rows in B. The product $(AB)_{m \times n}$ of the matrices $A_{m \times p}$ and $B_{p \times n}$ is the matrix with the element

$$c_{ij} = \sum_{k=1}^{p} a_{ik} b_{kj}$$

in the ith row and jth column. The element c_{ij} is found by adding the product of the first element a_{i1} in the ith row of A and the first element b_{1j} in the jth column of B, the product of the second element a_{i2} in the ith row of A and the second element b_{2j} in the jth column of B, . . . , and the product of the pth element a_{ip} in the ith row of A and the pth element a_{pj} in the jth column of B. If AB and BA are both defined, they may or may not be equal.

● **Example 2**   Find the product $(AB)_{2 \times 2}$ if

$$A_{2 \times 3} = \begin{bmatrix} a_{11} & a_{12} & a_{13} \\ a_{21} & a_{22} & a_{23} \end{bmatrix} \quad \text{and} \quad B_{3 \times 2} = \begin{bmatrix} b_{11} & b_{12} \\ b_{21} & b_{22} \\ b_{31} & b_{32} \end{bmatrix}$$

Solution   If we follow the procedure outlined above, we get

$$(AB)_{2 \times 2} = \begin{bmatrix} a_{11}b_{11} + a_{12}b_{21} + a_{13}b_{31} & a_{11}b_{12} + a_{12}b_{22} + a_{13}b_{32} \\ a_{21}b_{11} + a_{22}b_{21} + a_{23}b_{31} & a_{21}b_{12} + a_{22}b_{22} + a_{23}b_{32} \end{bmatrix}$$

● **Example 3**   Find the products AB and BA if

$$A = \begin{bmatrix} 3 & -2 \\ 1 & 4 \end{bmatrix} \quad \text{and} \quad B = \begin{bmatrix} 5 & 1 \\ -6 & 3 \end{bmatrix}$$

Solution

$$AB = \begin{bmatrix} (3)(5) + (-2)(-6) & (3)(1) + (-2)(3) \\ (1)(5) + (4)(-6) & (1)(1) + (4)(3) \end{bmatrix}$$

$$= \begin{bmatrix} 27 & -3 \\ -19 & 13 \end{bmatrix}$$

Similarly, we find that

$$BA = \begin{bmatrix} 16 & -6 \\ -15 & 24 \end{bmatrix}$$

● **Example 4**   If

$$A = \begin{bmatrix} 1 & 0 \\ 4 & 3 \end{bmatrix} \quad B = \begin{bmatrix} 2 & -5 \\ 1 & 3 \end{bmatrix} \quad \text{and} \quad C = \begin{bmatrix} 4 & 7 \\ -2 & 0 \end{bmatrix}$$

then

$$A(B + C) = \begin{bmatrix} 1 & 0 \\ 4 & 3 \end{bmatrix} \begin{bmatrix} 6 & 2 \\ -1 & 3 \end{bmatrix} = \begin{bmatrix} 6 & 2 \\ 21 & 17 \end{bmatrix}$$

and

$$AB + AC = \begin{bmatrix} 2 & -5 \\ 11 & -11 \end{bmatrix} + \begin{bmatrix} 4 & 7 \\ 10 & 28 \end{bmatrix} = \begin{bmatrix} 6 & 2 \\ 21 & 17 \end{bmatrix}$$

Identity matrix

If A is an $n \times n$ matrix and if each entry in the main diagonal (the diagonal from upper left to lower right) is one and all other elements are zero, the matrix is called an *identity matrix* since it acts as an identity for matrix multiplication. It is designated by I or I_n. Thus

$$I_2 = \begin{bmatrix} 1 & 0 \\ 0 & 1 \end{bmatrix} \quad \text{and} \quad I_3 = \begin{bmatrix} 1 & 0 & 0 \\ 0 & 1 & 0 \\ 0 & 0 & 1 \end{bmatrix}$$

are identity matrices.

Zero matrix   If each element of a matrix is zero, the matrix is called a *zero matrix* and is designated by O. It is the identity for matrix addition.

Transpose of a matrix   The matrix obtained by interchanging the rows and columns of a matrix A is called the *transpose* of A and is designated by A^T. Consequently, the transpose of

$$A = \begin{bmatrix} 3 & 1 & 2 \\ 4 & 0 & -5 \end{bmatrix} \quad \text{is} \quad A^T = \begin{bmatrix} 3 & 4 \\ 1 & 0 \\ 2 & -5 \end{bmatrix}$$

$A + B$ We have now defined matrix addition and multiplication. The sum $A + B$ is defined if and only if A and B have (i) the same number of rows, AB and (ii) the same number of columns. The product AB is defined if and only if the number of columns in A equals the number of rows in B. To talk about the field properties (Sec. 1.2), we will for the present restrict ourselves to $n \times n$ matrices where n is some fixed positive integer.

Thus we suppose that A, B, and C are $n \times n$ matrices. Then $A + B$ and AB are also $n \times n$ matrices. The associative laws

$$(A + B) + C = A + (B + C) \quad \text{and} \quad (AB)C = A(BC)$$

are true. For addition it is easy to prove (see Prob. 49), while for multiplication it is hard to prove. The matrix O is the identity for addition, and I is the identity for multiplication. The additive inverse of $A = (a_{ij})$ is $-A = (-a_{ij})$ since $A + (-A) = (a_{ij} - a_{ij}) = (0) = O$. Addition is commutative since $A + B = (a_{ij}) + (b_{ij}) = (a_{ij} + b_{ij}) = (b_{ij} + a_{ij}) = (b_{ij}) + (a_{ij}) = B + A$.

$A(B + C) = AB + AC$ The distributive law also holds: $A(B + C) = AB + AC$.

The two field properties which do not hold are the commutative law for multiplication and the existence of a multiplicative inverse. Example 3 $AB \neq BA$ shows that we may have $AB \neq BA$. Determinants will be presented later in this chapter. With their help, it can be shown that the $n \times n$ matrix A has an inverse if and only if the determinant of A is not zero. We will also present two methods for finding A^{-1} when it exists.

The matrix product $A \cdot A$ is written A^2. For A^3, we can write either $(A \cdot A) \cdot A$ or $A \cdot (A \cdot A)$. They have the same value because of the associative law for multiplication of matrices.

Matrices may be used in many ways in science, business, and government to organize data systematically. Usually the number of rows or columns is so large that, to handle them efficiently, computers must be used.

● **Example 5**   Suppose that a car dealer sells cars, vans, and trucks at two locations, A and B. Then his vehicles received for January, February, and March may be written as the matrices J, F, M.

$$J = \begin{bmatrix} C & V & T \\ 38 & 9 & 22 \\ 49 & 14 & 24 \end{bmatrix} \begin{matrix} \\ A \\ B \end{matrix} \qquad F = \begin{bmatrix} C & V & T \\ 33 & 7 & 21 \\ 45 & 9 & 21 \end{bmatrix} \begin{matrix} \\ A \\ B \end{matrix} \qquad M = \begin{bmatrix} C & V & T \\ 43 & 10 & 28 \\ 51 & 8 & 26 \end{bmatrix} \begin{matrix} \\ A \\ B \end{matrix}$$

For example, he received 49 cars at location B in January, and 10 vans at A in March. Then his total vehicles received for the 3 months is given by

$$J + F + M = \begin{array}{ccc} C & V & T \\ \begin{bmatrix} 114 & 26 & 71 \\ 145 & 31 & 71 \end{bmatrix} & & \end{array} \begin{array}{c} A \\ B \end{array}$$

Suppose further that his costs are given in dollars by the matrix D:

$$D = \begin{array}{cc} \text{purchase} & \text{overhead} \\ \begin{bmatrix} 5{,}300 & 300 \\ 7{,}400 & 400 \\ 11{,}900 & 550 \end{bmatrix} & \end{array} \begin{array}{c} C \\ V \\ T \end{array}$$

Then for January, the matrix product

$$JD = \begin{bmatrix} 38 & 9 & 22 \\ 49 & 14 & 24 \end{bmatrix} \begin{bmatrix} 5{,}300 & 300 \\ 7{,}400 & 400 \\ 11{,}900 & 550 \end{bmatrix} = \begin{array}{cc} \text{purchase} & \text{overhead} \\ \begin{bmatrix} 529{,}800 & 27{,}100 \\ 648{,}900 & 33{,}500 \end{bmatrix} \end{array} \begin{array}{c} A \\ B \end{array}$$

shows, for instance, that location A had \$529,800 in purchase prices and \$27,100 in overhead costs. Larger matrices would clearly be needed for a complete analysis. ●

Exercise 10.1 Find the values of a, b, c, and d so that the statement in each of Probs. 1 to 4 is true.

1 $\begin{bmatrix} 2 & 3 \\ -2 & a \end{bmatrix} = \begin{bmatrix} b & c \\ d & 1 \end{bmatrix}$ **2** $\begin{bmatrix} 1 & a & 3 \\ c & 4 & -2 \end{bmatrix} = \begin{bmatrix} b & -5 & d \\ 0 & 4 & -2 \end{bmatrix}$

3 $\begin{bmatrix} a & 3 & 2 \\ 5 & 0 & 6 \end{bmatrix} = \begin{bmatrix} -1 & 5 \\ 3 & 0 \\ 2 & 6 \end{bmatrix}$ **4** $\begin{bmatrix} 2 & a \\ -1 & 0 \\ b & 7 \end{bmatrix} = \begin{bmatrix} c & 3 \\ -1 & 0 \\ 6 & d \end{bmatrix}$

State the order and find the transpose of the matrix in each of Probs. 5 to 8.

5 $\begin{bmatrix} 2 & 1 & 5 \\ 3 & -6 & 2 \end{bmatrix}$ **6** $\begin{bmatrix} -1 & 4 & 0 \\ 4 & 5 & 2 \\ 0 & 2 & 3 \end{bmatrix}$ **7** $\begin{bmatrix} 3 & -2 \\ 1 & 5 \\ 0 & 4 \end{bmatrix}$ **8** $\begin{bmatrix} 3 & 4 \\ 5 & 8 \end{bmatrix}$

Use

$$A = \begin{bmatrix} 3 & -1 & 2 \\ 4 & 0 & 5 \end{bmatrix} \quad \text{and} \quad B = \begin{bmatrix} -2 & 3 & 0 \\ 1 & 5 & -4 \end{bmatrix}$$

to evaluate the matrices called for in Probs. 9 to 16.

9 $2A$ **10** $B^T + A$ **11** $-B$ **12** $-3A$ **13** $A + B$ **14** $4B - A$
15 $3A + 2B$ **16** $2A - 5B$

Find the matrix X in each of Probs. 17 to 20.

17 $X - \begin{bmatrix} 3 & 1 \\ 0 & -2 \end{bmatrix} = \begin{bmatrix} 4 & 1 \\ 0 & 3 \end{bmatrix}$ **18** $X - 2\begin{bmatrix} 2 & -1 \\ 3 & 5 \end{bmatrix} = 3\begin{bmatrix} 1 & 2 \\ -2 & -3 \end{bmatrix}$

19 $2X - 3\begin{bmatrix} 2 & 0 \\ 4 & -8 \end{bmatrix} = 2\begin{bmatrix} -3 & 5 \\ -7 & 12 \end{bmatrix}$ **20** $3X + \begin{bmatrix} 3 & -8 \\ 11 & 0 \end{bmatrix} = 3\begin{bmatrix} 1 & -3 \\ 4 & -2 \end{bmatrix}$

Verify the equations in Probs. 21 to 28 for the matrices

$$A = \begin{bmatrix} 3 & -1 \\ 2 & 1 \end{bmatrix} \quad B = \begin{bmatrix} 4 & 1 \\ 2 & 7 \end{bmatrix} \quad \text{and} \quad C = \begin{bmatrix} 6 & -6 \\ 2 & -3 \end{bmatrix}$$

21 $(A + 3B) + 4C = A + (3B + 4C)$
22 $A + (2B + 4C) = (A + 4C) + 2B$
23 $(3A)^T = 3(A^T)$ **24** $(B + C)^T = B^T + C^T$ **25** $(2A)(B) = A(2B)$
26 $(A - B)C = AC - BC$ **27** $(AC)B = A(CB)$ **28** $(B^T)^T = B$

Perform the indicated calculations to verify the statements in Probs. 29 to 36, if

$$A = \begin{bmatrix} 3 & 0 & 1 \\ 2 & -1 & 4 \\ -2 & 3 & 5 \end{bmatrix} \quad \text{and} \quad B = \begin{bmatrix} -1 & 4 & -3 \\ 3 & -2 & 5 \\ 0 & 1 & 2 \end{bmatrix}$$

29 $AB \neq BA$ **30** $A(B + A) = AB + A^2$ **31** $(AB)A = A(BA)$
32 $A(AB) = (A^2)B$ **33** $(A^T)^2 = (A^2)^T$ **34** $B^T A^T = (AB)^T$
35 $(AB)^2 \neq A^2 B^2$ **36** $(A + B)^2 \neq A^2 + 2AB + B^2$

Find the indicated products in each of Probs. 37 to 40.

37 $\begin{bmatrix} 2 & 1 & -3 \\ 3 & 0 & 1 \\ -2 & 2 & 5 \end{bmatrix}\begin{bmatrix} 2 \\ 0 \\ 1 \end{bmatrix}$ **38** $\begin{bmatrix} 1 & -2 \\ 4 & 3 \\ 3 & 5 \end{bmatrix}\begin{bmatrix} 3 \\ 2 \end{bmatrix}$

39 $\begin{bmatrix} 1 & 2 & 4 \\ 3 & -1 & 0 \end{bmatrix}\begin{bmatrix} 1 & 3 \\ 2 & -1 \\ 4 & 0 \end{bmatrix}$ **40** $\begin{bmatrix} 4 & -2 \\ 1 & 3 \\ 0 & 5 \end{bmatrix}\begin{bmatrix} 2 & -1 \\ 3 & 0 \end{bmatrix}$

Prove the statements in Probs. 41 to 44 for A, B, and C any 2×2 matrices.

41 $(AB)^T = B^T A^T$ **42** $(A + B)^T = A^T + B^T$
43 $A(B + C) = AB + AC$ **44** $A(BC) = (AB)C$

If $A_i = \begin{bmatrix} a_i & b_i & c_i \\ d_i & e_i & f_i \end{bmatrix}$ for $i = 1, 2, 3$ are three 2×3 matrices, and h and k are real, nonzero numbers, prove the statement in each of Probs. 45 to 52.

45 $A_1 + A_3 = A_3 + A_1$ **46** $h(A_1 + A_2) = hA_1 + hA_2$
47 $(h + k)A_1 = hA_1 + kA_1$ **48** $A_2 + (-A_2) = O$
49 $A_1 + (A_2 + A_3) = (A_1 + A_2) + A_3$ **50** $A_1 + O = A_1$
51 If $hA_1 + kA_2 = hA_1 + kA_3$, then $A_2 = A_3$, for $k \neq 0$.
52 If $hA_1 - kA_2 = hA_1 - kA_3$, then $A_2 = A_3$, for $k \neq 0$.

10.2 Matrices and systems of linear equations

If we have the system of linear equations

$$
\begin{aligned}
a_{11}x_1 + a_{12}x_2 + \cdots + a_{1n}x_n &= k_1 \\
a_{21}x_2 + a_{22}x_2 + \cdots + a_{2n}x_n &= k_2 \\
\cdots\cdots\cdots\cdots\cdots\cdots\cdots\cdots \\
a_{m1}x_m + a_{m2}x_2 + \cdots + a_{mn}x_n &= k_m
\end{aligned}
\tag{10.1}
$$

we can (1) interchange any two equations; (2) multiply any equation by a nonzero constant; (3) add a nonzero multiple of any equation to any other equation and have a system that is equivalent to the given system; i.e., we can have a system of equations with the same solution as the given system.

The matrix whose elements are the coefficients in the system of equations (10.1) and whose elements occur in the same relative position in the system is called the *coefficient matrix*. If the constant terms are adjoined on the right of the coefficient matrix, as an $(n + 1)$th column, the new matrix is called the *augmented matrix*. Therefore, if the system of equations is

Coefficient matrix

Augmented matrix

$$
\begin{aligned}
x + y + z &= 2 \\
2x + 5y + 3z &= 1 \\
3x - y - 2z &= -1
\end{aligned}
\tag{1}
$$

the matrix of the coefficient is

$$
A = \begin{bmatrix} 1 & 1 & 1 \\ 2 & 5 & 3 \\ 3 & -1 & -2 \end{bmatrix}
$$

and

$$
B = \begin{bmatrix} 1 & 1 & 1 & 2 \\ 2 & 5 & 3 & 1 \\ 3 & -1 & -2 & -1 \end{bmatrix}
$$

is the augmented matrix.

If we know the augmented matrix, we have each equation of the system just as clearly as if the variables and equality signs were written in.

In keeping with the relation between the system of equations and the augmented matrix and statements (1), (2), and (3) just after Eqs. (10.1), we can save time and space by performing the following *row operations* on the augmented matrix of the system of equations:

Row operations

1 *Interchange any two rows of the augmented matrix.*
2 *Multiply any row of the augmented matrix by a nonzero constant.*
3 *Add a nonzero multiple of any row to any other row.*

We will make use of these row operations to change the augmented matrix into one which represents an equivalent set of equations, but which is easier to solve. The new matrix will have 0s below the main diagonal.

● **Example 1**   We will illustrate the procedure by working with the system (1) above.

Solution   The matrix B is the augmented matrix for the system (1). We want to perform row operations on it so as to replace each element of the main diagonal by a 1 and the elements below the main diagonal by zeros.
   We normally begin by getting a 1 as the first element in the main diagonal, but it is already 1; hence we write

$$B_1 = \begin{bmatrix} 1 & 1 & 1 & 2 \\ 2 & 5 & 3 & 1 \\ 3 & -1 & -2 & -1 \end{bmatrix}$$

We now get zero as the second element in the first column by subtracting twice each element of row one from the corresponding element of row two. This operation is indicated by $R_2' = R_2 - 2R_1$. We get a zero in the third row of column one by performing $R_3' = R_3 - 3R_1$. Thus,

$$B_2 = \begin{bmatrix} 1 & 1 & 1 & 2 \\ 0 & 3 & 1 & -3 \\ 0 & -4 & -5 & -7 \end{bmatrix} \qquad \begin{array}{l} R_2' = R_2 - 2R_1 \\ R_3' = R_3 - 3R_1 \end{array}$$

We have now completed the first column since the main diagonal element is 1 and those below are 0. We now use row operations for the other columns also.

$$C_1 = \begin{bmatrix} 1 & 1 & 1 & 2 \\ 0 & 1 & \frac{1}{3} & -1 \\ 0 & -4 & -5 & -7 \end{bmatrix} \qquad R_2' = (\tfrac{1}{3})R_2$$

$$C_2 = \begin{bmatrix} 1 & 1 & 1 & 2 \\ 0 & 1 & \frac{1}{3} & -1 \\ 0 & 0 & -\frac{11}{3} & -11 \end{bmatrix} \qquad R_3' = R_3 + 4R_2$$

$$D_1 = \begin{bmatrix} 1 & 1 & 1 & 2 \\ 0 & 1 & \frac{1}{3} & -1 \\ 0 & 0 & 1 & 3 \end{bmatrix} \qquad R_3' = -\tfrac{3}{11}R_3$$

The matrix D_1 represents the system of equations $x + y + z = 2$, $y + z/3 = -1$, $z = 3$. Starting with the last equation, we see that $z = 3$. From the second equation, we have $y + \frac{3}{3} = -1$; hence $y = -1 - 1 = -2$. Now the first equation gives $x + (-2) + 3 = 2$, $x = 2 + 2 - 3 = 1$. The solution $(1, -2, 3)$ may be checked by substituting in the given system (1).   ●

The matrix method of solution is valuable and systematic, but doing it with pencil and paper is often slow and detailed. It, like many other Calculator   techniques, is ideal for the computer or a hand-held, programmable calculator.

● **Example 2**   Solve $x + y + 2z = 5$, $2x - y + 3z = 4$, and $5x - y + 8z = 10$.

Solution   We will use row operations on the augmented matrix B_1.

$$B_1 = \begin{bmatrix} 1 & 1 & 2 & 5 \\ 2 & -1 & 3 & 4 \\ 5 & -1 & 8 & 10 \end{bmatrix}$$

$$B_2 = \begin{bmatrix} 1 & 1 & 2 & 5 \\ 0 & -3 & -1 & -6 \\ 0 & -6 & -2 & -15 \end{bmatrix} \quad \begin{matrix} R'_2 = R_2 - 2R_1 \\ R'_3 = R_3 - 5R_1 \end{matrix}$$

$$C_2 = \begin{bmatrix} 1 & 1 & 2 & 5 \\ 0 & -3 & -1 & -6 \\ 0 & 0 & 0 & -3 \end{bmatrix} \quad R'_3 = R_3 - 2R_2$$

Notice that we did not calculate the matrix C_1 since it is unnecessary here to make $a_{22} = 1$. Now the last row of C_2 corresponds to the equation $0 \cdot x + 0 \cdot y + 0 \cdot z = -3$, which is not satisfied by any values for x, y, and z.
No solution   Hence the given system has *no solution*.   ●

Note   Solving linear equations by row operations may be used also when there is more than one solution. We must *take care* in interpreting the equation corresponding to the last row of the matrix, as in the next example.

● **Example 3**   Solve $x + 3y - z = 1$, $3x - y + z = 3$, and $5x - 5y + 3z = 5$.

Solution   We begin with the augmented matrix B_1.

$$B_1 = \begin{bmatrix} 1 & 3 & -1 & 1 \\ 3 & -1 & 1 & 3 \\ 5 & -5 & 3 & 5 \end{bmatrix}$$

$$B_2 = \begin{bmatrix} 1 & 3 & -1 & 1 \\ 0 & -10 & 4 & 0 \\ 0 & -20 & 8 & 0 \end{bmatrix} \quad \begin{matrix} R'_2 = R_2 - 3R_1 \\ R'_3 = R_3 - 5R_1 \end{matrix}$$

$$C_2 = \begin{bmatrix} 1 & 3 & -1 & 1 \\ 0 & -10 & 4 & 0 \\ 0 & 0 & 0 & 0 \end{bmatrix} \quad R'_3 = R_3 - 2R_2$$

The third row may be deleted now, since it represents the equation $0 \cdot x + 0 \cdot y + 0 \cdot z = 0$, which is satisfied by any x, y, and z. The second row represents $-10y + 4z = 0$, which may be solved for y by writing $y = 4z/10 = 2z/5$. The the first row corresponds to the equation

$$x + 3y - z = 1$$

$$x + 3\left(\frac{2z}{5}\right) - z = 1$$

$$x = 1 + z - \frac{6z}{5} = 1 - \frac{z}{5}$$

This means that for any value of z, there are values for x and y such that (x, y, z) is a solution of the given system of equations. For instance, if $z = 2$ then $y = \frac{4}{5}$ and $x = \frac{3}{5}$, and if $z = 5$ then $y = 2$ and $x = 0$. In fact, *any* solution has the form

$$\left(1 - \frac{z}{5}, \frac{2z}{5}, z\right)$$

Infinite number of solutions   for some value of z, and the system has an *infinite number of solutions.* ●

Exercise 10.2   Use row operations on matrices to solve the system of equations in each of Probs. 1 to 16.

1 $x - 2y = 0$
$\quad 2x + y = 5$

2 $2x + 3y = 0$
$\quad 3x - y = 11$

3 $2x + y = -9$
$\quad x - 3y = 13$

4 $5x - 4y = 7$
$\quad 4x - 3y = 5$

5 $3x + 2y + z = 8$
$\quad 2x + y + 3z = 7$
$\quad x + 3y + 2z = 9$

6 $x + 3y + z = 4$
$\quad x - 5y + 2z = 7$
$\quad 3x + y - 4z = -9$

7 $2x + 3y + z = -1$
$\quad 5x + 7y - z = 5$
$\quad 4x + 3y = 5$

8 $3x - 2y + 3z = 4$
$\quad 5x + 4z = 3$
$\quad 2x + 7y = -8$

9 $x - 2y + 3z = -5$
$\quad 3x + y = 1$
$\quad 2x - 3y + z = -4$

10 $3x - 4y + 2z = -6$
$\quad 4x + 3y - 2z = 18$
$\quad 2x - 3y + 4z = -10$

11 $2x + 3y - 6z = -3$
$\quad 4x - 3y = 1$
$\quad 4x + 3y + 12z = 13$

12 $x + 2z = 0$
$\quad 3x + 5y - 8z = 3$
$\quad 5x - 7y - 4z = 0$

13 $x + z - w = 0$
$\quad x - y - z = -2$
$\quad x - y - w = -3$
$\quad y + z - w = 1$

14 $x + 2z - w = 6$
$\quad y + z - w = 2$
$\quad z + 2w = -1$
$\quad y + z + w = 0$

15 $x - y - z - w = -1$
    $x + y - z + w = -5$
    $x + y + z + w = 1$
    $x + y + z - w = 9$

16 $x + y + z \quad\quad = 1$
    $x + y \quad\quad + w = 2$
    $x \quad\quad + z + w = 0$
    $\quad y + z + 3w = 0$

Use row operations on matrices to show that each problem has no solution.

17 $2x + y - 3z = 1$
    $2x - 3y - 2z = 2$
    $-2x + 11y \quad = -3$

18 $x - 3y + 5z = 6$
    $3x + 2y + 2z = 3$
    $-7x - 12y + 4z = 5$

19 $3x + 5y - z = 2$
    $4x + 3y + 2z = 5$
    $-6x + y - 8z = -1$

20 $2x - y - z = 0$
    $5x + y + 2z = 2$
    $11x + 5y + 8z = 4$

Each of Probs. 21 to 28 has an infinite number of solutions. Find them all.

21 $2x + 6y + z = 3$
    $x + 4y + 2z = 4$
    $4x + 10y - z = 1$

22 $4x - 2y + 3z = 5$
    $3x + y - z = -4$
    $18x - 4y + 7z = 7$

23 $4x + 6y + 5z = 8$
    $x + 2y + z = 6$
    $3x + 2y + 5z = -14$

24 $3x + 5y + 7z = 4$
    $x + y + z = 1$
    $11x + 15y + 19z = 13$

25 $x + 2y + z = 2$
    $3x + y + 2z = 3$
    $5x + 5y + 4z = 7$
    $10x + 5y + 7z = 11$

26 $3x + y + z = 1$
    $2x + 3y + 4z = 2$
    $12x + 11y + 14z = 8$
    $5x - 3y - 5z = -1$

27 $x + 2y + 2z + w = 4$
    $2x + 3y + 3z - w = 2$
    $x - y - 2z + 2w = 3$

28 $x + y + 2z + 3w = 3$
    $3x + 4y + 6z - w = 2$
    $2x + 3y + 5z - 2w = 6$

Geometry   29 For what values of a, b, and c does the parabola determined by $y = ax^2 + bx + c$ pass through $(-1, 0)$, $(0, -5)$, and $(3, 4)$?

10.3 Second- and third-order determinants

The solution of linear equations is enhanced by the use of determinants. We will later present Cramer's rule for solving linear equations, but we must first define determinants and show how to evaluate them efficiently.

If A is a 2×2 matrix $\begin{bmatrix} a & b \\ c & d \end{bmatrix}$, the value of its determinant is defined by

$|A|, n = 2$

$$\det A = |A| = ad - bc$$

● **Example 1**

$$\begin{vmatrix} 2 & 4 \\ 7 & 3 \end{vmatrix} = 2(3) - 4(7) = 6 - 28 = -22$$

$$\begin{vmatrix} 8 & -5 \\ -3 & 4 \end{vmatrix} = 8(4) - (-5)(-3) = 32 - 15 = 17$$

$$\begin{vmatrix} 6 & 3 \\ -5 & 4 \end{vmatrix} = 6(4) - 3(-5) = 24 + 15 = 39 \qquad ●$$

   To work systematically with determinants of order 3 in this section and order n in the next section, as well as finding the inverse of a square matrix in Sec. 10.6, we will write the matrices with the double subscript notation $A = (a_{ij})$. This means a_{ij} is the number in row i and column j. Thus a square matrix of order 2 is

$A = (a_{ij})$

$$A = \begin{bmatrix} a_{11} & a_{12} \\ a_{21} & a_{22} \end{bmatrix}$$

and its determinant is

$$\det(A) = |A| = a_{11}a_{22} - a_{12}a_{21}$$

● **Example 2**

$$\begin{vmatrix} 2 & -3 \\ 5 & 4 \end{vmatrix} = 2(4) - (-3)(5) = 8 + 15 = 23$$

$$\begin{vmatrix} -5 & 2 \\ -6 & 3 \end{vmatrix} = -5(3) - 2(-6) = -15 + 12 = -3 \qquad ●$$

The symbol for and value of a **third-order determinant** is

$|A|, n = 3$

$$D = \begin{vmatrix} a_{11} & a_{12} & a_{13} \\ a_{21} & a_{22} & a_{23} \\ a_{31} & a_{32} & a_{33} \end{vmatrix}$$

$$= a_{11}a_{22}a_{33} + a_{12}a_{23}a_{31} + a_{13}a_{21}a_{32}$$

$$- a_{11}a_{23}a_{32} - a_{12}a_{21}a_{33} - a_{13}a_{22}a_{31}$$

$$= a_{11}(a_{22}a_{33} - a_{23}a_{32}) + a_{12}(a_{23}a_{31} - a_{21}a_{33})$$

$$+ a_{13}(a_{21}a_{32} - a_{22}a_{31})$$

Hence, we may write this by using second-order determinants as

$$D = a_{11}\begin{vmatrix} a_{22} & a_{23} \\ a_{32} & a_{33} \end{vmatrix} - a_{12}\begin{vmatrix} a_{21} & a_{23} \\ a_{31} & a_{33} \end{vmatrix} + a_{13}\begin{vmatrix} a_{21} & a_{22} \\ a_{31} & a_{32} \end{vmatrix} \qquad (10.2)$$

A careful look at the subscripts in the definition shows that each of the six products consists of exactly one element from each row and exactly one from each column. Three of these products are preceded by a plus sign and the other three by a minus sign.

In Eq. (10.2), a_{11} is multiplied by the 2×2 determinant which remains from the given 3×3 determinant after crossing out the row and the column containing a_{11}. Similarly for a_{12} and a_{13}. Equation (10.2) is called the **expansion in terms of the first row** since the first row consists of a_{11}, a_{12}, and a_{13}. We may expand D in terms of any row or column. The expansion in terms of the second column is

Expansion

$$D = -a_{12}\begin{vmatrix} a_{21} & a_{23} \\ a_{31} & a_{33} \end{vmatrix} + a_{22}\begin{vmatrix} a_{11} & a_{13} \\ a_{31} & a_{33} \end{vmatrix} - a_{32}\begin{vmatrix} a_{11} & a_{13} \\ a_{21} & a_{23} \end{vmatrix}$$

The pattern of signs in front of the numbers a_{ij} is

$$\begin{matrix} + & - & + \\ - & + & - \\ + & - & + \end{matrix}$$

Another way of saying this is that the sign in front of a_{ij} is $(-1)^{i+j}$.

● **Example 3** We will expand the following determinant by the first row and also the second column.

First row
$$D = \begin{vmatrix} 2 & -4 & -5 \\ 1 & 0 & 4 \\ 2 & 3 & -6 \end{vmatrix} = +(2)\begin{vmatrix} 0 & 4 \\ 3 & -6 \end{vmatrix} - (-4)\begin{vmatrix} 1 & 4 \\ 2 & -6 \end{vmatrix} + (-5)\begin{vmatrix} 1 & 0 \\ 2 & 3 \end{vmatrix}$$

$$= 2(0 - 12) + 4(-6 - 8) - 5(3 - 0) = -24 - 56 - 15 = -95$$

Second column
$$D = -(-4)\begin{vmatrix} 1 & 4 \\ 2 & -6 \end{vmatrix} + 0\begin{vmatrix} 2 & -5 \\ 2 & -6 \end{vmatrix} - (3)\begin{vmatrix} 2 & -5 \\ 1 & 4 \end{vmatrix}$$

$$= 4(-6 - 8) + 0 - 3(8 + 5) = -56 - 39 = -95$$

The value, -95, is the same either way.   ●

● **Example 4**

$$D = \begin{vmatrix} 2 & 3 & 0 \\ 0 & -2 & 1 \\ -2 & 4 & 3 \end{vmatrix} = 2\begin{vmatrix} -2 & 1 \\ 4 & 3 \end{vmatrix} - 3\begin{vmatrix} 0 & 1 \\ -2 & 3 \end{vmatrix} + 0\begin{vmatrix} 0 & -2 \\ -2 & 4 \end{vmatrix}$$

$$= 2(-6 - 4) - 3(0 + 2) + 0(0 - 4)$$

$$= -20 - 6 + 0 = -26$$ ●

Minor The determinant that remains after eliminating the row and column in which an element lies is called the **minor** of that element. Thus, in Example 4, the minor of the element 4 is

$$\begin{vmatrix} 2 & 0 \\ 0 & 1 \end{vmatrix}$$

That is,

$$\begin{vmatrix} 2 & 3 & 0 \\ 0 & -2 & 1 \\ -2 & 4 & 3 \end{vmatrix}$$

Cofactor The **cofactor** A_{ij} of an element a_{ij} of a determinant is the minor preceded by a positive sign if $i + j$ is an even number and preceded by a negative sign if $i + j$ is an odd number. Consequently, the cofactor of 4 in D of Example 4 is $-\begin{vmatrix} 2 & 0 \\ 0 & 1 \end{vmatrix}$ since the element 4 is in the third row and second column and $3 + 2 = 5$ is an odd number.

By definition then, the *minor* M_{ij} of the element a_{ij} is a determinant, and the *cofactor* is

$$A_{ij} = (-1)^{i+j} M_{ij}$$

● **Example 5** Evaluate the determinant

$$\begin{vmatrix} 4 & 2 & -1 \\ 5 & 3 & 6 \\ -2 & -4 & 1 \end{vmatrix}$$

in terms of the third row.

Solution We have $-2 = a_{31}$, $-4 = a_{32}$, and $1 = a_{33}$. For the minors we get

$$M_{31} = \begin{vmatrix} 2 & -1 \\ 3 & 6 \end{vmatrix} = 12 - (-3) = 15$$

$$M_{32} = \begin{vmatrix} 4 & -1 \\ 5 & 6 \end{vmatrix} = 24 - (-5) = 29$$

$$M_{33} = \begin{vmatrix} 4 & 2 \\ 5 & 3 \end{vmatrix} = 12 - 10 = 2$$

The cofactors are

$$A_{31} = (-1)^{3+1} M_{31} = (+1)(15) = 15$$
$$A_{32} = (-1)^{3+2} M_{32} = (-1)(29) = -29$$
$$A_{33} = (-1)^{3+3} M_{33} = (+1)(2) = 2$$

The value of the determinant is

$$a_{31}A_{31} + a_{32}A_{32} + a_{33}A_{33} = (-2)(15) + (-4)(-29) + (1)(2)$$
$$= -30 + 116 + 2 = 88$$ ●

Function   For any square matrix A of order 2 or 3, we have defined a specific number $|A|$ called the determinant of A. In other words, we may define a **function** by setting

$$f(A) = |A|$$

Here A is a square matrix and $f(A)$ is a real number. We may find the value of $f(A)$ by expanding the determinant $|A|$ along any row or any column.

Exercise 10.3   Evaluate the determinant in each of Probs. 1 to 8.

1 $\begin{vmatrix} 3 & 2 \\ 5 & 4 \end{vmatrix}$ **2** $\begin{vmatrix} 4 & 1 \\ -7 & 0 \end{vmatrix}$ **3** $\begin{vmatrix} 7 & -3 \\ 2 & -4 \end{vmatrix}$ **4** $\begin{vmatrix} 6 & 7 \\ 8 & 9 \end{vmatrix}$

5 $\begin{vmatrix} 2 & a \\ 6 & 4 \end{vmatrix}$ **6** $\begin{vmatrix} -2 & -1 \\ a & -3 \end{vmatrix}$ **7** $\begin{vmatrix} 2 & b \\ -3 & a \end{vmatrix}$ **8** $\begin{vmatrix} -3 & 2 \\ b & a \end{vmatrix}$

Find the cofactors of the elements in the second column in each of Probs. 9 to 12.

9 $\begin{vmatrix} 3 & 2 & 1 \\ 2 & 3 & 6 \\ 5 & 7 & 2 \end{vmatrix}$ **10** $\begin{vmatrix} 3 & 0 & 4 \\ 1 & 1 & 4 \\ 4 & 4 & 1 \end{vmatrix}$ **11** $\begin{vmatrix} 2 & 2 & -1 \\ 3 & 0 & -3 \\ 1 & 0 & -2 \end{vmatrix}$ **12** $\begin{vmatrix} 2 & 5 & 4 \\ 1 & 4 & 3 \\ 2 & 3 & 1 \end{vmatrix}$

Expand the determinant in each of Probs. 13 to 20 by use of cofactors of the elements of some row or column.

13 $\begin{vmatrix} 2 & 5 & 0 \\ 3 & 1 & 6 \\ 0 & 4 & 2 \end{vmatrix}$ **14** $\begin{vmatrix} -3 & 0 & 0 \\ 0 & 2 & 0 \\ 0 & 0 & -4 \end{vmatrix}$ **15** $\begin{vmatrix} 1 & 2 & 3 \\ -3 & 2 & 4 \\ -2 & -3 & 2 \end{vmatrix}$

16 $\begin{vmatrix} 3 & -4 & 1 \\ 2 & 0 & -2 \\ -1 & 4 & 3 \end{vmatrix}$ **17** $\begin{vmatrix} 2 & 5 & 0 \\ 1 & 4 & 2 \\ 2 & 3 & 5 \end{vmatrix}$ **18** $\begin{vmatrix} 1 & 1 & 3 \\ 0 & 4 & 6 \\ 2 & 0 & -1 \end{vmatrix}$

19 $\begin{vmatrix} 2 & 3 & 4 \\ -4 & 1 & 3 \\ -2 & -3 & 2 \end{vmatrix}$ **20** $\begin{vmatrix} 2 & -5 & 0 \\ 3 & 1 & -1 \\ -2 & 3 & 6 \end{vmatrix}$

Solve the equation in each of Probs. 21 to 28 for x.

21 $\begin{vmatrix} 2 & x \\ 1 & 5 \end{vmatrix} = 7$

22 $\begin{vmatrix} x & 2 \\ -1 & 6 \end{vmatrix} = 4$

23 $\begin{vmatrix} 4 & 2 \\ x & -3 \end{vmatrix} = -14$

24 $\begin{vmatrix} 2 & -1 \\ -3 & x \end{vmatrix} = 5$

25 $\begin{vmatrix} 2 & 0 & 4 \\ 1 & x & 2 \\ 3 & -1 & 3 \end{vmatrix} = -12$

26 $\begin{vmatrix} 3 & 1 & 2 \\ 2 & 3 & 1 \\ 1 & 2 & x \end{vmatrix} = 11$

27 $\begin{vmatrix} x^2 & -3 & 2 \\ 3 & -2 & 1 \\ 2 & -1 & 0 \end{vmatrix} = 0$

28 $\begin{vmatrix} 5 & -3 & -1 \\ 3 & 4 & 2 \\ x^2 & -2 & 0 \end{vmatrix} = -6$

The area of the triangle with vertices (a, b), (c, d), and (e, f) is the absolute value of

$$\frac{1}{2} \begin{vmatrix} 1 & 1 & 1 \\ a & c & e \\ b & d & f \end{vmatrix}$$

Use this to find the areas of the triangles with the following vertices.

29 $(2, 3)$, $(4, -1)$, $(6, 6)$ **30** $(1, -1)$, $(2, 4)$, $(-8, 5)$
31 $(4, 5)$, $(6, 7)$, $(8, 9)$ **32** $(2, -1)$, $(-4, 3)$, $(5, 4)$

By direct calculation, verify the general properties in Probs. 33 to 36 for the particular matrices

$$A = \begin{bmatrix} 3 & 4 & 5 \\ 2 & 3 & 2 \\ 5 & 7 & 8 \end{bmatrix} \quad \text{and} \quad B = \begin{bmatrix} 4 & 1 & -2 \\ 3 & 4 & 1 \\ 2 & 5 & 3 \end{bmatrix}$$

33 $|AB| = |A| \cdot |B|$ **34** $|B| = |B^T|$ **35** $|2A| = 8|A|$
36 $|B^2| = |B|^2$

Prove the equations in Probs. 37 to 44.

37 $\begin{vmatrix} 1 & 1 & 1 \\ a & b & c \\ a^2 & b^2 & c^2 \end{vmatrix} = (a - b)(b - c)(c - a)$

38 $\begin{vmatrix} 1 & 1 & 1 \\ a & b & c \\ a^3 & b^3 & c^3 \end{vmatrix} = (a - b)(b - c)(c - a)(a + b + c)$

39 $\begin{vmatrix} a & x & y \\ 0 & b & z \\ 0 & 0 & c \end{vmatrix} = abc$

40 $\begin{vmatrix} ka & kb & kc \\ kd & ke & kf \\ kg & kh & ki \end{vmatrix} = k^3 \begin{vmatrix} a & b & c \\ d & e & f \\ g & h & i \end{vmatrix}$

41 $\begin{vmatrix} a & b \\ c & d \end{vmatrix} = - \begin{vmatrix} c & d \\ a & b \end{vmatrix}$

42 $\begin{vmatrix} a & b \\ ka & kb \end{vmatrix} = 0$

43 $\begin{vmatrix} a+A & b+B \\ c & d \end{vmatrix} = \begin{vmatrix} a & b \\ c & d \end{vmatrix} + \begin{vmatrix} A & B \\ c & d \end{vmatrix}$

44 $\begin{vmatrix} a & b \\ c+ka & d+kb \end{vmatrix} = \begin{vmatrix} a & b \\ c & d \end{vmatrix}$

45 Show that if a, b, and c are any real numbers, then the roots of the equation below are never complex numbers.

$$\begin{vmatrix} a-x & b \\ b & c-x \end{vmatrix} = 0$$

46 Let $A = \begin{bmatrix} 2 & 1 \\ 5 & 3 \end{bmatrix}$ and $I = \begin{bmatrix} 1 & 0 \\ 0 & 1 \end{bmatrix}$.

(a) Show that the value of the determinant $|A - xI|$ is $x^2 - 5x + 1$.
(b) Show that the matrix A satisfies the equation $A^2 - 5A + I = 0$.

47 Find the value of $\begin{vmatrix} \log 125 & 3 \\ \log 25 & 2 \end{vmatrix}$.

48 Verify that $|A + B| \neq |A| + |B|$ for

$$A = \begin{bmatrix} 3 & 8 \\ 6 & 1 \end{bmatrix} \quad \text{and} \quad B = \begin{bmatrix} 4 & 9 \\ 2 & 6 \end{bmatrix}$$

49 Show that

$$\begin{vmatrix} b+c & a-b & a \\ c+a & b-c & b \\ a+b & c-a & c \end{vmatrix} = 3abc - a^3 - b^3 - c^3$$

50 Show that

$$\begin{vmatrix} a-b-c & 2a & 2a \\ 2b & b-c-a & 2b \\ 2c & 2c & c-a-b \end{vmatrix} = (a+b+c)^3$$

51 If $w^3 = 1$, show that

$$\begin{vmatrix} 1 & w & w^2 \\ w & w^2 & 1 \\ w^2 & 1 & w \end{vmatrix} = 0$$

52 If $w^3 = 1$ and $w \neq 1$, show that

$$\begin{vmatrix} 1 & w^3 & w^2 \\ w^3 & 1 & w \\ w^2 & w & 1 \end{vmatrix} = 3$$

Hint: $0 = w^3 - 1 = (w - 1)(w^2 + w + 1)$.

10.4 Expansion of determinants of order n

We will give some definitions and notation for use in connection with determinants of order n.

Minor   The *minor* of the element a_{ij} of a determinant D_n is the determinant that remains after deleting the ith row and then jth column of D_n and is designated by M_{ij}.

Cofactor   The *cofactor* of a_{ij} is the minor of a_{ij} if $i + j$ is an even number, and the negative of the minor if $i + j$ is an odd number. We designate the cofactor of a_{ij} by A_{ij}.

According to this definition,

$$A_{ij} = (-1)^{i+j} M_{ij} \tag{10.3}$$

Expansion of a
determinant by minors   The expansion of a determinant D_n of order n can be expressed in terms of the minors of the elements of the ith row in the following way:

$$D_n = (-1)^{i+1} a_{i1} M_{i1} + (-1)^{i+2} a_{i2} M_{i2} + (-1)^{i+3} a_{i3} M_{i3}$$
$$+ \cdots + (-1)^{i+j} a_{ij} M_{ij} + \cdots + (-1)^{i+n} a_{in} M_{in} \tag{10.4}$$

Expansion by the ith
row   By use of (10.3) we can express (10.4) in terms of cofactors as follows:

Cofactors
$$D_n = a_{i1} A_{i1} + a_{i2} A_{i2} + a_{i3} A_{i3} + \cdots + a_{ij} A_{ij} + \cdots + a_{in} A_{in} \tag{10.5}$$

Similarly, the expansion of D_n in terms of the minors and cofactors of the elements of the jth column is, respectively,

Expansion by
jth column
$$D_n = (-1)^{1+j} a_{1j} M_{1j} + (-1)^{2+j} a_{2j} M_{2j} + (-1)^{3+j} a_{3j} M_{3j}$$
$$+ \cdots + (-1)^{i+j} a_{ij} M_{ij} + \cdots + (-1)^{n+j} a_{nj} M_{nj} \tag{10.6}$$

and

Cofactors
$$D_n = a_{1j} A_{1j} + a_{2j} A_{2j} + a_{3j} A_{3j} + \cdots + a_{ij} A_{ij} + \cdots + a_{nj} A_{nj} \tag{10.7}$$

These two ways of expansion of D_n give the same value, but the proof is difficult and will be omitted.

● **Example 1**   Obtain the value of

$$D = \begin{vmatrix} 2 & 5 & 4 & 3 \\ 3 & 2 & 5 & 1 \\ 4 & 0 & 2 & 1 \\ 3 & 0 & 3 & 2 \end{vmatrix}$$

Solution   We expand in terms of the second column and obtain

Second column

$$D = (-1)^{1+2}(5)\begin{vmatrix} 3 & 5 & 1 \\ 4 & 2 & 1 \\ 3 & 3 & 2 \end{vmatrix} + (-1)^{2+2}(2)\begin{vmatrix} 2 & 4 & 3 \\ 4 & 2 & 1 \\ 3 & 3 & 2 \end{vmatrix} + 0 + 0$$

We next expand each of the third-order determinants in terms of the elements of the first row and obtain

First row

$$
\begin{aligned}
D &= -5[3(4-3) - 5(8-3) + 1(12-6)] \\
&\quad + 2[2(4-3) - 4(8-3) + 3(12-6)] \\
&= -5(3 - 25 + 6) + 2(2 - 20 + 18) \\
&= -5(-16) + 2(0) = 80
\end{aligned}
$$

● **Example 2**   Find the value of

$$D = \begin{vmatrix} 4 & 2 & 5 & 0 & 3 \\ 1 & 0 & 2 & 5 & 0 \\ 0 & 3 & 2 & 3 & 1 \\ 0 & 0 & 3 & 2 & 0 \\ 5 & 2 & 4 & 4 & 3 \end{vmatrix}$$

Solution   We expand by the fourth row since it has three zeros.

Fourth row

$$D = -0 + 0 + (-1)^{3+4}(3)\begin{vmatrix} 4 & 2 & 0 & 3 \\ 1 & 0 & 5 & 0 \\ 0 & 3 & 3 & 1 \\ 5 & 2 & 4 & 3 \end{vmatrix} + (-1)^{4+4}(2)\begin{vmatrix} 4 & 2 & 5 & 3 \\ 1 & 0 & 2 & 0 \\ 0 & 3 & 2 & 1 \\ 5 & 2 & 4 & 3 \end{vmatrix} - 0$$

We now expand each 4×4 determinant by the second row.

Second row

$$D = -3\left\{(-1)^{2+1}(1)\begin{vmatrix} 2 & 0 & 3 \\ 3 & 3 & 1 \\ 2 & 4 & 3 \end{vmatrix} + (-1)^{2+3}(5)\begin{vmatrix} 4 & 2 & 3 \\ 0 & 3 & 1 \\ 5 & 2 & 3 \end{vmatrix}\right\}$$

$$+ 2\left\{(-1)^{2+1}(1)\begin{vmatrix} 2 & 5 & 3 \\ 3 & 2 & 1 \\ 2 & 4 & 3 \end{vmatrix} + (-1)^{2+3}(2)\begin{vmatrix} 4 & 2 & 3 \\ 0 & 3 & 1 \\ 5 & 2 & 3 \end{vmatrix}\right\}$$

Expanding each of the four 3×3 determinants now gives

$$
\begin{aligned}
D &= -3\{(-1)(28) - 5(-7)\} + 2\{(-1)(-7) - 2(-7)\} \\
&= -3(-28 + 35) + 2(7 + 14) \\
&= -3(7) + 2(21) = -21 + 42 = 21
\end{aligned}
$$

Introduce zeros   As the two examples above illustrate, it is a decided advantage to have lots of zeros in a determinant. We will soon see how to make every element but one equal 0 in any row or column without changing the value of the determinant. See property 6 below.

If an $n \times n$ determinant is expanded completely, there will be $n!$ terms. Each of these terms has n factors, and the terms may be found (except for the sign) by forming every possible product by taking one and only one factor from each row and each column.

Properties of determinants

Although the expansion of a determinant by minors enables us to express the determinant in terms of determinants of lower order, the computation in calculating the value of a determinant of order 4 or more by use of this method would be very tedious. The computation can be greatly simplified if we use the following seven properties.

See Probs. 1 and 2   1 *If the rows of one determinant are the same as the columns of another, and in the same order, the two determinants are equal. That is,*

$$|A| = |A^T|$$

Proof   We will prove the theorem for determinants of the second and third order. If we expand the two determinants

$$D_2 = \begin{vmatrix} a_{11} & a_{12} \\ a_{21} & a_{22} \end{vmatrix} \quad \text{and} \quad D_2' = \begin{vmatrix} a_{11} & a_{21} \\ a_{12} & a_{22} \end{vmatrix}$$

we get $a_{11}a_{22} - a_{12}a_{21}$ in both cases since multiplication is commutative. Hence, the theorem is true for $n = 2$. We will now prove it for $n = 3$ by considering

$$D_3 = \begin{vmatrix} a_{11} & a_{12} & a_{13} \\ a_{21} & a_{22} & a_{23} \\ a_{31} & a_{32} & a_{33} \end{vmatrix} \quad \text{and} \quad D_3' = \begin{vmatrix} a_{11} & a_{21} & a_{31} \\ a_{12} & a_{22} & a_{32} \\ a_{13} & a_{23} & a_{33} \end{vmatrix}$$

If we expand D_3 in terms of the elements of the first row and D_3' in terms of the elements of the first column, we obtain $a_{11}A_{11} + a_{12}A_{12} + a_{13}A_{13}$ in each case. Hence the theorem is true for $n = 3$.   ○

True for order n We will prove that the properties stated in the remainder of this section hold for determinants of order 4. The arguments used, however, are general, and can be applied to *a determinant of any given order.*

See Probs. 3 and 4   2 *If two columns (or rows) of a determinant are interchanged, the value of the determinant is equal to the negative of the value of the given determinant.*

Proof   The proof of this statement is as follows. By (10.7), with $n = 4$, the expansion of the determinant D in terms of the cofactors of the elements of the jth column is

$$D = a_{1j}A_{1j} + a_{2j}A_{2j} + a_{3j}A_{3j} + a_{4j}A_{4j}$$

Recall that $A_{ij} = (-1)^{i+j}M_{ij}$.

Now if we interchange the jth column with the column immediately to the left, we obtain a new determinant D'. This operation changes neither the elements of the original jth column nor the cofactors of the elements. It does, however, decrease the number of the original column by 1; hence, the jth column of D becomes the $(j-1)$th column of D'. Therefore, the expansion of D' will be the same as the expansion of D except that the exponent of -1 in each term will be decreased by 1. Since $(-1)^{i+j-1} = -(-1)^{i+j}$, then $D = -D'$.

If the two columns interchanged are not adjacent, we can prove that this interchange can be accomplished by an odd number of interchanges of adjacent columns. For example, to interchange the first and fourth columns of D, we interchange the fourth column successively with the third, the second, and the first, placing it immediately at the left of the first column. Next we interchange the first column successively with the second and third, placing it in the position vacated by the fourth. Hence, we have made $3 + 2 = 5$ interchanges, and therefore have five changes in sign. Consequently, if two nonadjacent columns of a determinant are interchanged, the value of the determinant obtained will be the negative of the value of the original determinant.

See Probs. 5 and 6   **3** *If two columns (or rows) of a determinant are identical, the value of the determinant is zero.*

Proof   If any two columns of the determinant D are identical, and if we obtain the determinant D' by interchanging these two columns, then by property 2, $D = -D'$. On the other hand, since the two columns interchanged are identical, $D = D'$. Therefore, $D = -D$, and it follows that $D = 0$.

See Probs. 7 and 8   **4** *If the elements of a column (or row) of a determinant are multiplied by k, the value of the determinant is multiplied by k.*

Proof   To prove this statement, we multiply the elements of the jth column of the determinant D by k and obtain the determinant D''. If we expand D'' in terms of the minors of the elements of the jth column, we obtain

$$D'' = (ka_{1j})A_{1j} + (ka_{2j})A_{2j} + (ka_{3j})A_{3j} + (ka_{4j})A_{4j}$$
$$= k(a_{1j}A_{1j} + a_{2j}A_{2j} + a_{3j}A_{3j} + a_{4j}A_{4j}) = kD$$

See Probs. 9 and 10   **5** *If the elements of the jth column of a determinant D are of the form $a_{ij} + b_{ij}$, then D is the sum of the determinants D' and D'' in which all the columns of D, D', and D'' are the same except the jth; furthermore, the jth column of D' is a_{ij}, $i = 1, 2, 3, 4, \ldots, n$, and the jth column of D'' is b_{ij}, $i = 1, 2, 3, 4, \ldots, n$. Similarly for rows.*

Proof   We will prove that this property is valid for a determinant of the fourth order in which the third column is of the form $a_{ij} + b_{ij}$. If

$$D = \begin{vmatrix} a_{11} & a_{12} & a_{13} + b_{13} & a_{14} \\ a_{21} & a_{22} & a_{23} + b_{23} & a_{24} \\ a_{31} & a_{32} & a_{33} + b_{33} & a_{34} \\ a_{41} & a_{42} & a_{43} + b_{43} & a_{44} \end{vmatrix}$$

and we expand D in terms of the cofactors of the elements of the third column, we obtain

$$\begin{aligned} D &= (a_{13} + b_{13})A_{13} + (a_{23} + b_{23})A_{23} + (a_{33} + b_{33})A_{33} \\ &\quad + (a_{43} + b_{43})A_{43} \\ &= [a_{13}A_{13} + a_{23}A_{23} + a_{33}A_{33} + a_{43}A_{43}] \\ &\quad + [b_{13}A_{13} + b_{23}A_{23} + b_{33}A_{33} + b_{43}A_{43}] \end{aligned}$$

By (10.7) the first and second bracketed expressions are, respectively, the expansions of

$$D' = \begin{vmatrix} a_{11} & a_{12} & a_{13} & a_{14} \\ a_{21} & a_{22} & a_{23} & a_{24} \\ a_{31} & a_{32} & a_{33} & a_{34} \\ a_{41} & a_{42} & a_{43} & a_{44} \end{vmatrix} \quad \text{and} \quad D'' = \begin{vmatrix} a_{11} & a_{12} & b_{13} & a_{14} \\ a_{21} & a_{22} & b_{23} & a_{24} \\ a_{31} & a_{32} & b_{33} & a_{34} \\ a_{41} & a_{42} & b_{43} & a_{44} \end{vmatrix}$$

Therefore, $D = D' + D''$.   ◯

See Probs. 11 and 12   6 *If in a given determinant D the elements a_{ik} of the kth column, i = 1, 2, 3, . . . , n, are replaced by $a_{ik} + ta_{ij}$, where a_{ij}, i = 1, 2, 3, . . . , n are the elements of the jth column, the determinant obtained is equal to D. Similarly for rows. In other words, the value of a determinant is not changed if a column (or row) is replaced by that column plus a multiple of another column (or row).*

Proof   We will prove that this property is true for a determinant of order 4 in which we multiply each element of the fourth column by t and add the product to the corresponding element of the second column. If

$$D = \begin{vmatrix} a_{11} & a_{12} & a_{13} & a_{14} \\ a_{21} & a_{22} & a_{23} & a_{24} \\ a_{31} & a_{32} & a_{33} & a_{34} \\ a_{41} & a_{42} & a_{43} & a_{44} \end{vmatrix} \quad \text{and} \quad D' = \begin{vmatrix} a_{11} & a_{12} + ta_{14} & a_{13} & a_{14} \\ a_{21} & a_{22} + ta_{24} & a_{23} & a_{24} \\ a_{31} & a_{32} + ta_{34} & a_{33} & a_{34} \\ a_{41} & a_{42} + ta_{44} & a_{43} & a_{44} \end{vmatrix}$$

then, by properties 4 and 5,

$$D' = \begin{vmatrix} a_{11} & a_{12} & a_{13} & a_{14} \\ a_{21} & a_{22} & a_{23} & a_{24} \\ a_{31} & a_{32} & a_{33} & a_{34} \\ a_{41} & a_{42} & a_{43} & a_{44} \end{vmatrix} + t \begin{vmatrix} a_{11} & a_{14} & a_{13} & a_{14} \\ a_{21} & a_{24} & a_{23} & a_{24} \\ a_{31} & a_{34} & a_{33} & a_{34} \\ a_{41} & a_{44} & a_{43} & a_{44} \end{vmatrix}$$

Now the first determinant in the above pair is equal to D, and the second is equal to zero, since two columns are identical. Therefore, $D' = D$.   ◯

By property 1, the above property is true if the word "column" is replaced by "row."

By a repeated application of property 6 to a determinant D of order n, we can obtain a determinant in which all the elements except one of some row or column are zeros, and the determinant thus obtained will be equal to D. We may then expand the determinant in terms of the minors of the row or column that contains the zeros and thus have the given determinant equal to the product of a constant and a determinant of order $n - 1$. We will illustrate the method with two examples.

Simplify a row or a column by introducing zeros.

We will use the *notation* $C_k + tC_j$ to indicate that each element in the *j*th column of D is multiplied by t and the product is added to the corresponding element of the *k*th column, and we will use a similar notation for rows.

● **Example 3** Find the value of

$$D = \begin{vmatrix} 2 & 1 & 4 & 1 \\ 4 & 2 & 6 & 2 \\ 3 & 5 & 2 & 3 \\ 7 & 3 & 1 & 3 \end{vmatrix}$$

Solution We first notice that the second and fourth columns have three elements in common; hence, we replace C_2 by $C_2 - C_4$ and get

$$D = \begin{vmatrix} 2 & 0 & 4 & 1 \\ 4 & 0 & 6 & 2 \\ 3 & 2 & 2 & 3 \\ 7 & 0 & 1 & 3 \end{vmatrix}$$

Now we expand by the second column and get

$$D = -2 \begin{vmatrix} 2 & 4 & 1 \\ 4 & 6 & 2 \\ 7 & 1 & 3 \end{vmatrix}$$

Finally, we replace R_2 by $R_2 - 2R_1$, expand the determinant thus obtained in terms of the elements of the second row, and get

$$D = -2 \begin{vmatrix} 2 & 4 & 1 \\ 0 & -2 & 0 \\ 7 & 1 & 3 \end{vmatrix} = (-2)(-2) \begin{vmatrix} 2 & 1 \\ 7 & 3 \end{vmatrix} = 4(6 - 7) = -4 \quad ●$$

● **Example 4** Obtain the value of the determinant

$$D = \begin{vmatrix} 2 & 3 & 5 & 1 \\ 4 & 2 & 3 & 5 \\ 3 & 1 & 4 & 2 \\ 5 & 4 & 2 & 3 \end{vmatrix}$$

Solution   If we examine the above determinant, we see that we cannot obtain a determinant with three zeros in either one row or one column by adding or subtracting corresponding terms in either rows or columns. Since $a_{14} = 1$, we can, however, obtain a determinant in which the elements in the first row are 0, 0, 0, 1 by performing successively the operations $C_1 - 2C_4$, $C_2 - 3C_4$, and $C_3 - 5C_4$ and then writing column 4 unchanged. Thus we get

$$D = \begin{vmatrix} 0 & 0 & 0 & 1 \\ -6 & -13 & -22 & 5 \\ -1 & -5 & -6 & 2 \\ -1 & -5 & -13 & 3 \end{vmatrix}$$

If we now expand in terms of the elements of the first row, we get

$$D = -1 \begin{vmatrix} -6 & -13 & -22 \\ -1 & -5 & -6 \\ -1 & -5 & -13 \end{vmatrix}$$

Now we notice that the first two terms in the second and third rows are the same; so we perform the operation $R_2 - R_3$ and get

$$D = -1 \begin{vmatrix} -6 & -13 & -22 \\ 0 & 0 & 7 \\ -1 & -5 & -13 \end{vmatrix}$$

$$= (-1)(-7) \begin{vmatrix} -6 & -13 \\ -1 & -5 \end{vmatrix} = 7(30 - 13) = 119 \qquad \bullet$$

7 *If each element of a row is multiplied by the cofactor of the corresponding element of another row and the products are added, the sum is zero.*

Proof   We will prove this theorem for a third-order determinant, but the procedure can be applied to one of any order. We will consider the elements of the first row of

$$D = \begin{vmatrix} a_{11} & a_{12} & a_{13} \\ a_{21} & a_{22} & a_{23} \\ a_{31} & a_{32} & a_{33} \end{vmatrix} = a_{11}A_{11} + a_{12}A_{12} + a_{13}A_{13}$$

and the cofactors of the elements of the third row and prove that

$$a_{11}A_{31} + a_{12}A_{32} + a_{13}A_{33} = 0 \qquad (1)$$

Since

$$A_{31} = \begin{vmatrix} a_{12} & a_{13} \\ a_{22} & a_{23} \end{vmatrix} \qquad A_{32} = - \begin{vmatrix} a_{11} & a_{13} \\ a_{21} & a_{23} \end{vmatrix} \qquad A_{33} = \begin{vmatrix} a_{11} & a_{12} \\ a_{21} & a_{22} \end{vmatrix}$$

the left member of (1) is the expansion of

$$\begin{vmatrix} a_{11} & a_{12} & a_{13} \\ a_{11} & a_{12} & a_{13} \\ a_{21} & a_{22} & a_{23} \end{vmatrix}$$

in terms of the elements of the first row. The value of the determinant is zero since two rows are identical. $\bigcirc$

Exercise 10.4   By use of the properties of determinants, prove the statement in each of Probs. 1 to 20 without expanding.

1 $\begin{vmatrix} b & a & d \\ b & o & y \\ r & u & n \end{vmatrix} = \begin{vmatrix} b & b & r \\ a & o & u \\ d & y & n \end{vmatrix}$

2 $\begin{vmatrix} 8 & a & m \\ g & e & t \\ o & u & t \end{vmatrix} = \begin{vmatrix} 8 & g & o \\ a & e & u \\ m & t & t \end{vmatrix}$

3 $\begin{vmatrix} 1 & 4 & 3 \\ 2 & 1 & 8 \\ 3 & 2 & 5 \end{vmatrix} = - \begin{vmatrix} 1 & 4 & 3 \\ 3 & 2 & 5 \\ 2 & 1 & 8 \end{vmatrix}$

4 $\begin{vmatrix} 3 & 1 & 7 \\ 2 & 0 & 3 \\ 4 & 5 & 6 \end{vmatrix} = - \begin{vmatrix} 3 & 7 & 1 \\ 2 & 3 & 0 \\ 4 & 6 & 5 \end{vmatrix}$

5 $\begin{vmatrix} 1 & 3 & 1 \\ 5 & 0 & 5 \\ 2 & 7 & 2 \end{vmatrix} = 0$

6 $\begin{vmatrix} 8 & 1 & 4 \\ 6 & 7 & 3 \\ 6 & 7 & 3 \end{vmatrix} = 0$

7 $2 \begin{vmatrix} b & a & t \\ b & o & y \\ s & i & d \end{vmatrix} = \begin{vmatrix} 2b & a & t \\ 2b & o & y \\ 2s & i & d \end{vmatrix}$

8 $6 \begin{vmatrix} s & h & o \\ v & e & r \\ h & i & g \end{vmatrix} = \begin{vmatrix} 6s & 6h & 6o \\ v & e & r \\ h & i & g \end{vmatrix}$

9 $\begin{vmatrix} 2 & 1 & 3 \\ 1 & 5 & 7 \\ 5 & 2 & 4 \end{vmatrix} = \begin{vmatrix} 2 & 1 & 3 \\ 1 & 1 & 2 \\ 5 & 2 & 4 \end{vmatrix} + \begin{vmatrix} 2 & 1 & 3 \\ 0 & 4 & 5 \\ 5 & 2 & 4 \end{vmatrix}$

10 $\begin{vmatrix} 1 & 2 & 3 \\ 4 & 5 & 6 \\ 7 & 8 & 9 \end{vmatrix} = \begin{vmatrix} 1 & 2 & 2 \\ 4 & 5 & 2 \\ 7 & 8 & 2 \end{vmatrix} + \begin{vmatrix} 1 & 2 & 1 \\ 4 & 5 & 4 \\ 7 & 8 & 7 \end{vmatrix}$

11 $\begin{vmatrix} 3 & 6 & 2 \\ 1 & 1 & 5 \\ 4 & 3 & 8 \end{vmatrix} = \begin{vmatrix} 3 & 6 & 20 \\ 1 & 1 & 8 \\ 4 & 3 & 17 \end{vmatrix}$

12 $\begin{vmatrix} 4 & 3 & 2 \\ 6 & 1 & 6 \\ 5 & 9 & 4 \end{vmatrix} = \begin{vmatrix} 4 & 3 & 2 \\ 22 & 13 & 14 \\ 5 & 9 & 4 \end{vmatrix}$

13 $\begin{vmatrix} a & b & c \\ d & e & f \\ g & h & i \end{vmatrix} = - \begin{vmatrix} a & g & d \\ b & h & e \\ c & i & f \end{vmatrix}$

14 $\begin{vmatrix} a & b & c \\ d & e & f \\ g & h & i \end{vmatrix} = \begin{vmatrix} a+b-c & b & c \\ d+e-f & e & f \\ g+h-i & h & i \end{vmatrix}$

15 $\begin{vmatrix} x_1 & y_1 & z_1 \\ x_2 & y_2 & z_2 \\ x_3 & y_3 & z_3 \end{vmatrix} = \begin{vmatrix} x_1 + ax_2 & y_1 + ay_2 & z_1 + az_2 \\ x_2 & y_2 & z_2 \\ x_3 & y_3 & z_3 \end{vmatrix}$

16 $\begin{vmatrix} a & b \\ c & d \end{vmatrix} = \begin{vmatrix} a + 2b & b \\ c + 2d & d \end{vmatrix} = \begin{vmatrix} a & b - 7a \\ c & d - 7c \end{vmatrix}$

In Probs. 17 to 20, assume that

$$\begin{vmatrix} L & S & U \\ T & I & G \\ E & R & Z \end{vmatrix} = 7$$

and find the value of each determinant.

17 $\begin{vmatrix} L & U & S \\ E & Z & R \\ T & G & I \end{vmatrix}$ **18** $\begin{vmatrix} L & -2U & S \\ T & -2G & I \\ E & -2Z & R \end{vmatrix}$

19 $\begin{vmatrix} L & S & U \\ T & I & G \\ E - T & R - I & Z - G \end{vmatrix}$ **20** $\begin{vmatrix} 2L & 2S & 2U \\ 3T & 3I & 3G \\ -E & -R & -Z \end{vmatrix}$

Use property 6 (properties of determinants) to find the value of the determinant in each of Probs. 21 to 28.

21 $\begin{vmatrix} 1 & 8 & 1 \\ 1 & 3 & 3 \\ 8 & 3 & 24 \end{vmatrix}$ **22** $\begin{vmatrix} 1 & -1 & 1 \\ 4 & 4 & -7 \\ -2 & -3 & 4 \end{vmatrix}$

23 $\begin{vmatrix} 2 & 9 & 4 \\ 8 & 2 & -3 \\ -2 & 1 & 2 \end{vmatrix}$ **24** $\begin{vmatrix} 6 & 2 & 10 \\ 2 & 0 & 5 \\ 2 & 3 & -4 \end{vmatrix}$

25 $\begin{vmatrix} 1 & 1 & 2 \\ 2 & 2 & 1 \\ 3 & 1 & 3 \end{vmatrix}$ **26** $\begin{vmatrix} 2 & 1 & 2 \\ 3 & 2 & 4 \\ 5 & 1 & 2 \end{vmatrix}$ **27** $\begin{vmatrix} a & b & c \\ 0 & a & b \\ a & 0 & b \end{vmatrix}$ **28** $\begin{vmatrix} x & y & 0 \\ 0 & x & y \\ x & 0 & y \end{vmatrix}$

Find the value or values of x so that the statement in each of Probs. 29 to 32 is true.

29 $\begin{vmatrix} 3 & 5 & -1 \\ 2 & 1 & 1 \\ 4 & x & 2 \end{vmatrix} = 0$ **30** $\begin{vmatrix} 5 & 2 & x \\ 1 & 3 & -1 \\ x & 2 & 1 \end{vmatrix} = -4$

31 $\begin{vmatrix} 2 & 0 & x \\ 3 & 1 & x \\ x & 2 & 1 \end{vmatrix} = -13$

32 $\begin{vmatrix} x & 1 & x \\ 1 & x & 2 \\ 3 & x & 1 \end{vmatrix} = -7$

In each of Probs. 33 to 36, select the row (or column) that contains the most zeros, then expand the determinant in terms of this row (or column), and find the value of the determinant.

33 $\begin{vmatrix} 2 & 0 & 0 & 0 \\ 3 & 0 & 1 & 0 \\ 4 & 2 & 0 & 1 \\ 0 & 0 & 5 & 3 \end{vmatrix}$

34 $\begin{vmatrix} 4 & 1 & 0 & 2 \\ 0 & 3 & 0 & 0 \\ 2 & 0 & 1 & 0 \\ 4 & 0 & 1 & 2 \end{vmatrix}$

35 $\begin{vmatrix} 2 & 0 & 3 & 2 \\ 0 & 1 & 0 & 4 \\ 0 & 0 & 5 & 0 \\ 1 & 2 & 0 & 0 \end{vmatrix}$

36 $\begin{vmatrix} 3 & 2 & 0 & 5 \\ 0 & 2 & 0 & 1 \\ 1 & 6 & 3 & 4 \\ 0 & 1 & 0 & 2 \end{vmatrix}$

Verify these equations.

37 $\begin{vmatrix} 2 & 1 & 0 & 0 \\ 3 & 4 & 0 & 0 \\ 0 & 0 & 1 & 2 \\ 0 & 0 & 4 & 5 \end{vmatrix} = \begin{vmatrix} 2 & 1 \\ 3 & 4 \end{vmatrix} \cdot \begin{vmatrix} 1 & 2 \\ 4 & 5 \end{vmatrix}$

38 $\begin{vmatrix} 0 & 0 & 2 & 4 \\ 0 & 0 & 3 & 1 \\ 5 & 4 & 0 & 0 \\ 3 & 2 & 0 & 0 \end{vmatrix} = \begin{vmatrix} 2 & 4 \\ 3 & 1 \end{vmatrix} \cdot \begin{vmatrix} 5 & 4 \\ 3 & 2 \end{vmatrix}$

39 $\begin{vmatrix} a & 0 & b & 0 \\ 0 & x & 0 & y \\ x & 0 & b & 0 \\ 0 & a & 0 & y \end{vmatrix} = \begin{vmatrix} a & b \\ x & b \end{vmatrix} \cdot \begin{vmatrix} x & y \\ a & y \end{vmatrix}$

40 $\begin{vmatrix} a & 0 & 0 & b \\ 0 & a & b & 0 \\ 0 & b & a & 0 \\ b & 0 & 0 & a \end{vmatrix} = \begin{vmatrix} a & b \\ b & a \end{vmatrix}^2$

Find the value of the determinants in Probs. 41 to 52.

41 $\begin{vmatrix} 2 & 5 & 2 & 3 \\ 3 & 2 & 3 & 4 \\ 1 & 5 & 1 & 3 \\ 4 & 1 & 2 & 2 \end{vmatrix}$

42 $\begin{vmatrix} 2 & 4 & 1 & 3 \\ 1 & 3 & 5 & 2 \\ 3 & 1 & 2 & 1 \\ 1 & 3 & 5 & 3 \end{vmatrix}$

43 $\begin{vmatrix} 3 & 2 & 1 & 1 \\ 2 & 4 & 3 & 4 \\ 1 & 2 & 4 & 2 \\ 1 & 2 & 1 & 2 \end{vmatrix}$

44 $\begin{vmatrix} 2 & 5 & 1 & 3 \\ 1 & 3 & 2 & -1 \\ -2 & 1 & -1 & 3 \\ 2 & 5 & 1 & 6 \end{vmatrix}$

45 $\begin{vmatrix} 2 & 1 & 0 & 3 & 4 \\ 0 & 2 & 0 & 0 & 0 \\ 0 & 4 & 2 & 1 & 0 \\ 0 & 5 & 1 & 3 & 2 \\ 0 & 2 & 3 & 2 & 1 \end{vmatrix}$

46 $\begin{vmatrix} 3 & 5 & 0 & 0 & 1 \\ 1 & 2 & 5 & 0 & 0 \\ 0 & 1 & 0 & 0 & 0 \\ 2 & 3 & 1 & 4 & 2 \\ 1 & 3 & 4 & 0 & 2 \end{vmatrix}$

47 $\begin{vmatrix} 1 & 2 & 1 & 3 & 4 & 1 \\ 0 & 0 & 0 & 0 & 1 & 0 \\ 3 & 0 & 1 & 0 & 2 & 0 \\ 1 & 0 & 2 & 4 & 0 & 2 \\ 2 & 0 & 3 & 1 & 4 & 0 \\ 1 & 0 & 1 & 0 & 2 & 1 \end{vmatrix}$
48 $\begin{vmatrix} 2 & 1 & 0 & 0 & 0 & 0 \\ 1 & 2 & 0 & 0 & 0 & 0 \\ 0 & 0 & 1 & 1 & 0 & 0 \\ 0 & 0 & 2 & 1 & 0 & 0 \\ 0 & 0 & 0 & 0 & 3 & 1 \\ 0 & 0 & 0 & 0 & 1 & 3 \end{vmatrix}$

49 $\begin{vmatrix} 0 & 0 & Q & 0 \\ S & 0 & 0 & 0 \\ 0 & 0 & 0 & R \\ 0 & P & 0 & 0 \end{vmatrix}$
50 $\begin{vmatrix} D & x & y & z \\ 0 & I & p & q \\ 0 & 0 & A & r \\ 0 & 0 & 0 & G \end{vmatrix}$

51 $\begin{vmatrix} a & b & w & x \\ c & d & y & z \\ 0 & 0 & e & f \\ 0 & 0 & g & h \end{vmatrix}$
52 $\begin{vmatrix} 0 & x & a & g \\ y & 0 & d & p \\ 0 & 0 & 0 & u \\ 0 & 0 & t & 0 \end{vmatrix}$

53 Prove that if two rows of a determinant are proportional, then its value is zero.

54 Show that

$$\begin{vmatrix} x & y & z \\ x^2 & y^2 & z^2 \\ yz & zx & xy \end{vmatrix} = (x - y)(y - z)(z - x)(xy + yz + zx)$$

55 Show that

$$\begin{vmatrix} a & b & c & d \\ b & a & d & c \\ c & d & a & b \\ d & c & b & a \end{vmatrix} = \begin{array}{l}(a + b + c + d)(a - b + c - d) \\ \cdot (a - b - c + d)(a + b - c - d)\end{array}$$

56 Show that

$$\begin{vmatrix} 1 & 1 & 1 & 1 \\ a & b & c & d \\ a^2 & b^2 & c^2 & d^2 \\ a^3 & b^3 & c^3 & d^3 \end{vmatrix} = (a - b)(a - c)(a - d)(b - c)(b - d)(c - d)$$

10.5 Cramer's rule

In the eighteenth century the Swiss mathematician Cramer devised a rule for solving a system of n linear equations in n unknowns if the determinant of the coefficient matrix is not zero. We will now prove Cramer's rule in two different ways for $n = 2$ and $n = 3$.

For $n = 2$, suppose the system of equations is

$$ax + by = c$$
$$dx + ey = f$$

Multiplying the first by e and the second by b gives

$$aex + bey = ce$$
$$bdx + bey = bf$$

Subtracting gives

$$(ae - bd)x = ce - bf$$

If the determinant

$$\begin{vmatrix} a & b \\ d & e \end{vmatrix} = ae - bd$$

is not zero, we can divide by it and get

Cramer's rule for
$n = 2$

$$x = \frac{ce - bf}{ae - bd} = \frac{\begin{vmatrix} c & b \\ f & e \end{vmatrix}}{\begin{vmatrix} a & b \\ d & e \end{vmatrix}} \qquad (10.8)$$

In the same way

$$y = \frac{\begin{vmatrix} a & c \\ d & f \end{vmatrix}}{\begin{vmatrix} a & b \\ d & e \end{vmatrix}}$$

The two denominators are the same, the determinant of the coefficient matrix $\begin{bmatrix} a & b \\ d & e \end{bmatrix}$. The *numerator* in each case can be obtained from the denominator by replacing the coefficients of the variable for which we are solving by the constant terms.

● **Example 1** Find the simultaneous solution of the equations

$$3x - 6y - 2 = 0 \qquad (1)$$

$$4x + 7y + 3 = 0 \qquad (2)$$

by use of Cramer's rule.

Solution We first add 2 to each member of (1) and -3 to each member of (2) and get

$$3x - 6y = 2$$
$$4x + 7y = -3$$

We now obtain the solution by the following steps:

Step 1   Form the determinant D whose elements are the coefficients of the unknowns in the order in which they appear, and get

$$D = \begin{vmatrix} 3 & -6 \\ 4 & 7 \end{vmatrix} = 21 + 24 = 45$$

Step 2   Replace the column of coefficients of x in D by the column of constant terms and get

$$N(x) = \begin{vmatrix} 2 & -6 \\ -3 & 7 \end{vmatrix} = 14 - 18 = -4$$

Step 3   Replace the column of coefficients of y in D by the column of constant terms and get

$$N(y) = \begin{vmatrix} 3 & 2 \\ 4 & -3 \end{vmatrix} = -9 - 8 = -17$$

Step 4   By Cramer's rule,

$$x = \frac{N(x)}{D} = \frac{-4}{45} = -\frac{4}{45}$$

$$y = \frac{N(y)}{D} = \frac{-17}{45} = -\frac{17}{45}$$

Proof for $n = 3$ We will now prove Cramer's rule for a system of three linear equations by using properties of determinants. The method used is applicable regardless of the number of equations. We will consider the system

$$a_1 x + b_1 y + c_1 z = d_1 \tag{3}$$
$$a_2 x + b_2 y + c_2 z = d_2 \tag{4}$$
$$a_3 x + b_3 y + c_3 z = d_3 \tag{5}$$

The determinant of the coefficients is

$$D = \begin{vmatrix} a_1 & b_1 & c_1 \\ a_2 & b_2 & c_2 \\ a_3 & b_3 & c_3 \end{vmatrix}$$

We define the determinants formed when the constant terms are substituted for the coefficients of x, y, and z in D as

$$N(x) = \begin{vmatrix} d_1 & b_1 & c_1 \\ d_2 & b_2 & c_2 \\ d_3 & b_3 & c_3 \end{vmatrix} \qquad N(y) = \begin{vmatrix} a_1 & d_1 & c_1 \\ a_2 & d_2 & c_2 \\ a_3 & d_3 & c_3 \end{vmatrix} \qquad N(z) = \begin{vmatrix} a_1 & b_1 & d_1 \\ a_2 & b_2 & d_2 \\ a_3 & b_3 & d_3 \end{vmatrix}$$

We multiply the first column of D by x, and by property 4, we have

$$xD = \begin{vmatrix} a_1x & b_1 & c_1 \\ a_2x & b_2 & c_2 \\ a_3x & b_3 & c_3 \end{vmatrix} \qquad (6)$$

Next, we multiply the elements of the second column of the above determinant by y, and the elements of the third column by z, and add the products to the elements of the first column. Thus by property 6 we get

$$xD = \begin{vmatrix} a_1x + b_1y + c_1z & b_1 & c_1 \\ a_2x + b_2y + c_2z & b_2 & c_2 \\ a_3x + b_3y + c_3z & b_3 & c_3 \end{vmatrix}$$

$$= \begin{vmatrix} d_1 & b_1 & c_1 \\ d_2 & b_2 & c_2 \\ d_3 & b_3 & c_3 \end{vmatrix} \qquad \text{since } a_ix + b_iy + c_iz = d_i, \text{ for } i = 1, 2, 3$$

$$xD = N(x) \qquad (10.9)$$

Cramer's rule for $n = 3$ Consequently, $x = N(x)/D$, if $D \neq 0$.
By a similar argument, we can show that $y = N(y)/D$ and $z = N(z)/D$.

Note If the determinant of coefficients $D = 0$, we do *not* use Cramer's rule. If $D = 0$, the system of equations is not independent, and we may solve the system (if there is a solution) by row operations as in Sec. 10.2.

$D = 0$ If the determinant D in the denominator is 0, the equations are dependent if $N(x) = N(y) = N(z) = 0$, and the equations are inconsistent if at least one of them is not 0.

● **Example 2** Use Cramer's rule to solve the system of equations

$$\begin{aligned} 3x + y - 2z &= -3 \\ 2x + 7y + 3z &= 9 \\ 4x - 3y - z &= 7 \end{aligned}$$

Solution The terms in the left members are arranged in the proper order, and only the constant terms appear in the right members. Hence, we proceed as follows:

Step 1

$$D = \begin{vmatrix} 3 & 1 & -2 \\ 2 & 7 & 3 \\ 4 & -3 & -1 \end{vmatrix}$$

$$= 3\begin{vmatrix} 7 & 3 \\ -3 & -1 \end{vmatrix} - 1\begin{vmatrix} 2 & 3 \\ 4 & -1 \end{vmatrix} + (-2)\begin{vmatrix} 2 & 7 \\ 4 & -3 \end{vmatrix}$$

$$= 3(-7 + 9) - 1(-2 - 12) - 2(-6 - 28)$$

$$= 6 + 14 + 68 = 88$$

Step 2
$$N(x) = \begin{vmatrix} -3 & 1 & -2 \\ 9 & 7 & 3 \\ 7 & -3 & -1 \end{vmatrix}$$

$$= -3 \begin{vmatrix} 7 & 3 \\ -3 & -1 \end{vmatrix} - 1 \begin{vmatrix} 9 & 3 \\ 7 & -1 \end{vmatrix} + (-2) \begin{vmatrix} 9 & 7 \\ 7 & -3 \end{vmatrix}$$

$$= -3(-7 + 9) - 1(-9 - 21) - 2(-27 - 49)$$

$$= -6 + 30 + 152 = 176$$

Step 3
$$N(y) = \begin{vmatrix} 3 & -3 & -2 \\ 2 & 9 & 3 \\ 4 & 7 & -1 \end{vmatrix}$$

$$= 3 \begin{vmatrix} 9 & 3 \\ 7 & -1 \end{vmatrix} - (-3) \begin{vmatrix} 2 & 3 \\ 4 & -1 \end{vmatrix} + (-2) \begin{vmatrix} 2 & 9 \\ 4 & 7 \end{vmatrix}$$

$$= 3(-9 - 21) + 3(-2 - 12) - 2(14 - 36)$$

$$= -90 - 42 + 44 = -88$$

Step 4
$$N(z) = \begin{vmatrix} 3 & 1 & -3 \\ 2 & 7 & 9 \\ 4 & -3 & 7 \end{vmatrix}$$

$$= 3 \begin{vmatrix} 7 & 9 \\ -3 & 7 \end{vmatrix} - 1 \begin{vmatrix} 2 & 9 \\ 4 & 7 \end{vmatrix} + (-3) \begin{vmatrix} 2 & 7 \\ 4 & -3 \end{vmatrix}$$

$$= 3(49 + 27) - 1(14 - 36) - 3(-6 - 28)$$

$$= 228 + 22 + 102 = 352$$

Step 5
$$x = \frac{N(x)}{D} = \frac{176}{88} = 2$$

$$y = \frac{N(y)}{D} = \frac{-88}{88} = -1$$

$$z = \frac{N(z)}{D} = \frac{352}{88} = 4$$

by Cramer's rule. Hence the solution is $(2, -1, 4)$; it can be checked in the original equations. ●

● **Example 3** Show that the following equations are not independent:

$$5x + 4y + 11z = 3$$
$$6x - 4y + 2z = 1$$
$$x + 3y + 5z = 2$$

Solution

$$D = \begin{vmatrix} 5 & 4 & 11 \\ 6 & -4 & 2 \\ 1 & 3 & 5 \end{vmatrix} = 5\begin{vmatrix} -4 & 2 \\ 3 & 5 \end{vmatrix} - 4\begin{vmatrix} 6 & 2 \\ 1 & 5 \end{vmatrix} + 11\begin{vmatrix} 6 & -4 \\ 1 & 3 \end{vmatrix}$$

$$= 5(-20 - 6) - 4(30 - 2) + 11(18 + 4)$$

$$= -130 - 112 + 242 = 0$$

Hence, since $D = 0$, the equations are not independent and Cramer's rule does not apply. Row operations show that no solution exists. ●

Cramer's rule for n variables

Cramers rule can be extended to n linear equations in the n unknowns, $x_1, x_2, \ldots, x_n$ where each equation has the form

$$a_1 x_1 + a_2 x_2 + \cdots + a_n x_n = b_1$$

The solution, with D equal to the determinant of the coefficient matrix, is

$$x_1 = \frac{N(x_1)}{D} \qquad x_2 = \frac{N(x_2)}{D} \qquad \cdots \qquad x_n = \frac{N(x_n)}{D}$$

where, as for $n = 3$, the determinant for $N(x_i)$ is obtained by replacing the ith column of D by the column of constants.

Cramer's rule is not an efficient procedure if there are a large number of variables and equations. It can not be used if the number of variables differs from the number of equations or if the determinant of the matrix of the coefficients is zero.

Method of least squares

As one example of the use of linear equations and determinants, we will present the **method of least squares** for finding the line which "best fits" n points, found usually by an experiment. If the points $(x_1, y_1), (x_2, y_2), \ldots, (x_n, y_n)$ are given, we will find the line $y = mx + b$ which gives the smallest value for the sum

$$[y_1 - (mx_1 + b)]^2 + [y_2 - (mx_2 + b)]^2 + \cdots + [y_n - (mx_n + b)]^2$$

To find this line, we must find m and b. We will get two equations in these two unknowns. Suppose that the points are $A(5, 14)$, $B(4, 9)$, $C(6, 16)$, and $D(8, 23)$. See Fig. 10.1. We first substitute the coordinates of

FIGURE 10.1

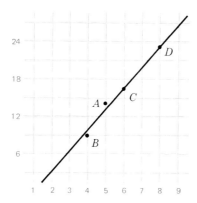

each point in $y = mx + b$. We then (1) multiply each equation by the coefficient of m and (2) add these two groups of equations. The two resulting equations are then solved for m and b.

Substitute given points	Multiply by coefficient of m
$14 = 5m + b$	$70 = 25m + 5b$
$9 = 4m + b$	$36 = 16m + 4b$
$16 = 6m + b$	$96 = 36m + 6b$
$23 = 8m + b$	$184 = 64m + 8b$
Add $62 = 23m + 4b$	$386 = 141m + 23b$

By Cramer's rule, the values for m and b are

$$m = \frac{\begin{vmatrix} 386 & 23 \\ 62 & 4 \end{vmatrix}}{\begin{vmatrix} 141 & 23 \\ 23 & 4 \end{vmatrix}} = \frac{1544 - 1426}{564 - 529} = \frac{118}{35} \approx 3.37$$

$$b = \frac{\begin{vmatrix} 141 & 386 \\ 23 & 62 \end{vmatrix}}{\begin{vmatrix} 141 & 23 \\ 23 & 4 \end{vmatrix}} = \frac{8742 - 8878}{564 - 529} = \frac{-136}{35} \approx -3.89$$

The least squares line of best fit for the four given points is

$$y = 3.37x - 3.89$$

The same type of procedure may be used to fit a parabola $y = ax^2 + bx + c$ to n given points. The method requires solving three equations in the three unknowns a, b, and c.

If there is a logarithmic relationship (see Chap. 8) of the form $w = c + d \log t$, we may let $y = w$ and $x = \log t$, then proceed as above to find c and d.

Exercise 10.5   Use Cramer's rule to find the solutions of each of the following systems of linear equations.

1 $2x - y = 3$
     $x + y = 3$

2 $3x + y = 2$
     $2x + 3y = -1$

3 $2x + 3y = 3$
     $3x + 2y = 7$

4 $3x - y = 9$
     $4x + 3y = -1$

5 $7x + 3y = 1$
     $3x - 4y = 11$

6 $5x + 3y = 7$
     $4x + 5y = 3$

7 $5x - 4y = 2$
     $3x - 5y = 9$

8 $6x + 7y = -11$
     $3x + 5y = -4$

9 $\quad x + 5y = 3$
$\quad 3x - 2y = 9$

10 $7x + 2y = 1$
$\quad 5x + 3y = 7$

11 $2x - 3y = 4$
$\quad 3x - 4y = 5$

12 $4x - 5y = 7$
$\quad 3x + 4y = -18$

13 $2x - y = 2a - b$
$\quad 3x + 2y = 3a + 2b$

14 $5x + 3y = 10a - 2b$
$\quad 4x - 5y = 8a - 9b$

15 $ax + by = a^2 + b^2$
$\quad bx - 2ay = -ab$

16 $ax + by = a^2 + b^2$
$\quad 2ax - by = 2a^2 + 3ab - b^2$

17 $3x + 2y + z = 8$
$\quad 2x + y + 3z = 7$
$\quad 5x - 3y + 4z = 3$

18 $2x - 3y - 2z = 10$
$\quad 3x - 4y + 3z = 8$
$\quad 4x - 5y + 4z = 10$

19 $7x + 3y + 4z = 15$
$\quad 2x + 6y + 3z = 1$
$\quad 7x + 3y + 2z = 13$

20 $2x - 5y - 8z = 12$
$\quad 5x - y + 3z = 7$
$\quad 3x + 4y + 5z = 7$

21 $6x + 5y + 4z = 8$
$\quad 7x - 5y + 3z = 26$
$\quad 5x - 2y - 6z = -9$

22 $3x + y + 4z = 13$
$\quad 6x + 2y - 3z = -29$
$\quad 2x + 3y + 2z = 3$

23 $3x + 5y + 4z = 2$
$\quad 5x + 4y + 3z = 7$
$\quad 4x + 2y + 5z = 3$

24 $7x + 5y + 9z = 10$
$\quad 6x + 4y + 7z = 7$
$\quad 5x + 3y + 4z = 1$

25 $3x + y = 9$
$\quad 2x + 3z = -2$
$\quad 3y - z = 11$

26 $3x + z = 7$
$\quad 2x - 3y = -1$
$\quad 2y + 3z = -9$

27 $5x + 3z = 1$
$\quad 3x + 2y = 4$
$\quad 4y - 5z = 11$

28 $ y + 4z = -14$
$\quad 3x + 2y = 11$
$\quad 2x + 3z = 1$

29 $x + y + z = 0$
$\quad y + z - 2w = -5$
$\quad 2x - y + 3w = 9$
$\quad x + 3z - w = -1$

30 $2x - y + z - w = 5$
$\quad 2y - z + 2w = -4$
$\quad x + y - 3w = 0$
$\quad 3y + 2z - 2w = 1$

31 $x + 2y + 4z + 6w = -5$
$\quad x - y - 2z + w = 0$
$\quad 2x + 3y + 2z + 2w = -8$
$\quad x - y - 2z - 3w = 4$

32 $x + 2y - z + w = -1$
$\quad x + 4y - z + 2w = -1$
$\quad 3x - 2y - 2z - w = 1$
$\quad 2x + y - z + w = 1$

Solve for any one variable in Probs. 33 to 36.

33 $x + y + z + w + t = 12$
$\quad x + y - t = 1$
$\quad y + z + w + t = 10$
$\quad x - y - w - t = -4$
$\quad x - z + w - t = -1$

34 $x + y - z - w + t = 5$
$\quad x + z + w - t = 3$
$\quad x + y - w - t = 0$
$\quad x - y + z + t = 5$
$\quad x + y + z - w = 4$

35 $x + 2y - w = 9$
$\quad 2x + y + w = 12$
$\quad y + z + w = 6$
$\quad y - z + 2t = 1$
$\quad z - 2w + t = 0$

36 $x + z + t = 9$
$\quad 2x - z - 3t = -6$
$\quad y + 2z - 4w = 3$
$\quad z - 3w + 3t = 9$
$\quad w + 2t = 7$

Use the method of least squares to find the line which best fits the following points. A calculator is useful here.

37 (6, 16), (7, 20), (8, 21), (9, 24)
38 (6, 16), (7, 14), (8, 8), (9, 6)
39 (5.8, 19.2), (6.4, 20.2), (7.7, 22.4), (8.1, 23.2)
40 (18.3, 48.2), (18.7, 47.9), (19.1, 47.7), (19.9, 47.2)

Show that Cramer's rule does not apply in Probs. 41 to 44 by verifying that the determinant of the coefficient matrix is 0.

41
$$3x - 7y + z = 10$$
$$4x + 2y - 5z = -7$$
$$-6x - 20y + 17z = 8$$

43 $6x + y - 2z = 11$
$$3x + 5y + 5z = 1$$
$$3x - y - 3z = -4$$

42 $6x + y - z = 10$
$$2x + 11y + 9z = 17$$
$$4x - 2y - 3z = 15$$

44 $-7x - 11y + 4z = 3$
$$4x - 4y + 5z = 10$$
$$5x + y + 2z = 12$$

10.6 Inverse of a square matrix

The use of the inverse of a square matrix enables us to find the solution of a system of linear equations in a way that is different from any studied heretofore.

Inverse   If there is a square matrix B such that the product of it and the square matrix A is the multiplicative identity I, then B is the *multiplicative* **inverse** of A, and B is written A^{-1}. We have by definition

$$AA^{-1} = I \quad \text{and} \quad A^{-1}A = I$$

We will now find the multiplicative inverse of the matrix

$$A = \begin{bmatrix} a_{11} & a_{12} \\ a_{21} & a_{22} \end{bmatrix}$$

If it is

$$A^{-1} = \begin{bmatrix} b & c \\ d & e \end{bmatrix}$$

we must have

$$\begin{bmatrix} a_{11} & a_{12} \\ a_{21} & a_{22} \end{bmatrix} \begin{bmatrix} b & c \\ d & e \end{bmatrix} = \begin{bmatrix} 1 & 0 \\ 0 & 1 \end{bmatrix}$$

Therefore,

$$\begin{bmatrix} a_{11}b + a_{12}d & a_{11}c + a_{12}e \\ a_{21}b + a_{22}d & a_{21}c + a_{22}e \end{bmatrix} = \begin{bmatrix} 1 & 0 \\ 0 & 1 \end{bmatrix}$$

Consequently, using matrix equality gives

$$a_{11}b + a_{12}d = 1 \qquad a_{11}c + a_{12}e = 0$$
$$a_{21}b + a_{22}d = 0 \qquad a_{21}c + a_{22}e = 1$$

If these two pairs of equations are solved for b, c, d, and e, we get (provided that det $A \neq 0$)

$$b = \frac{a_{22}}{a_{11}a_{22} - a_{12}a_{22}} = \frac{a_{22}}{\det A} \qquad c = \frac{-a_{12}}{\det A}$$

$$d = \frac{-a_{21}}{\det A} \qquad\qquad e = \frac{a_{11}}{\det A}$$

Therefore,

A^{-1}

$$A^{-1} = \begin{bmatrix} b & c \\ d & e \end{bmatrix} = \frac{1}{\det A} \begin{bmatrix} a_{22} & -a_{12} \\ -a_{21} & a_{11} \end{bmatrix}$$

The form of this matrix can be changed by making use of the fact that it is the transpose of the matrix obtained by replacing each element by its cofactor. Since $A_{11} = a_{22}$, $A_{12} = -a_{21}$, $A_{21} = -a_{12}$, and $A_{22} = a_{11}$, then

Inverse of a
2 × 2 matrix

$$A^{-1} = \frac{1}{\det A} \begin{bmatrix} A_{11} & A_{21} \\ A_{12} & A_{22} \end{bmatrix} = \frac{1}{\det A} \begin{bmatrix} A_{11} & A_{12} \\ A_{21} & A_{22} \end{bmatrix}^{T} \qquad (10.9)$$

● **Example 1**   To get the inverse of

$$A = \begin{bmatrix} 3 & -2 \\ 4 & 5 \end{bmatrix}$$

we find its determinant to be $15 - (-8) = 23$ and make use of the fact that the cofactors of the elements, 3, -2, 4, 5 are 5, -4, $-(-2) = 2$, 3, respectively. Now replacing each element by its cofactor gives $\begin{bmatrix} 5 & -4 \\ 2 & 3 \end{bmatrix}$

and the transpose of this is $\begin{bmatrix} 5 & 2 \\ -4 & 3 \end{bmatrix}$; hence

$$A^{-1} = \frac{1}{23} \begin{bmatrix} 5 & 2 \\ -4 & 3 \end{bmatrix} = \begin{bmatrix} \frac{5}{23} & \frac{2}{23} \\ -\frac{4}{23} & \frac{3}{23} \end{bmatrix}$$

This can be verified as the inverse of A by evaluating AA^{-1} or $A^{-1}A$ and finding the product to be

$$\begin{bmatrix} 1 & 0 \\ 0 & 1 \end{bmatrix} = I$$

Note   Notice that if det $A = 0$, then A does not have an inverse.

● **Example 2**   The matrix $\begin{bmatrix} 6 & 9 \\ 2 & 3 \end{bmatrix}$ does not have an inverse since its determinant is

$$6(3) - 9(2) = 0$$ ●

We will not prove it, but it is true that if the determinant of a square matrix A of order n is different from zero, then the matrix has an inverse. Furthermore, the inverse of A is $(1/\det A)$ times the transpose of the matrix obtained from A by replacing each element by its cofactor. Symbolically, the inverse of $A = (a_{ij})$ is $\dfrac{1}{\det A}(A_{ij})^T$, and so

A^{-1} The inverse of $\begin{bmatrix} a_{11} & a_{12} & \cdots & a_{1n} \\ a_{21} & a_{22} & \cdots & a_{2n} \\ \cdots\cdots\cdots\cdots\cdots \\ a_{n1} & a_{n2} & \cdots & a_{nn} \end{bmatrix}$ is $\left(\dfrac{1}{\det A}\right) \begin{bmatrix} A_{11} & A_{21} & \cdots & A_{n1} \\ A_{12} & A_{22} & \cdots & A_{n2} \\ \cdots\cdots\cdots\cdots\cdots \\ A_{1n} & A_{2n} & \cdots & A_{nn} \end{bmatrix}$

● **Example 3**   Find A^{-1} if

$$A = \begin{bmatrix} 1 & 2 & 5 \\ 2 & 3 & 8 \\ -1 & 1 & 2 \end{bmatrix}$$

Solution   The determinant of A is $-1 \neq 0$; hence, A^{-1} exists. Now replacing each element in A by its cofactor we get

$$\begin{bmatrix} -2 & -12 & 5 \\ 1 & 7 & -3 \\ 1 & 2 & -1 \end{bmatrix}$$

whose transpose is

$$\begin{bmatrix} -2 & 1 & 1 \\ -12 & 7 & 2 \\ 5 & -3 & -1 \end{bmatrix}$$

Consequently

$$A^{-1} = \frac{1}{-1} \begin{bmatrix} -2 & 1 & 1 \\ -12 & 7 & 2 \\ 5 & -3 & -1 \end{bmatrix} = \begin{bmatrix} 2 & -1 & -1 \\ 12 & -7 & -2 \\ -5 & 3 & 1 \end{bmatrix}$$

We can verify this by multiplying A by A^{-1}. Thus,

$$AA^{-1} = \begin{bmatrix} 1 & 2 & 5 \\ 2 & 3 & 8 \\ -1 & 1 & 2 \end{bmatrix} \times \begin{bmatrix} 2 & -1 & -1 \\ 12 & -7 & -2 \\ -5 & 3 & 1 \end{bmatrix} = \begin{bmatrix} 1 & 0 & 0 \\ 0 & 1 & 0 \\ 0 & 0 & 1 \end{bmatrix}$$ ●

In the system of equations (10.1), suppose that $m = n$ and that the coefficient matrix A has an inverse. Let the $n \times 1$ matrix of unknowns be $X = [x_1 x_2 x_3 \cdots x_n]^T$, and the $n \times 1$ matrix of constants be

$$K = [k_1 k_2 k_3 \cdots k_n]^T$$

Both X and K are column matrices with n elements. Then (10.1) may be written in the compact form $AX = K$, whose left member is the product of an $n \times n$ and $n \times 1$ matrix, and whose right member is an $n \times 1$ matrix. With the help of A^{-1} and properties of matrices, we may solve the system (10.1).

Solution by
matrix inverse

$$
\begin{array}{ll}
AX = K & \text{Eq. (10.1)} \\
A^{-1}(AX) = A^{-1}K & \text{multiply on left by } A^{-1} \\
(A^{-1}A)X = A^{-1}K & \text{associative law} \\
IX = A^{-1}K & A^{-1}A = I = \text{identity} \\
X = A^{-1}K & IX = X
\end{array}
$$

● **Example 4** Solve $x + 2y + 5z = 2$, $2x + 3y + 8z = 3$, and $-x + y + 2z = 3$.

Solution The matrix of coefficients A is the same as the matrix M of Example 3. Since we found its inverse there, we may use the above procedure to solve the equations.

$$
\begin{bmatrix} x \\ y \\ z \end{bmatrix} = X = A^{-1}K = \begin{bmatrix} 2 & -1 & -1 \\ 12 & -7 & -2 \\ -5 & 3 & 1 \end{bmatrix} \begin{bmatrix} 2 \\ 3 \\ 3 \end{bmatrix} = \begin{bmatrix} -2 \\ -3 \\ 2 \end{bmatrix}
$$

●

Exercise 10.6 Find the inverse of the matrix in each of Probs. 1 to 12

1 $\begin{bmatrix} 2 & 4 \\ 3 & 7 \end{bmatrix}$ **2** $\begin{bmatrix} -1 & 0 \\ 2 & 3 \end{bmatrix}$ **3** $\begin{bmatrix} 3 & 5 \\ -2 & -1 \end{bmatrix}$ **4** $\begin{bmatrix} 0 & -5 \\ 2 & 6 \end{bmatrix}$

5 $\begin{bmatrix} 1 & 3 & 2 \\ 2 & -1 & 3 \\ -2 & 4 & 1 \end{bmatrix}$ **6** $\begin{bmatrix} 0 & 1 & -1 \\ 2 & -2 & 3 \\ -3 & -1 & 1 \end{bmatrix}$

7 $\begin{bmatrix} 1 & 7 & -3 \\ 2 & 0 & 1 \\ -3 & 3 & 2 \end{bmatrix}$ **8** $\begin{bmatrix} 0 & 2 & 0 \\ 3 & 0 & 1 \\ 2 & -1 & 3 \end{bmatrix}$

9 $\begin{bmatrix} 1 & 0 & 0 & 2 \\ 0 & 2 & 1 & 0 \\ 0 & 1 & 0 & 1 \\ 0 & 0 & -1 & 0 \end{bmatrix}$ **10** $\begin{bmatrix} 0 & 0 & 1 & 2 \\ 1 & 0 & 0 & 2 \\ 0 & 1 & 2 & 0 \\ 2 & 2 & 0 & 0 \end{bmatrix}$

11 $\begin{bmatrix} 1 & 0 & 1 & -1 \\ 1 & -1 & -1 & 0 \\ 1 & -1 & 0 & -1 \\ 0 & 1 & 1 & -1 \end{bmatrix}$ **12** $\begin{bmatrix} 1 & 1 & 0 & 1 \\ 0 & -1 & -1 & 0 \\ -1 & -1 & 1 & -1 \\ -1 & 0 & 1 & 0 \end{bmatrix}$

Use the matrix inverse method to solve Probs. 13 to 24.

13 $5x + 7y = -11$
      $3x + 4y = -6$

14 $x + 2y = 6$
      $2x + 5y = 13$

15 $-3x + y = -7$
      $-8x + 5y = -14$

16 $6x + 7y = -1$
      $-4x - 5y = 1$

17 $x + 2y + 2z = 18$
      $2x + 3y + 3z = 29$
      $x - y - 2z = -6$

18 $x + 2y - z = 2$
      $-3x - 4y - 4z = -37$
      $2x + 3y + 2z = 22$

19 $2x + 2y + 3z = 9$
      $3x + 3y + 4z = 13$
      $x + 2y + 4z = 9$

20 $3x + 2y + 2z = 12$
      $4x + 3y + 3z = 17$
      $5x + 2y + 3z = 17$

21 $x + y + 2z = 7$
      $3x + 4y + 6z = 21$
      $2x + 3y + 5z = 16$

22 $x + 3y + 2z = 5$
      $2x + 7y + 6z = 11$
      $2x + 6y + 5z = 10$

23 $x + z - w = -2$
      $x - y - z = -4$
      $x - y - w = -10$
      $y + z - w = 0$

24 $x + y + w = 6$
      $-y - z = -5$
      $-x - y + z - w = -3$
      $-x + z = 0$

In Probs. 25 to 32, use the matrices

$$A = \begin{bmatrix} 1 & 2 & 3 \\ 3 & 5 & 7 \\ -1 & 2 & 4 \end{bmatrix} \quad B = \begin{bmatrix} 5 & 12 & 6 \\ 3 & 7 & 4 \\ 5 & 13 & 2 \end{bmatrix} \quad C = \begin{bmatrix} 7 & 4 & -8 \\ 4 & 2 & -5 \\ -6 & -3 & 7 \end{bmatrix}$$

to verify the stated property.

25 $(AB)^{-1} = B^{-1}A^{-1}$

26 $(BC)^{-1} = C^{-1}B^{-1}$

27 $(A^T)^{-1} = (A^{-1})^T$

28 $(B^{-1})^T = (B^T)^{-1}$

29 The inverse of A^{-1} is A.

30 $(A^2)^{-1} = (A^{-1})^2$

31 $(B^2)^{-1} = (B^{-1})^2$

32 $(A + B)^{-1} \neq A^{-1} + B^{-1}$

33 (a) Show that if M has an inverse, and $MP = MQ$, then $P = Q$.
    (b) Use (a) to show that if there is an inverse of a matrix, then there is only one. *Hint:* If P and Q are inverses for M, then $MP = I = MQ$. Now use part (a).

10.7   ## Key concepts

Be certain that you understand and can use each of the following important words and ideas.

Augmented matrix $A + B$
Row operations kA
Matrix inverse $A \cdot B$
Determinant   Identity
Minor   Transpose
Cofactor   Coefficient matrix
Cramer's rule

(10.3) $A_{ij} = (-1)^{i+j} M_{ij}$

(10.4) $D_n = a_{i1}A_{i1} + a_{i1}A_{i2} + \cdots + a_{ij}A_{ij} + \cdots + a_{in}A_{in}$

(10.8) $x = \dfrac{\begin{vmatrix} c & b \\ f & e \end{vmatrix}}{\begin{vmatrix} a & b \\ d & e \end{vmatrix}}$

(10.9) $x = \dfrac{N(x)}{D}$

Exercise 10.7 review

1 Find x and y so that $\begin{bmatrix} 2x+y & -4 \\ 14 & 5 \end{bmatrix} = \begin{bmatrix} 7 & x-2y \\ 14 & 5 \end{bmatrix}$.

2 Find x, y, and z so that $\begin{bmatrix} x+y & 7 & 0 \\ 2 & y-z & 6 \\ -1 & 5 & 8 \end{bmatrix} = \begin{bmatrix} 3 & 7 & 0 \\ 2 & 3 & 6 \\ x+2z & 5 & 8 \end{bmatrix}$.

Perform the operations indicated in Probs. 3 to 6.

3 $\begin{bmatrix} 2 & x-y & 4 \\ 3 & y & -6 \end{bmatrix} + \begin{bmatrix} -2 & y+x & x-4 \\ 0 & 1-y & x+6 \end{bmatrix}$

4 $\begin{bmatrix} 2 & 4 \\ 0 & 3 \\ 1 & -2 \end{bmatrix} \begin{bmatrix} 0 & 1 & 4 & 2 \\ -3 & 2 & 7 & 5 \end{bmatrix}$

5 $\begin{bmatrix} a & 5 & b \\ b & -2 & 0 \\ c & 1 & 3 \end{bmatrix} + \begin{bmatrix} 1-a & -5 & 1-b \\ 2-b & 2 & a \\ 3-c & -2 & -4 \end{bmatrix}$

6 $\begin{bmatrix} 1 & 0 & 2 \\ 3 & -1 & 4 \end{bmatrix} \begin{bmatrix} 2 & 3 & -1 & 5 \\ 1 & 0 & 2 & -3 \\ -3 & 4 & 1 & 0 \end{bmatrix}$

Use $A = \begin{bmatrix} 3 & 2 & -2 \\ 0 & -1 & 3 \\ 1 & 4 & 5 \end{bmatrix}$ and $B = \begin{bmatrix} 2 & -1 & -1 \\ 12 & -7 & -2 \\ -5 & 3 & 1 \end{bmatrix}$ to find the quantities

called for in Probs. 7 to 9.

7 $(AB)^T$ **8** $A^T B^T$ **9** B^{-1}

10 Prove that the value of B^{-1} found in Prob. 9 is correct by showing that $BB^{-1} = I$.

11 Show that I is its own inverse.

Find the value of the determinant in each of Probs. 12 to 17.

12 $\begin{vmatrix} 3 & 2 \\ 1 & 4 \end{vmatrix}$ **13** $\begin{vmatrix} 5 & 0 \\ -1 & 3 \end{vmatrix}$ **14** $\begin{vmatrix} -2 & 3 \\ 4 & 6 \end{vmatrix}$

15 $\begin{vmatrix} 1 & 2 & 3 \\ 1 & 2 & 3 \\ 3 & 2 & 1 \end{vmatrix}$ **16** $\begin{vmatrix} 1 & 3 & 5 \\ 3 & 1 & 3 \\ 5 & 3 & 1 \end{vmatrix}$ **17** $\begin{vmatrix} 0 & 2 & 1 \\ 0 & 1 & 0 \\ 1 & 2 & 2 \end{vmatrix}$

Prove the statements in Probs. 18 to 23 by use of the properties of determinants or by expanding.

18 $\begin{vmatrix} 2 & 5 & 0 \\ 1 & 4 & -2 \\ 3 & 0 & 6 \end{vmatrix} = \begin{vmatrix} 2 & 1 & 3 \\ 5 & 4 & 0 \\ 0 & -2 & 6 \end{vmatrix}$

19 $\begin{vmatrix} 3 & 4 & -2 \\ 1 & 0 & 7 \\ 6 & 5 & 8 \end{vmatrix} = -1 \begin{vmatrix} 1 & 3 & 6 \\ 0 & 4 & 5 \\ 7 & -2 & 8 \end{vmatrix}$

20 $\begin{vmatrix} 3 & 0 & 6 \\ -1 & 4 & -2 \\ 5 & 0 & 10 \end{vmatrix} = \begin{vmatrix} 3 & 3 & 9 \\ -1 & 3 & -3 \\ 5 & 5 & 15 \end{vmatrix}$

21 $6 \begin{vmatrix} 3 & 5 & -1 \\ 2 & 4 & 3 \\ 1 & 0 & 2 \end{vmatrix} = \begin{vmatrix} 9 & 15 & -3 \\ 2 & 4 & 3 \\ 2 & 0 & 4 \end{vmatrix}$

22 $\begin{vmatrix} 3 & 5 & 2 \\ 1 & 3 & -4 \\ 0 & 2 & 6 \end{vmatrix} = \begin{vmatrix} 3 & 4 & 2 \\ 1 & 1 & -4 \\ 0 & -1 & 6 \end{vmatrix} + \begin{vmatrix} 3 & 1 & 2 \\ 1 & 2 & -4 \\ 0 & 3 & 6 \end{vmatrix}$

23 $\begin{vmatrix} 2 & 3 & 1 \\ 1 & -1 & 2 \\ -3 & 2 & 3 \end{vmatrix} = \begin{vmatrix} 2 & 9 & 1 \\ 1 & 2 & 2 \\ -3 & -7 & 3 \end{vmatrix}$

Find values for x, y, and z by use of determinants, so that the statements in Probs. 24 to 30 are true.

24 $\begin{vmatrix} x & 5 \\ 4 & 3 \end{vmatrix} = 1$ **25** $\begin{vmatrix} x & 1 & 3 \\ 2 & -1 & 5 \\ 2 & -3 & 4 \end{vmatrix} = 1$

26 $\begin{vmatrix} x & 0 & 3 \\ -1 & 2 & x \\ 4 & -2 & 1 \end{vmatrix} = 6$

27 $3x + y = 3$
$2x - 3y = 13$

28 $2x - 3y = -9$
$3x + y = 14$

29 $x + y + z = 4$
$x \quad\;\; + 2z = 11$
$2x - y \quad\;\; = 4$

30 $2x + 3y - z = 10$
$3x - 2y + 4z = -7$
$x + 5y - 2z = 11$

31 Solve the system of Prob. 29 by use of row operations on matrices.

32 Solve the system of Prob. 30 by use of a matrix inverse.

33 Is it true that $AB = BA$ if $A = \begin{bmatrix} 1 & 2 \\ 3 & 4 \end{bmatrix}$ and $B = \begin{bmatrix} 5 & 4 \\ 6 & 11 \end{bmatrix}$?

34 Let $A = \begin{bmatrix} a & b \\ c & d \end{bmatrix}$ and $B = \begin{bmatrix} w & x \\ y & z \end{bmatrix}$. Show that $|AB| = |A| \cdot |B|$.

35 Define A as in Prob. 34. Show that $|A|^2 = |A^2|$.

36 Show that if $A = \begin{bmatrix} \frac{3}{5} & \frac{4}{5} \\ \frac{4}{5} & -\frac{3}{5} \end{bmatrix}$, then $A^{-1} = A^T$.

37 Show that if $A^2 - 3A - 18I = 0$, then $A^{-1} = (\frac{1}{18})(A - 3I)$.

38 If

$$A = \begin{bmatrix} 4 & 4 & 7 \\ 8 & 5 & 2 \\ 3 & 6 & 6 \end{bmatrix} \quad \text{and} \quad B = \begin{bmatrix} 4 & 5 & 9 \\ 11 & 6 & 1 \\ 3 & 7 & 8 \end{bmatrix}$$

then A and B are magic squares since each row and column has the same sum (although a different sum for each matrix). Show that the product AB also has the same row sum and column sum.

39 Show that the equation of the line through the points (a, b) and (c, d) is

$$\begin{vmatrix} 1 & 1 & 1 \\ x & a & c \\ y & b & d \end{vmatrix} = 0$$

40 The complex number $a + bi$ can be identified with the matrix

$$\begin{bmatrix} a & b \\ -b & a \end{bmatrix}$$

by writing $\begin{bmatrix} a & b \\ -b & a \end{bmatrix} = aI + bJ$, where $I = \begin{bmatrix} 1 & 0 \\ 0 & 1 \end{bmatrix}$ and $J = \begin{bmatrix} 0 & 1 \\ -1 & 0 \end{bmatrix}$.

(a) Show that $\begin{bmatrix} a & b \\ -b & a \end{bmatrix} = aI + bJ$.

(b) Show that $J^2 = -I$. (Note the parallel between this and $i^2 = -1$.)

(c) Show that $(3 + 2i)(1 - i)$ is identified with $(3I + 2J)(I - J)$, by multiplying out each product.

41 Show that the equation of the circle through the points (x_1, y_1), (x_2, y_2), (x_3, y_3) is

$$\begin{vmatrix} x^2 + y^2 & x & y & 1 \\ x_1^2 + y_1^2 & x_1 & y_1 & 1 \\ x_2^2 + y_2^2 & x_2 & y_2 & 1 \\ x_3^2 + y_3^2 & x_3 & y_3 & 1 \end{vmatrix} = 0$$

42 Use the method of least squares to find the line which best fits the points $(-2, -10)$, $(0, -4)$, $(1, -1)$, $(4, 8)$.

43 If

$$A = \begin{bmatrix} 0.9 & 0.1 \\ 0.2 & 0.8 \end{bmatrix}$$

then it can be shown that A^n gets closer and closer to the matrix

$$\begin{bmatrix} \frac{2}{3} & \frac{1}{3} \\ \frac{2}{3} & \frac{1}{3} \end{bmatrix}$$

as n gets larger and larger. Calculate A^2, $A^4 = (A^2)^2$, $A^8 = (A^4)^2$, and $A^{16} = (A^8)^2$, using two decimal places in the calculations.

Management **44** When considering the inputs in parts and assemblies that go into an output assembly, we are led to a square matrix N and the expression

$$T = I + N + N^2 + N^3$$

for the total requirement matrix T at stage 3. Find T for

$$I = \begin{bmatrix} 1 & 0 & 0 \\ 0 & 1 & 0 \\ 0 & 0 & 1 \end{bmatrix} \quad \text{and} \quad N = \begin{bmatrix} 0 & 4 & 1 \\ 0 & 0 & 0 \\ 3 & 2 & 0 \end{bmatrix}$$

Management **45** For the final total requirements matrix T and the assembly matrix N,

$$T = NT + I$$

Show that $T = (I - N)^{-1}$.

CHAPTER 11 · POLYNOMIAL EQUATIONS

In this chapter we will discuss equations of the type $f(x) = 0$, where $f(x)$ is a polynomial of degree greater than 2. Unfortunately, no general method exists for obtaining the solution set of such equations if the degree of the polynomial is greater than 4, and the direct methods for solving equations in which the polynomial is of degree 3 or 4 are long and tedious. We will present several theorems, however, that will enable us to obtain pertinent information about the roots. We will also discuss methods for obtaining an approximation of any irrational root to the desired degree of accuracy. Roots of polynomial equations can be approximated quickly by a computer or a calculator. However, exact solutions (such as $7 - \sqrt{2}$) are often preferable to approximations (such as 2.7573593), since they allow factorization and simplification of expressions and they allow as much accuracy as is desired.

11.1 The remainder and factor theorems

An equation of the type

Polynomial equation

$$a_0 x^n + a_1 x^{n-1} + a_2 x^{n-2} + \cdots + a_{n-1}x + a_n = 0 \qquad (11.1)$$

where n is a positive integer and the coefficients $a_0, a_1, a_2, \ldots, a_{n-1}, a_n$ are constants, is a *polynomial equation*. The coefficients a_i will often be rational in this chapter, but they may also be any real or complex numbers.

Function · The left member of (11.1) is called a *polynomial function*.

● Example 1   Some examples of polynomial equations are

$$3x^4 + 2x^3 - 4x = 0$$

$$x^3 + \sqrt{3}x^2 - 31\pi x + \sqrt[3]{4} - \frac{1}{\sqrt{2}} = 0$$

$$2ix^3 - (3i - 4)x^2 + x + 1 = 0$$

We will make extensive use of functional notation. For example, if $f(x) = 2x^3 + x^2 - 2x + 4$, then $f(-3) = 2(-3)^3 + (-3)^2 - 2(-3) + 4 = -54 + 9 + 6 + 4 = -35$ and $f(r) = 2r^3 + r^2 - 2r + 4$.

The remainder theorem

The computation involved in finding the solution set of Eq. (11.1) is greatly simplified if the left member is factored into linear factors, or even if one or more linear factors are found. The **remainder theorem** stated and proved below is useful for this purpose and is also essential in finding the solutions of Eq. (11.1) when the left member is not readily factorable.

○ **Theorem**   If a polynomial $f(x)$ is divided by $x - r$ until a remainder independent of x
Remainder theorem   is obtained, then the remainder is equal to $f(r)$.   ○

Before proving the remainder theorem, we will illustrate its meaning in the following example.

● Example 2   If we divide $x^3 - 2x^2 - 4x + 5$ by $x - 3$, using the method of Sec. 2.2, we obtain $x^2 + x - 1$ as the quotient and 2 as the remainder. Note that the remainder is independent of x. In this problem, $x - r = x - 3$. Hence $r = 3$. Since

$$f(x) = x^3 - 2x^2 - 4x + 5$$

we have

$$f(3) = 3^3 - 2(3)^2 - 4(3) + 5 = 27 - 18 - 12 + 5 = 2$$

and this is equal to the remainder obtained above.   ●

In a division process, we have the following relation between the dividend, the divisor, the quotient, and the remainder:

Proof of   Dividend = (quotient)(divisor) + remainder
remainder theorem

Hence, if the quotient obtained by dividing $f(x)$ by $x - r$ is $Q(x)$ and if the remainder is R, we have

$$f(x) = Q(x)(x - r) + R \qquad (11.2)$$

Equation (11.2) is true for all values of x including $x = r$. Hence, if we substitute r for x in Eq. (11.2), we have

$$f(r) = Q(r)(r - r) + R = Q(r) \cdot (0) + R = R$$

This proves the remainder theorem.

The factor theorem and its converse

If r is a root of $f(x) = 0$, then $f(r) = 0$. Hence, by the remainder theorem, R in Eq. (11.2) is zero, and Eq. (11.2) then becomes

$$f(x) = Q(x)(x - r) \qquad (11.3)$$

Therefore, $x - r$ is a factor of $f(x)$, and we have proved the **factor theorem:**

○ **Theorem**   If r is a root of the polynomial equation $f(x) = 0$,

Factor theorem   then $x - r$ is a factor of $f(x)$ ○

Conversely, if $x - r$ is a factor of $f(x)$, then the remainder obtained by dividing $f(x)$ by $x - r$ is equal to zero. Hence, by the remainder theorem, $f(r) = 0$. Therefore, r is a root of $f(x) = 0$. Hence we have the converse of the factor theorem:

Converse of factor   If $x - r$ is a factor of the polynomial $f(x)$, then r is a root of $f(x) = 0$.
theorem

The factor theorem and its converse may be combined by saying that

$x - r$ is a factor of the polynomial $f(x)$

if and only if r is a root of $f(x) = 0$

● **Example 3**   We may show that $x - 2$ is a factor of $f(x) = -x^3 + 2x^2 - 2x + 4$ by calculating $f(2) = -8 + 8 - 4 + 4 = 0$. By the factor theorem, $x - 2$ is a factor since $f(2) = 0$.

Notice that $x + 3$ is not a factor of $f(x)$, since $x + 3 = x - (-3)$ and $f(-3) = -(-27) + 2(9) - 2(-3) + 4 = 55$. The converse of the factor theorem guarantees that $x + 3$ is not a factor since $f(-3) \neq 0$.   ●

Synthetic division   The remainder theorem as well as the factor theorem and its converse deal with function values. We may use *synthetic division, which was presented fully in Sec. 2.2,* to find these function values quickly. We will now give a summary of synthetic division. The general comments below are illustrated by the following example.

$$
\begin{array}{rrrrr|r}
3 & 0 & -7 & 6 & 5 & \underline{-2} \\
& -6 & 12 & -10 & 8 & \\
\hline
3 & -6 & 5 & -4 & 13 &
\end{array}
$$

Dividing $f(x)$ by $x - r$ To divide a polynomial $f(x)$ (above: $3x^4 - 7x^2 + 6x + 5$) synthetically by $x - r$ (above: $x - r = x + 2$, and so $r = -2$):

1 With the polynomial in *decreasing* powers of x,

First row *write the coefficients*, including any zeros, in the first row *followed* by the r from the divisor $x - r$.

$$\text{Above we have } 3 \quad 0 \quad -7 \quad 6 \quad 5 \ \underline{|-2}$$

2 Bring down the first coefficient to the third row,

Multiply and add *multiply* it by r [above: $(3)(-2) = -6$] and put it in the second row, *add* this product to the number in the first row (above: $0 - 6 = -6$), *continue* this "multiply and add" procedure

$$\text{Above: } (-6)(-2) = 12 \text{ and } 12 + (-7) = 5$$
$$(5)(-2) = -10 \text{ and } -10 + 6 = -4$$
$$(-4)(-2) = 8 \text{ and } 8 + 5 = 13$$

Third row 3 *Interpret* the third row correctly. The last number in the third row is the remainder (above: 13); the other numbers, from left to right, are the coefficients of the quotient (above: $3 \quad -6 \quad 5 \quad -4$)

As a result of the synthetic division, we have

$$3x^4 - 7x^2 + 6x + 5 = (x + 2)(3x^3 - 6x^2 + 5x + 4) + 13$$

By the remainder theorem, $f(r) = $ remainder, and so, above, $f(-2) = 13$.

● **Example 4** Find $f(2)$ and $f(-1)$ if $f(x) = 2x^3 - 5x^2 + 6x - 11$.

Solution By synthetic division we find that the remainders are

$$
\begin{array}{rrrr|r}
2 & -5 & 6 & -11 & 2 \\
  & 4 & -2 & 8 & \\
\hline
2 & -1 & 4 & -3 &
\end{array}
\qquad
\begin{array}{rrrr|r}
2 & -5 & 6 & -11 & -1 \\
  & -2 & 7 & -13 & \\
\hline
2 & -7 & 13 & -24 &
\end{array}
$$

-3 and -24. Hence by the remainder theorem, we have $f(2) = -3$ and $f(-1) = -24$.   ●

● **Example 5** For $f(x) = x^4 + 3x^3 + x^2 - 7x - 30$, we divide synthetically by $x + 3$, $x + 1$, $x - 2$, and $x - 4$.

$$
\begin{array}{rrrrr|r}
1 & 3 & 1 & -7 & -30 & -3 \\
  & -3 & 0 & -3 & 30 & \\
\hline
1 & 0 & 1 & -10 & 0 &
\end{array}
\qquad
\begin{array}{rrrrr|r}
1 & 3 & 1 & -7 & -30 & -1 \\
  & -1 & -2 & 1 & 6 & \\
\hline
1 & 2 & -1 & -6 & -24 &
\end{array}
$$

$$
\begin{array}{rrrrr|r}
1 & 3 & 1 & -7 & -30 & 2 \\
  & 2 & 10 & 22 & 30 & \\
\hline
1 & 5 & 11 & 15 & 0 &
\end{array}
\qquad
\begin{array}{rrrrr|r}
1 & 3 & 1 & -7 & -30 & 4 \\
  & 4 & 28 & 116 & 436 & \\
\hline
1 & 7 & 29 & 109 & 406 &
\end{array}
$$

By the remainder theorem, $f(-3) = 0$, $f(-1) = -24$, $f(2) = 0$, and $f(4) = 406$. Thus by the factor theorem and its converse, we know that $x + 3$ and $x - 2$ are factors of $f(x)$, while $x + 1$ and $x - 4$ are not factors.

Synthetic division may be used for complex numbers as well as real numbers.

Example 6 Find $f(i)$ and $f(2i)$ if $f(x) = 2x^3 + x^2 + 8x + 4$.

Solution

$$
\begin{array}{cccc|c}
2 & 1 & 8 & 4 & i \\
  & 2i & -2+i & -1+6i & \\
\hline
2 & 1+2i & 6+i & 3+6i &
\end{array}
\qquad
\begin{array}{cccc|c}
2 & 1 & 8 & 4 & 2i \\
  & 4i & -8+2i & -4 & \\
\hline
2 & 1+4i & 2i & 0 &
\end{array}
$$

Thus $f(i) = 3 + 6i$, $f(2i) = 0$ and $x - 2i$ is a factor of $f(x)$.

**Exercise 11.1
remainder and
factor theorems**

By use of the remainder theorem, find the remainder obtained by dividing the polynomial in each of Probs. 1 to 12 by the binomial $x - r$ which follows it. Use synthetic division as desired.

1 $x^3 + 5x^2 - 3x + 2$, $x - 1$ 2 $x^3 - 9x^2 + 3x - 9$, $x - 3$
3 $x^3 - 3x - 1$, $x + 2$ 4 $x^3 + 5x^2 + 4x + 3$, $x + 4$
5 $2x^3 + 3x^2 - 5x - 8$, $x + 2$ 6 $3x^3 - 2x^2 + 2x - 3$, $x + 3$
7 $2x^3 + 5x^2 - 6x + 4$, $x + 1$ 8 $2x^3 - 4x^2 + 4x - 7$, $x - 2$
9 $2x^4 + 7x^3 + 4x^2 - 2x + 7$, $x + 3$
10 $-3x^4 + 8x^3 + 2x^2 - 3x + 4$, $x + 1$
11 $x^5 - 2x^4 + 2x^3 - 3x^2 - x + 5$, $x - 2$
12 $16x^5 + 8x^4 + 4x^3 + 2x^2 + 4x + 1$, $x + \frac{1}{2}$

Use the factor theorem to show that the binomial $x - r$ is a factor of the polynomial in each of Probs. 13 to 36. Use synthetic division as desired.

13 $2x^3 + 3x^2 - 6x + 1$, $x - 1$
14 $3x^3 - 9x^2 - 4x + 12$, $x - 3$
15 $5x^4 + 8x^3 + x^2 + 2x + 4$, $x + 1$
16 $3x^4 + 9x^3 - 4x^2 - 9x + 9$, $x + 3$
17 $-2x^5 + 11x^4 - 12x^3 - 5x^2 + 22x - 8$, $x - 4$
18 $3x^5 + 17x^4 + 17x^3 + 35x^2 - 4x - 20$, $x + 5$
19 $x^6 - x^5 - 7x^4 + x^3 + 8x^2 + 5x + 2$, $x + 2$
20 $2x^6 - 5x^5 + 4x^4 + x^3 - 7x^2 - 7x + 2$, $x - 2$
21 $x^3 - 4ax^2 + 2a^2x + a^3$, $x - a$
22 $x^3 - (2a + b)x^2 + (3a + 2ab)x - 3ab$, $x - b$
23 $x^{2n} - a^{2n}$, $x + a$ 24 $x^{2n+1} + a^{2n+1}$, $x + a$
25 $3x^3 + 2x^2 - 4x + 1$, $x - \frac{1}{3}$ 26 $4x^3 + 3x^2 - 5x + 1$, $x - \frac{1}{4}$
27 $4x^5 - 7x^4 - 5x^3 + 2x^2 + 11x - 6$, $x - \frac{3}{4}$
28 $3x^4 - 4x^3 + 5x^2 - 4$, $x + \frac{2}{3}$
29 $x^3 - 3x^2 + x - 3$, $x - i$ 30 $x^4 - 2x^2 - 3$, $x + i$
31 $x^3 - 2x^2 + 4x - 8$, $x - 2i$ 32 $x^4 + 12x^2 + 27$, $x + 3i$
33 $x^3 - 7x^2 + 25x - 39$, $x - 2 - 3i$ 34 $x^3 + x - 10$, $x + 1 - 2i$
35 $2x^3 - 19x^2 + 48x + 29$, $x - 5 + 2i$
36 $3x^3 + 22x^2 + 59x - 50$, $x + 4 + 3i$

In Probs. 37 to 40, find the value of k for which the binomial $x - r$ is a factor of the polynomial.

37 $x^3 + 2x^2 + 4x + k, x + 1$
38 $-x^3 + 3x^2 + kx - 4, x - 1$
39 $2x^4 - 5x^3 + kx^2 - 6x + 8, x - 2$
40 $-3x^4 + kx^3 + 6x^2 - 9x + 3, x - 1$

In Probs. 41 to 44, use the converse of the factor theorem to show that $x - r$ is not a factor of the polynomial.

41 $-2x^3 + 4x^2 - 4x + 9, x - 2$
42 $-3x^3 - 9x^2 + 5x + 12, x + 3$
43 $3x^4 - 8x^3 + 5x^2 + 7x - 3, x - 3$
44 $4x^4 + 9x^3 + 3x^2 + x + 4, x + 2$

Show that the given value of r is a root of the equation in each of Probs. 45 to 52.

45 $r = 5, (x - 5)(x + 1)(x + 2) = 0$
46 $r = -\frac{3}{2}, (x - 5)(x + 2)(2x + 3) = 0$
47 $r = -\frac{2}{3}, (x + 3)(3x + 2)(3x + 4) = 0$
48 $r = -\frac{4}{5}, (2x + 3)(10x + 8)(x - 4) = 0$
49 $r = 1, x^3 + 3x^2 - 4x = 0$
50 $r = -1, x^3 + 5x^2 - 6x - 10 = 0$
51 $r = -3, 5x^4 + 17x^3 + 6x^2 + 9x + 27 = 0$
52 $r = 1, 7x^4 - 8x^3 - 9x^2 + 6x + 4 = 0$

In Probs. 53 to 56, show that the polynomial has no factor $x - r$ if r is real.

53 $x^2 + 1$ **54** $x^4 + 3x^2 + 2$
55 $x^4 + 5x^2 + 3$ **56** $x^6 + 3x^2 + 5$

In Probs. 57 to 60, find the function values by using synthetic division. A calculator is suggested.

57 $f(1.32)$ if $f(x) = 1.72x^2 - 6.03x + 3.10$
58 $f(2.47)$ if $f(x) = 2.43x^3 - 5.94x - 4.11$
59 $f(3.1)$ if $f(x) = 2.6x^3 + 4.9x^2 - 9.1x - 99$
60 $f(1.7)$ if $f(x) = 0.63x^3 - 4.7x^2 + 11.4x + 5.10081$
61 Show that

$$2 + 4x + 7x^2 - 8x^3 - 5x^4 = 2 + x\{4 + x[7 + x(-8 - 5x)]\}$$

This special form simplifies evaluation of polynomial function values in computers and calculators.

11.2 Number of roots, identical polynomials, and conjugate roots

Fundamental theorem of algebra

The **fundamental theorem of algebra** states that *every polynomial equation with complex coefficients has at least one complex root.*

We will consider the polynomial equation

$$f(x) = a_0 x^n + a_1 x^{n-1} + \cdots + a_{n-1}x + a_n = 0 \qquad (1)$$

and let r_1 be a root. Then, by the factor theorem,

$$f(x) = Q_1(x)(x - r_1)$$

where $Q_1(x)$ is of degree $n - 1$. The equation $Q_1(x) = 0$ also has at least one root, and we shall let it be r_2. Then $x - r_2$ is a factor of $Q_1(x)$ and $Q_1(x) = Q_2(x)(x - r_2)$, where $Q_2(x)$ is of degree $n - 2$. Thus $f(x) = (x - r_1)(x - r_2)Q_2(x)$. If we continue this process, we can find $n - 2$ additional factors, $(x - r_3), (x - r_4), \ldots, (x - r_n)$, of $f(x)$ and have

$$f(x) = (x - r_1)(x - r_2)(x - r_3) \cdots (x - r_n)Q_n(x) \qquad (2)$$

where $Q_n(x)$ is of degree $n - n = 0$ and is therefore a constant. Obviously, $Q_n(x)$ is the coefficient of x^n in the right member of (2). Hence it is equal to a_0. Then the factored form of $f(x)$ is

$$f(x) = a_0(x - r_1)(x - r_2)(x - r_3) \cdots (x - r_n) \qquad (11.4)$$

By the converse of the factor theorem, $r_1, r_2, r_3, \ldots, r_n$ are roots of $f(x) = 0$ and so the equation $f(x) = 0$ has at least n roots. Furthermore, $f(x) = 0$ has no other roots, for no one of the factors in (11.4) is equal to zero for any value of x not equal to $r_1, r_2, r_3, \ldots,$ or r_n, since $r_{n+1} - r_i \neq 0$. Thus $f(x) = 0$ has at most n roots.

If s values of $r_1, r_2, r_3, \ldots, r_n$ are equal, say if $r_1 = r_2 = r_3 = \cdots = r_s$, then (11.4) becomes

$$f(x) = a_0(x - r_s)^s(x - r_{s+1}) \cdots (x - r_n)$$

Multiplicity

and r_s is called a *root of multiplicity s* of $f(x) = 0$. Hence we have the following theorem on the number of roots.

○ **Theorem**

Theorem on the number of roots

A polynomial equation of degree n with complex coefficients has exactly n complex roots, where a root of multiplicity s is counted as s roots.   ○

● **Example 1**

The roots of the equation

$$(x + 1)(x - 3)^2(x + \tfrac{3}{2})^2 = 0$$

are -1, 3 with multiplicity 2, and $-\tfrac{3}{2}$ with multiplicity 2.   ●

● **Example 2**

Find the four roots of $f(x) = x^4 - 8x^3 + 20x^2 - 32x + 64 = 0$ if 4 is a root of multiplicity 2.

Solution   Since 4 is a root of $f(x) = 0$, then $x - 4$ is a factor of $f(x)$. We find the quotient by synthetic division:

$$1 \quad -8 \quad 20 \quad -32 \quad 64 \underline{\lfloor 4}$$
$$\underline{\qquad 4 \quad -16 \quad 16 \quad -64}$$
$$1 \quad -4 \quad 4 \quad -16 \quad 0$$

Thus $f(x) = (x - 4)(x^3 - 4x^2 + 4x - 16)$. Since 4 is a root of multiplicity 2, we divide synthetically again by $x - 4$.

$$1 \quad -4 \quad 4 \quad -16 \underline{\lfloor 4}$$
$$\underline{\qquad 4 \quad 0 \quad 16}$$
$$1 \quad 0 \quad 4 \quad 0$$

We now have $f(x) = (x - 4)^2(x^2 + 4)$. The other two roots of $f(x) = 0$ are found by setting $x^2 + 4 = 0$. The roots are $2i$ and $-2i$. Hence the four roots of $f(x) = 0$ are 4, 4, $2i$, and $-2i$.   ●

Identical polynomials

One way in which a polynomial $f(x)$ may be zero for all x is for each of the coefficients to be zero. The following theorem shows that, in fact, this is the only way. It is also useful in dealing with partial fractions, which allow a complicated fraction to be expressed as a sum of simpler fractions. See Sec. 9.6.

○ **Theorem**   If two polynomials of degree n with complex coefficients are equal for more than n distinct values, then they are identical; that is, the coefficients of equal powers of the variable are equal.

Proof   Let $f(x) = a_0x^n + a_1x^{n-1} + \cdots + a_{n+1}x + a_n$ and $g(x) = b_0x^n + b_1x^{n-1} + \cdots + b_{n-1}x + b_n$. The hypothesis says that $f(x_i) = g(x_i)$ for $i = 1, 2, \ldots, n, n + 1$. Consider.

$$h(x) = f(x) - g(x) = (a_0 - b_0)x^n + (a_1 - b_1)x^{n-1} + \cdots + (a_n - b_n)$$

Now $h(x_i) = f(x_i) - g(x_i) = 0$ if $i = 1, 2, \ldots, n$; so by the factor theorem

$$h(x) = (a_0 - b_0)(x - x_1)(x - x_2) \cdots (x - x_n) \tag{3}$$

Furthermore, $h(x_{n+1}) = f(x_{n+1}) - g(x_{n+1}) = 0$; so by Eq. (3)

$$0 = h(x_{n+1}) = (a_0 - b_0)(x_{n+1} - x_1)(x_{n+1} - x_2) \cdots (x_{n+1} - x_n) \tag{4}$$

Now $x_1, x_2, \ldots, x_n, x_{n+1}$ are all distinct; so $x_{n+1} - x_i \neq 0$ if $i = 1, 2, \ldots, n$, and thus by Eq. (4) we must have $a_0 - b_0 = 0$; that is, $a_0 = b_0$.

We thus have $h(x) = (a_1 - b_1)x^{n-1} + \cdots + (a_n - b_n)$, and may repeat the above argument with $x_1, x_2, \ldots, x_{n-1}, x_n$ to show that $a_1 = b_1$. Similarly $a_2 = b_2, \ldots, a_n = b_n$.   ○

● **Example 3**   We will have $2x^3 + 5x^2 - 4x + 1 = Ax^3 + Bx^2 + Cx + D$ for all x if and only if $2 = A$, $5 = B$, $-4 = C$, and $1 = D$.   ●

● **Example 4**   Find a third-degree polynomial $f(x)$ such that

$$f(2) = 0 \qquad f(5) = 0 \qquad f(-3) = 0 \qquad f(4) = 4$$

Solution   If $f(r) = 0$, then $x - r$ is a factor of $f(x)$. Hence

$$f(x) = k(x - 2)(x - 5)(x + 3)$$

Furthermore, $4 = f(4) = k(4 - 2)(4 - 5)(4 + 3)$, and so $4 = k(2)(-1)(7) = -14k$. It follows that $k = -\frac{4}{14} = -\frac{2}{7}$, and

$$f(x) = (-\tfrac{2}{7})(x - 2)(x - 5)(x + 3)$$

This is the only polynomial satisfying the given conditions.   ●

Conjugate roots

It can be readily verified that $x = 2$ is one root of $x^3 - 4x^2 + 9x - 10 = 0$, and the other two roots are $x = 1 + 2i$ and $x = 1 - 2i$. Also, the roots of the quadratic equation $ax^2 + bc + c = 0$ are

$$x = -\frac{b}{2a} + \frac{1}{2a}\sqrt{b^2 - 4ac} \qquad \text{and} \qquad x = -\frac{b}{2a} - \frac{1}{2a}\sqrt{b^2 - 4ac}$$

which are conjugates if they are complex numbers and a, b, and c are real.

Conjugate complex roots
    This illustrates the fact that the complex roots of a polynomial equation with real coefficients occur in pairs, and that the members of each pair are complex conjugates.

○ **Theorem on conjugate roots**   If the complex number $a + bi$, $b \neq 0$, is a root of the polynomial equation $f(x) = 0$ and f has real coefficients, then its conjugate $a - bi$ is also a root of $f(x) = 0$.

Proof   Suppose $f(x) = a_0x^n + a_1x^{n-1} + \cdots + a_n = 0$ (5)

where $a_0, a_1, \ldots, a_n$ are real numbers, and $x = a + bi$ is a root. By Exercise 4.4, Probs. 37 and 39, the conjugate of a sum is the sum of the conjugates, and the conjugate of a product is the product of the conjugates. Since a_i is real, then $\bar{a}_i = a_i$ for $i = 0, 1, 2, \ldots, n$. Thus, taking the conjugate of each member of (5) gives

$$\overline{f(x)} = \overline{a_0x^n + a_1x^{n-1} + \cdots + a_n} = \bar{0}$$

or $f(\bar{x}) = a_0\bar{x}^n + a_1\bar{x}^{n-1} + \cdots + a_n = 0$ (6)

Equation (6) says that $f(\bar{x}) = 0$; hence the conjugate $\bar{x} = a - bi$ is also a root of $f(x) = 0$.   ○

● **Example 5**   Find a third-degree polynomial with real coefficients which has roots $x = 3$ and $x = 3 - 2i$.

Solution   By the theorem on conjugate roots, we may take $x = 3 + 2i$ as a root along with $x = 3 - 2i$ and $x = 3$. By the factor theorem the desired polynomial is

$$(x - 3)[x - (3 - 2i)][x - (3 + 2i)] = (x - 3)(x - 3 + 2i)(x - 3 - 2i)$$
$$= (x - 3)[(x - 3)^2 + 4]$$
$$= x^3 - 9x^2 + 31x - 39 \qquad (7)$$

Any nonzero constant times this polynomial would also be an answer. If real coefficients were not required, then $(x - 3)[x - (3 - 2i)](x - r)$, where r is any complex number, would be an answer.   ●

Conjugate real roots   If a polynomial has real coefficients, then we have just seen that its complex roots occur in pairs. If the coefficients are rational, then certain real roots occur in pairs. The following theorem may be proved in a similar way, although we will not do so.

○ **Theorem**   If the real number $a + b\sqrt{c}$, $c > 0$ and c not a perfect square, is a root of the polynomial equation $f(x) = 0$ and f has rational coefficients, then $a - b\sqrt{c}$ is also a root of $f(x) = 0$.   ○

For instance, $2 - \sqrt{5}$ is a root of $x^4 - 4x^3 + 2x^2 - 12x - 3 = 0$ since $2 + \sqrt{5}$ is a root, and the coefficients are rational.

As a consequence of Eq. (11.4) and the theorem on conjugate roots, we can now state the following corollary.

Corollary   A polynomial with real coefficients can be expressed as the product of linear and irreducible quadratic factors, each factor having real coefficients.

The proof is simply that if any factor in Eq. (11.4) has $r = a + bi$, then another factor will have $r = a - bi$; and, as in Eq. (7) above,

$$[x - (a + bi)][x - (a - bi)] = x^2 - 2ax + a^2 + b^2$$

which has real coefficients.

Exercise 11.2   State the degree of the equation in each of Probs. 1 to 8, find each root,
identical   and give its multiplicity.
polynomials and
complex roots

1 $(x - 1)^3(x + 2)(x - 3)^2 = 0$
2 $(x + 1)^5(x - 2)^3(x + 3) = 0$
3 $(x + 6)^2(x - 3)^4(x - 1)(x + 5) = 0$
4 $(x - 1)^5(x - 2)^3(x + 2)^2 (x + 3) = 0$
5 $(2x - 7)^2(3x + 2) = 0$ **6** $(3x + 5)^4(4x - 1)^3 = 0$
7 $(3x + 2)^5(x - 3)^3(2x + 7) = 0$ **8** $(5x - 4)^4(3x + 5)^3(x - 8) = 0$

In Probs. 9 to 28, find all roots of $f(x) = 0$.

9 $f(x) = x^3 - 2x^2 - x + 2 = 0$, one root is -1
10 $f(x) = x^3 - 13x + 12 = 0$, one root is 3
11 $f(x) = x^3 - 3x^2 - 4x + 12 = 0$, one root is -2
12 $f(x) = x^3 + 7x^2 + 16x + 12 = 0$, one root is -2
13 $f(x) = x^4 + x^3 - 3x^2 - x + 2$; 1 is a root of multiplicity 2
14 $f(x) = x^4 + 2x^3 + 2x^2 + 2x + 1$; -1 is a root of multiplicity 2
15 $f(x) = x^4 - 3x^3 + 6x^2 - 12x + 8$; two roots are 1 and 2
16 $f(x) = x^4 - 2x^3 + 6x^2 - 18x - 27$; two roots are -1 and 3
17 $f(x) = x^3 + x^2 + x + 1$; one root is i
18 $f(x) = 2x^3 - 3x^2 + 8x - 12$; one root is $-2i$
19 $f(x) = 3x^3 + 5x^2 + 12x + 20$; one root is $2i$
20 $f(x) = 4x^3 - 3x^2 + 24x - 18$; one root is $i\sqrt{6}$
21 $f(x) = 2x^3 - x^2 - 2x + 6$; one root is $1 + i$
22 $f(x) = 4x^3 + 9x^2 + 22x + 5$; one root is $-1 + 2i$
23 $f(x) = 4x^3 - 19x^2 + 32x - 15$; one root is $2 - i$
24 $f(x) = 2x^3 + 7x^2 - 4x - 65$; one root is $-3 - 2i$
25 $f(x) = x^4 + 2x^3 + 9x^2 + 8x + 20$; $2i$ and $-1 + 2i$ are roots
26 $f(x) = x^5 + 5x^4 + 2x^3 - 38x^2 - 95x - 75$; $-2 + i$ is a root of multiplicity 2
27 $f(x) = x^6 + 4x^5 + 15x^4 + 24x^3 + 39x^2 + 20x + 25$; $-1 + 2i$ is a root of multiplicity 2
28 $f(x) = x^5 - 7x^4 + 24x^3 - 32x^2 + 64$; $2 + 2i$ is a root of multiplicity 2

Find a polynomial equation of least possible degree with rational (and thus real) coefficients and with the given roots.

29 $1, -1, \frac{2}{3}$ **30** $2, 1, \frac{1}{4}$ **31** $i, -i, \frac{2}{3}$ **32** $2i, -2i, \frac{2}{5}$
33 $-\frac{1}{3}, 1 + 2i$ **34** $\frac{7}{2}, -3 + i$ **35** $1, 2 + i$ **36** $3, 4 - 2i$
37 $-\frac{1}{3}, 1 + \sqrt{2}$ **38** $\frac{7}{2}, -3 + \sqrt{5}$ **39** $1, 2 - \sqrt{2}$ **40** $3, 4 - \sqrt{3}$
41 $-2 - i$, multiplicity 2   **42** $-1 + \sqrt{2}$, multiplicity 2
43 $2 - 2i$, multiplicity 2   **44** $2i$, multiplicity 3

In Probs. 45 to 56, find a polynomial function of smallest possible degree which satisfies the given conditions.

45 $f(1) = f(3) = f(6) = 0, f(4) = -12$
46 $f(-2) = f(0) = f(3) = 0, f(1) = -18$
47 $f(-3) = f(1) = f(3) = f(4) = 0, f(0) = -36$
48 $f(-2) = f(-1) = f(1) = f(2) = 0, f(0) = 16$
49 $f(-\sqrt{2}) = f(\sqrt{2}) = f(1) = 0, f(-1) = 16$
50 $f(-\sqrt{3}) = f(\sqrt{3}) = f(2) = 0, f(0) = 24$
51 $f(2 + \sqrt{5}) = f(2 - \sqrt{5}) = f(4) = 0, f(3) = 20$
52 $f(1 + \sqrt{5}) = f(1 - \sqrt{5}) = f(3 - \sqrt{3}) = f(3 + \sqrt{3}) = 0, f(2) = 40$
53 $f(2i) = f(-2i) = f(1) = 0, f(4) = 60$
54 $f(3i) = f(-3i) = f(-2) = 0, f(-1) = 30$

55 $f(2 + 3i) = f(2 - 3i) = f(3 + 2i) = f(3 - 2i) = 0, f(1) = 160$

56 $f(5 + i) = f(5 - i) = f(4 + 2i) = f(4 - 2i) = 0, f(3) = 75$

57 Show that if

$$x^n + a_1 x^{n-1} + a_2 x^{n-2} + \cdots + a_{n-1}x + a_n = 0$$

has roots $r_1, r_2, \ldots, r_n$, *then the sum of the roots satisfies*

$$r_1 + r_2 + \cdots + r_n = -a_1 \tag{S}$$

and the product of the roots satisfies

$$r_1 r_2 \cdots r_n = (-1)^n a_n \tag{P}$$

Hint: Factor the polynomial; then multiply it out using $r_1, r_2, \ldots, r_n$, and compare coefficients with the original polynomial.

58 Find a polynomial equation with roots $2 + 3i$ and -5 and verify Eqs. (S) and (P) in Prob. 57 for this polynomial.

59 Repeat Prob. 58 for roots $3 - 2i$ and $2 + \sqrt{3}$.

60 Repeat Prob. 58 for roots $4 - i$, $1 + 2i$, and $\frac{3}{2}$.

61 Show that a quadratic equation with real coefficients has either two real roots or none, counting multiplicity.

62 Show that a cubic equation with real coefficients has either three or one real roots, counting multiplicity.

63 Show that a fourth-degree equation with real coefficients has four, two, or no real roots, counting multiplicity.

64 Show that a fifth-degree equation with real coefficients has five, three, or one real roots, counting multiplicity.

11.3   Locating the real roots

The first theorem given in this section enables us to find a number that is greater than or equal to the largest real root of an equation, and another number that is smaller than or equal to the least root of the equation. Thus we can restrict the range in which the real roots are known to lie. Any

Upper bound   number that is larger than or equal to the greatest root of an equation is called *an upper bound of the roots;* any number that is smaller than or equal

Lower bound   to the least root of an equation is called a *lower bound of the roots.*

We will now state and then prove a theorem that enables us to determine upper and lower bounds.

○ **Theorem on bounds**   If the coefficient of x^n in the polynomial equation $f(x) = 0$ is positive and if there are no negative terms in the third line of the synthetic division of $f(x)$ by $x - k$, $k > 0$, then k is an upper bound of the real roots of $f(x) = 0$. Furthermore, if the signs in the third line of the synthetic division of $f(x)$ by $x - (-k) = x + k$ are alternately plus and minus,† then $-k$ is a lower bound of the real roots.

---

† A zero may be given either a plus or a minus sign to help the signs alternate.

Proof   To prove the first part of the theorem, we use Eq. (11.2) with $r = k$ and have

$$f(x) = Q(x)(x - k) + R \qquad (1)$$

By synthetic division the coefficients in $Q(x)$ and the value of R are the numbers in the third row of the division of $f(x)$ by $x - k$. If these numbers are positive or zero and if x is greater than k, then $Q(x)(x - k) + R > 0$ since $x > k > 0$ and $x - k > 0$. Hence there are no real roots of $f(x) = 0$ that are greater than k; that is k is an upper bound of the real roots, as stated in the first part of the theorem. We omit the proof of the second part, but see Prob. 45.   ○

● **Example 1**   Find an upper and a lower bound of the real roots of the equation $x^3 - 2x^2 + 3x + 3 = 0$.

Solution   The synthetic division of the left member of $x^3 - 2x^2 + 3x + 3 = 0$ by $x - 3$ and by $x + 1$ is given below:

$$
\begin{array}{c}
1 - 2 + 3 + 3\lfloor 3 \\
\underline{\quad 3 + 3 + 18 \quad} \\
1 + 1 + 6 + 21
\end{array}
\qquad
\begin{array}{c}
1 - 2 + 3 + 3\lfloor -1 \\
\underline{\quad -1 + 3 - 6 \quad} \\
1 - 3 + 6 - 3
\end{array}
$$

In the first case, the terms in the third row are all positive. Hence, 3 is an upper bound. In the second case, the terms in the third row are alternately plus and minus. Therefore, -1 is a lower bound of the roots. Hence, all real roots are between -1 and 3.   ●

● **Example 2**   Find upper and lower bounds for the real roots of the equation $x^4 - x^3 - 12x^2 - 2x + 3 = 0$.

Solution   We first divide synthetically by -3:

$$
\begin{array}{c}
1 - 1 - 12 - 2 + 3\lfloor -3 \\
\underline{\quad -3 + 12 \quad 0 + 6 \quad} \\
1 - 4 \quad\ 0 - 2 + 9
\end{array}
$$

If we replace the 0 in the middle of the third line by $+$, then the third-row signs alternate. Thus -3 is a lower bound of the roots.

If we now divide by 4 and 5, we get

$$
\begin{array}{c}
1 - 1 - 12 - 2 + 3\lfloor 4 \\
\underline{\quad 4 \quad 12 \quad 0 - 8 \quad} \\
1 + 3 \quad\ 0 - 2 - 5
\end{array}
\qquad
\begin{array}{c}
1 - 1 - 12 - 2 + \quad 3\lfloor 5 \\
\underline{\quad 5 + 20 + 40 + 190 \quad} \\
1 + 4 + \quad 8 + 38 + 193
\end{array}
$$

In the division by 4, not every sign in the third row is positive or zero; so we cannot say that 4 is an upper bound of the roots. However, in the division by 5 all third-row entries are positive; so 5 is an upper bound of the roots.   ●

Note   The theorem on bounds does not always give the *best* upper bound for positive roots, even among integers, but it does give an upper bound. See the next example.

● **Example 3**   For $(x - 4)(x - 5) = x^2 - 9x + 20 = 0$, we know that the roots are 4 and 5, and it follows that 6 is an upper bound. However, as the synthetic divisions below show,

$$
\begin{array}{rrr|r}
1 & -9 & 20 & 6 \\
  & 6 & -18 & \\
\hline
1 & -3 & 2 &
\end{array}
\qquad
\begin{array}{rrr|r}
1 & -9 & 20 & 7 \\
  & 7 & -14 & \\
\hline
1 & -2 & 6 &
\end{array}
$$

$$
\begin{array}{rrr|r}
1 & -9 & 20 & 8 \\
  & 8 & -8 & \\
\hline
1 & -1 & 12 &
\end{array}
\qquad
\begin{array}{rrr|r}
1 & -9 & 20 & 9 \\
  & 9 & 0 & \\
\hline
1 & 0 & 20 &
\end{array}
$$

the theorem on bounds alone does not show that 6 (or 7 or 8) is an upper bound. The number 9 is the smallest integral upper bound that is revealed by the theorem on bounds.   ●

Descartes' rule of signs

We will now present a criterion or rule that enables us to determine a number that is greater than or equal to the number of positive roots of a polynomial equation, and another that is equal to or greater than the number of negative roots.

If the terms of a polynomial are arranged according to the ascending or descending powers of the variable, we say that a *variation of signs* occurs when the signs of two consecutive terms differ. For example, in the

Variation of signs   polynomial $2x^4 - 5x^3 - 6x^2 + 7x + 3$, the signs of the terms are $+ \ - \ -$ $+ \ +$. Hence, there are two variations of sign, since the sign changes from positive to negative and back again to positive. Furthermore, there are three variations of sign in $x^4 - 2x^3 + 3x^2 + 6x - 4$.

We will now state Descartes' rule of signs and then illustrate it. The proof is omitted.

Statement of   *The number of positive roots of a polynomial equation $f(x) = 0$ with real*
Descartes' rule of signs   *coefficients is equal to the number of variations in sign of $f(x)$ or is less than this number by an even integer. The number of negative roots equals the number of variations in sign of $f(-x)$ or is less than this number by an even integer.*

● **Example 4**   Find the maximum number of positive roots and the maximum number of negative roots that can be in the solution set of $x^4 - 3x^3 - 5x^2 + 7x - 3 = 0$.

Solution   There are three variations of sign in $f(x) = x^4 - 3x^3 - 5x^2 + 7x - 3$. Hence, the number of positive roots of the equation $x^4 - 3x^3 - 5x^2 + 7x - 3 = 0$ does not exceed three. Furthermore, $f(-x) = x^4 + 3x^3 - 5x^2 - 7x - 3$, and this polynomial has one variation of sign. Therefore, the number of negative roots of the above equation is one.   ●

Size of the roots
Number of roots

Using 0

The theorem on bounds gives information about *upper and lower bounds* for the values of roots. Descartes' theorem gives information about *the number of positive and negative roots* without saying how large any of them may be. In the application of either theorem a 0 may occur. A 0 in the third line of a synthetic division may be given either a plus sign or a minus sign in the process of determining the bounds of the root values. A 0 that occurs as a coefficient in the application of Descartes' rule of signs should be ignored.

● **Example 5**   What information can be given about the positive and negative roots of $f(x) = x^3 - 4x^2 + 3x + 2 = 0$?

Solution   There are two variations in sign of $f(x)$; so there are either two positive roots or none. Since $f(-x) = -x^3 - 4x^2 - 3x + 2$, there is one variation of sign in $f(-x)$, and so there is exactly one negative root. The synthetic divisions

$$
\begin{array}{r}
1 - 4 + 3 + 2\,\lfloor{-1} \\
\underline{- 1 + 5 - 8} \\
1 - 5 + 8 - 6
\end{array}
\qquad
\begin{array}{r}
1 - 4 + 3 + 2\,\lfloor 4 \\
\underline{4 0 + 12} \\
1 0 + 3 + 14
\end{array}
$$

show that there is no negative root less than -1 and no positive root greater than 4.   ●

Locating the roots

The roots of
$$2x^3 - 3x^2 - 12x + 6 = 0$$

are approximately -2.1, 0.5, and 3.1, which may be checked by synthetic division. Furthermore, for $x = -3$, the curve is below the x axis since it passes through the point $(-3, -39)$; for $x = -2$, the curve is above the x axis since it passes through the point $(-2, 2)$. This is an illustration of the intermediate value theorem: If the graph of $y = f(x)$ passes through two points on opposite sides of the x axis and if there are no gaps or breaks in the curve, it must cross the x axis between the two points. Curves that contain no gaps or that are not made up of separate or disjointed parts are called *continuous curves*. It can be proved that the graph of any polynomial is continuous, although it is beyond the scope of this book to do so. We have the following rule for locating the roots:

Location theorem   *If $f(a)$ and $f(b)$ differ in sign, there is an odd number of real roots of $f(x) = 0$ between $x = a$ and $x = b$. Therefore, there is at least one such root.*

● **Example 6**   Locate the roots of $2x^3 - x^2 - 6x + 3 = 0$.

Solution   In the equation $2x^3 - x^2 - 6x + 3 = 0$, $f(x) = 2x^3 - x^2 - 6x + 3$. Furthermore, it can be verified by substitution or synthetic division that $f(-2) = -5$, $f(-1) = 6$, $f(0) = 3$, $f(1) = -2$, and $f(2) = 3$. Hence, there is an odd number of roots between -2 and -1, between 0 and 1, and between 1 and 2. Since the degree of the equation is 3, it has only three roots. Hence, there is exactly one root in each of the above intervals.   ●

● **Example 7**   What can we say about the roots of

$$x^5 - 10x^4 + 33x^3 - 38x^2 + 6x + 4 = 0 \qquad (2)$$

Solution
5 roots
Descartes' rule of signs

Since $f(x)$ has degree 5, there are 5 roots counting real and complex roots and multiplicities. In $f(x)$, there are 4 variations in sign; hence there are 4, 2, or 0 positive roots. In $f(-x) = -x^5 - 10x^4 - 33x^3 - 38x^2 - 6x + 4$, there is one variation in sign; hence there is precisely one negative root. ●

Location theorem

The following table of values may be found either by calculating function values directly from Eq. (2) or by using synthetic division. We thus see that $x = 2$ is a root, and there also are roots between -1 and 0, between 0 and 1, between 1 and 3, between 3 and 4, and between 4 and 5—this gives a total of four positive roots, and one negative root.

x	-1	0	1	2	3	4	5
$f(x)$	-84	4	-4	0	4	-4	84

Some of the synthetic divisions are shown below.

```
1  -10   33  -38    6    4 |-1      1  -10   33  -38    6    4 | 2
      -1   11  -44   82  -88                 2  -16   34   -8   -4
1  -11   44  -82   88  -84          1   -8   17   -4   -2    0
```

```
1  -10   33  -38    6    4 | 5      1  -10   33  -38    6    4 | 4
       5  -25   40   10   80                 4  -24   36   -8   -8
1   -5    8    2   16   84          1   -6    9   -2   -2   -4
```

Theorem on bounds

We know that -1 is a lower bound by the synthetic division—or by the fact that the one negative root is between -1 and 0. We know that 5 is an upper bound since we have located the four positive roots between 0 and 5. The third line in synthetic division will not yield an upper bound until we try $x = 10$, since the second number will be negative for $x = 5, 6, 7, 8, 9$.

Exercise 11.3
real roots

In Probs. 1 to 12 use the theorem on bounds to determine upper and lower bounds for the real roots of the given equation.

1 $3x^3 + 2x^2 - 2x + 3 = 0$ 2 $2x^2 + 2x - 8x - 5 = 0$
3 $4x^3 - 3x^2 - 3x - 13 = 0$ 4 $3x^3 + 2x^2 - 14x - 3 = 0$
5 $x^4 + 3x^3 - 15x^2 - 9x + 31 = 0$
6 $3x^4 - 20x^3 + 28x^2 + 19x - 13 = 0$
7 $6x^4 + 11x^3 - 25x^2 - 33x + 21 = 0$
8 $6x^4 + x^3 - 43x^2 - 7x + 7 = 0$
9 $2x^3 + x^2 + 8x + 4 = 0$ 10 $x^3 + 3x^2 + x - 4 = 0$
11 $6x^4 + 23x^3 + 25x^2 - 9x - 5 = 0$ 12 $x^4 + 5x^3 + 7x^2 - 3x - 9 = 0$

Determine the number of positive roots and number of negative roots of the equation in each of Probs. 13 to 24 by use of Descartes' rule of signs.

13 $x^3 - 4x^2 - 5x + 2 = 0$ 14 $-2x^3 - 3x^2 + 5x + 7 = 0$
15 $5x^3 + 3x^2 + 6x + 1 = 0$ 16 $2x^3 + 5x^2 - x + 2 = 0$
17 $2x^4 + 5x^3 - 3x^2 + x + 2 = 0$
18 $x^4 - 4x^3 + 12x^2 + 24x + 24 = 0$
19 $-3x^4 - 5x^3 + 8x^2 - 2x + 6 = 0$
20 $3x^4 + 8x^3 - 2x^2 + 5x - 1 = 0$
21 $4x^5 + 3x^4 - 2x^3 + x^2 - x + 4 = 0$
22 $3x^5 - 3x^3 + 2x^2 + 5x - 1 = 0$
23 $3x^5 + 2x^2 + 5x - 1 = 0$
24 $2x^6 - 3x^4 + x^3 - 3 = 0$

In Probs. 25 to 36, use the location theorem to locate each real root between two consecutive integers.

25 $3x^3 + 10x^2 - 2x - 4 = 0$ 26 $3x^3 - 19x^2 + 21x - 4 = 0$
27 $2x^3 - 19x^2 + 50x - 28 = 0$ 28 $3x^3 - 28x^2 + 54x + 20 = 0$
29 $x^4 - 8x^3 + 12x^2 + 16x - 16 = 0$
30 $x^4 + 4x^3 - 15x^2 - 66x - 54 = 0$
31 $3x^4 - 12x^3 - 9x^2 + 16x + 4 = 0$
32 $3x^4 - 12x^3 - 6x^2 + 36x - 9 = 0$
33 $2x^3 - 11x^2 + 18x - 14 = 0$ 34 $3x^3 - 13x^2 + 19x - 5 = 0$
35 $x^4 - 4x^2 - 8x - 4 = 0$ 36 $x^4 - 8x^3 + 21x^2 - 20x - 6 = 0$

The equation in each of Probs. 37 to 40 has two roots between consecutive integers. Locate these roots by use of a value halfway between the consecutive integers. Also locate all other roots between two consecutive integers.

37 $3x^3 - 4x^2 - 7x - 2 = 0$ 38 $3x^3 - 5x^2 - 6x + 10 = 0$
39 $9x^3 - 15x^2 - 12x + 20 = 0$ 40 $9x^3 - 24x^2 - 48x + 128 = 0$

Use Descartes' rule of signs in Probs. 41 to 44.

41 Show that $x^6 + 2x^4 + 3x^2 + 4 = 0$ has six imaginary roots.
42 Show that $5x^5 + 3x^3 + x + 1 = 0$ has four imaginary roots.
43 Show that $x^5 + 2x - 3 = 0$ has four imaginary roots.
44 Show that $x^3 + 2x + 3 = 0$ has two imaginary roots.
45 The following example illustrates the method used to prove the theorem on bounds in general for the case $k < 0$. Give reasons for each step. We use $f(x) = x^4 + 3x^3 - 5x^2 - 8x + 4$ and $k = -5$.

Theorem on bounds

(a)

$$
\begin{array}{r|rrrr}
1 \quad 3 \quad -5 \quad -8 \quad 4 & -5 \\
\quad\quad -5 \quad 10 \quad -25 \quad 165 \\
\hline
1 \quad -2 \quad 5 \quad -33 \quad 169 &
\end{array}
$$

signs alternate

(b) $f(x) = (x + 5)(x^3 - 2x^2 + 5x - 33) + 169$
(c) If $x < -5$, then
$$f(x) = (\text{negative})(\text{negative}) + 169$$
$$= \text{positive} + \text{positive} = \text{positive}$$
(d) And thus if $x < -5$, then $f(x) \neq 0$.

11.4 The rational roots

If a polynomial equation has one or more rational roots, the work involved in finding the others is greatly reduced if the rational roots are found first. The process of identifying a rational root is a matter of trial. Hence, the following theorem on rational roots is very useful because it enables us to find a set of numbers which includes the rational roots.

○ **Theorem on rational roots**

If the coefficients of

$$a_0x^n + a_1x^{n-1} + \cdots + a_{n-1}x + a_n = 0 \qquad (1)$$

are integers, then each of the rational roots, after being reduced to lowest terms, has a factor of a_n for its numerator and a factor of a_0 for its denominator.

Proof We begin our proof of the theorem by assuming that q/p is a rational root of (1) and that q and p do not have a common factor greater than 1. Now, if we substitute q/p for x in (1) and multiply by p^n, we obtain

$$a_0q^n + a_1q^{n-1}p + \cdots + a_{n-1}qp^{n-1} + a_np^n = 0 \qquad (2)$$

By adding $-a_np^n$ to each member and then dividing by q, we have

$$a_0q^{n-1} + a_1q^{n-2}p + \cdots + a_{n-1}p^{n-1} = -\frac{a_np^n}{q} \qquad (3)$$

The left member of (3) is made up of the sum, product, and integral powers of integers; hence, it is an integer. Therefore, the right member must be an integer. Consequently, q is a factor of a_n since by hypothesis q and p have no common factor greater than 1.

If we add $-a_0 q^n$ to each member of (2) and divide by p, we obtain

$$a_1 q^{n-1} + \cdots + a_{n-1} q p^{n-2} + a_n p^{n-1} = -\frac{a_0 q^n}{p} \tag{4}$$

Hence, p is a factor of a_0, since q and p have no common factor greater than 1 and the left member of (4) is an integer.

If $a_0 = 1$ in (1), we get the following corollary:

Corollary   *Each rational root of an equation*

Integer roots if
integer coefficients

$$x^n + a_1 x^{n-1} + \cdots + a_{n-1} x + a_n = 0 \tag{5}$$

with integral coefficients is an integer and a factor of a_n.

● **Example 1**   Find the set of possible rational roots of $2x^4 + x^3 - 9x^2 - 4x + 4 = 0$.

Solution   In the equation $2x^4 + x^3 - 9x^2 - 4x + 4 = 0$, the numerators of the rational roots must be factors of 4, and the denominators factors of 2. Hence the possibilities for the rational roots are $\pm 1, \pm 2, \pm 4, \pm\frac{1}{2}, \pm\frac{2}{2}, \pm\frac{4}{2}$. If we

Possible rational
roots

eliminate repetitions, this set of quotients becomes $\{-4, -2, -1, -\frac{1}{2}, \frac{1}{2}, 1, 2, 4\}$. We can use synthetic division and the remainder theorem to determine which of these are roots; see the top half of page 386.   ●

The depressed equation

If r_1 is a root of $F(x) = 0$, then by the factor theorem,

$$F(x) = F_1(x)(x - r_1) = 0$$

Depressed equation   and $F_1(x) = 0$ is called the *depressed equation* corresponding to r_1. Furthermore, the degree of $F_1(x)$ is one less than the degree of $F(x)$. Now if r_2 is a root of $F_1(x) = 0$, then r_2 is also a root of $F(x) = 0$ and we have

$$F(x) = (x - r_1)(x - r_2)F_2(x) = 0$$

We can use $F_2(x) = 0$ in seeking the remaining roots.

This operation is continued after each rational root is found; when and if we obtain a depressed equation that is a quadratic, then the remaining roots can be found by factoring or the quadratic formula.

Note   The depressed equation $F_1(x) = 0$ is used to find roots of the given equation $F(x) = 0$. The depressed equation may *not* be used to find function values of $y = F(x)$ directly.

We illustrate the use of the depressed equation by obtaining all roots of the equation $2x^4 + x^3 - 9x^2 - 4x + 4$ of Example 1. We first use synthetic division to determine whether 2 is a root and get

$$
\begin{array}{r}
2 + 1 - \ \ 9 - 4 + 4\underline{|2} \\
+ 4 + 10 + 2 - 4 \\
\hline
2 + 5 + \ \ 1 - 2 \quad \ 0
\end{array}
$$

Since the remainder is zero, 2 is a root. Furthermore, all roots of the given equation $2x^4 + x^3 - 9x^2 - 4x + 4 = 0$, except possibly $x = 2$, are roots of the depressed equation $2x^3 + 5x^2 + x - 2 = 0$. If 2 is a multiple root of the given equation, it is also a root of the depressed equation. We next try $x = -2$ in the depressed equation above and obtain

$$
\begin{array}{r}
2 + 5 + 1 - 2\underline{|-2} \\
- 4 - 2 + 2 \\
\hline
2 + 1 - 1 \quad \ 0
\end{array}
$$

Since the remainder is zero, -2 is a root. The depressed equation, corresponding to $x = 2$ and $x = -2$, is the quadratic $2x^2 + x - 1 = 0$, which we may solve by factoring and get

$$(2x - 1)(x + 1) = 0$$
$$x = \tfrac{1}{2}, -1$$

Consequently, the solutions of the given equation are $2, -2, -1, \tfrac{1}{2}$. It should be noticed that the degree of the equation is 4, and there are four solutions.

Process of obtaining all rational roots

We now outline the steps that should be followed in determining the rational roots of a polynomial equation.

Steps in process of obtaining all rational roots

1 Use the theorem on rational roots to write the set of rational numbers that contains the rational roots, and put them in the order of the magnitude of their numerical values.
2 Test the smallest positive integer in the set, then the next larger, and so on, until each integral root or a bound of the roots is found. (*a*) If an upper bound is found, discard all larger numbers in the set. (*b*) If a root is found, use the depressed equation in further calculations, and check first for multiplicity.
3 Test the fractions that remain in the set after considering any bound that has been found.
4 Repeat steps 2 and 3 for negative roots, beginning with the negative integer nearest 0.

Reminder   If a quadratic is obtained by use of the depressed equation, its roots can be found by use of any of the methods for solving quadratics.

● **Example 2**   Find all rational roots of $4x^4 - 4x^3 - 25x^2 + x + 6 = 0$.

Solution   The possible numerators of rational roots are ± 6, ± 3, ± 2, ± 1, and the possible denominators are ± 4, ± 2, ± 1. The set of possible rational roots is $\{\pm\frac{1}{4}, \pm\frac{1}{2}, \pm\frac{3}{4}, \pm 1, \pm\frac{3}{2}, \pm 2, \pm 3, \pm 6\}$. We now test the positive integral possibilities.

$$
\begin{array}{rrrrr|l}
4 & -\,4 & -\,25 & +\,1 & +\,6 & 1 \\
  & +\,4 & 0 & -\,25 & -\,24 & \\
\hline
4 & 0 & -\,25 & -\,24 & -\,18 &
\end{array}
$$

Since the remainder is not zero and the signs are neither all plus in the third line nor alternately plus and minus, 1 is neither a root nor a bound.

$$
\begin{array}{rrrrr|l}
4 & -\,4 & -\,25 & +\,1 & +\,6 & 2 \\
  & +\,8 & +\,8 & -\,34 & -\,66 & \\
\hline
4 & +\,4 & -\,17 & -\,33 & -\,60 &
\end{array}
$$

Therefore, 2 is neither a root nor a bound.

$$
\begin{array}{rrrrr|l}
4 & -\,4 & -\,25 & +\,1 & +\,6 & 3 \\
  & +\,12 & +\,24 & -\,3 & -\,6 & \\
\hline
4 & +\,8 & -\,1 & -\,2 & 0 &
\end{array}
$$

3 is a root   Consequently, 3 is a root and the corresponding depressed equation is $4x^3 + 8x^2 - x - 2 = 0$, which has all the roots of the original equation, with the possible exception of 3. Since the constant term in the depressed equation is -2 and the coefficient of x^3 is 4, the set of possible rational *Depressed equation* roots of the depressed equation is $\{\pm\frac{1}{4}, \pm\frac{1}{2}, \pm 1, \pm 2\}$. Not all these need be considered, since we found that 1 and 2 are not roots of the original equation.

It is readily seen that $\frac{1}{4}$ is not a root, but $\frac{1}{2}$ is a root with $4x^2 + 10x + 4 = 0$ as the corresponding depressed equation. This is a quadratic that can be solved by factoring. Its roots are -2 and $-\frac{1}{2}$. Hence, the solutions of the original equation are $3, \frac{1}{2}, -\frac{1}{2}, -2$.   ●

● **Example 3**   Find all roots of $4x^4 - 12x^3 + 17x^2 - 24x + 18 = 0$.

Solution   The possible numerators of rational roots are ± 18, ± 9, ± 6, ± 3, ± 2, ± 1, and the possible denominators are ± 4, ± 2, ± 1. Hence the possible rational roots are $+$ or $-$ the following numbers:

$$\tfrac{1}{4}, \tfrac{1}{2}, \tfrac{3}{4}, 1, \tfrac{3}{2}, 2, \tfrac{9}{4}, 3, \tfrac{9}{2}, 6, 9, 18.$$

There are 24 possible rational roots, which is a lot. The only one which will work is $x = \frac{3}{2}$. We are led to try $\frac{3}{2}$ since $f(1) = 3$, and $f(2) = 6$, and 3 and 6 are fairly close to 0 compared to $f(0) = 18$ and $f(3) = 99$; hence we can at least hope that a nearby number will produce exactly 0.

$$
\begin{array}{rrrrr|l}
4 & -\,12 & 17 & -\,24 & 18 & \tfrac{3}{2} \\
  & 6 & -\,9 & 12 & -\,18 & \\
\hline
4 & -\,6 & 8 & -\,12 & 0 &
\end{array}
\qquad
\begin{array}{rrrr|l}
4 & -\,6 & 8 & -\,12 & \tfrac{3}{2} \\
  & 6 & 0 & 12 & \\
\hline
4 & 0 & 8 & 0 &
\end{array}
$$

We try $x = \frac{3}{2}$ in the depressed equation $4x^3 - 6x^2 + 8x - 12 = 0$ and find that it is a root of multiplicity 2. The resulting depressed equation is $4x^2 + 8 = 0$; hence $x^2 = -2$. The four roots of the original equation are thus $\frac{3}{2}$, $\frac{3}{2}$, $i\sqrt{2}$, and $-i\sqrt{2}$. ●

Exercise 11.4 rational roots

Find all roots of the equation in each of Probs. 1 to 28.

1 $x^3 - x^2 - 4x + 4 = 0$ **2** $x^3 - 4x^2 + x + 6 = 0$

3 $x^3 - 7x + 6 = 0$ **4** $x^3 + 2x^2 - 5x - 6 = 0$

5 $2x^3 + x^2 - 13x + 6 = 0$ **6** $3x^3 + 19x^2 + 16x - 20 = 0$

7 $4x^3 - 7x - 3 = 0$ **8** $6x^3 + 13x^2 + x - 2 = 0$

9 $12x^3 - 4x^2 - 3x + 1 = 0$ **10** $8x^3 - 12x^2 - 18x + 27 = 0$

11 $4x^3 - 8x^2 - 15x + 9 = 0$ **12** $6x^3 - x^2 - 31x - 10 = 0$

13 $x^3 - 4x^2 + 3x + 2 = 0$ **14** $x^3 + x^2 - 8x - 6 = 0$

15 $2x^3 + 3x^2 - 4x - 5 = 0$ **16** $2x^3 + 3x^2 - 6x - 8 = 0$

17 $2x^3 - x^2 + 2x - 1 = 0$ **18** $3x^3 + 7x^2 + 8x + 2 = 0$

19 $2x^3 - 3x^2 + 2x - 3 = 0$ **20** $2x^3 + x^2 + 8x + 4 = 0$

21 $6x^4 + x^3 - 22x^2 - 11x + 6 = 0$

22 $6x^4 + 17x^3 - 14x^2 - 27x + 18 = 0$

23 $x^4 + x^3 - 15x^2 + 23x - 10 = 0$

24 $x^4 - x^3 - 13x^2 + x + 12 = 0$

25 $x^4 - x^3 - x^2 - x - 2 = 0$

26 $x^4 - 2x^3 + x^2 + 2x - 2 = 0$

27 $2x^4 + 2x^3 + \frac{1}{2}x^2 + 2x - \frac{3}{2} = 0$

Hint: Multiply by the lowest common denominator.

28 $x^4 - \frac{17}{6}x^3 - \frac{7}{6}x^2 + \frac{20}{3}x - 2 = 0$.

29 Show that $\frac{1}{3}$ is a root of multiplicity 2 of $9x^3 + 12x^2 - 11x + 2 = 0$.

30 Show that $\frac{2}{5}$ is a root of multiplicity 2 of $25x^3 + 5x^2 - 16x + 4 = 0$.

31 Show that $-\frac{1}{2}$ is a root of multiplicity 2 of $4x^4 + 4x^3 - 3x^2 - 4x - 1 = 0$.

32 Show that $-\frac{1}{3}$ is a root of multiplicity 3 of $27x^4 - 27x^3 - 45x^2 - 17x - 2 = 0$.

33 Show that there are no rational roots of $x^3 - 7x^2 + 2x - 1 = 0$.

34 Show that there are no rational roots of $x^3 + 13x^2 - 6x - 2 = 0$.

35 Show that there are no rational roots of $x^4 - 2x^3 + 10x^2 - x + 1 = 0$.

36 Show that there are no rational roots of $3x^4 - 9x^3 - 2x^2 - 15x - 5 = 0$.

Show that the number in each of Probs. 37 to 44 is irrational. *Hint:* To show that $x = 1 + \sqrt{2}$ is irrational, form the equations $x - 1 = \sqrt{2}$, $(x - 1)^2 = (\sqrt{2})^2 = 2$, and $x^2 - 2x - 1 = 0$ and then show that the last equation has no rational roots.

37 $\sqrt{6}$ **38** $\sqrt[3]{5}$ **39** $2 - \sqrt{5}$ **40** $1 + \sqrt[3]{6}$ **41** $\sqrt{2} + \sqrt{7}$

42 $\sqrt{5} - \sqrt{2}$ **43** $6^{2/3}$ **44** $1 + \sqrt{2} + \sqrt{3}$

Note that if $x = 1 + \sqrt{3} + \sqrt{2}$, then $x - 1 = \sqrt{3} + \sqrt{2}$.

45 Solve $(x + 3)^4 + (x - 1)^4 = 82$.
46 Solve $(x^2 + x + 1)(x^2 + x + 3) = 15$.
47 Solve $(x + 1)(x - 1)(x - 2)(x - 4) = -8$.
48 Solve $(x - 2)^4 + (x - 1)^4 = 1$.
49 Show that the sum of the roots of $ax^4 + bx^2 + c = 0$ is 0. *Hint:* Treat it as a quadratic in x^2. You might also use Prob. 57, Exer. 11.2.

 ## The irrational roots

There are formulas for solving polynomial equations of degree 3 and degree 4. However, it was proved, about 1825, that no formula can possibly exist which works for every fifth-degree equation. And no formula can exist for the general polynomial equation of degree n, if $n \geq 5$. Hence we are forced to find approximations.

Cubic equation   A *third-degree equation* always has at least one real root, but in general we need to use complicated formulas and trigonometry to find them. At our level, it is best to find the rational roots (if any), then use the depressed

Successive equation, which will be a quadratic. If there are no rational roots, we must
approximation approximate the irrational roots by the location theorem. Thus we first find consecutive integers whose function values have opposite signs; then we narrow the root between consecutive tenths, and then between consecutive hundredths. If more accuracy is needed, then techniques from calculus are used, usually with the aid of a computer (or perhaps a hand calculator).

Problem 37 shows how the general **cubic equation** $x^3 + ax^2 + bx + c = 0$ can be put in the form $y^3 + py + q = 0$. Furthermore, it can be shown that

$$\text{If} \quad \frac{p^3}{27} + \frac{q^2}{4} < 0, \quad \text{then there are 3 real roots}$$

$$\text{If} \quad \frac{p^3}{27} + \frac{q^2}{4} > 0, \quad \begin{array}{l}\text{then there is 1 real root} \\ \text{and a pair of complex conjugate roots}\end{array}$$

● **Example 1**   Approximate the largest root of $2x^3 - 3x^2 - 12x + 6 = 0$.

Solution   Calculation of function values gives $f(-3) = -39$ and $f(-2) = 2$; so there is a root between -3 and -2. Also there is a root between 0 and 1, since $f(0) = 6$ and $f(1) = -7$. Finally, there is a root between 3 and 4, since $f(3) = -3$ and $f(4) = 38$. We want the root between 3 and 4. Since $f(3)$ is much closer to 0 than $f(4)$ is, the root is much closer to 3 than it is to 4. The following synthetic divisions

$$
\begin{array}{rrrr|r}
2 & -3 & -12 & +6 & \underline{3.1} \\
  & 6.2 & 9.92 & -6.448 & \\
\hline
2 & +3.2 & -2.08 & -0.448 &
\end{array}
\qquad
\begin{array}{rrrr|r}
2 & -3 & -12 & +6 & \underline{3.2} \\
  & 6.4 & 10.88 & -3.584 & \\
\hline
2 & +3.4 & -1.12 & +2.416 &
\end{array}
$$

show that the root is between 3.1 and 3.2, since $f(3.1) < 0$ and $f(3.2) > 0$. The root is closer to 3.1 than to 3.2, since $f(3.1)$ is closer to 0. Now synthetic division gives

$$
\begin{array}{rrrr|l}
2 & -3 & -12 & 6 & \underline{3.11} \\
  & 6.22 & 10.0142 & -6.175838 & \\
\hline
2 & 3.22 & -1.9858 & -0.175838 &
\end{array}
$$

$$
\begin{array}{rrrr|l}
2 & -3 & -12 & 6 & \underline{3.12} \\
  & 6.24 & 10.1088 & -5.900544 & \\
\hline
2 & 3.24 & -1.8912 & 0.099456 &
\end{array}
$$

so the root is between 3.11 and 3.12, and closer to 3.12. Further synthetic divisions would show the root is between 3.116 and 3.117.   ●

● **Example 2**   Approximate the roots of $x^3 - 4x^2 + 7x - 5 = 0$.

Solution   The following table of values narrows the root within smaller and smaller limits.

x	1	2	1.5	1.6	1.56	1.57
$f(x)$	-1	1	-0.125	0.056	-0.018	0.003

We will thus use 1.57 as our *approximation* to the real root. The following synthetic division

One real root

$$
\begin{array}{rrrr|l}
1 & -4 & 7 & -5 & \underline{1.57} \\
  & 1.57 & -3.8151 & 5.000293 & \\
\hline
1 & -2.43 & 3.1849 & 0.000293 &
\end{array}
$$

gives us the approximate depressed equation $x^2 - 2.43x + 3.1849 = 0$. The quadratic formula gives

Two complex roots

$$
\begin{aligned}
2x &= 2.43 \pm \sqrt{(2.43)^2 - 4(1)(3.1849)} \\
&= 2.43 \pm \sqrt{5.9049 - 12.7396} \\
&= 2.43 \pm \sqrt{-6.8347} \\
&= 2.43 \pm 2.614i \\
x &= 1.215 \pm 1.307i
\end{aligned}
$$

Hence there is one real root and two complex roots.   ●

Quartic equations have degree 4

As we have seen, the general method for solving third-degree equations is very involved, even for approximations. The method for *fourth-degree equations*, surprisingly, is somewhat easier to apply, though still formidable. It usually involves completing the square twice and solving a cubic equation, thereby reducing the problem to solving two quadratic equations. It will be illustrated in the next two examples.

● **Example 3**   Solve $x^4 - 8x^3 + 7x^2 + 30x - 25 = 0$.

Solution
$$x^4 - 8x^3 = -7x^2 - 30x + 25 \qquad \text{transposing}$$
$$x^4 - 8x^3 + 16x^2 = 16x^2 - 7x^2 - 30x + 25 \qquad \text{completing the square}$$
$$\text{of the left member}$$

$$(x^2 - 4x)^2 = 9x^2 - 30x + 25$$
$$(x^2 - 4x)^2 = (3x - 5)^2 \qquad \text{factoring right member}$$

Now if $a^2 = b^2$, then either $a = b$ or $a = -b$. Hence

$x^2 - 4x = 3x - 5$	$x^2 - 4x = -(3x - 5)$
$x^2 - 7x + 5 = 0$	$x^2 - x - 5 = 0$
$x = \dfrac{7 \pm \sqrt{49 - 20}}{2}$	$x = \dfrac{1 \pm \sqrt{1 + 20}}{2}$
$= \dfrac{7 \pm \sqrt{29}}{2}$	$= \dfrac{1 \pm \sqrt{21}}{2}$

Decimal approximations to these four roots are 6.1926, 0.8074, 2.7913, and -1.7913.

● **Example 4** Solve $x^4 - 2x^3 - 18x^2 + 34x - 7 = 0$.

Solution
$$x^4 - 2x^3 = 18x^2 - 34x + 7$$
$$x^4 - 2x^3 + x^2 = x^2 + 18x^2 - 34x + 7 \qquad \text{completing the square}$$
$$(x^2 - x)^2 = 19x^2 - 34x + 7$$

If we now add $(x^2 - x)(2y) + y^2$ to both members, the left member remains a perfect square for any y. We will shortly choose a value for y so that the right member is also a perfect square.

$$(x^2 - x)^2 + 2y(x^2 - x) + y^2 = 19x^2 - 34x + 7 + 2y(x^2 - x) + y^2$$
$$[(x^2 - x) + y]^2 = x^2(19 + 2y) + x(-34 - 2y) + (7 + y^2) \quad (1)$$

For any value of y, the left member is a perfect square, and we will now choose y so that the right member is a perfect square also. This requires that its discriminant $b^2 - 4ac = 0$.

$$(-34 - 2y)^2 - 4(19 + 2y)(7 + y^2) = 0 \qquad \text{collecting terms}$$
$$1156 + 136y + 4y^2 - 4(133 + 19y^2 + 14y + 2y^3) = 0 \qquad \text{and dividing}$$
$$y^3 + 9y^2 - 10y - 78 = 0 \qquad \text{by } -8$$

We now find *any* solution of this last equation. By the theorem on rational roots, we see that $y = 3$ is a solution since $27 + 81 - 30 - 78 = 108 - 108 = 0$. Using $y = 3$ in both members of Eq. (1) gives

$$(x^2 - x + 3)^2 = 25x^2 - 40x + 16$$

$$(x^2 - x + 3)^2 = (5x - 4)^2$$

$x^2 - x + 3 = 5x - 4$	$x^2 - x + 3 = -5x + 4$
$x^2 - 6x + 7 = 0$	$x^2 + 4x - 1 = 0$
$x = \dfrac{6 \pm \sqrt{36 - 28}}{2}$	$x = \dfrac{-4 \pm \sqrt{16 + 4}}{2}$
$= 3 \pm \sqrt{2}$	$= -2 \pm \sqrt{5}$

This method of solving quartic equations reduces the original equation to two quadratic equations. We not only can find the exact solutions by the quadratic formula, but we also get both irrational and complex roots.

Exercise 11.5 irrational roots

In Probs. 1 to 8, find the real root to two decimal places.

1 $x^3 - 6x^2 + 12x - 12 = 0$ 2 $x^3 - 3x^2 + 3x - 4 = 0$
3 $x^3 + 9x^2 + 27x + 7 = 0$ 4 $x^3 + 12x^2 + 48x + 55 = 0$
5 $x^3 + x - 4 = 0$ 6 $x^3 + 3x - 18 = 0$
7 $x^3 + 5x + 16 = 0$ 8 $x^3 + 9x + 13 = 0$

In Probs. 9 to 16, find the value of the least positive root to two decimal places.

9 $x^3 + 8x^2 + 15x - 2 = 0$ 10 $x^3 - x^2 + 4x - 2 = 0$
11 $x^3 - 8x + 5 = 0$ 12 $3x^3 + 9x^2 - 3x - 16 = 0$
13 $x^3 + 4x^2 - x - 6 = 0$ 14 $2x^3 - 9x^2 + x + 11 = 0$
15 $x^3 + 3x^2 - 6x - 3 = 0$ 16 $x^3 - 4x^2 + 3x + 1 = 0$

In Probs. 17 to 24, find the value of each irrational root to three decimal places.

17 $x^4 - 2x^3 - 4x - 4 = 0$
18 $x^4 - 8x^3 + 12x^2 - 8x + 11 = 0$
19 $x^4 - 6x^3 + 4x^2 - 6x + 3 = 0$
20 $x^4 + 4x^3 - 2x^2 + 4x - 3 = 0$
21 $x^3 + 3x^2 - 3x - 7 = 0$ 22 $x^3 + 6x^2 - 28 = 0$
23 $x^3 - 6x^2 + 6x + 2 = 0$ 24 $2x^3 - 11x^2 + 12x + 7 = 0$

In Probs. 25 to 36, use the method of Examples 3 and 4 to find all four exact roots of the equation. In each problem, find the largest real root to three decimal places.

25 $x^4 - 4x^3 - 2x^2 + 12x + 8 = 0$
26 $x^4 - 2x^3 - 6x^2 - 14x - 3 = 0$
27 $x^4 - 2x^3 - 16x^2 + 22x + 7 = 0$
28 $x^4 + 2x^3 - 35x^2 - 22x + 36 = 0$
29 $x^4 - 8x^3 + 14x^2 - 8x + 13 = 0$
30 $x^4 - 4x^3 + x^2 - 12x - 6 = 0$
31 $x^4 + 6x^3 + 5x^2 + 24x + 4 = 0$
32 $x^4 + 4x^3 + 2x^2 + 20x - 15 = 0$
33 $x^4 - 10x^3 + 21x^2 - 44x - 121 = 0$
34 $x^4 + 12x^3 + 20x^2 + 24x - 9 = 0$
35 $x^4 + 6x^3 + 24x - 16 = 0$
36 $x^4 - 8x^3 - 8x - 1 = 0$
37 Show that if, in the general cubic equation $x^3 + ax^2 + bx + c = 0$, we replace x by $t - a/3$, then we obtain a *reduced cubic equation* of the form $t^3 + pt + q = 0$, which has no t^2 term.

38 Suppose we have $y^3 + py + q = 0$, and we replace y by $z - (p/3z)$. Show that the resulting equation can be put in the form $z^6 + qz^3 - p^3/27 = 0$, which is a quadratic in z^3.

39 Obtain a reduced cubic from $x^3 + 3x^2 - 2x + 4 = 0$ by making use of the substitution given in Prob. 37.

40 Find the quadratic in z^3 obtained from $y^3 + 3y + 2 = 0$ by use of the substitution given in Prob. 38.

11.6 Key concepts

Make sure you understand the following important words and ideas.

Polynomial equation	Number of positive and negative
Remainder theorem	roots
Factor theorem and its converse	Rule for locating real roots
Synthetic division	Rational roots
Fundamental theorem of algebra	Depressed equation
Multiplicity	Irrational roots
Conjugate complex roots	Cubic equation
Upper and lower bounds	Quartic equation
of the real roots	

Exercise 11.6 review

1 Find the remainder if $x^3 + 3x^2 - x - 4$ is divided by $x - 2$.

2 Find the remainder if $-2x^3 + 3x^2 + 7x - 40$ is divided by $x + 3$.

3 What is the remainder when $x^4 - 2x^3 + 5x + 2$ is divided by $x - d$?

4 Show that $x + 4$ is a factor of $x^3 + 4x^2 - x - 4$.

5 Show that $x^3 - 3x^2 - 2x + 6$ is divisible by $x - 3$.

6 Show that $\frac{3}{2}$ is a root of $2x^3 - 3x^2 + 6x - 9 = 0$.

7 Show that $\sqrt{2}$ is a root of $x^3 + 4x^2 - 2x - 8 = 0$.

8 Show that $3i$ is a root of $x^3 - 4x^2 + 9x - 36 = 0$.

9 Find the quotient and remainder if $2x^3 + 5x - 6$ is divided by $x - 2$.

10 Find the quotient and remainder if $3x^4 + 5x^3 + 2x^2 + 6x - 3$ is divided by $x + 2$.

11 Find all roots of $x^3 - x^2 - 8x + 12 = 0$ if 2 is a root of multiplicity 2.

12 Find the other two roots of $x^4 + 6x^3 + 11x^2 + 12x + 18 = 0$ if -3 is a root of multiplicity 2.

13 Find a polynomial of least possible degree with real coefficients whose zeros include 2 and $3 + i$.

14 Find a polynomial of least possible degree with rational coefficients whose zeros include 3 and $1 + \sqrt{3}$.

15 Find a polynomial of degree 4 with real coefficients that has $x - 1 + 2i$ and $x + 2 - i$ as factors.

16 Find a polynomial of degree 4 with real coefficients which has the zeros $2 + \sqrt{3}$ and $2 + i\sqrt{3}$.

17 What is the multiplicity of the root $x = 3$ in $x^4 - 7x^3 + 9x^2 + 27x - 54 = 0$?

18 Find the multiplicity of the root $x = i$ in $x^5 - 2x^4 + 2x^3 - 4x^2 + x - 2 = 0$.

19 What are the sum and the product of the roots of $x^3 - x^2 + x - 1 = 0$?

20 What are the sum and the product of the roots of

$$x^4 - 4x^3 + 8x^2 - 16x + 16 = 0$$

21 Show that all real roots of $2x^3 - x^2 + 4x + 3 = 0$ are between -1 and 1.

22 Show that there is exactly one positive root of $x^4 + 2x^3 + 4x - 5 = 0$.

23 Find the roots of

$$(x - 1)(x - 2)(x - 3)(x + 4) = (x - 1)(x - 2)(x - 3)(x + 5)$$

24 Show that $x = \sqrt[3]{4}$ is irrational by checking the equation $x^3 - 4 = 0$ for rational roots. Show by substitution that the three roots of this equation are

$$x = \sqrt[3]{4} \quad \text{and} \quad x = -\frac{\sqrt[3]{4}}{2}(1 \pm i\sqrt{3})$$

25 Show that the graphs of

$$y = x^3 + 3x^2 - 5x + 4 \quad \text{and} \quad y = x^3 + 2x^2 - 9x - 3$$

do not intersect each other.

26 Find all roots of the equation $x^3 - 2x^2 + x - 2 = 0$.

27 Find all roots of the equation $4x^4 + 12x^3 + 25x^2 + 48x + 36 = 0$.

28 Find all roots of $6x^5 - 17x^4 - 7x^3 + 58x^2 - 78x + 20 = 0$.

29 Show that the arithmetic mean of the roots of

$$9x^4 - 24x^3 - 35x^2 - 4x + 4 = 0$$

equals the arithmetic mean of the roots of

$$36x^3 - 72x^2 - 70x - 4 = 0$$

30 Show that $\sqrt[4]{6}$ is irrational.

31 Show that $\sqrt{3} - \sqrt{2}$ is irrational.

32 Is $\sqrt{3} + \sqrt{12} - \sqrt{27}$ irrational?

33 Find the real root of $x^3 + 6x^2 + 12x + 2 = 0$ to two decimal places.

34 Find both real roots of $x^4 + 2x^3 + x^2 + 6x - 6 = 0$ to two decimal places.

35 Find any real root of $x^4 - x^3 - 4x^2 + 3x + 3 = 0$ to three decimal places.

36 Find the exact value of each root of $x^4 - x^3 - 4x^2 + 3x + 3 = 0$.

37 Make use of the substitution suggested in Prob. 37 of Exer. 11.5 to find the reduced cubic corresponding to $x^3 + x^2 - 3x - 1 = 0$.

38 Make use of the substitution given in Prob. 38 of Exer. 11.5 to find the quadratic in z^3 corresponding to $y^3 - 3y + 2 = 0$.

CHAPTER 12·
PROGRESSIONS AND ANNUITIES

The inventor of chess, so it is said, asked that he be rewarded with one grain of wheat for the first square of the board, two grains for the second, four for the third, and so on for the 64 squares. Fortunately, this apparently modest request was examined before it was granted. By the twentieth square, the reward would have amounted to more than a million grains of wheat; by the sixty-fourth square the number called for would have been astronomical and the amount would have far exceeded all the grain in the kingdom.

Sequence

This story deals with a *sequence* of numbers, and such relationships have a great many important applications. If a sequence of numbers is such that each term can be obtained from the preceding one by the operation of some law, the sequence is also called a *progression*.

Progression

12.1   Arithmetic progressions

Sequences

Earlier in the book, we studied functions such as

$$f(x) = x^3 + 3x^2 + 8x - 4 \qquad \text{(polynomials)}$$
$$g(x) = 10^x \qquad \text{(exponentials)}$$
$$h(x) = \ln x \qquad \text{(logarithms)}$$
$$r(x) = \frac{2x - 1}{x^2 - 1} \qquad \text{(rational)}$$

Various domains

and others whose *domains* were a set of real numbers. We also looked at functions of two variables like $F(x, y) = 3x + 5y - 9$, and at functions of matrices such as $G(A) = |A| =$ the determinant of the matrix A.

Sequence   In this chapter we will study functions whose *domain is the set of positive integers*. A function of this type is called a **sequence.** Thus the domain of a sequence is the set $\{1, 2, 3, \ldots, n, \ldots\}$, and the value of the sequence, or the value of the function, at n is $f(n)$. Standard notation is to let $f(n) = a_n$. For instance, if $f(n) = n^2$, then the values of the sequence are

$$f(1) = 1, f(2) = 4, f(3) = 9, \ldots, f(n) = n^2, \ldots$$

or

$$a_1 = 1, a_2 = 4, a_3 = 9, \ldots, a_n = n^2, \ldots$$

or even just

$$1, 4, 9, 16, 25, 36, \ldots, n^2, \ldots$$

Arithmetic sequence   An **arithmetic** (arithMETic) **sequence** is one in which each term is d more than the preceding one, where d is a fixed real number, positive, negative, or zero. In symbols

$$f(n + 1) = d + f(n)$$

and $f(1)$ is given. Thus if $f(1) = -8$ and $d = 3$, then $f(2) = 3 + f(1) = 3 + (-8) = -5, f(3) = 3 + f(2) = 3 - 5 = -2, f(4) = 3 + f(3) = 3 - 2 = 1, f(5) = 3 + f(4) = 3 + 1 = 4$, etc. Hence the values of the sequence are

$$-8, -5, -2, 1, 4, \ldots$$

We will soon be able to show that $f(n) = -8 + (n - 1)(3)$.

Arithmetic progressions   We are interested not only in *infinite* arithmetic sequences but in *finite* ones as well. Historically these are called **arithmetic progressions,** and the following special notation is used.

a = first term = $f(1) = a_1$
l = last term = nth term = $f(n) = a_n$
d = common difference = $f(2) - f(1) = f(3) - f(2)$, etc.
n = the number of terms
s = the sum of the n terms
$\quad = f(1) + f(2) + \cdots + f(n) = a_1 + a_2 + \cdots + a_n$

Lands   Note that these letters can be rearranged to make the word "lands."

● **Example 1**   (a)   If the first term is $a = 2$ and the common difference is $d = 5$, then the first eight terms of the arithmetic progression are 2, 7, 12, 17, 22, 27, 32, 37.

(b)   In the sequence 16, $14\frac{1}{2}$, 13, $11\frac{1}{2}$, 10, $8\frac{1}{2}$, each term after the first is $1\frac{1}{2}$ less than the preceding. Hence, this is an arithmetic progression with the common difference equal to $-1\frac{1}{2} = d$, with $a = 16$, and $n = 6$.   ●

Most problems in arithmetic progressions deal with three or more of the five quantities defined above: the first term, the last term, the number of terms, the common difference, and the sum of the terms. We will now derive formulas which enable us to determine the remaining two of these five quantities if we know the values of three of them.

Last term of an arithmetic progression

In terms of the above notation, the first four terms of an arithmetic progression are

$$a \qquad a + d \qquad a + 2d \qquad a + 3d$$

We notice that d enters with the coefficient 1 in the second term and that this coefficient increases by 1 as we move from each term to the next. Hence the coefficient of d in any term is 1 less than the number of that term in the progression. Therefore, the sixth term is $a + 5d$, the ninth is $a + 8d$, and finally the last, or nth, term is $a + (n - 1)d$. Hence we have the formula

$$l = a + (n - 1)d \tag{12.1}$$

● **Example 2**   If the first three terms of an arithmetic progression are 2, 6, and 10, find the eighth term.

Solution   Since the first and second terms, as well as the second and third, differ by 4, it follows that $d = 4$. Furthermore, $a = 2$ and $n = 8$. Hence if we substitute these values in Eq. (12.1), we have

$$l = 2 + (8 - 1)4 = 2 + 28 = 30 \qquad ●$$

● **Example 3**   If the first term of an arithmetic progression is -3 and the eighth and last term is 11, find d and write the eight terms of the progression.

Solution   In this problem, $a = -3$, $n = 8$, and $l = 11$. If these values are substituted in Eq. (12.1), we have

$$
\begin{aligned}
11 &= -3 + (8 - 1)d \\
11 &= -3 + 7d & &\text{performing indicated operations} \\
-7d &= -14 & &\text{adding } -11 - 7d \text{ to each member} \\
d &= 2 & &\text{solving for } d
\end{aligned}
$$

Therefore, since $a = -3$, the first eight terms of the desired progression are $-3, -1, 1, 3, 5, 7, 9, 11$.   ●

Sum of an arithmetic progression

The sum of the arithmetic progression 1, 5, 9, 13 may be written as $1 + (1 + 4) + (1 + 2 \times 4) + (1 + 3 \times 4)$ or, starting with the last term, $13 + (13 - 4) + (13 - 2 \times 4) + (13 - 3 \times 4)$. Similarly we may write the sum of any arithmetic progression as

$$s = a + (a + d) + (a + 2d) + \cdots + [a + (n - 2)d] + [a + (n - 1)d]$$

or as

$$s = l + (l - d) + (l - 2d) + \cdots + [l - (n - 2)d] + [l - (n - 1)d]$$

Adding corresponding terms of these two equations gives

$$2s = (a + l) + [(a + d) + (l - d)] + \cdots + [a + (n - 1)d + l - (n - 1)d]$$
$$= (a + l) + (a + l) + \cdots + (a + l) = n(a + l)$$

Hence, dividing by 2, we obtain the formula

s using *l, a, n*
$$s = \frac{n}{2}(a + l) \qquad (12.2)$$

By (12.1), $a + l = a + [a + (n - 1)d] = 2a + (n - 1)d$, and thus

(s) using *d, a, n*
$$s = \frac{n}{2}[2a + (n - 1)d] \qquad (12.3)$$

is a second formula for *s*.

Our proof of (12.2) is essentially what Gauss did as a child when his teacher made him find the sum $1 + 2 + 3 + \cdots + 100$.

● **Example 4** Find the sum of all the even integers from 2 to 1000, inclusive.

Solution Since the even integers 2, 4, 6, etc., taken in order, form an arithmetic progression with $d = 2$, we can use Eq. (12.2), with $a = 2$, $n = 500$, and $l = 1000$, to obtain the desired sum. The substitution of these values in Eq. (12.2) yields

$$s = \tfrac{500}{2} \times (2 + 1000)$$
$$= 250 \times 1002 = 250{,}500$$

If any three of the quantities *l, a, n, d*, and *s* are known, the other two can be found by use of Eqs. (12.1) to (12.3), either separately or by solving two of them simultaneously.   ●

● **Example 5** If $a = 4$, $n = 10$, and $l = 49$, find *d* and *s*.

Solution Since each of Eqs. (12.1) and (12.2) contains *a, n*, and *l*, we can find *d* and *s* by using the formulas separately. If we substitute the given values for *a*, *n*, and *l* in Eq. (12.1), we get

$$49 = 4 + (10 - 1)d$$
$$49 = 4 + 9d \qquad \text{performing indicated operations}$$
$$-9d = -45 \qquad \text{adding } -49 - 9d \text{ to each member}$$
$$d = 5 \qquad \text{solving for } d$$

Similarly, by substituting in Eq. (12.2), we have

$$s = \tfrac{10}{2} \times (4 + 49)$$
$$= 5 \times 53$$
$$= 265$$

● **Example 6**   If *l* = 23, *d* = 3, and *s* = 98, find *a* and *n*.

Solution   If we substitute these values in Eqs. (12.1) and (12.2), we obtain

$$23 = a + (n - 1)3 \qquad\qquad (1)$$

from the former, and

$$98 = \frac{n}{2}(a + 23) \qquad\qquad (2)$$

from the latter. Each of these equations contains the two desired variables *a* and *n*. Hence we complete the solution by solving (1) and (2) simultaneously. If we solve (1) for *a*, we get

$$a = 23 - (n - 1)3$$
$$= 26 - 3n \qquad\qquad (3)$$

Upon substituting this expression for *a* in (2), we obtain

$$98 = \frac{n}{2}(26 - 3n + 23)$$

$$196 = n(49 - 3n) \qquad \text{multiplying by 2 and combining}$$
$$196 = 49n - 3n^2 \qquad \text{performing indicated operations}$$
$$3n^2 - 49n + 196 = 0 \qquad \text{adding } 3n^2 - 49n \text{ to each member}$$
$$n = 9\tfrac{1}{3} \text{ and } 7 \qquad \text{solving for } n$$

Since *n* cannot be a fraction, we discard $9\tfrac{1}{3}$ and have

$$n = 7$$

If we substitute 7 for *n* in (3), we obtain

$$a = 26 - (3 \times 7)$$
$$= 5$$

Hence, the progression consists of the seven terms 5, 8, 11, 14, 17, 20, and 23.

Arithmetic means

The terms between the first and last terms of an arithmetic progression are called *arithmetic means*. If the progression contains only three terms, the middle term is called *the arithmetic mean* of the first and last term. If the progression consists of three terms a, m, and l, then by the definition of an arithmetic progression, $m - a = l - m$, $2m = a + l$, and

$$m = \frac{a + l}{2}$$

Rule for finding arithmetic mean

Therefore, *the arithmetic mean of two numbers is equal to one-half their sum.*

● **Example 7**   Insert five arithmetic means between 6 and -10.

Solution   Since we are to find five means between 6 and -10, we have seven terms in all. Hence, $n = 7$, $a = 6$, and $l = -10$. Thus, by Eq. (12.1), we have

$$-10 = 6 + (7 - 1)d$$
$$-6d = 16 \qquad \text{adding } -6d + 10 \text{ to each member and combining terms}$$
$$d = -\tfrac{16}{6} = -\tfrac{8}{3} \qquad \text{solving for } d$$

and the progression is $\tfrac{18}{3} = 6, \tfrac{10}{3}, \tfrac{2}{3}, -\tfrac{6}{3}, -\tfrac{14}{3}, -\tfrac{22}{3}, -\tfrac{30}{3} = -10$.   ●

Harmonic progressions

The series formed by the reciprocals of the terms of an arithmetic progression is called a *harmonic progression*. For example, since two quantities are reciprocals if and only if their product is 1 and since $-5, -3, -1, 1, 3, 5, 7$ is an arithmetic series, it follows that $-\tfrac{1}{5}, -\tfrac{1}{3}, -1, 1, \tfrac{1}{3}, \tfrac{1}{5}, \tfrac{1}{7}$ is a harmonic progression. In general, if $w, x, y, \ldots, z$ is an arithmetic progression with no term equal to zero, then

$$\frac{1}{w}, \frac{1}{x}, \frac{1}{y}, \ldots, \frac{1}{z}$$

is a harmonic progression.

From the above definition, we can derive the following rule:

Rule for finding *n*th term of harmonic progression

To determine the nth term of a harmonic progression, we write the corresponding arithmetic progression, find the nth term of the arithmetic progression, and take its reciprocal.

The terms between any two terms of a harmonic progression are called

Harmonic means   *harmonic means.*

● **Example 8**   What is the tenth term of a harmonic progression if the first and third terms are $\frac{1}{2}$ and $\frac{1}{6}$? What is the harmonic mean of $\frac{1}{2}$ and $\frac{1}{6}$?

Solution   The first and third terms of the corresponding arithmetic progression are 2 and 6. Hence, $l = a + (n - 1)d$ becomes $6 = 2 + 2d$, and consequently, $d = 2$. Therefore, when $n = 10$, $l = 2 + (10 - 1)2 = 20$. Taking the reciprocal of 20, we find that the tenth term of the harmonic progression is $\frac{1}{20}$. Since $d = 2$, the first three terms of the arithmetic progression are 2, 4, 6; so the harmonic mean of $\frac{1}{2}$ and $\frac{1}{6}$ is $\frac{1}{4}$.   ●

Exercise 12.1

Write the n terms of the arithmetic progression described in each of Probs. 1 to 12.

1 $a = 1, d = 3, n = 4$ **2** $a = 4, d = 1, n = 3$
3 $a = 13, d = -2, n = 5$ **4** $a = 18, d = -3, n = 7$
5 $a = 2$, second term 4, $n = 4$ **6** $a = -3$, third term 3, $n = 5$
7 $a = -7$, third term -3, $n = 6$ **8** $a = -12$, second term -7, $n = 4$
9 Second term 6, third term 2, $n = 5$
10 Second term 5, fourth term -7, $l = -19$
11 Third term 4, fifth term 6, $n = 6$
12 Second term 9, fifth term 0, $l = -6$

In Probs. 13 to 24, find the quantity at the right by using the given values in the formulas for the last term and the sum of an arithmetic progression.

13 $a = 1, n = 6, d = 2; l$ **14** $l = -5, a = 7, n = 7; d$
15 $n = 5, d = 4, l = 5; a$ **16** $a = 13, d = -4, l = -7; n$
17 $a = 2, l = 14, n = 7; s$ **18** $a = 14, l = 2, s = 56; n$
19 $s = 51, n = 6, a = 1; l$ **20** $n = 7, l = 20, s = 77; a$
21 $a = 9, d = -3, s = 0; n$ **22** $s = 15, n = 6, a = 10; d$
23 $n = 7, d = -3, s = 28; a$ **24** $a = 17, s = 35, d = -4; l$

Find the two of the quantities l, a, n, d, and s that are missing in each of Probs. 25 to 36.

25 $a = -6, d = 3, n = 7$ **26** $a = 12, d = -3, n = 7$
27 $a = 18, n = 6, l = -2$ **28** $a = 19, l = -11, d = -5$
29 $a = 12, l = 3, s = 52.5$ **30** $n = 8, l = \frac{17}{3}, s = \frac{80}{3}$
31 $a = -\frac{4}{5}, d = -\frac{3}{5}, s = -\frac{116}{5}$ **32** $a = \frac{19}{7}, d = -\frac{2}{7}, s = 13$
33 $d = -3, l = -7, s = 3$ **34** $d = 5, l = 7, s = -2$
35 $l = 16, d = 3, s = 51$ **36** $l = -10, d = -4, s = 0$
37 Find the sum of all even integers between 5 and 29.
38 Find the sum of all multiples of 3 between 2 and 43.
39 Find the sum of the first n positive multiples of 4.
40 Find the sum of the first n positive multiples of 5.
41 Find the value of x if $2x + 1$, $x - 2$, and $3x + 4$ are consecutive terms of an arithmetic progression.

42 Find the value of x if $3x - 1$, $1 - 2x$, and $2x - 5$ are consecutive terms of an arithmetic progression.

43 Find the values of x and y if $3x - y$, $2x + y$, $4x + 3$, and $3x + 3y$ are consecutive terms of an arithmetic progression.

44 Show that if a, b, c and x, y, z are two arithmetic progressions, then $a + x$, $b + y$, $c + z$ is an arithmetic progression.

45 How many times will a clock strike in 24 h if it strikes only at the hours?

Physics   **46** If a compact body falls vertically 16 ft during the first second, 48 ft during the next second, 80 ft during the third, and so on, how far will it fall during the seventh second? During the first seven seconds?

Engineering   **47** A bomb was dropped from an altitude of 10,000 ft. Neglecting air resistance, find the time required for it to reach the ground. (See Prob. 46.)

48 A student made a grade of 64 on the first test, and did 7 points better on each succeeding test than on the previous one. What score was made on the fifth test, and what was the average grade of the five tests?

Economics   **49** A machine that cost $5800 depreciated 15 percent the first year, 13.5 percent the second, 12 percent the third, and so on. What was its value at the end of 9 years if all percentages apply to the original cost?

50 If Robin buys a painting on June 14, 1980 for $7000 and sells it on June 14, 1988 for $15,400, and the increase in value each year is $100 more than that of the previous year, find the value of the painting on June 14, 1986.

Film   **51** Find the approximate length of a motion picture film 0.01 cm thick if it is wound on a reel 6 cm in diameter that has a central core 2 cm in diameter. Consider the film as being wound in concentric circles.

Marketing   **52** A display of cans has 18 cans on the bottom row, 17 on the row above, 16 on the next row, and so on. If there are 12 rows of cans, how many cans are there in all?

53 Find three arithmetic means between 3 and 15.

54 Find five arithmetic means between 3 and 15.

55 Insert four arithmetic means between 10 and -10.

56 Insert six arithmetic means between 18 and 7.5.

57 Find the sixth term of the harmonic progression $\frac{1}{4}$, $\frac{1}{8}$, $\frac{1}{12}$, ...

58 Find the eighth term of the harmonic progression $\frac{3}{5}$, $\frac{3}{7}$, $\frac{1}{3}$, ...

59 Find the seventh term of the harmonic progression $\frac{2}{3}$, $\frac{4}{9}$, $\frac{1}{3}$, ...

60 Find the sixth term of the harmonic progression 3, 1, $\frac{3}{5}$, ...

61 What is the first term of a harmonic progression whose third term is $\frac{1}{5}$ and ninth term is $\frac{1}{8}$?

62 What is the eighth term of a harmonic progression whose second term is 2 and fifth term is -2?

63 What is the sixth term of a harmonic progression whose third term is -1 and eighth term is $\frac{1}{9}$?

64 What is the thirteenth term of a harmonic progression whose third term is 12 and eighth term is 2?

65 Show that the harmonic mean between a and b is $2ab/(a + b)$.

66 Show that log 2, log 6, log 18, log 54, log 162 is an arithmetic progression.

67 If $\sqrt{3}$, $\sqrt{6}$ is an arithmetic progression, what is d?

68 If $\sqrt{2}$, a_2, $\sqrt{18}$ is a harmonic progression, find a_2.

69 The second term of an arithmetic progression is equal to -5, and the difference between the sixth and the fourth one is 6. Show that the sum of the first 10 terms of the progression is 55.

70 Show that the sum $1 - 3 + 5 - 7 + 9 - 11 + \cdots - 171 + 173 = 87$.

71 Prove that if the numbers $\dfrac{2}{b+c}$, $\dfrac{2}{a+c}$, $\dfrac{2}{a+b}$ form an arithmetic progression, then the numbers a^2, b^2, c^2 constitute an arithmetic progression.

72 Show that if a, b, c, d, e form an arithmetic progression, then

$$a - 4b + 6c - 4d + e = 0$$

In Probs. 73 to 76, show that the arithmetic mean of the roots of $f(x) = 0$ is the same as the arithmetic mean of the roots of $g(x) = 0$.

73 $f(x) = x^3 + 4x^2 + x - 6 = (x-1)(x+2)(x+3) = 0$
 $g(x) = 3x^2 + 8x + 1 = 0$

74 $f(x) = 2x^3 + x^2 - 18x - 9 = (2x+1)(x-3)(x+3) = 0$
 $g(x) = 6x^2 + 2x - 18 = 0$

75 $f(x) = 12x^3 - 8x^2 - 3x + 2 = (3x-2)(2x-1)(2x+1) = 0$
 $g(x) = 36x^2 - 16x - 3 = 0$

76 $f(x) = 6x^3 + 5x^2 - 7x - 4 = (3x+4)(2x+1)(x-1) = 0$
 $g(x) = 18x^2 + 10x - 7$

12.2 Geometric progressions

Geometric sequence   A **geometric sequence** is one in which each term is r times the preceding one, where r is a fixed real number. In symbols

$$f(n + 1) = r \cdot f(n)$$

and $f(1)$ is given. For instance if $f(1) = a_1 = 5$ and $r = 2$, then the geometric sequence is

$$5, 10, 20, 40, 80, \ldots$$

In the next section, we will look at *infinite* geometric sequences. Here we will consider geometric sequences with a *finite* number of terms. Historically these are called **geometric progressions,** and the following special notation is used.

Geometric progression

a = first term = $f(1) = a_1$
l = last term = nth term = $f(n) = a_n$
r = common ratio = $a_2/a_1 = a_3/a_2$, etc.
n = the number of terms
s = the sum of the n terms
   = $f(1) + f(2) + \cdots + f(n) = a_1 + a_2 + \cdots + a_n$

Snarl   These letters can be rearranged to make the word "snarl." Note that the symbols for geometric progressions are the same as those for arithmetic progressions except for the use of r, the common ratio, in place of d, the common difference.

● **Example 1**   2,   6,   18,   54,   162   common ratio 3
                   3, -3,   3, -3,   3   common ratio -1
                  96,   24,   6, $\frac{3}{2}$, $\frac{3}{8}$ common ratio $\frac{1}{4}$ ●

Last term of a geometric progression

In terms of the above notation, the first six terms of a geometric progression in which the first term is a and the common ratio is r are

$$a,\ ar,\ ar^2,\ ar^3,\ ar^4,\ ar^5$$

We notice here that the exponent of r in the second term is 1 and that this exponent increases by 1 as we proceed from each term to the next. Hence, the exponent of r in any term is 1 less than the number of that term in the progression. Therefore, the nth term is ar^{n-1}. Thus, we have the formula

$$l = ar^{n-1} \tag{12.4}$$

● **Example 2**   Find the seventh term of the geometric progression $36, -12, 4, \ldots$.

Solution   In this progression, each term after the first is obtained by multiplying the preceding term by $-\frac{1}{3}$.† Hence, $r = -\frac{1}{3}$. Also $a = 36$, $n = 7$, and the seventh term is l. If we substitute these values in Eq. (12.4), we have

$$l = 36 \times (-\tfrac{1}{3})^{7-1} = \frac{36}{(-3)^6} = \frac{36}{729} = \frac{4}{81} \qquad ●$$

Sum of a geometric progression

If we add the terms of the geometric progression $a,\ ar,\ ar^2,\ \ldots,\ ar^{n-2}$, ar^{n-1}, we have

$$s = a + ar + ar^2 + \cdots + ar^{n-2} + ar^{n-1} \tag{1}$$

However, by use of an algebraic device, we can obtain a more compact formula for s. First, we multiply each member of (1) by r and get

$$rs = ar + ar^2 + ar^3 + \cdots + ar^{n-1} + ar^n \tag{2}$$

---

† Given any two consecutive terms of a geometric progression, we find r by dividing the second of the two by the first; in this case $-12 \div 36 = -\frac{1}{3}$.

Next, we notice that if we subtract the corresponding members of (1) and (2) and combine like terms, we have

$$s - rs = a + ar + \cdots + ar^{n-1} - (ar + ar^2 + \cdots + ar^{n-1} + ar^n)$$
$$= a - ar^n$$

or $$s(1 - r) = a - ar^n$$

By solving this equation for s, we obtain

s using *r*, *a*, *n* $$s = \frac{a - ar^n}{1 - r} \qquad r \neq 1 \tag{12.5}$$

If we multiply each member of Eq. (12.4) by r, we get $rl = ar^n$. Now, if we replace ar^n by rl in Eq. (12.5), we have

s using *l*, *a*, *r*, *n* $$s = \frac{a - rl}{1 - r} \qquad r \neq 1 \tag{12.6}$$

● **Example 3**   Find the sum of the first six terms of the progression 2, -6, 18, ...

Solution   In this progression, $a = 2$, $r = -3$, and $n = 6$. Hence, if we substitute these values in Eq. (12.5), we have

$$s = \frac{2 - [2 \times (-3)^6]}{1 - (-3)} = \frac{2 - (2 \times 729)}{1 + 3} = -364$$ ●

Note   If any three of the quantities s, n, a, r, and l are known, the other two can be found by using Eqs. (12.4) through (12.6).

● **Example 4**   The first term of a geometric progression is 3; the fourth term is 24. Find the tenth term and the sum of the first 10 terms.

Solution   To find either the tenth term or the sum, we must have the value of r. We can obtain this value by considering a progression made up of the first four terms defined above. Then we have $a = 3$, $n = 4$, and $l = 24$. If we substitute these values in Eq. (12.4), we get

$$24 = 3r^{4-1}$$
$$3r^3 = 24 \qquad \text{performing indicated operations}$$
$$r^3 = 8 \qquad \text{dividing both members by 3}$$
$$r = 2 \qquad \text{solving for } r$$

Now, by using Eq. (12.4) again with $a = 3$, $r = 2$, and $n = 10$, we get

$$l = 3 \times 2^{10-1} = 3 \times 512 = 1536$$

Hence, the tenth term is 1536.
To obtain s, we use Eq. (12.5), with $a = 3$, $r = 2$, and $n = 10$, and get

$$s = \frac{3 - (3 \times 2^{10})}{1 - 2} = \frac{3 - (3 \times 1024)}{-1} = \frac{3 - 3072}{-1} = 3069$$ ●

● **Example 5**   If $s = 61$, $l = 81$, and $n = 5$, find a and r.

Solution   If $s = 61$, $l = 81$, and $n = 5$, we can find a and r by use of any two of Eqs. (12.4) to (12.6). However, the work is easier if we use Eqs. (12.4) and (12.6). By substituting the given values in these, we get

$$81 = ar^4 \tag{3}$$

$$61 = \frac{a - 81r}{1 - r} \tag{4}$$

We solve (3) and (4) simultaneously by first solving (4) for a in terms of r and then substituting in (3).

$$
\begin{aligned}
61 - 61r &= a - 81r && \text{multiplying (4) by } 1 - r \\
a &= 61 + 20r && \text{solving for } a \tag{5} \\
81 &= (61 + 20r)r^4 && \text{substituting the value of } a \text{ in (3)} \\
20r^5 + 61r^4 - 81 &= 0 && \text{removing parentheses and adding} \\
& && -81 \text{ to each member}
\end{aligned}
$$

Now, by use of the methods of Chap. 11, we find the rational roots of this equation to be -3 and 1. The solution $r = 1$ must be discarded since, for this value, (4) is meaningless. However, if $r = -3$, then by (5), $a = 1$. Hence, the solution is $a = 1$, $r = -3$.   ●

Geometric means

The terms between the first and last terms of a geometric progression are called *geometric means*. If the progression contains only three terms, the middle term is called *the geometric mean* of the other two. If the three terms in the progression are a, g, and l, then by the definition of a geometric progression, $g/a = l/g$, so $g^2 = al$. Thus the second term, or the geometric mean between a and l, is

$$g = \pm\sqrt{al}$$

Rule for finding geometric mean   Hence, *the geometric mean between two quantities is either the square root of their product or its negative.*

● **Example 6**   Find the five geometric means between 3 and 192.

Solution   A geometric progression starting with 3, ending with 192, and containing five intermediate terms has seven terms. Hence, $n = 7$, $a = 3$, and $l = 192$. Therefore, by Eq. (12.4),

$$
\begin{aligned}
192 &= 3(r^{7-1}) \\
r^6 &= \tfrac{192}{3} \\
&= 64
\end{aligned}
$$

$$r = \sqrt[6]{64} = \pm 2 \qquad \text{solving for } r$$

Consequently, the two sets of geometric means of five terms each between 3 and 192 are 6, 12, 24, 48, 96, and -6, 12, -24, 48, -96. ●

● **Example 7** Find the two geometric means of $\frac{1}{2}$ and $\frac{1}{8}$.

Solution The geometric mean of $\frac{1}{2}$ and $\frac{1}{8}$ is $g = \pm \sqrt{\frac{1}{2} \times \frac{1}{8}} = \pm \sqrt{\frac{1}{16}} = \pm \frac{1}{4}$. ●

Exercise 12.2
geometric
progressions

Write the n terms of each geometric progression described in Probs. 1 to 8.

1 $a = 1, r = 2, n = 4$ **2** $a = 3, r = 3, n = 5$
3 $a = 2, r = -2, n = 5$ **4** $a = -2, r = -3, n = 4$
5 $a = 81$, fourth term 3, $n = 6$ **6** $a = 128$, second term -64, $n = 8$
7 Second term 1, fourth term $\frac{16}{9}$, last term $\frac{64}{27}$
8 Third term $\frac{2}{3}$, fourth term $\frac{4}{9}$, last term $\frac{16}{81}$

In Probs. 9 to 20, use Eqs. (12.4) to (12.6) to find the quantity on the right.

9 $a = 2, r = 3, n = 5, l$ **10** $a = 3, r = 2, l = 96, n$
11 $r = -1, n = 7, l = -1, a$ **12** $l = 81, a = 1, r = 3, n$
13 $a = 2, r = 4, n = 4, s$ **14** $a = \frac{1}{3}, r = 3, n = 5, s$
15 $r = -4, n = 6, s = -819, a$ **16** $a = -1, s = 182, r = -3, n$
17 $a = 64, r = \frac{1}{2}, l = 1, s$ **18** $a = 27, l = \frac{16}{3}, r = \frac{2}{3}, s$
19 $a = 125, r = -\frac{1}{5}, s = 104\frac{1}{5}, l$ **20** $a = 64, l = 1, s = 43, r$

In Probs. 21 to 36, find the two of the quantities s, n, a, r, and l that are missing.

21 $a = 2, r = 3, n = 5$ **22** $a = 3, r = 2, n = 5$
23 $a = 8, r = -\frac{3}{2}, l = \frac{81}{2}$ **24** $a = 162, r = -\frac{1}{3}, l = 2$
25 $a = 343, r = \frac{1}{7}, l = 1$ **26** $n = 4, r = 3, l = 3$
27 $n = 9, a = 256, l = 1$ **28** $n = 7, a = \frac{1}{8}, l = 8$
29 $s = 242, a = 2, r = 3$ **30** $s = 781, n = 5, r = \frac{1}{5}$
31 $s = 400, r = \frac{1}{7}, l = 1$ **32** $s = 1, n = 7, l = 1$
33 $l = 12, s = 9, n = 3$ **34** $r = -\frac{3}{2}, l = \frac{81}{2}, s = \frac{55}{2}$
35 $a = 5, n = 3, s = 65$ **36** $a = 8, l = \frac{1}{8}, s = \frac{127}{8}$
37 Find the sum of all of the integral powers of 2 between 5 and 500.
38 Find the sum of the first seven integral powers of 3 beginning with 3.
39 Find the sum $1 + 2 + 4 + 8 + \cdots + 2^n$.
40 Find the sum $\dfrac{1}{4} + \dfrac{1}{16} + \dfrac{1}{64} + \cdots + \dfrac{1}{4^n}$.
41 For which values of k are $2k$, $5k + 2$, and $20k - 4$ consecutive terms of a geometric progression?
42 If 1, 4, and 19 are added to the first, second, and third terms, respectively, of an arithmetic progression with $d = 3$, a geometric progression is obtained. Find the arithmetic progression and the common ratio of the geometric progression.

43 Show that if

$$\frac{1}{y-x} \qquad \frac{1}{2y} \qquad \text{and} \qquad \frac{1}{y-z}$$

form an arithmetic progression, then x, y, and z form a geometric progression.

44 Show that if a, b, c and x, y, z are two geometric progressions, then ax, by, cz is also a geometric progression.

Fish bait **45** Twelve men are fishing. If the first is worth $1000, the second $2000, the third $4000, and so on, how many of the men are millionaires?

Probate **46** Mrs. Timken willed one-third of her estate to one person, one-third of the remainder to a second, and so on, until the fifth received $1600. What was the value of the estate?

Biology **47** The number of bacteria in a culture doubles every 2 h. If there were n bacteria present at noon one day, how many were there at noon the next day?

Physics **48** The first stroke of a pump removes one-fourth of the air from a bell jar, and each stroke thereafter removes one-fourth of the remaining air. What part of the original amount is left after six strokes?

Gambling **49** Mr. Hughes bets $1 on the first poker hand, $2 on the second, $4 on the third, and so on. If he loses nine hands in a row, how much does he lose? If he then wins the tenth hand, what is his net profit or loss?

Finance **50** If $100 is put in a savings account at the end of each 6 months, how much money is in the account at the end of 6 years if the bank pays 6 percent interest per year compounded semiannually?

Genealogy **51** If there were no duplications, how many ancestors did Jennifer have in the immediately preceding seven generations?

Management **52** Each year, a $30,000 machine depreciates by 20 percent of its value at the beginning of that year. Find its value at the end of the fifth year.

53 What are the two geometric means between 2 and 54?

54 What are the four geometric means between $\frac{2}{9}$ and $\frac{27}{16}$?

55 Find two sets of three geometric means between 2 and 32.

56 Find two sets of five geometric means between 2 and $\frac{1}{32}$.

Classify the progression in each of Probs. 57 to 64 as arithmetic, geometric, or harmonic and give the next two terms.

57 $\frac{2}{3}, \frac{4}{9}, \frac{8}{27}, \frac{16}{81}$ **58** $\frac{2}{3}, \frac{4}{9}, \frac{1}{3}, \frac{4}{15}$ **59** $\frac{2}{3}, \frac{4}{9}, \frac{2}{9}, 0$ **60** $\frac{1}{20}, \frac{3}{10}, \frac{11}{20}, \frac{4}{5}$

61 $\frac{1}{4}, 1, -\frac{1}{2}, -\frac{1}{5}$ **62** $2, 3, 4\frac{1}{2}, \frac{27}{4}$ **63** $\frac{1}{6}, \frac{2}{3}, \frac{7}{6}, \frac{5}{3}$ **64** $\frac{1}{7}, \frac{2}{11}, \frac{1}{4}, \frac{2}{5}$

65 Show that if $0 < a < b$ and the plus sign is used for the geometric mean, then the harmonic mean of a and b is less than their geometric mean, and their geometric mean is less than their arithmetic mean.

66 Suppose the distance between two houses is d mi, and that the trip in one direction is made with an average speed of v mi/h. Show that if w is the average speed in the other direction, then the average speed for the round trip is at most $(v+w)/2$. *Hint:* The total distance is clearly $2d$.

Show that the total time is $d/v + d/w$, and then use

$$\frac{\text{Distance}}{\text{Time}} = \text{speed}$$

67 Three numbers form a geometric progression. If 8 is added to the second number, then these numbers will constitute an arithmetic progression. If 64 is then added to the third number, the resulting numbers will form a geometric progression once again. Find these three numbers.

68 Prove that

$$\log \sqrt{ab} = \frac{\log a + \log b}{2}$$

In words, this says that "the logarithm of the geometric mean" is equal to "the arithmetic mean of the logarithms."

12.3 Infinite geometric series

Sequence The **infinite geometric sequence** with first term a and ratio r is

$$a, \ ar, \ ar^2, \ ar^3, \ ar^4, \ \ldots, \ ar^{n-1}, \ \ldots$$

Series The associated **infinite geometric series** is the indicated sum

$$a + ar + ar^2 + ar^3 + ar^4 + \cdots$$

We cannot literally "add" an infinite number of terms, but we can most certainly add a finite number of the terms. The sum of the first n terms is denoted by S_n, and by Eq. (12.5) that sum is

$$S_n = \frac{a - ar^n}{1 - r} = \frac{a(1 - r^n)}{1 - r} = \frac{a}{1 - r}(1 - r^n)$$

Now if $|r| < 1$, then r^n approaches 0 as n becomes larger and larger (see Prob. 49), and it follows that S_n approaches

$$\frac{a}{1 - r}(1 - 0) = \frac{a}{1 - r}$$

These ideas are treated more fully in a calculus course. The standard notation for this is

$$S = \lim_{n \to \infty} S_n = \frac{a}{1 - r} \qquad (12.7)$$

and we call $S = a/(1 - r)$ the "sum" of the infinite geometric series.

● **Example 1** Find the sum of the infinite geometric series $1 + \frac{1}{2} + \frac{1}{4} + \cdots$, where the dots indicate that there is no end to the series.

Solution   In this series, $a = 1$ and $r = \frac{1}{2}$. Hence, by Eq. (12.7),

$$S = \frac{a}{1 - r} = \frac{1}{1 - \frac{1}{2}} = \frac{1}{\frac{1}{2}} = 2$$

●

A nonterminating, repeating decimal fraction is an illustration of an infinite geometric series with $-1 < r < 1$. For example,

$$0.232323 \cdots = 0.23 + 0.0023 + 0.000023 + \cdots$$

The terms on the right form a geometric series with $a = 0.23$ and $r = \frac{1}{100}$

By use of Eq. (12.7), we can express any repeating decimal fraction as a common fraction by the method illustrated in the following example. In fact, a decimal is repeating (or terminates) if and only if it is equal to a rational number.

● **Example 2**   Show that $0.333 \cdots = \frac{1}{3}$. (1)

Solution   The decimal fraction $0.333 \cdots$ can be expressed as the series

$$0.3 + 0.03 + 0.003 + \cdots$$

in which $a = 0.3$ and $r = 0.1$. Hence, by Eq. (12.7), the sum S is

$$S = \frac{a}{1 - r} = \frac{0.3}{1 - 0.1} = \frac{0.3}{0.9} = \frac{3}{9} = \frac{1}{3}$$

●

Problems 29 to 32 are related to Example 2.

● **Example 3**   Express $3.2181818 \cdots$ as a fraction.

Solution   The given number can be expressed as 3.2 plus a geometric progression with $a = 0.018$ and $r = 0.01$. Thus, from Eq. (12.7),

$$3.2181818 \cdots = 3.2 + 0.018 + 0.00018 + 0.0000018 + \cdots$$
$$= 3.2 + \frac{0.018}{1 - 0.01} = 3.2 + \frac{0.018}{0.99}$$
$$= 3 + \frac{1}{5} + \frac{1}{55}$$
$$= 3\frac{12}{55} = \frac{177}{55}$$

●

Exercise 12.3   Find the sum of each infinite geometric series determined by the data in Probs. 1 to 12.

1 $a = 7, r = \frac{6}{7}$ **2** $a = 7, r = \frac{1}{3}$ **3** $a = 7, r = -\frac{6}{7}$

4 $a = 7, r = -\frac{1}{3}$ **5** $a = 5, r = -\frac{4}{5}$ **6** $a = 15, r = -\frac{2}{3}$

7 $a = 9, r = -\frac{1}{8}$ **8** $a = 12, r = \frac{1}{4}$

9 $a = 11$, second term $-\frac{11}{6}$ **10** $a = 15$, third term $\frac{135}{16}$

11 $a = 32$, fourth term -4 **12** Second term 27, fifth term 1.

Find a or r, whichever is missing.

13 $r = \frac{1}{3}, s = \frac{9}{2}$ **14** $r = -\frac{2}{5}, s = \frac{25}{7}$ **15** $a = 16, s = 10$ **16** $a = 7, s = 9$

If n is a positive integer, find the sum of all the numbers of the form given in each of Probs. 17 to 20.

17 $(\frac{1}{3})^n$ **18** $(\frac{2}{3})^n$ **19** $(-\frac{2}{5})^n$ **20** $(-\frac{3}{4})^n$

Express the repeating decimal in each of Probs. 21 to 28 as a rational number in lowest terms.

21 $0.444\cdots$ **22** $0.2424\cdots$ **23** $2.343434\cdots$ **24** $1.414141\cdots$
25 $4.1222\cdots$ **26** $2.2111\cdots$ **27** $6.54848\cdots$ **28** $2.0124124\cdots$

Problems 29 to 32 give four different ways of showing that $0.999\cdots = 1$.

29 Treat $0.999\cdots$ as in Probs. 21 to 28.
30 Let $x = 0.999\cdots$, so $10x = 9.999\cdots$. Then subtract and solve for x.
31 Show that if $(a + b)/2 = a$, then $a = b$. Apply this with $a = 0.999\cdots$ and $b = 1.000\cdots$.
32 Multiply both members of (1) in Example 2 by 3.

In each of Probs. 33 to 36, find the sum of the series. Verify by calculation that the sum of an odd number of terms is larger than S, and the sum of an even number of terms is less than S.

33 $1 - \dfrac{1}{3} + \dfrac{1}{3^2} - \dfrac{1}{3^3} + \cdots$ **34** $1 - \dfrac{1}{4} + \dfrac{1}{4^2} - \dfrac{1}{4^3} + \cdots$

35 $\frac{2}{5} - (\frac{2}{5})^2 + (\frac{2}{5})^3 - (\frac{2}{5})^4 + \cdots$ **36** $2 - 1 + \frac{1}{2} - \frac{1}{4} + \cdots$

Physics **37** If the first arc made by the tip of a pendulum is 12 cm and each arc thereafter is 0.995 as long as the one just before it, how far does the tip move before coming to rest?

Physics **38** If a ball rebounds $\frac{3}{5}$ as far as it falls, how far will it travel before coming to rest if it is dropped from a height of 30 m?

Engineering **39** The motion of a particle through a certain medium is such that it moves $\frac{2}{3}$ as far each second as in the preceding second. If it moves 6 m the first second, how far will it move before coming to rest?

Finance **40** An alumna gave an oil field to her university. If the university received $230,000 from the field the first year and $\frac{2}{3}$ as much each year thereafter as during the immediately preceding year, how much did the college realize?

Agriculture **41** Assume that potatoes shrink one-half as much each week as during the previous week. If a dealer stores 1000 kg when the price is p cents per kilogram, and if the weight decreases to 950 kg during the first week, for which values of p can he afford to hold the potatoes until the price rises to $p + 1$ cents per kilogram?

Biology   **42** A hamster receives a dose of 3 mg of a compound and then $\frac{2}{3}$ as much as the previous dose at the end of every 3 h. What is the maximum amount of the compound it will receive?

Geometry   **43** A series of squares is drawn by connecting the midpoints of the sides of a given square, then the midpoints of the square thus drawn, and so on. Find the sum of the areas of all the squares if the original square had sides of 10 in.

Geometry   **44** Find the sum of the perimeters of the squares of Prob. 43.

Infinite series   Find the values for x in each of Probs. 45 to 48 in order for the sum to exist. Also find the sum.

45 $\dfrac{1}{2x-1} + \dfrac{1}{(2x-1)^2} + \dfrac{1}{(2x-1)^3} + \cdots$

46 $\dfrac{2}{3x-2} + \dfrac{4}{(3x-2)^2} + \dfrac{8}{(3x-2)^3} + \cdots$

47 $\dfrac{1}{x^2+2} + \dfrac{1}{(x^2+2)^2} + \dfrac{1}{(x^2+2)^3} + \cdots$

48 $\dfrac{3}{x^2+1} + \dfrac{3}{(x^2+1)^2} + \dfrac{3}{(x^2+1)^3} + \cdots$

49 We stated in this section that if $|r| < 1$, then r^n approaches 0 as n becomes larger and larger. Verify the following steps, which show that if $n > 52$, then $(0.7)^n < 10^{-8}$.

$$(0.7)^n < 10^{-8}$$

if and only if $\quad\quad\quad \log(0.7)^n < \log(10^{-8})$

if and only if $\quad\quad\quad\quad\quad n \log 0.7 < -8$

if and only if $\quad\quad\quad\quad n(\log 7 - 1) < -8$

if and only if $\quad\quad\quad\quad\quad\quad n > \dfrac{-8}{\log 7 - 1} \approx 51.6$

12.4   Simple and compound interest

Interest   A person who uses another's money pays a sum, called *interest,* for the use. The amount of interest paid depends on the sum borrowed, the interest rate per period, and the number of periods used. The sum borrowed is

Principal   called the *principal,* and the length of time for which the principal is

Term   borrowed is called the *term.* If the interest is paid as due at the end of each interest period, we say that *simple interest* is being used. If, however, the interest is not paid when due but is added to the principal to form a new principal, we say that *compound interest* is being used.

● **Example 1**   If $1000 is borrowed at 8 percent per year, then the interest to be paid at the end of each year is $1000(0.08) = $80 if simple interest is being used.

If, however, compound interest is being used, the $80 is added to the principal and the principal for the second year is $1080. Consequently, the interest for the second year is paid on $1080 and is $1080(0.08) = $86.40. Hence, the principal for the third year is $1080 + $86.40 = $1166.40 and the interest is $1166.40(0.08) = $93.31. Consequently, the amount that should be repaid at the end of 3 years is $1166.40 + $93.31 = $1259.71. The reader should verify the fact that this is $1000(1.08)^3 by using multiplication, logarithms, a calculator, or Eq. (12.8). ●

Values by calculator or table   The values called for in interest and annuity problems in this section and the next can be found by using either a calculator or the tables in the Appendix.

Interest period   The time between consecutive compoundings is called the *interest period*. *It may be a year but need not be.* We will now find the sum to which a principal of P accumulates if invested for n years at a rate of i per year compounded annually.

Period	Principal	Interest	Accumulated value
1	P	Pi	$P + Pi = P(1 + i)$
2	$P(1 + i)$	$P(1 + i)i$	$P(1 + i) + P(1 + i)i = P(1 + i)^2$
3	$P(1 + i)^2$	$P(1 + i)^2 i$	$P(1 + i)^2 + P(1 + i)^2 i = P(1 + i)^3$
.			
.			
n	$P(1 + i)^{n-1}$	$P(1 + i)^{n-1} i$	$P(1 + i)^{n-1} + P(1 + i)^{n-1} i = P(1 + i)^n$

Consequently, if we designate the *accumulated value* by S, we have

Accumulated value
$$S = P(1 + i)^n \qquad (12.8)$$

● **Example 2**   Find the accumulated value if $1700 is invested at 6 percent compounded annually for 13 years.

Solution   If we substitute $P = \$1700$, $i = 6$ percent, and $n = 13$ in (12.8), we get

$$\begin{aligned} S &= \$1700(1.06)^{13} \\ &= \$1700(2.1329) \qquad \text{from Table A.3} \\ &= \$3625.93 \end{aligned}$$
●

Note   If the investment had been at simple interest, the yearly interest would have been $1700(0.06) = $102; hence the interest for 13 years would have been 13($102) = $1326 and the accumulated value would have been $1700 + $1326 = $3026.

If the interest is compounded other than annually, we call the stated

Nominal rate   annual rate the *nominal rate*. If the nominal rate is j and if it is compounded

Periodic rate m times per year, then the rate per period or the *periodic rate* is j/m, and by (12.8), 1 invested at this rate will in 1 year be $1(1 + j/m)^m$. Hence the

Effective rate   interest for 1 year is $(1 + j/m)^m - 1$, and it is called the *effective rate e.*
                  Thus the rate e such that

$$1 + e = \left(1 + \frac{j}{m}\right)^m$$

(12.9)

is called the *effective rate* where j is the nominal rate and m is the number
of times per year it is compounded. The effective rate is the simple interest
rate which gives the same interest after 1 year as the nominal rate j
compounded m times per year.

● **Example 3**   The effective rate corresponding to the rate 10 percent compounded quarterly
                   is determined by $m = 4$, $j = 0.10$, $j/m = 0.10/4 = 0.025$. Hence by (12.9)

$$1 + e = 1.025^4 = 1.1038$$

and e is $0.1038 = 10.38$ percent.
    If we replace $1 + i$ in (12.8) by $(1 + j/m)^m$, we find that

$$S = P\left(1 + \frac{j}{m}\right)^{mn}$$

(12.8a)

is the amount to which P will accumulate in n years at the nominal rate j
compounded m times per year.   ●

● **Example 4**   To what sum will $1000 accumulate in 5 years if invested at 8 percent
                   compounded (a) semiannually? (b) quarterly?

Solution   The principal is $1000, the term is $n = 5$ years, the nominal rate $j = 8$
            percent.
            (a) Here $m = 2$ since the interest is compounded each 6 months; hence
                 $j/m = 0.08/2 = 0.04$, and by use of (12.8a) and Table A.3, the
                 accumulated value is

$$S = \$1000(1.04)^{10} = \$1000(1.4802) = \$1480.20$$

            (b) Here $j/m = 0.08/4 = 0.02$, $m = 4$, and hence

$$S = \$1000(1.02)^{20} = \$1000(1.4859) = \$1485.90$$

●

        Equation (12.8a) is solved for S, but it involves P, j, m, and n; any one
of them can be found if the other three and S are known.

● **Example 5**   How much must be invested at 6 percent compounded quarterly in order
                   to accumulate to $2763 in 3 years?

Solution   We are given that $S = \$2763$, $n = 3$, $j = 6$ percent, $m = 4$ and want to
            find P. By use of (12.8a), we have

$$\$2763 = P(1.015)^{12}$$
$$P = \$2763(1.015)^{-12}$$
$$= \$2763(0.83639) \qquad \text{by Table A.4}$$
$$= \$2310.95$$

●

Example 6   How long will it take \$616.63 to accumulate to \$2000 if compounded semiannually at 8 percent?

Solution   We substitute $P = 616.63$, $S = 2000$, $j = 8$ percent, and $m = 2$ in (12.8a), and have

$$2000 = 616.63(1.04)^{2n}$$

$$1.04^{2n} = \frac{2000}{616.63}$$

$$= 3.2434$$

Now using Table A.2, we get $2n = 30$; hence $n = 15$. Consequently, the required time is 15 years.   ●

Example 7   If \$1500 accumulates to \$3305.70 in 8 years, find the nominal rate compounded quarterly and the simple interest rate.

Solution   If we substitute the given values in (12.8a), we get

$$3305.70 = 1500\left(1 + \frac{j}{4}\right)^{32}$$

since $S = \$3305.70$, $P = \$1500$, $m = 4$, and $n = 8$. Therefore

$$\left(1 + \frac{j}{4}\right)^{32} = \frac{3305.70}{1500} = 2.2038$$

Now, from Table A.2, we get

$$\frac{j}{4} = 2.5 \text{ percent}$$

Consequently, the nominal rate is $j = 4(2.5$ percent$) = 10$ percent. Since $S = \$3305.70$, $P = \$1500$, and the term is 8 years, the annual simple interest is ($\$3305.70 - \$1500)/8 = \$1805.70/8 = \$225.71$ and the simple interest rate is \$225.71/\$1500 = 15.05 percent.   ●

Management   It is interesting to compare three different ways of figuring interest. If P dollars are invested at a rate i per period for n periods, then during the nth period, $n \le t < n + 1$, we have

$$S = P(1 + i)^n$$

The graph is a step function. See Fig. 12.1. If we replace the discrete variable n by the continuous variable t, allowing fractional parts of a year, we have

$$S = P(1 + i)^t$$

This graph connects the corners of the above function.
    If compounding occurs m times per year, then after n years we have

$$S = \left(1 + \frac{i}{m}\right)^{mn} \qquad \text{by Eq. (12.8a)}$$

FIGURE 12.1

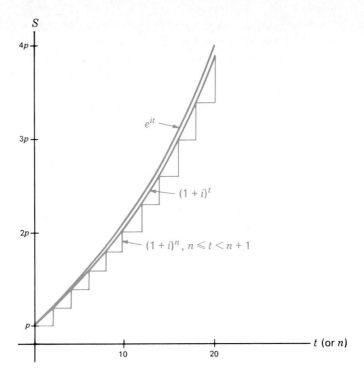

As m becomes infinite, S approaches e^{in} since we have

$$S = \left[\left(1 + \frac{i}{m}\right)^{m/i}\right]^{in}$$

$$= \left[\left(1 + \frac{1}{k}\right)^{k}\right]^{in} \qquad \text{with } k = \frac{m}{i}$$

See Sec. 8.1 after Example 2, where $e \approx 2.718$. We call this situation *continuous compounding,* and in this case have, letting $t = n$ to allow fractional parts of a year,

$$S = e^{it} \qquad \text{approximately}$$

This graph is also in Fig. 12.1. Notice that it is always above the other two. We have used $i = 7$ percent in Fig. 12.1

Exercise 12.4   Find the interest and accumulated value if the principal in each of Probs. 1 to 4 is invested at simple interest at the given rate for the given time.

1 $1200 at 11 percent for 1 year   **2** $700 at 12 percent for 2 years
3 $5762 at 9 percent for 10 months   **4** $8888 at 8 percent for 8 months
5 Mrs. Schaffer borrowed some money for 8 months at 12 percent simple interest and had to repay $903.39. How much did she borrow?

6 How much can one borrow so as to repay it with $1000 after 6 months if the simple interest rate is 11 percent?

7 What simple interest rate did Reverend Alverson pay if $2558.60 was required to repay a loan of $2326 after 15 months?

8 How long did Mrs. Bradley keep the loan of $1500 if $1815 was required to repay it at 12 percent simple interest?

Find the value of the one of S, P, j, m, and n that is missing in each of Probs. 9 to 24. The answers were found by using values from the appendix tables in this book. Answers by calculator should be very close.

9 P = $560, j = 6 percent, m = 2, n = 17 years
10 P = $2350, j = 5 percent, m = 2, n = 20 years
11 P = $4600, j = 6 percent, m = 4, n = 12 years
12 P = $7500, j = 8 percent, m = 4, n = 9 years
13 S = $5200, j = 5 percent, m = 2, n = 8 years
14 S = $2100, j = 8 percent, m = 2, n = 23 years
15 S = $1120, j = 6 percent, m = 4, n = 11 years
16 S = $480, j = 8 percent, m = 4, n = 7 years
17 P = $1000, S = $1194.10, j = 6 percent, m = 2
18 P = $1800, S = $2771.10, j = 8 percent, m = 2
19 P = $540, S = $2836.78, j = 10 percent, m = 2
20 P = $770, S = $1768.84, j = 4 percent, m = 2
21 S = $819.30, P = $500, m = 2, n = 10 years
22 S = $512.07, P = $300, m = 3, n = 9 years
23 P = $750, S = $1206.30, m = 3, n = 8 years
24 P = $470, S = $1204.75, m = 2, n = 12 years

Find the effective rate that is equivalent to the nominal rate indicated in each of Probs. 25 to 32.

25 j = 6 percent, m = 2   26 j = 4 percent, m = 2
27 j = 5 percent, m = 2   28 j = 3 percent, m = 2
29 j = 8 percent, m = 4   30 j = 6 percent, m = 4
31 j = 10 percent, m = 4   32 j = 9 percent, m = 3

33 On the day of his son's birth, a father deposited $1000 to the credit of his boy in a bank that pays 4 percent compounded semiannually. How much will be to the son's credit when he is ready for college on his eighteenth birthday?

34 Mr. Nobl offered Marie the choice between $13,000 cash now for a boat and $17,450 to be paid in 5 years. Which should she choose, if money is worth 6 percent compounded semiannually? How much better off will she be 5 years after the sale as a result of her choice?

35 A store offered a refrigerator for $600 cash now or $638 at the end of a year. Which should a buyer choose if money is worth 6 percent compounded quarterly? How much better off will the buyer be in a year as a result of his choice?

36 The Hofbauers leased a car and were told the cost could be $1080 at the beginning of the year or $1169 at the end. Which should the Hofbauers choose if money is worth 8 percent compounded quarterly? How much better off at the end of the year will they be as a result of their choice?

37 A rancher sold his holdings for $80,000 cash, $70,000 due in 5 year, and $90,000 due in 10 years. What was the equivalent cash price if money is worth 5 percent?

38 To be able to buy a lot, Ms. Jager deposited $1200 in a savings bank on June 1 for 3 consecutive years. How much was available 1 year after the last payment if money is worth 4 percent?

39 Mr. Johnson bought a truck by paying $7000 cash and $3000 at the end of each year for 4 consecutive years. What was the cash value of the truck if money is worth 4 percent compounded semiannually?

40 As payment for a lot, Tompkins accepted $8000 in cash and $5000 at the end of each year for 3 years. What price payable 3 years after the sale would have been equivalent if money was worth 6 percent compounded quarterly?

In Probs. 41 to 44, use the appendix tables and interpolate, or use a calculator.

41 How long will it take for a sum of money to double at 5 percent effective?

42 How long will be required for a sum of money to double at 4 percent compounded semiannually?

43 How much longer does it take for money to double at 6 percent effective than at 6 percent, $m = 4$?

44 How much longer is required for money to triple at 5 percent effective than at 5 percent, $m = 2$?

45 At what effective rate will $1000 accumulate to $1665.10 in 13 years?

46 At what rate compounded semiannually will $1400 accumulate to $1792.14 in 5 years?

47 Jones invested $800 for 9 years at an effective rate and then had $1351.60 to his credit. What rate was earned?

48 What rate compounded semiannually must be earned in order for $1600 to accumulate to $2754.56 in 11 years?

12.5 Annuities

Annuity

Payment interval
Annuity due
Ordinary annuity

Interest period

A sequence of equal payments at equal intervals is called an *annuity*. Thus the payments for rent and on house notes are both annuities. The time between consecutive payments is the *payment interval* or period. If each payment is made at the beginning of a payment interval, we have an *annuity due;* if made at the end of the payment period, we have an *ordinary annuity*. The length of time between consecutive additions of interest to the then current principal is called the *conversion period* or interest period. We will consider only annuities in which the payment period and interest period coincide. The term of an annuity is obtained by multiplying the number of payments by the payment interval.

● **Example 1**   If a payment is made at the end of each quarter until 17 payments have been made, the term is $17(3) = 51$ months and we have an ordinary annuity. If the rate is 8 percent compounded quarterly, the periodic interest rate is 2 percent.   ●

   We will now develop a formula for the value at the end of the term of an annuity of 1 at the end of each period for n periods with a rate of i per period where a period may be a year but need not be. The first payment draws interest for $(n - 1)$ periods, the second for $(n - 2)$, . . . , the next to last for 1 period, and the last for no time. Consequently, the sum of the values of the payments at the end of the term is

Geometric
progression
$$(1 + i)^{n-1} + (1 + i)^{n-2} + (1 + i)^{n-3} + \cdots$$
$$+ (1 + i)^2 + (1 + i)^1 + (1 + i)^0$$

This is a geometric progression with

$$a = (1 + i)^{n-1} \qquad r = (1 + i)^{-1} \qquad l = (1 + i)^0 = 1 \qquad \text{and} \qquad n = n$$

We will designate the sum of this progression by $s_{\overline{n}|i}$. By use of $s = (a - rl)/(1 - r)$, we get

$$s_{\overline{n}|i} = \frac{(1 + i)^{n-1} - (1 + i)^{-1} \cdot 1}{1 - (1 + i)^{-1}}$$

$$= \frac{(1 + i)^n - 1}{1 + i - 1} \qquad \text{multiplying by } \frac{1 + i}{1 + i}$$

Therefore, we know that

Accumulated value
$$s_{\overline{n}|i} = \frac{(1 + i)^n - 1}{i} \tag{12.10}$$

is the accumulated value of an ordinary annuity of 1 per period for n periods at rate i per period.

   If the periodic payment is R in place of 1, we have

$$S = R \cdot s_{\overline{n}|i} \tag{12.11}$$

● **Example 2**   How much did Grandpa accumulate by depositing $150 at the end of each quarter for 9 years into a fund that pays 8 percent compounded quarterly?

Solution   These payments form an ordinary annuity with $R = \$150$, $i = 2$ percent, and $n = 9 \times 4 = 36$. We now substitute in (12.11) and have

$$S = \$150 \cdot s_{\overline{36}|2\%} = (\$150)\frac{1.02^{36} - 1}{0.02}$$

$$= \$150(51.9944) \qquad \qquad \text{by use of Table A.4}$$

$$= \$7799.16$$ ●

We can find the accumulated value of an *annuity due* by multiplying the accumulated value of an *ordinary annuity* by $1 + i$ since each payment is made one period earlier for an annuity due than for an ordinary annuity. Therefore, the accumulated value of an annuity due of R per period for n periods at rate i per period is

$$S(\text{due}) = (1 + i)Rs_{\overline{n}|i} \qquad (12.12)$$

● **Example 3**   If, in Example 2, Grandpa had made each deposit at the beginning of the period, what would the accumulated value have been?

Solution   This annuity differs from the one in Example 2 only in that this is an annuity due and that one is an ordinary annuity. Therefore, the accumulated value is

$$S(\text{due}) = \$7799.16(1.02)$$
$$= \$7955.14 \qquad ●$$

The value of an annuity at the beginning of its term is referred to as the
Present value   *present value* of the annuity. The wording is unfortunately used because present value does not always refer to "now." If the accumulated value of an annuity is S, then its present value is the sum of money A which would grow to S if it were accumulated at compound interest over the term of the annuity.

Thus, by (12.8), $S = A(1 + i)^n$. Consequently, we have

$$A = S(1 + i)^{-n}$$

$$a_{\overline{n}|i} = \dfrac{1 - (1 + i)^{-n}}{i} \qquad\qquad = R\dfrac{(1 + i)^n - 1}{i} \cdot (1 + i)^{-n} \qquad \text{by (12.11)}$$

$$= R \cdot \dfrac{1 - (1 + i)^{-n}}{i} = R \cdot a_{\overline{n}|i}$$

Therefore

$$A = R \cdot \dfrac{1 - (1 + i)^{-n}}{i} = R \cdot a_{\overline{n}|i} \qquad (12.13)$$

is the present value of an ordinary annuity of R per period for n periods at rate i per period.

● **Example 4**   On the day that Jo Beth graduated from college, Grandma gave her the choice between $500 at the end of each 6 months for 10 years or a single payment of $6900 on graduation day. Which should Jo Beth choose if money is worth 8 percent compounded semiannually?

Solution   The payments form an ordinary annuity with $R = \$500$, $j = 8$ percent, $m = 2$, and $n = 10$. Consequently, the present value of the annuity is

$$A = \$500 \cdot a_{\overline{20}|4\%}$$
$$= \$500(13.5903) \qquad \text{by Table A.6}$$
$$= \$6795.15$$

She should choose the single payment of $6900.   ●

Example 5   Find the sale price of a house if it was paid for by a cash payment of $75,000 and $2000 at the end of each quarter for 12 years. Assume that the interest rate is 10 percent compounded quarterly.

Solution   The sale price was the down payment of $75,000 plus the present value of the ordinary annuity formed by the payments. In it, $R = \$2000$, $i = 2\frac{1}{2}$ percent, and $n = 4 \times 12 = 48$. Therefore,

$$A = \$2000 \cdot a_{\overline{48}|2.5\%}$$
$$= \$2000(27.7732) \qquad \text{by use of Table A.5}$$
$$= \$55,546.40$$

Consequently, the sale price was $\$55,546.40 + \$75,000 = \$130,546.40$. ●

Determining the periodic payment, the term, and the rate

Quite often it is desirable to find the periodic payment that must be made for a certain time at a given rate in order to pay off a debt or to produce a desired sum. At times, the periodic rate or term may have to be determined. Problems of these types can be solved by use of the accumulated-value formula (12.11) and the present-value formula (12.13), as will be demonstrated by working three examples.

Example 6   How much must one pay at the end of each 6 months to accumulate $8000 in 10 years if money is worth 6 percent compounded semiannually?

Solution   We must use the accumulated-value formula (12.11), since $8000 is the value
Use Eq. (12.11)   at the **end of the term.** Substituting $i = \frac{1}{2}(6 \text{ percent}) = 3$ percent, $n = 10(2) = 20$, and $S = \$8000$ gives

$$\$8000 = Rs_{\overline{20}|0.03}$$

$$R = \frac{\$8000}{s_{\overline{20}|0.03}} \qquad \text{solving for } R$$

$$= \frac{\$8000}{26.8704} \qquad \text{from Table A.4, under 3 and opposite 20}$$

$$= \$297.73 \qquad\qquad\qquad\qquad ●$$

Example 7   If a debt of $7300 can be paid off with quarterly payments of $237.99 at the ends of the quarters for 12 years, what rate of interest compounded quarterly is used?

Solution   Since $7300 is the value at the **beginning of the term,** we must use the

Use Eq. (12.13)   present-value formula (12.13). By substituting the given values in it, we
have

$$7300 = 237.99a_{\overline{48}|i}$$

$$a_{\overline{48}|i} = \frac{7300}{237.99} \qquad \text{solving for } a_{\overline{48}|i}$$

$$= 30.6736$$

Table A.5 lists the values of $a_{\overline{n}|i}$. If we look in Table A.5 across from
$n = 48$, we discover that 30.6731 is under 2 percent. Hence $j/4 = 2$ percent
and $j = 8$ percent.   ●

● Example 8   How many payments of $70.67 are required to accumulate $4268.61 if the
interest rate is 2 percent per period?

Solution   Substituting $R = \$70.67$, $S = \$4268.61$, and $i = 2$ percent $= 0.02$ in (12.11)
gives

Use Eq. (12.11) $$4268.61 = 70.67s_{\overline{n}|0.02}$$

$$s_{\overline{n}|0.02} = \frac{4268.61}{70.67} = 60.4020$$

We now look in Table A.4 and find 60.4020 under 2 percent and across
from 40; hence 40 payments are needed.   ●

Exercise 12.5   In Probs. 1 to 12, R represents the periodic payment, n the term in years,
annuities   and j the nominal rate that is compounded m times per year; furthermore,
the payment period and interest period coincide. Find the accumulated and
present value of each ordinary annuity described in Probs. 1 to 12.

1 $R = \$500$, $n = 7$ years, $j = 4$ percent, $m = 1$
2 $R = \$600$, $n = 5$ years, $j = 5$ percent, $m = 1$
3 $R = \$300$, $n = 17$ years, $j = 6$ percent, $m = 1$
4 $R = \$700$, $n = 25$ years, $j = 4$ percent, $m = 1$
5 $R = \$200$, $n = 19$ years, $j = 6$ percent, $m = 2$
6 $R = \$720$, $n = 23$ years, $j = 5$ percent, $m = 2$
7 $R = \$1250$, $n = 24$ years, $j = 8$ percent, $m = 2$
8 $R = \$540$, $n = 16$ years, $j = 4$ percent, $m = 2$
9 $R = \$270$, $n = 11$ years, $j = 8$ percent, $m = 4$
10 $R = \$460$, $n = 7$ years, $j = 10$ percent, $m = 4$
11 $R = \$950$, $n = 12$ years, $j = 6$ percent, $m = 4$
12 $R = \$180$, $n = 5$ years, $j = 8$ percent, $m = 4$
13 The Williams bought a lot for a down payment of $3000 and a payment
of $1500 at the end of each year for 9 years. If money is worth 6 percent,
what is the equivalent cash price?

14 Linda made a down payment of $1000 on a car and agreed to pay $400 at the end of each 3-month period for 2.5 years. If money is worth 8 percent compounded quarterly, what was the cash price of the car?

15 Jean, the beneficiary of an insurance policy received $2000 in cash and is to receive $1000 at the end of each 6 months for 10 years. What was the cash value of the policy if money is worth 5 percent compounded semiannually?

16 Carol, the purchaser of an insurance policy is to receive $3200 on her sixty-fifth birthday and $800 at the end of each 3 months from then until and including the day she is 70. What is the cash equivalent of this policy on her sixty-fifth birthday provided money is worth 6 percent compounded quarterly?

17 To buy a house in 5 years, a couple decided to invest $2000 at the end of each 6 months. If they live up to their intentions and can get 6 percent compounded semiannually on their funds, how much will they have for a down payment?

18 If a business can be bought for $5000 at the end of each quarter for 8 years or for a lump sum of $222,000 due in 8 years, which should a buyer choose provided money is worth 8 percent compounded quarterly?

19 The machinery in a plant can be replaced in 10 years by depositing $3300 at the end of each 6 months into a fund that accumulates at 6 percent compounded semiannually. What is the cost of replacement?

20 A father deposited $1000 at the end of each 6 months for 14 years in a fund to enable his son to go into business. How much was available from this source if the money was invested at 5 percent compounded semiannually?

Find the periodic payment for each ordinary annuity described in Probs. 21 to 24.

21 $A = \$7750, n = 13$ years, $j = 6$ percent, $m = 1$
22 $A = \$3960, n = 5$ years, $j = 10$ percent, $m = 2$
23 $S = \$9600, n = 10$ years, $j = 8$ percent, $m = 4$
24 $S = \$29,860, n = 12$ years, $j = 10$ percent, $m = 4$

Find the nominal rate in Probs. 25 to 28.

25 $R = \$162.89, A = \$2420, n = 23$ years, $m = 1$
26 $R = \$383.44, A = \$7000, n = 25$ years, $m = 2$
27 $R = \$130, S = \$5785.26, n = 12$ years, $m = 2$
28 $R = \$565.79, S = \$44,444, n = 11$ years, $m = 4$

Find the term in years in Probs. 29 to 32.

29 $A = \$3086.42, R = \$220, j = 4$ percent, $m = 1$
30 $A = \$3073.12, R = \$170, j = 10$ percent, $m = 2$
31 $S = \$4626.67, R = \$75, j = 6$ percent, $m = 4$
32 $S = \$7483.60, R = \$308, j = 8$ percent, $m = 4$

33 Meredith bought a piece of property for $37,500 and made a cash payment of $7500. How much must be paid at the end of each 6 months for 3 years to complete payment if money is worth 6 percent compounded semiannually?

34 The Hornsbys anticipate that they will need to replace their stove, refrigerator, and water heater in 6 years at a cost of $2300. How much must they set aside at the end of each 3 months for the next 6 years to be able to pay cash if the savings draw 6 percent compounded quarterly?

35 Ms. Gundersen wants to be able to pay $7200 cash for a car in 3 years. If she can invest her funds at 8 percent compounded quarterly, how much must she invest at the end of each 3 months?

36 Mr. Hulin bought a lot for $13,300 and paid $1300 in cash. In order to finish paying for it in 30 months, how much must he pay at the end of each 6 months if money is worth 6 percent with $m = 2$?

37 If Klein pays off a debt of $3900 in 7 years by means of payments of $649.77 at the end of each year, what effective rate of interest is used?

38 Winchel accumulated $9300 in 8 years by investing $939.63 at the end of each year. What effective rate was earned?

39 The Daigles paid off a debt of $7200 by making a payment of $812.34 at the end of each 6 months for 6 years. What nominal rate compounded semiannually was used?

40 The Boones accumulated $6750 in 4 years by depositing $732.56 at the end of each 6 months in a savings bank. What nominal rate compounded semiannually was paid by the bank?

Find the present and accumulated values of the annuities due described in Probs. 41 to 44.

41 $R =$ $325.13, $i = 6$ percent, $n = 15$ years
42 $R =$ $576.23, $j = 8$ percent, $m = 4$, $n = 12$ years
43 $R =$ $1234, $j = 10$ percent, $m = 2$, $n = 25$ years
44 $R =$ $1208.78, $j = 18$ percent, $m = 12$, $n = 4$ years

12.6   Key concepts

Make sure that you understand the following important words and ideas.

Arithmetic progression   Harmonic means
Common difference   Simple interest
Geometric progression   Compound interest
Common ratio   Nominal rate
Infinite geometric series   Effective rate
Arithmetic means   Accumulated value
Geometric means   Annuity
Harmonic progression   Present value

$$(12.1) \qquad l = a + (n - 1)d$$

$$(12.2) \qquad s = \frac{n}{2}(a + l)$$

$$(12.3) \qquad s = \frac{n}{2}[2a + (n - 1)d]$$

$$(12.4) \qquad l = ar^{n-1}$$

$$(12.5) \qquad s = \frac{a(1 - r^n)}{1 - r}, r \neq 1$$

$$(12.6) \qquad s = \frac{a - rl}{1 - r}, r \neq 1$$

$$(12.7) \qquad s = \frac{a}{1 - r}, r \neq 1$$

$$(12.8) \qquad S = P(1 + i)^n$$

$$(12.8a) \qquad S = P\left(1 + \frac{j}{m}\right)^{mn}$$

$$(12.9) \qquad 1 + e = \left(1 + \frac{j}{m}\right)^{m}$$

$$(12.10) \qquad s_{\overline{n}|i} = \frac{(1 + i)^n - 1}{i}$$

$$(12.11) \qquad S = R(s_{\overline{n}|i})$$

$$(12.12) \qquad S(\text{due}) = (1 + i)Rs_{\overline{n}|i}$$

$$(12.13) \qquad A = R\frac{1 - (1 + i)^{-n}}{i} = Ra_{\overline{n}|i}$$

Exercise 12.6 review

1 Write the seven terms of an arithmetic progression with $a = 3$ and $d = \frac{1}{2}$.

2 Write the six terms of a geometric progression with $a = \frac{3}{4}$ and $r = \frac{2}{3}$.

3 Write the six terms of a harmonic progression with the first two terms being 12 and 3.

4 If a, b, c, d is an arithmetic progression, then $1/a$, $1/b$, $1/c$, $1/d$ is by definition a harmonic progression. Show that if x, y, z, w is a geometric progression, then $1/x$, $1/y$, $1/z$, $1/w$ is also a geometric progression.

5 Find the sixth term and the sum of the six terms of the geometric progression that begins 2, $\frac{2}{3}$, ...

6 Find the seventh term and the sum of the seven terms of the arithmetic progression that begins $\frac{1}{12}$, $\frac{5}{24}$, ...

7 If the second and fourth terms of an arithmetic progression are $\frac{3}{4}$ and $\frac{7}{4}$ and $n = 5$, find a, d, l, and s.

8 If the second and fifth terms of a geometric progression are $\frac{1}{2}$ and 32 and $s = \frac{341}{8}$, find a, r, l, and n.

9 Find the sum of all integers between 8 and 800 that are multiples of 7.

10 Find all values of k so that $2k + 2$, $5k - 11$, and $7k - 13$ is a geometric progression.

11 If $1000 is put in a savings account and left for 6 years, how much is in the account at the end of that time if the bank pays 6 percent interest compounded semiannually?

12 Express $0.5151 \cdots$ as a rational number.

13 Find the sum of all numbers of the form $(-\frac{1}{6})^n$, where n is a positive integer.

14 Show that if a, b, c, d is an arithmetic progression and x is a real number, then $a + x$, $b + x$, $c + x$, $d + x$ is also an arithmetic progression.

15 Repeat Prob. 14 for ax, bx, cx, dx.

16 Show that p^2, q^2, r^2 is a geometric progression if p, q, r is.

17 Show that $\frac{1}{4}$, $\frac{3}{4}$, $\frac{9}{4}$, $\frac{27}{4}$ is a geometric progression and $\log \frac{1}{4}$, $\log \frac{3}{4}$, $\log \frac{9}{4}$, $\log \frac{27}{4}$ is an arithmetic progression.

18 Show that if a^2, b^2, c^2 is an arithmetic progression, then $a + b$, $a + c$, $b + c$ is a harmonic progression.

19 Prove that if each term of a geometric progression is added to nine times the term, if any, that follows it, the resulting sums form a geometric progression.

20 If the sum of n terms of an arithmetic progression is equal to the sum of m terms, $m \neq n$, show that the common difference is zero.

21 Find a geometric progression of 3 terms whose sum is 26 and whose product is 216.

22 Find r for an infinite geometric progression in which each term after the first is twice the sum of all terms that follow it.

23 Find three geometric means between 4 and 16.

24 Find three arithmetic means between 4 and 16.

25 Find three harmonic means between 4 and 16.

26 In Prob. 65 of Exercise 12.2, a proof was requested that the arithmetic mean exceeds the geometric mean for positive, unequal numbers a and b. Show that the difference is $\frac{1}{2}(\sqrt{a} - \sqrt{b})^2$.

27 Let $f(x) = x^3 - 2x^2 - x + 2$ and $g(x) = 3x^2 - 4x - 1$. Show that the arithmetic average of the roots of $f(x) = 0$ equals the arithmetic average of the roots of $g(x) = 0$.

28 Find the infinite geometric progression whose sum is $\frac{4}{3}$ and the sum of whose first three terms is $\frac{3}{2}$.

Classify the progression in each of Probs. 29 to 33, and give the next two terms.

29 $\frac{1}{12}$, $\frac{1}{4}$, $\frac{5}{12}$, $\frac{7}{12}$ **30** $\frac{3}{5}$, $\frac{2}{5}$, $\frac{4}{15}$, $\frac{8}{45}$ **31** $\frac{1}{9}$, $\frac{5}{18}$, $\frac{4}{9}$, $\frac{11}{18}$

32 $\frac{3}{2}, \frac{6}{5}, 1, \frac{6}{7}$ **33** $\frac{4}{3}, \frac{8}{7}, 1, \frac{8}{9}$

34 What effective rate corresponds to 12 percent compounded quarterly?

35 What sum must be repaid if $13,750 is borrowed for 5 years at 10 percent compounded semiannually?

36 What is the present value of $9470 due in 27 months if money is worth 8 percent compounded quarterly?

37 How much could the Thrifties accumulate toward their daughter's educational expenses by depositing $300 at the end of each 6 months into a fund that accumulates at 8 percent compounded semiannually? Assume that the first payment is made when she is 6 months old and the last when she is 18 years of age.

38 The Senoj family made a down payment of 10 percent of the cash value of a house and finished paying for it with payment of $2000 at the end of each 6 months for 20 years. What was the cash price if money is worth 8 percent, $m = 2$?

39 If $2000 is invested for a year at the nominal rate of 12 percent, find the accumulated value if interest is compounded annually, semiannually, quarterly, and continuously.

40 The outcome of a certain experiment that uses mice is age-dependent. A first group of mice is 4 weeks old and a second group is 6 weeks old. Find the ages of two more groups if the ages of all four groups must form a geometric progression.

41 About 1200 Leonardo of Pisa introduced the Fibonacci sequence, defined by $a_1 = 1$, $a_2 = 1$, and for $n \geq 3$, $a_n = a_{n-1} + a_{n-2}$. Find the first eight terms in this sequence. Verify by calculation that

a_4 is a factor of a_8
a_7 is prime
$(a_6)^2 + 1 = a_7 a_5$
$(a_7)^2 - 1 = a_8 a_6$

42 Show that if the real numbers a, b, c form

   (a)   An arithmetic progression, then $b^2 > ac$
   (b)   A geometric progression, then $b^2 = ac$
   (c)   A harmonic progression, then $b^2 < ac$

43 The formula $C = \frac{5}{9}(F - 32)$ converts degrees Fahrenheit to degrees Celsius. Show that if F_1, F_2, F_3 is an arithmetic progression in degrees Fahrenheit, then the corresponding numbers C_1, C_2, C_3 in degrees Celsius also form an arithmetic progression.

CHAPTER 13·
MATHEMATICAL INDUCTION

All sciences, including mathematics, proceed by establishing generalizations, or laws. The usefulness of a generalization is that it frees us from the particular case; we can handle the new case as it arises simply by identifying it as falling in the class covered by the generalization.

One way to arrive at a generalization is to work first with the general case itself, just as we did in arriving at the quadratic formula. Then in applying the formula we need only identify the equation as being in the general form. That is, we work from the general to the particular, a method of logic called *deduction*.

Another way to arrive at a generalization is to examine a number of specific cases to see in what way they are related; the relation, once found, is stated as a generalization, or law. That is, we work from the particular to the general, a method of logic known is *induction*.

The danger of induction is that the specific cases, however large in number, may be special ones, and the generalization based on them may therefore be erroneous. Science must often be content to work with this limitation, called *incomplete induction*, but in mathematics we are able to avoid the pitfall by a method of proof called *mathematical induction*, or *complete induction*.

13·1 Method of mathematical induction

If we let $n = 1, 2, 3$ in $q(n) = n^2 - n + 41$, we find that $q(n)$ becomes 41, 43, and 47, respectively. These numbers are primes; that is, no one of them is divisible by any integer other than itself and unity and each is greater than 1. If we calculate the value of $q(n)$ for each integral value of n up to

$1 \leq n \leq 40$ and including 40, we will see that $q(n)$ is always prime, and this surely suggests that $q(n)$ represents a prime number for every integral value assigned to n. However, if n is equal to 41, $q(n) = 41^2 - 41 + 41 = 41^2$,

$n = 41$ which is not a prime.

428

On the other hand, $1 + 3 = 4 = 2^2$, $1 + 3 + 5 = 9 = 3^2$, $1 + 3 + 5 + 7 = 16 = 4^2$. These results suggest that the sum of the first n odd integers is n^2; that is,

$$1 + 3 + 5 + \cdots + (2n - 1) = n^2 \tag{13.1}$$

Since repeated verification of the truth of Eq. (13.1) for particular values of n does not constitute a proof, we must find some other means of demonstrating its general validity. The type of reasoning involved in mathematical induction is illustrated by the following hypothetical example: Suppose that a certain goal can be reached by a sequence of steps that are successive but of unknown number. Suppose, further, that a person in the process of achieving this goal can be assured that it will always be possible for him to take the next step. Then, regardless of all other circumstances, he knows that he can ultimately attain the goal.

To apply this method of reasoning to the proof of the statement in Eq. (13.1), we assume that the statement is true for some definite but unknown integral value of n, that is, for $n = k$, where k is a number for which the statement can be verified. Then we show that it *necessarily* follows that the statement is true for the *next* integer, $k + 1$. Then, if we can show that the statement is true for some number, say $k = 3$, we know that it is true for the next integer, $k = 4$, and, by proceeding in the same manner, for all following integers. We will show how this is done in Example 1.

● **Example 1**   Prove Eq. (13.1) by mathematical induction.

Solution   We begin with Eq. (13.1), which is

$$1 + 3 + 5 + \cdots + (2n - 1) = n^2 \tag{1}$$

We then assume that (1) is true for $n = k$ and obtain

$n = k$ $$1 + 3 + 5 + \cdots + (2k - 1) = k^2 \tag{2}$$

Next we write (1) with $n = k + 1$ and get

$n = k + 1$ $$1 + 3 + 5 + \cdots + (2k + 1) = (k + 1)^2 \tag{3}$$

Now we will prove that the truth of (3) necessarily follows from (2). The last term in the left member of (3) is the $(k + 1)$th term of (1); hence the next to the last is the kth term and, therefore, is $(2k - 1)$. Consequently, we can write (3) in the form

$$1 + 3 + 5 + \cdots + (2k - 1) + (2k + 1) = (k + 1)^2 \tag{4}$$

To prove that (4) is true, provided we assume the truth of (2), we notice that the left member of (4) is the corresponding member of (2) increased by $(2k + 1)$, that is, by the $(k + 1)$th term of (1). Hence, we add $2k + 1$ to each member of (2), thus obtaining

$$1 + 3 + 5 + \cdots + (2k + 1) = k^2 + 2k + 1 = (k + 1)^2$$

which is the same as (4). Therefore, if (2) is true, (4) is true. That is, (1) is true for $n = k + 1$ if it is true for $n = k$.

$n = 1$ Evidently, (1) is true for $n = 1$, since $1 = 1^2$.

Since (4) has been proved true if (2) is true, (1) holds for $n = 1, 2, 3, 4, 5$, and so on. ●

The formal process of a proof by mathematical induction consists of the following five steps:

Steps in proof by
mathematical induction

1 Verify the theorem for the least integral value q of n for which it has a meaning (often $q = 1$).
2 Assume that the theorem or statement to be proved is true for $n = k$, a particular but unspecified integer, and express this assumption in symbolic form, usually an equation or an inequality.
3 Obtain a symbolic statement of the theorem for $n = k + 1$.
4 Prove that if the statement in step 2 is true, then the statement in step 3 is true also.
5 Using the conclusion in step 4, observe that the theorem is true for $n = q + 1$ since it is true for the integer q of step 1; furthermore, that it is true for $n = q + 2$ since it is true for $n = q + 1$; . . . ; and finally, that it is true regardless of the positive integral value of n if $n \geq q$.

(1) $n = q$
(often $n = 1$)
(4) If true for $n = k$,
then true for
$n = k + 1$

Notice that the basic steps are 1 and 4. No general directions can be given for carrying out the work of step 4. However, the following additional examples illustrate procedures that can frequently be followed.

● **Example 2**   Prove that

$$(1)(2) + (3)(4) + (5)(6) + \cdots + (2n - 1)(2n) = \frac{n(n + 1)(4n - 1)}{3} \quad (5)$$

Notice that the left side of (5) consists of the product of the first odd and even natural numbers, plus the product of the second odd and even natural numbers, and so on to the product of the nth odd and even natural numbers.

Solution   For $n = 1$, the left member of (5) is $(1)(2) = 2$, and the right member is
$n = 1$ $(1)(2)(3)/3 = 2$. Hence Eq. (5) is true for $n = 1$.

We now assume it is true for $n = k$. That is, we assume that

$n = k$ $$1(2) + 3(4) + 5(6) + \cdots + (2k - 1)(2k) = \frac{k(k + 1)(4k - 1)}{3} \quad (6)$$

We are trying to prove Eq. (5) for $n = k + 1$, that is,

$n = k + 1$ $$1(2) + 3(4) + 5(6) + \cdots + (2k + 1)(2k + 2) = \frac{(k + 1)(k + 2)(4k + 3)}{3} \quad (7)$$

In comparing Eqs. (6) and (7), we see that (6) is a sum of k terms and (7)

is a sum of $k + 1$ terms. Hence if we add $(2k + 1)(2k + 2)$ to both members of (6), we have

$$1(2) + 3(4) + \cdots + (2k - 1)(2k) + (2k + 1)(2k + 2)$$

$$= \frac{k(k + 1)(4k - 1)}{3} + (2k + 1)(2k + 2) \quad (8)$$

whose left member is the same as that of (7). Hence we need only show that the right members of (7) and (8) are equal, and we will have completed the proof. Now the right member of (8) is

$$\frac{k(k + 1)(4k - 1)}{3} + (2k + 1)(2k + 2)$$

$$= \frac{1}{3}[k(k + 1)(4k - 1) + 3(2k + 1)(2)(k + 1)]$$

$$= \frac{k + 1}{3}[(4k^2 - k) + (12k + 6)] \qquad \text{factoring}$$

$$= \frac{k + 1}{3}[4k^2 + 11k + 6]$$

$$= \frac{(k + 1)(k + 2)(4k + 3)}{3}$$

which is also the right member of (7).

Example 3 Prove by mathematical induction that $4^n - 1$ is divisible by 3.

Solution Supposing that the statement is true for $n = k$, we have $4^k - 1 = 3q$ for some integer q. Now for $n = k + 1$ we may write

$$4^{k+1} - 1 = 4^{k+1} - 4^k + (4^k - 1) \qquad \text{adding and subtracting } 4^k$$
$$= 4^k(4 - 1) + 3q \qquad \text{using the induction hypothesis}$$
$$= 3(4^k + q)$$

which shows that the statement is true for $n = k + 1$ if it is true for $n = k$. For $n = 1$, the statement is that $4 - 1$ is divisible by 3, and since this is certainly true, the proof by induction is complete.

Example 4 Show that for $n \geq 7$, $\qquad\qquad 3^n < n!$ $\qquad\qquad\qquad$ (9)

Solution For $n = 7$, we have $3^7 = 3 \cdot 3 \cdot 3 \cdot 3 \cdot 3 \cdot 3 \cdot 3 = 9 \cdot 9 \cdot 9 \cdot 3 = 81 \cdot 27 = 2187$, and $7! = 7 \cdot 6 \cdot 5 \cdot 4 \cdot 3 \cdot 2 \cdot 1 = 42(20)(6) = 42(120) = 5040$, and
$n = 7$ so the inequality is true for $n = 7$. It is, incidentally, false for $n = 1, 2, 3,$
$n = k$ 4, 5, and 6.
$n = k + 1$ Suppose now that (9) is true for $n = k$, and hence $3^k < k!$. We must show that $3^{k+1} < (k + 1)!$. Now

$$3^{k+1} = 3(3^k) \qquad \text{by laws of exponents}$$
$$< 3(k!) \qquad \text{true for } n = k \text{ by assumption}$$
$$< (k + 1)(k!) \qquad \text{if } 3 < k + 1, \text{ i.e., for } k > 2$$
$$= (k + 1)! \qquad \text{definition of factorial}$$

Note   In all mathematical induction proofs, we must know clearly what the statement says for $n = k$ and for $n = k + 1$, either in an equation or in words. We then must figure out some way to use the $n = k$ statement (which we are assuming is true) to help prove the $n = k + 1$ statement.

We must always do two basic things: step 1 and step 4. See Probs. 57 to 64 to see how false statements may satisfy one but not the other of these two steps.

Exercise 13.1 mathematical induction   By use of mathematical induction, show that the equation in each of Probs. 1 to 40 is true for all positive integers [for step 4, add the $(k + 1)$th term of the formula under consideration to each member of the equation].

1 $1 + 2 + 3 + \cdots + n = n(n + 1)/2$

2 $3 + 5 + 7 + \cdots + (2n + 1) = n(n + 2)$

3 $1 + 4 + 7 + \cdots + (3n - 2) = n(3n - 1)/2$

4 $5 + 9 + 13 + \cdots + (4n + 1) = n(2n + 3)$

5 $3 + 7 + 11 + \cdots + (4n - 1) = n(2n + 1)$

6 $7 + 13 + 19 + \cdots + (6n + 1) = n(3n + 4)$

7 $2 + 9 + 16 + \cdots + (7n - 5) = n(7n - 3)/2$

8 $a + (a + d) + (a + 2d) + \cdots + [a + (n - 1)d]$
$$= (n/2)[2a + (n - 1)d]$$

9 $2 + 6 + 18 + \cdots + 2(3^{n-1}) = 3^n - 1$

10 $\tfrac{1}{2} + \tfrac{1}{4} + \tfrac{1}{8} + \cdots + 1/2^n = 1 - 1/2^n$

11 $7 + 7^2 + 7^3 + \cdots + 7^n = 7(7^n - 1)/6$

12 $1 + 6 + 6^2 + \cdots + 6^{n-1} = (6^n - 1)/5$

13 $\dfrac{1}{3} + \dfrac{2}{9} + \dfrac{4}{27} + \cdots + \dfrac{1}{3}\left(\dfrac{2}{3}\right)^{n-1} = 1 - \left(\dfrac{2}{3}\right)^n$

14 $\dfrac{1}{4} + \dfrac{3}{16} + \dfrac{9}{64} + \cdots + \dfrac{1}{4}\left(\dfrac{3}{4}\right)^{n-1} = 1 - \left(\dfrac{3}{4}\right)^n$

15 $1 - \dfrac{1}{2} + \dfrac{1}{4} - \dfrac{1}{8} + \cdots + \left(-\dfrac{1}{2}\right)^{n-1} = \dfrac{2}{3}\left[1 - \left(-\dfrac{1}{2}\right)^n\right]$

16 $a + ar + ar^2 + \cdots + ar^{n-1} = \dfrac{a(1 - r^n)}{1 - r}$

17 $2 + 6 + 12 + \cdots + n(n + 1) = n(n + 1)(n + 2)/3$

18 $4 + 10 + 18 + \cdots + n(n + 3) = n(n + 1)(n + 5)/3$

19 $2 + 14 + 36 + \cdots + n(5n - 3) = n(n + 1)(5n - 2)/3$

20 $-3 - 2 + 3 + \cdots + n(2n - 5) = n(n + 1)(4n - 13)/6$

21 $(1)(3) + (2)(4) + (3)(5) + \cdots + (n)(n + 2) = n(n + 1)(2n + 7)/6$

22 $(1)(2)(3) + (2)(3)(4) + (3)(4)(5) + \cdots + n(n + 1)(n + 2)$
$$= n(n + 1)(n + 2)(n + 3)/4$$

23 $2(4) + 5(7) + 8(10) + \cdots + (3n - 1)(3n + 1) = \dfrac{n}{2}(6n^2 + 9n + 1)$

24 $3(5) + 7(9) + 11(13) + \cdots + (4n - 1)(4n + 1)$
$$= n(4n + 1)(4n + 5)/3$$

25 $\dfrac{1}{(1)(2)} + \dfrac{1}{(2)(3)} + \dfrac{1}{(3)(4)} + \cdots + \dfrac{1}{n(n+1)} = \dfrac{n}{n+1}$

26 $\dfrac{1}{(1)(2)(3)} + \dfrac{1}{(2)(3)(4)} + \dfrac{1}{(3)(4)(5)} + \cdots$

$$+ \dfrac{1}{(n)(n+1)(n+2)} = \dfrac{n(n+3)}{4(n+1)(n+2)}$$

27 $\dfrac{1}{1(3)} + \dfrac{1}{3(5)} + \dfrac{1}{5(7)} + \cdots + \dfrac{1}{(2n-1)(2n+1)} = \dfrac{n}{2n+1}$

28 $\dfrac{3}{4} + \dfrac{5}{36} + \dfrac{7}{144} + \cdots + \dfrac{2n+1}{[n(n+1)]^2} = 1 - \dfrac{1}{(n+1)^2}$

29 $1^2 + 2^2 + 3^2 + \cdots + n^2 = n(n+1)(2n+1)/6$

30 $1^2 + 3^2 + 5^2 + \cdots + (2n-1)^2 = n(2n-1)(2n+1)/3$

31 $1^3 + 2^3 + 3^3 + \cdots + n^3 = [n(n+1)/2]^2$

32 $1^3 + 3^3 + 5^3 + \cdots + (2n-1)^3 = 2n^4 - n^2$

33 $1^4 + 2^4 + 3^4 + \cdots + n^4 = n(n+1)(2n+1)(3n^2 + 3n - 1)/30$

34 $1^4 + 3^4 + 5^4 + \cdots + (2n-1)^4 = n(2n-1)(2n+1)(12n^2 - 7)/15$

35 $n^2 - (n-1)^2 + (n-2)^2 - (n-3)^2 + \cdots + (-1)^{n-3} \cdot 3^2$

$$+ (-1)^{n-2} \cdot 2^2 + (-1)^{n-1} \cdot 1^2 = \dfrac{n(n+1)}{2}$$

36 $n^3 - (n-1)^3 + (n-2)^3 - (n-3)^3 + \cdots + (-1)^{n-3} \cdot 3^3$

$$+ (-1)^{n-2} \cdot 2^3 + (-1)^{n-1} \cdot 1^3 = \tfrac{1}{8}\{2(n+1)^2(2n-1) + [1 + (-1)^n]\}$$

37 $1 \cdot 2^1 + 2 \cdot 2^2 + 3 \cdot 2^3 + 4 \cdot 2^4 + \cdots + n \cdot 2^n = 2[1 + (n-1) \cdot 2^n]$

38 $1 \cdot 3^1 + 2 \cdot 3^2 + 3 \cdot 3^3 + 4 \cdot 3^4 + \cdots + n \cdot 3^n$

$$= \tfrac{3}{4}[1 + (2n-1) \cdot 3^n]$$

39 $1 \cdot 4^1 + 2 \cdot 4^2 + 3 \cdot 4^3 + 4 \cdot 4^4 + \cdots + n \cdot 4^n$

$$= \tfrac{4}{9}[1 + (3n-1) \cdot 4^n]$$

40 $1 \cdot 5^1 + 2 \cdot 5^2 + 3 \cdot 5^3 + 4 \cdot 5^4 + \cdots + n \cdot 5^n$

$$= \tfrac{5}{16}[1 + (4n-1) \cdot 5^n]$$

41 Show that $5^n - 1$ is divisible by 4.

42 Show that $3^{2n+1} + 1$ is divisible by 4.

43 Show that $4^{2n} - 1$ is divisible by 3.

44 Show that $2^{3n} - 1$ is divisible by 7.

45 Show that $x^{2n+1} + y^{2n+1}$ is divisible by $x + y$.

46 Show that $x^{2n-1} - y^{2n-1}$ is divisible by $x - y$.

47 Show that $x^{2n} - y^{2n}$ is divisible by $x - y$.

48 Show that $x^{2n} - y^{2n}$ is divisible by $x + y$.

49 Show that $n < 2^n$

50 Show that $n^2 < 2^n$ if $n \geq 5$.

51 Show that $2^{n-1} \leq n!$

52 $2n + 1 < n^2$ for $n \geq 3$.

53 $\left(1 + \dfrac{1}{1}\right)\left(1 + \dfrac{1}{2}\right)\left(1 + \dfrac{1}{3}\right) \cdots \left(1 + \dfrac{1}{n}\right) = n + 1$

54 For $n \geq 2$, $\left(1 - \dfrac{1}{2^2}\right)\left(1 - \dfrac{1}{3^2}\right)\left(1 - \dfrac{1}{4^2}\right) \cdots \left(1 - \dfrac{1}{n^2}\right) = \dfrac{n+1}{2n}$

55 $\dfrac{1}{n+1} + \dfrac{1}{n+2} + \dfrac{1}{n+3} + \cdots + \dfrac{1}{2n} \geq \dfrac{1}{2}$

56 $\dfrac{1}{\sqrt{1}} + \dfrac{1}{\sqrt{2}} + \dfrac{1}{\sqrt{3}} + \cdots + \dfrac{1}{\sqrt{n}} > \sqrt{n}$

Verify for $n = 1$ **57** Show that if

$$2 + 5 + 8 + \cdots + (3n - 1) = \frac{(3n + 4)(n - 1)}{2}$$

is true for $n = k$, then it is true for $n = k + 1$. Nevertheless, try any positive integer n and observe that the formula fails.

58 Repeat Prob. 57 for

$$2 + 7 + 12 + \cdots + (5n - 3) = \frac{(5n + 4)(n - 1)}{2}$$

59 Repeat Prob. 57 for

$$6 + 11 + 16 + \cdots + (5n + 1) = \frac{(5n + 2)(n + 1)}{2}$$

60 For the statement $n = n + 1$, show that if it is true for $n = k$, then it is also true for $n = k + 1$.

Must show if true   **61** Show that $1 + 3 + 5 + \cdots + (2n - 1) = 3n - 2$ holds for $n = 1$
for $n = k$, then true   and $n = 2$, but not for all n.
for $n = k + 1$ **62** Let $p_1 = 2, p_2 = 3, p_3 = 5, \ldots$ be the primes. Show that $p_1 p_2 \cdots p_n + 1$ is prime for $n = 1, 2, 3$, but not for all n. For $n = 6$, try 59.

63 Show that $n^3 > 2^n$ for $n = 2, 3, 4, 5, 6, 7, 8$, and 9, but not for $n = 10$.

64 Let $f(n) = n^2 - n + 2$. Show that $f(n) = 2^n$ for $n = 1, 2$, and 3, but not for $n = 4$.

65 Show that the number of subsets (counting the empty set) of a set with n elements is 2^n.

66 Show that

$$\frac{1}{n} + \frac{1}{n+1} + \frac{1}{n+2} + \cdots + \frac{1}{2n-1} = 1 - \frac{1}{2} + \frac{1}{3} - \frac{1}{4} + \cdots + \frac{1}{2n-1}$$

67 Show by mathematical induction that for any real numbers a and b, $(ab)^n = a^n b^n$, $n = 1, 2, 3, \ldots$.

68 Show that for $a > 0$, $(1 + a)^n \geq 1 + na$.

CHAPTER 14·
PERMUTATIONS
AND COMBINATIONS

In the theory of probability, in statistics, in industry, in science, and in government, it is often desirable or necessary to determine the number of ways in which the elements of a set can be arranged or combined into subsets. For example, a telephone company must provide each subscriber with a unique number, and a state government must provide each car with a unique number or combination of numbers and letters. We will study problems of this nature in this chapter.

14.1 Definitions and the fundamental principle

In this chapter we will deal with collections and arrangements of symbols, objects, or events. Each symbol, object, or event will be called an *element;* furthermore, each set of elements will be called a **combination,** and each unique arrangement of the elements in a combination will be called a **permutation.** Note that a permutation is distinguished by the *order* of the elements that form it; a combination is a set of elements *without regard to order.* In either case, all the elements may be of the same kind, but they need not be. All the elements may be used in any permutation or combination, but it is not necessary to use all of them.

Elements may be drawn from a set with or without replacement. Depending on the situation, we sometimes use the word replacement and sometimes use the word repetition—for us, they will mean the same thing. If they are drawn *without replacement,* we have a combination or permutation. *If each element is replaced* before the next is drawn, we have neither a combination nor a permutation, and the elements withdrawn are called a **sample** if order matters and a **selection** if order does not matter. In the example below, we choose two elements from the letters *A, B, C, D* according to whether or not there is **replacement** and whether or not **order** matters.

Margin terms: Element · Combination · Permutation · Sample · Selection

● **Example 1**   (a)   If there is replacement (that is, if repetition is allowed) and if order matters, we get the following 16 samples of *A, B, C, D* two at a time:

16 samples   *AA, AB, AC, AD, BA, BB, BC, BD, CA, CB, CC, CD, DA, DB, DC, DD*

(b)   If there is replacement and if order does not matter, we get the following 10 *selections* of *A, B, C, D* two at a time:

10 selections   *AA, AB, AC, AD, BB, BC, BD, CC, CD, DD*

Note that *BA* and *AB* are the same selection, but different samples.

(c)   If there is no replacement and if order does matter, we get the following 12 *permutations* of *A, B, C, D* two at a time:

12 permutations   *AB, AC, AD, BA, BC, BD, CA, CB, CD, DA, DB, DC*

Note that *BA* and *AB* are different permutations.

(d)   If there is no replacement and if order does not matter, we get the following 6 *combinations* of *A, B, C, D* two at a time:

6 combinations   *AB, AC, AD, BC, BD, CD*

Note that *BA* and *AB* are the same combination, and that *BB* is neither a permutation nor a combination.   ●

In the flowchart below, we use repetition and replacement of an element interchangeably.

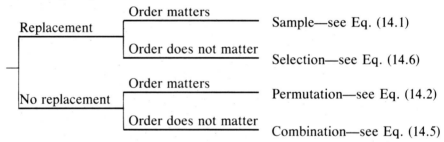

For brevity, we use *A, B, C, D* to represent things that can happen. One possible interpretation is to let *A, B, C,* and *D* represent the event that a child's illness is measles, mumps, chicken pox, or scarlet fever, respectively. Many, many others are possible.

The fundamental principle of counting

Suppose that a woman flies from New Orleans to Los Angeles and back and must leave on Monday, Tuesday, or Wednesday of one week, and must return on Friday or Saturday of the next week. Then her possible trips are: (1) leave Monday, return Friday; (2) leave Monday, return Saturday; (3) leave Tuesday, return Friday; (4) leave Tuesday, return Saturday; (5) leave Wednesday, return Friday; (6) leave Wednesday, return Saturday. She has three possibilities going out and two returning, and thus she has $3 \cdot 2 = 6$ possibilities altogether.

The **tree diagram** below illustrates the possibilities.

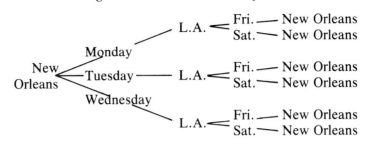

The above travel situation illustrates the following fundamental principle of counting:

Fundamental principle   **If an event can occur in h_1 ways and if another event, which is not influenced by the first event, can occur in h_2 ways, then the two events can occur in $h_1 h_2$ ways.**

● Example 2   In how many ways can a boy and a girl be selected from a group of five boys and six girls?

Solution   The boy can be selected in five ways, and independently the girl can be selected in six ways; hence a boy and a girl can be selected in $5 \times 6 = 30$ ways.   ●

We can extend the fundamental principle by thinking of the first two events as a single one that can happen in $(h_1 h_2)$ ways; hence, if after the first two events have occurred, a third can happen in h_3 ways, then the three can happen in $(h_1 h_2) h_3 = h_1 h_2 h_3$ ways. If then a fourth can occur in h_4 ways, the four can happen in $(h_1 h_2 h_3) h_4 = h_1 h_2 h_3 h_4$ ways. This procedure can be continued, event by event, until we find that n events can happen in $h_1 h_2 h_3 \cdots h_n$ ways in the order indicated by the subscripts.

● Example 3   How many automobile license plates can be made if the inscription on each contains two different letters followed by three different digits?

Solution   (a)   There are 26 letters; therefore, the first of the two letters can be chosen in 26 ways. Since the two letters must be different, there are only 25 ways in which the second letter can be chosen. The first digit can be selected in 10 ways; and since the three digits must be different, the second can be chosen in 9 ways and the third in 8. Consequently such license plates can be made up in $26 \times 25 \times 10 \times 9 \times 8 = 468,000$ ways.

(b)   If, in Example 3, we allow repeated digits, the number of license plates is then $26 \times 25 \times 10 \times 10 \times 10 = 650,000$.   ●

The fundamental principle of counting is used later in this chapter to prove the formulas for both the number of samples and the number of permutations.

Using set terminology at this point may help to clear up a possible point of confusion. The fundamental principle of counting may be written as

$$n(A \times B) = n(A) \cdot n(B)$$

where $A \times B$ is the set $\{(a, b) \mid a \in A \text{ and } b \in B\}$ = "the set of all ordered pairs from A and B," and

$$n(A) = \text{the number of elements in } A$$

Multiply and similarly for $n(B)$ and $n(A \times B)$. For instance if $A = \{1,4\}$ and $B = \{w, x, y, z\}$, then $A \times B = \{(1,w), (1,x), (1,y), (1,z), (4,w), (4,x), (4,y), (4,z)\}$, which has eight elements, and also

$$n(A \times B) = n(A) \cdot n(B) = 2 \cdot 4 = 8$$

On the other hand, if C and D are *disjoint* sets, then

$$n(C \cup D) = n(C) + n(D)$$

Add For $C = \{j,4\}$ and $D = \{m,a,t,h\}$, we have $n(C \cup D) = 2 + 4 = 6$. Hence in some cases we multiply and sometimes we add.

Sample If a set has n elements and we choose a **sample** of r elements, we begin by choosing any one of the n elements, recording that it was chosen first, and then replacing it in the set. The second element is chosen similarly; so by the *fundamental principle* we may choose the first two elements in $nn = n^2$ ways. This may be continued for r drawings, showing that

Number of samples **The number of samples of r elements that can be drawn from a set of n elements is**

$$n^r \tag{14.1}$$

if each element is replaced before the next is drawn, and order counts.

● **Example 4** How many three-symbol "words" may be formed for a code which uses the letters A, E, I, O, and U and the numbers 2, 3, and 5 if any symbol may be repeated?

Solution (a) We have $n = 8$ elements and use $r = 3$ of them for each "word," so that the answer is $8^3 = 512$.

(b) If we require that the first two symbols be letters and the last one be a number, the answer is $5 \times 5 \times 3 = 5^2 \times 3 = 75$.

(c) If we only require that the first symbol be a letter, then the answer is $5 \cdot 8 \cdot 8 = 5 \cdot 8^2 = 320$.   ●

Exercise 14.1

1 How many license plates can be made if each has three different digits followed by two different letters?

2 How many sets of three texts consisting of a math book, one in English, and one on history can be chosen from five math books, seven English texts, and eight history books?

3 How many odd integers between 10 and 99, inclusive, can be formed from the odd digits if there are no repetitions?

4 In how many ways can a coach choose a basketball team from four men who can play guard, three who can play center, and six who can play forward?

5 How many license plates can be made if each has three letters followed by three digits?

6 How many integers between 1000 and 9000 may be formed from the digits 2, 4, 5, 6, and 7?

7 How many three-letter "words" may be formed from the alphabet, excluding A, B, X, Y, and Z?

8 If each person has three initials, how large a population may a city have if everybody has different initials?

9 How many five-letter "words" may be made using the 12 letters of the Hawaiian alphabet?

10 How many four-letter "words" may be made using the last 10 letters of the Hawaiian alphabet if no letter is repeated?

11 How many integers between 2000 and 8000 may be formed from the digits 2, 4, 6, 7, and 9 if the second digit is a 2?

12 If a man has 4 suits, 6 shirts, 13 ties, 8 pairs of socks, 2 hats, and 5 pairs of shoes, in how many ways can he dress?

13 In how many ways can a Japanese, two Chinese, and an Australian be chosen from eight of each nationality to be put into single hotel rooms numbered 1135, 1241, 1382, and 1364?

14 If a music store has 14 salespersons, in how many ways may it choose a salesperson of the month for 6 months?

15 How many four-digit even numbers can be formed from 7, 6, 4, and 2 if no digit is repeated?

16 How many four-digit even numbers can be formed from 7, 6, 4, and 2?

17 Five people always buy their food from any of three stores. In how many ways may they shop on a given day if each person shops in just one store?

18 How many results are possible if two standard dice are thrown?

19 In how many ways may a fortune-teller draw and replace a card three times from a standard deck?

20 In how many ways may a true-false test of 12 questions be answered?

In Probs. 21 to 24, show that the first number is larger than the second.

21 3^4, 4^3 22 2^5, 5^2 23 3^8, 8^3 24 4^{16}, 16^4

In Probs. 25 to 36, classify the situation as a sample, selection, permutation, or combination.

25 Choose six letters from the English alphabet.
26 Choose six different letters from the alphabet.
27 Make a six-letter "word" using any letters.
28 Make a six-letter "word" using different letters.
29 Two dice, red and green, are thrown and all outcomes recorded.
30 Two red dice are thrown and distinct outcomes are recorded (1, 3 is the same as 3, 1).
31 A bridge hand is dealt—13 cards are chosen from a standard deck of 52 cards.
32 Six students are seated in a row of six chairs.
33 Four different digits are chosen.
34 A four-digit number is written, all digits being different.
35 George arranges three of his six reference books on his desk.
36 Marie chooses two of her projects to work on today.

14.2 Permutations

n different elements taken *r* at a time

A **permutation** of *n* distinct elements consists of *r* elements arranged in some order. We do not allow repetition and we do consider order.

The arrangements *abc, acb, bac, bca, cab,* and *cba* constitute the six permutations of the letters *a*, *b*, and *c* taken three at a time. Furthermore, *ab, ba, ac, ca, bc,* and *cb* are the six permutations of the same three letters taken two at a time.

We use here a notation which allows a more compact form for certain expressions. The product of any positive integer *n* and all positive integers *n*! less than *n* is called factorial *n* or *n* factorial, and is written *n*!. For example,

$$3! = 3 \cdot 2 \cdot 1 = 6$$
$$5! = 5 \cdot 4 \cdot 3 \cdot 2 \cdot 1 = 120$$

Notice that $4! = 4 \cdot 3 \cdot 2 \cdot 1 = 4 \cdot 3 \cdot 2$. It is convenient to define $0! = 1$.

We will let $P(n, r)$ represent the number of permutations of *n* elements taken *r* at a time, and we will develop a formula for evaluating this number. We can fill the position 1 in the arrangement in *n* ways. After position 1 has been filled, we have $n - 1$ choices for position 2, then $n - 2$ choices for position 3, and finally, $n - (r - 1)$ choices for position *r*. Hence, since $n - (r - 1) = n - r + 1$, we have

$$P(n, r) = n(n - 1)(n - 2) \cdots (n - r + 1)$$

Note   Note that this is the *product of r consecutive integers.* If we multiply the right member of this equation by $(n - r)!/(n - r)!$, where $(n - r)!$ is the product of the integer $n - r$ and all the positive integers less than $n - r$, we get

$$\frac{n(n - 1)(n - 2) \cdots (n - r + 1)(n - r)!}{(n - r)!} = \frac{n!}{(n - r)!}$$

Permutation of *n* elements taken *r* at a time   Therefore,

$$P(n, r) = \frac{n!}{(n - r)!} \tag{14.2}$$

To obtain the number of permutations of *n* elements taken *n* at a time, we let $r = n$ in Eq. (14.2). Thus, we obtain

Permutation of *n* elements taken *n* at a time

$$P(n, n) = n! \tag{14.3}$$

since, by definition, $0! = 1$.

● **Example 1**   Find the number of four-digit numbers that can be formed from the digits 1, 2, 3, and 4 if no digit is repeated.

Solution   Here we have four elements to be taken four at a time; by Eq. (14.3), the number is 4! or $4 \times 3 \times 2 \times 1 = 24$.   ●

● **Example 2**   Six people enter a room that contains 10 chairs. In how many ways can they be seated?

Solution   Since only six of the chairs are to be occupied, the number of different seating arrangements is equal to the number of permutations of 10 elements taken 6 at a time and, by Eq. (14.2), is

$$P(10, 6) = \frac{10!}{(10 - 6)!} = \frac{10 \times 9 \times 8 \times 7 \times 6 \times 5 \times (4!)}{(4!)}$$
$$= 10 \times 9 \times 8 \times 7 \times 6 \times 5 = 151{,}200 \quad ●$$

n elements, not all different, taken *n* at a time

If the *n* elements of a set are not all different, the problem of determining the number of permutations of the set presents a new aspect. For example, there are two permutations of the letters *a* and *b*, namely, *ab* and *ba*, but there is only one permutation of the letters *a* and *a*, since neither letter can be distinguished from the other and the two can therefore be put in only one *unique* arrangement. We will suppose that *s* members of the set are alike and then designate the $n - s$ different elements by $t_1, t_2, \ldots, t_{n-s}$. The permutations, that is, the distinguishable arrangements of the *n* elements, will then depend only on the arrangement of the elements $t_1, t_2, \ldots, t_{n-s}$.

We can therefore obtain all the permutations of the n elements by distributing $t_1, t_2, \ldots, t_{n-s}$ in the n positions in all possible ways and then filling in the vacant places with the s identical elements.† This amounts to distributing $n - s$ elements in the positions $1, 2, 3, \ldots, n$ in as many ways as possible, that is, to finding the number of permutations of n elements taken $n - s$ at a time. By Eq. (14.2), we have

$$P(n, n - s) = \frac{n!}{[n - (n - s)]!} = \frac{n!}{s!}$$

Hence, if s members of a set of n elements are alike, the number of permutations of the n elements taken n at a time is equal to the number of permutations of n things taken n at a time divided by the number of permutations of s things taken s at a time. By a repeated application of this principle, we arrive at the following theorem:

○ **Theorem**

Permutations of
like elements

If, in a set of n elements, there are g groups, the first containing n_1 members all of which are alike; the second containing n_2 which are alike; the third, n_3 which are alike; and so on to the gth group, which has n_g members alike; then, the number of permutations of the n elements taken all at a time is given by

$$\frac{n!}{n_1! n_2! \cdots n_g!} \qquad (14.4)$$

where $n_1 + n_2 + \cdots + n_g = n$.   ○

● **Example 3** In how many ways can the letters of the word "abracadabra" be arranged?

Solution The solution of this problem involves two factors: (1) the number of permutations of 11 letters taken 11 at a time and (2) the letters b and r two times and the letter a five times. Hence, the number of unique arrangements is given by

$$\frac{11!}{5! 2! 2! 1! 1!} = \frac{11 \times 10 \times 9 \times 8 \times 7 \times 6 \times 5!}{5! \times 2 \times 2} = 83,160 \qquad ●$$

Exercise 14.2 Find the value of the symbol in each of Probs. 1 to 8.

1 $P(5, 3)$ **2** $P(7, 5)$ **3** $P(8, 4)$ **4** $P(6, 2)$
5 $P(9, 1)$ **6** $P(4, 4)$ **7** $P(7, 2)$ **8** $P(8, 5)$

† The reasoning here is the same as that in finding the number of ways in which two people can be seated in a row of five chairs: since when one person has been seated in any one of the five chairs, the other may be seated in any one of four, the permutations are $5 \times 4 = 20$. It makes no difference which chairs are *empty;* that is, it makes no difference where the particular members of the s like elements are located.

Show that the equation in each of Probs. 9 to 12 is an identity.

9 $3P(n, n - 3) = P(n, n - 2)$ **10** $P(n, n - 2) = 12P(n, n - 4)$
11 $P(n, r) = (n - r + 1)P(n, r - 1)$
12 $P(n, n) - P(n - 1, n - 2) = (n - 1)^2 P(n - 2, n - 3)$

Solve the equation in each of Probs. 13 to 16.

13 $(n + 3)! = 30(n + 1)!$ **14** $P(n, 2) = 56$
15 $3P(n, 4) = P(n - 1, 5)$ **16** $P(n, 3) = 336$

How many permutations can be made from the letters in the words in Probs. 17 to 20 if all letters are used in each permutation?

17 Analysis   **18** Referee
19 Company   **20** Bookkeeper
21 How many three-letter "words" can be made from the first eight consonants if no letter is repeated?
22 In how many ways can five people be arranged in a lineup?
23 In how many ways may an inspector spot-check 3 of 10 tax returns?
24 In how many ways may a student visit four of eight fraternity houses?
25 A Scrabble player tests all possible seven-letter words that can be made with his letters, A E E M S S S. If he tests a word every second, how long will it take him?
26 Repeat Prob. 25 for six-letter words.
27 In how many ways can six Russians, four Germans, and four Libyans be seated in a row so that all the Russians are on the left and all the Germans are in the middle?
28 A plane has four double seats on one side and four single seats on the other. In how many ways may 12 people be seated?
29 How many nine-digit social-security numbers may be assigned to an area where the first and second digits are 4 and 3 respectively?
30 After he has selected his nine players, how many batting orders can a little-league coach have if his pitcher bats fourth?
31 How many four-digit numbers may be made using 1, 3, 5, 7, and 8 if no digit is repeated?
32 How many ways are there to choose four of the 12 face cards and give one to each of four poker players?
33 How many television call signals of four different letters may be formed if the first letter is W or K?
34 In how many ways may seven people be seated for a picture of five of them if there are only five chairs?
35 In how many ways may seven people line up for a picture if Mr. Kitchen insists on being on the left end?
36 In how many ways may n people be arranged about a round table?

Derangement   A *derangement* of 1, 2, 3, . . . , n is a permutation of all n elements which leaves no element in its original position. The number of derangements is

$$D_n = n!\left[1 - \frac{1}{1!} + \frac{1}{2!} - \frac{1}{3!} + \cdots + \frac{(-1)^n}{n!}\right]$$

37 Write out all derangements of 1, 2, 3, 4.

38 Calculate D_4, D_5, and D_7.

39 Chefs from China, Italy, Greece, New Zealand, Mexico, and Sweden each prepare one national speciality. In how many ways may they each eat a national speciality they did not prepare?

40 Six people are scheduled for one operation each, all different operations. In how many ways may the operations be performed so that at least one person gets the correct operation?

14.3 Combinations

The definition of a combination is given in Sec. 14.1, but it will be repeated here for ready reference: A set of elements (taken without replacement and without regard to the order in which the elements are arranged) is called a *combination*. According to this definition, the six permutations *xyz*, *xzy*, *yzx*, *yxz*, *zxy*, and *zyx* are all the same combination of the letters *x*, *y*, and *z*.

There is only one combination of n elements taken n at a time, but the problem of determining the number of combinations of n elements taken r at a time, where $r < n$, is both interesting and important. We will designate the combination of n elements taken r at a time by $C(n, r)$. Since a permutation involves choosing r elements (a combination) and then arranging these r elements in one of the $r!$ possible ways, the fundamental principle gives $P(n, r) = C(n, r) \cdot r!$. Hence, by (14.2),

$$C(n, r) = \frac{P(n, r)}{r!} = \frac{1}{r!} \cdot \frac{n!}{(n - r)!}$$

Combinations of n elements taken r at a time

$$C(n, r) = \frac{n!}{(n - r)!r!} \tag{14.5}$$

● **Example 1**   How many committees of 7 women can be formed from a group of 25 women?

Solution   The number of committees is equal to the number of combinations of 25 elements taken 7 at a time. Hence it is

$$C(25, 7) = \frac{25!}{18!7!} = \frac{25 \times 24 \times 23 \times 22 \times 21 \times 20 \times 19 \times (18!)}{(18!) \times 7 \times 6 \times 5 \times 4 \times 3 \times 2 \times 1}$$
$$= 480{,}700 \qquad\qquad ●$$

● **Example 2**   A business firm wishes to employ six men and three boys. In how many ways can the selection be made if nine men and five boys are available?

Solution   The six men can be selected from the nine in $C(9, 6)$ ways, and the three boys from the five in $C(5, 3)$. Hence by the fundamental principle the number of ways in which the selection of the employees can be made is

$$C(9, 6)C(5,3) = \frac{9!}{(9 - 6)!6!}\frac{5!}{(5 - 3)!3!}$$

$$= \frac{9 \times 8 \times 7 \times (6!)}{3 \times 2 \times (6!)}\frac{5 \times 4 \times (3!)}{2 \times (3!)}$$

$$= 840 \qquad ●$$

Selection   If we choose r elements from a set of n elements with replacement after each choice and without order being taken into account, we have a *selection*. We begin by choosing an element from the given n elements and then replacing it. Repeating this to a total of $r - 1$ choices means that, before making the next choice, we replace a chosen element $r - 1$ times. This effectively increases our original n elements to $n + (r - 1)$ elements. We then make the rth or last choice. Since order does not matter, the number of selections is

Number of selections

$$C(n + r - 1, r) \qquad (14.6)$$

● **Example 3**   How may distinct throws of two dice are there?

Solution   This is a selection of two numbers from six numbers, since order does not matter and repetition is allowed. Since $n = 6$ and $r = 2$, we see that the number of selections desired is

$$C(6 + 2 - 1, 2) = C(7, 2) = \frac{7!}{2!5!} = \frac{7 \cdot 6}{2} = 21 \qquad ●$$

Ordered partitions

In this section we will give another application of Eq. (14.4). An example will illustrate the idea.

● **Example 4**   In how many ways may an English class split up so that four people discuss Bacon, six discuss Johnson, five discuss Emerson, and four discuss Pope?

Solution   There are 19 people in the class, so the 4 may be chosen to discuss Bacon in $C(19, 4)$ ways. Out of the 15 remaining, the 6 to discuss Johnson may be chosen in $C(15, 6)$ ways, and the 5 to discuss Emerson in $C(9, 5)$ ways, leaving 4 to discuss Pope. By the fundamental principle, the answer is

$$C(19, 4)C(15, 6)C(9, 5) = \frac{19!}{4!15!}\frac{15!}{6!9!}\frac{9!}{5!4!} = \frac{19!}{4!6!5!4!} = 2{,}444{,}321{,}880 \qquad ●$$

In the example we were dividing a set of n elements into g subsets with n_1 in the first subset, n_2 in the second, and so on to n_g in the gth or last subset, where $n_1 + n_2 + \cdots + n_g = n$. The subsets were distinguished before the division of the set was made. This gives an *ordered partition*, and a simple extension of the argument in the example shows that the number of ordered partitions is given by

Ordered partition

$$\frac{n!}{n_1!n_2! \cdots n_g!} \tag{14.7}$$

Exercise 14.3
Combinations,
ordered
partitions

Calculate the numbers in Probs. 1 to 4.

1 $C(10, 2)$ **2** $C(11, 4)$ **3** $C(8, 5)$ **4** $C(12, 8)$

5 It is true that $C(n, r) = C(n, n - r)$. Verify this for $n = 7, r = 2$ and for $n = 9, r = 3$.

6 It is true that $C(n + 1, r) = C(n, r) + C(n, r - 1)$. Verify this for $n = 8, r = 4$ and for $n = 6, r = 4$.

7 It is true that $C(n, 0) + C(n, 1) + \cdots + C(n, n) = 2^n$. Verify this for $n = 4$ and for $n = 7$.

8 Verify that $C(7, 4) = C(6, 4) + C(5, 3) + C(4, 2) + C(3, 1) + C(2, 0)$.

9 How many groups of four letters may be chosen without replacement from nine letters?

10 How many subsets of 3 points may be chosen from 10 points?

11 How many different bridge hands are there?

12 In how many ways may 5 finalists be chosen from 50 contestants?

13 How many five-card poker hands are there?

14 In how many ways may a buyer choose 12 dresses from a group of 18?

15 In how many ways may a teacher give 6 A's to a class of 22 students?

16 In how many ways may a bear catch half of a school of 10 fish?

17 In how many ways may two men and three women be chosen from five men and seven women?

18 Ten people are to ride in two identical cars, with three in one car and seven in the other. How many ways are there to do this?

19 In how many ways may a committee of two Jews, two Catholics, and two Protestants be chosen from eight of each?

20 Out of four Jerseys, five Holsteins, and seven orioles, how many groups of three cows and two birds may be chosen?

21 In how many ways may a teacher give six A's, five B's, seven C's, three D's, and one F in a class of 22?

22 In how many ways may nine women split into three groups of two, three, and four women, to meet in the red, green, and blue rooms, respectively?

23 How many ways are there for a football team to win seven games, lose two, and tie two?

24 If 13 people go to a play in three different cars, how many ways are there to put six in the Ford, four in the Chevrolet, and three in the Chrysler? How many if the owners drive their own cars?

25 Show that the number of combinations of n things taken r at a time is the same as the number of ordered partitions of n things into two groups with r in the first group.

26 In how many ways may 27 people get five different diseases if at least five people get each disease?

27 Show that $C(8, 4) = C^2(4, 0) + C^2(4, 1) + C^2(4, 2) + C^2(4, 3) + C^2(4, 4)$.

28 How many groups of at most 3 flowers may be made from 10 flowers?

29 (a) In how many ways may three letters be chosen from B, E, A, D?
(b) In how many ways may three different letters be chosen from B, E, A, D?

30 A small firm has four trucks available each week. If Jimmy gets first choice of all the trucks for each of three weeks, in how many ways may Jimmy make his three choices if order does not matter?

31 In how many ways may five dwarfs be chosen to be "most handsome" by five judges if the dwarfs are Doc, Grumpy, Happy, Sneezy, Sleepy, Dopey, and Bashful?

32 There are seven danger signals for cancer. In how many ways may each of five people have exactly one of the danger signals?

33 A commodities trader deals in wheat, corn, pork bellies, gold, and silver. In how many ways may she buy four contracts on Monday if order does not matter?

34 Shizophrenia has five main characteristics. In how many ways may three people each exhibit precisely one such characteristic?

35 A car has eight basic systems. A test is run on five similar cars to see which system breaks down first. How many results are possible?

36 A light bulb has five essential components. If any of the five fail, the bulb goes out. In how many ways may four bulbs fail?

14.4 Key concepts

Make sure you understand the following important words and ideas.

Order	Sample
Replacement	Selection
Permutation	Fundamental principle of counting
Combination	Ordered partition

(14.1)   Number of samples $= n^r$
(14.2)   Number of permutations $= P(n, r) = n!/(n - r)!$
(14.3) $P(n, n) = n!$
(14.4) $n!/(n_1!n_2! \cdots n_g!)$
(14.5)   Number of combinations $= C(n, r) = n!/[r!(n - r)!]$
(14.6)   Number of selections $= C(n + r - 1, r)$

Exercise 14.4 In Probs. 1 to 4, find the value indicated.
review
    1 $P(12, 3)$ **2** $P(8, 3)$ **3** $C(12, 3)$ **4** $C(8, 5)$

Verify the equation in Probs. 5 to 8.

5 $3^5 - 5^3 = C(10, 3) - 2$
6 $P(9, 4) - P(15, 3) = C(13, 3) + C(8, 1)$
7 $C(4, 1) + 2C(4, 2) + 3C(4, 3) + 4C(4, 4) = (4)2^3$
8 $C(4, 1) + 4C(4, 2) + 9C(4, 3) + 16C(4, 4) = (4)(5)2^2$
9 How many ways are there to choose three numbers from the set $\{\frac{1}{2}, \frac{1}{3},$ $\frac{2}{3}, \frac{1}{4}, \frac{3}{4}, \frac{1}{5}\}$ if numbers may be repeated?
10 Repeat Prob. 9 for the case in which a number may not be repeated.
11 Repeat Prob. 10 for the case in which numbers are chosen and then placed, one each, on a red, blue, or green circle.
12 In how many ways may a coach keep two centers, four forwards, and four guards from a group of tryouts of five centers, seven forwards, and nine guards?
13 How many meals may a person order from a menu with five appetizers, three salads, six entrees, six vegetables, and four desserts if two vegetables and one of everything else is allowed?
14 Two cards are drawn from the 12 face cards without replacement. In how many ways may this be done with three decks?
15 Solve the equation $P(n, 3) = 6C(n, 3)$.
16 How many permutations of the letters in "seventeenth" are there?
17 In how many ways may eight diamonds be placed on the edge of a round silver tray?
18 In how many ways may a board of directors of eight people reach a majority decision?
19 Verify that $\dfrac{9!}{2!3!4!} = \dfrac{8!}{1!3!4!} + \dfrac{8!}{2!2!4!} + \dfrac{8!}{2!3!3!}$.
20 How many triangles are formed by n lines in a plane if no two of the lines are parallel and no three pass through a common point?
21 Calculate (14.1), (14.2), (14.5), and (14.6) for $n = 8$, $r = 3$.
22 How many three-digit numbers are there using the digits 1, 3, 5, 7, 9?
23 How many three-digit numbers are there with different digits using the digits 1, 3, 5, 7, 9?
24 How many ways are there of choosing three different odd digits?
25 How many ways are there of choosing three odd digits?
26 In how many ways may five coins be placed so that one is at each vertex of a pentagon?
27 Verify the formula

$$n! = C(n, 0)n^n - C(n, 1)(n - 1)^n + C(n, 2)(n - 2)^n - + - + \cdots$$

$$+ (-1)^{n-1}C(n, n - 1)1^n \qquad \text{for } n = 3 \quad \text{and} \quad n = 4$$

28 Show that $2 \cdot 4 \cdot 6 \cdots \cdots (2n) = 2^n \cdot (n!)$.

29 Show that $1 \cdot 3 \cdot 5 \cdot 7 \cdot \cdots \cdot (2n - 1) = \dfrac{(2n)!}{2^n(n!)}$.

30 Show that if $f(n) = \dfrac{1 \cdot 3 \cdot 5 \cdot \cdots \cdot (2n - 1)}{2 \cdot 4 \cdot 6 \cdot \cdots \cdot (2n)}$, then $\dfrac{f(n + 1)}{f(n)} = \dfrac{2n + 1}{2n + 2}$.

31 Show that if $f(n) = \dfrac{n^n}{n!}$, then $\dfrac{f(n + 1)}{f(n)} = \left(1 + \dfrac{1}{n}\right)^n$.

Psychology **32** In making paired judgments, an approximation to the chi-square test of significance is

$$K = \frac{8}{n - 4}\left[\frac{1}{4}C(n, 3) - d + \frac{1}{2}\right] + f$$

where n = number of stimuli

d = number of circular triads

f = number of degrees of freedom = $\dfrac{n(n - 1)(n - 2)}{(n - 4)^2}$

Calculate f and K for $d = 9$ and $n = 7$.

CHAPTER 15 ·
THE BINOMIAL THEOREM

We have examined the general expression for the square of a binomial and later made use of the expression to solve equations. Obviously, a binomial can be raised not only to the second power but to any power, and there are many occasions to deal with such an expansion. It is the purpose of this chapter to develop a general formula for obtaining any integral power of a binomial. Newton developed such a formula in *Principia Mathematica*, which he then used to help develop calculus.

15.1 The binomial formula

In this section we will develop a formula that will enable us to express any positive integral power of a binomial as a polynomial. This polynomial is

Expansion   called the *expansion* of the power of the binomial.

By actual multiplication, we obtain the following expansions of the first, second, third, fourth, and fifth powers of $x + y$:

$$(x + y)^1 = x + y$$
$$(x + y)^2 = x^2 + 2xy + y^2$$
$$(x + y)^3 = x^3 + 3x^2y + 3xy^2 + y^3$$
$$(x + y)^4 = x^4 + 4x^3y + 6x^2y^2 + 4xy^3 + y^4$$
$$(x + y)^5 = x^5 + 5x^4y + 10x^3y^2 + 10x^2y^3 + 5xy^4 + y^5$$

By referring to these expansions, we can readily verify the fact that the following properties of $(x + y)^n$ hold for $n = 1, 2, 3, 4,$ and 5:

Properties of the   1 The first term in the expansion is x^n.
expansion of $(x + y)^n$ 2 The second term is $nx^{n-1}y$.
  3 The exponent of x decreases by 1 and the exponent of y increases by 1 as we proceed from term to term.
  4 There are $n + 1$ terms in the expansion.
  5 The nth, or next to the last, term of the expansion is nxy^{n-1}.
  6 The $(n + 1)$th, or last, term is y^n.

7 If we multiply the coefficient of any term by the exponent of x in that term and then divide the product by the number of the term in the expansion, we obtain the coefficient of the next term.

8 The sum of the exponents of x and y in any term is n.

If we assume that these eight properties hold for all positive integral n, we can write the first five terms in the expansion of $(x + y)^n$ as follows:

$$\begin{aligned}
&\text{First term} = x^n &&\text{by property 1}\\
&\text{Second term} = nx^{n-1}y &&\text{by property 2}\\
&\text{Third term} = \frac{n(n-1)}{2}x^{n-2}y^2 &&\text{by properties 7 and 3}\\
&\text{Fourth term} = \frac{n(n-1)(n-2)}{3 \times 2}x^{n-3}y^3 &&\text{by properties 7 and 3}\\
&\text{Fifth term} = \frac{n(n-1)(n-2)(n-3)}{4 \times 3 \times 2}x^{n-4}y^4 &&\text{by properties 7 and 3}
\end{aligned}$$

We can continue this process until we have

$$\begin{aligned}
&n\text{th term} = nxy^{n-1} &&\text{by property 5}\\
&(n+1)\text{th term} = y^n &&\text{by property 6}
\end{aligned}$$

We are now in a position to form the sum of the above terms and obtain the binomial formula. Notice that $4 \times 3 \times 2 = 4 \times 3 \times 2 \times 1 = 4!$, $3 \times 2 = 3 \times 2 \times 1 = 3!$, and $2 = 2 \times 1 = 2!$, and write

Binomial formula

$$(x + y)^n = x^n + nx^{n-1}y + \frac{n(n-1)}{2!}x^{n-2}y^2$$

$$+ \frac{n(n-1)(n-2)}{3!}x^{n-3}y^3$$

$$+ \frac{n(n-1)(n-2)(n-3)}{4!}x^{n-4}y^4$$

$$+ \cdots + nxy^{n-1} + y^n \tag{15.1}$$

Binomial theorem   Equation (15.1) is called the *binomial formula*, and the statement that it is true is called the *binomial theorem*.

● **Example 1**   Use the binomial formula to obtain the expansion of $(2a + b)^6$.

Solution   We first apply Eq. (15.1) with $x = 2a$, $y = b$, and $n = 6$. Thus,

$$(2a + b)^6 = (2a)^6 + 6(2a)^5b + \frac{6 \times 5}{2!}(2a)^4b^2 + \frac{6 \times 5 \times 4}{3!}(2a)^3b^3$$

$$+ \frac{6 \times 5 \times 4 \times 3}{4!}(2a)^2b^4 + \frac{6 \times 5 \times 4 \times 3 \times 2}{5!}(2a)b^5$$

$$+ \frac{6 \times 5 \times 4 \times 3 \times 2 \times 1}{6!}b^6$$

Now simplifying the coefficients, and raising $2a$ to the indicated powers, we obtain

$$(2a + b)^6 = 64a^6 + 6(32a^5)b + 15(16a^4)b^2 + 20(8a^3)b^3$$
$$+ 15(4a^2)b^4 + 6(2a)b^5 + b^6$$

Finally, we perform the indicated multiplication in each term and get

$$(2a + b)^6 = 64a^6 + 192a^5b + 240a^4b^2 + 160a^3b^3 + 60a^2b^4 + 12ab^5 + b^6$$

⬤

The computation of the coefficients can, in most cases, be performed mentally by use of property 7, and thus we can avoid writing the first step in the expansion in the above example.

⬤ **Example 2**   Expand $(a - 3b)^5 = [a + (-3b)]^5$.

Solution   In (15.1), we use $x = a$, $y = -3b$, and $n = 5$. The first term in the expansion is a^5, and the second is $5a^4(-3b)$. To get the coefficient of the third, we multiply 5 by 4 and divide the product by 2, thus obtaining 10. Hence, the third term is $10a^3(-3b)^2$. Similarly, the fourth term is

$$\tfrac{30}{3} a^2(-3b)^3 = 10a^2(-3b)^3$$

By continuing this process, we obtain the following expansion:

$$(a - 3b)^5 = a^5 + 5a^4(-3b) + 10a^3(-3b)^2 + 10a^2(-3b)^3$$
$$+ 5a(-3b)^4 + (-3b)^5$$
$$= a^5 - 15a^4b + 90a^3b^2 - 270a^2b^3 + 405ab^4 - 243b^5 \quad ⬤$$

Note   It should be noted that we carry the second term of the binomial, $-3b$, through the first step of the expansion as a single term. Then we raise $-3b$ to the indicated power and simplify the result.

⬤ **Example 3**   Expand $(2x - 5y)^4$.

Solution   We will carry through the expansion with $2x$ as the first term and $-5y$ as the second and get

$$(2x - 5y)^4 = (2x)^4 + 4(2x)^3(-5y) + 6(2x)^2(-5y)^2 + 4(2x)(-5y)^3 + (-5y)^4$$
$$= 16x^4 + 4(8x^3)(-5y) + 6(4x^2)(25y^2) + 4(2x)(-125y^3) + 625y^4$$
$$= 16x^4 - 160x^3y + 600x^2y^2 - 1000xy^3 + 625y^4 \quad ⬤$$

The *r*th term of the binomial formula

In the preceding examples, we explained the method for obtaining any term of a binomial expansion from the term just before it. However, by use of this method, it is impossible to obtain any specific term of the expansion without first computing all the terms which precede it. We will now develop a formula for finding the general rth term without using the other terms.

To find the rth term, we will prove the binomial theorem. By the definition of a product,

$$(x + y)^n = (x + y)(x + y) \cdots (x + y) \qquad n \text{ factors of } (x + y) \qquad (1)$$

Proof using combinations To multiply out the right member of (1), we must take either x or y from each of the n factors $(x + y)$. If y is chosen r times, and x chosen $n - r$ times, the resulting term will be $x^{n-r}y^r$. But by the definition of a combination (see Chap. 14), there are $C(n, r)$ ways of choosing r of the y's. Thus the term involving $x^{n-r}y^r$ is $C(n, r)x^{n-r}y^r$. This shows that

$$(x + y)^n = C(n, 0)x^n + C(n, 1)x^{n-1}y + C(n, 2)x^{n-2}y^2 + \cdots$$
$$+ C(n, r)x^{n-r}y^r + \cdots + C(n, n - 1)xy^{n-1} + C(n, n)y^n \qquad (15.2)$$

The student should show that (15.2) is the same as (15.1).

If we call $C(n, 0)x^n$ the first term, $C(n, 1)x^{n-1}y$ the second term, and so on, then the rth term will involve $C(n, r - 1)$. Using (14.5), we find the rth term to be

$$C(n, r - 1)x^{n-r+1}y^{r-1} = \frac{n!}{(r - 1)!(n - r + 1)!} x^{n-r+1}y^{r-1}$$
$$= \frac{n(n - 1)(n - 2) \cdots (n - r + 2)}{(r - 1)!} x^{n-r+1}y^{r-1} \qquad (15.3)$$

Note Notice that the numerator and denominator of the coefficient in (15.3) are each the product of $r - 1$ consecutive integers; the largest factor in the numerator is n, the smallest in the denominator is 1.

● **Example 4** Find the fourth term in the expansion of $(2a - b)^9$.

Solution In this problem, $x = 2a$, $y = -b$, $n = 9$, and $r = 4$. Therefore, $r - 1 = 3$, $n - r + 1 = 9 - 4 + 1 = 6$, and $n - r + 2 = 7$. Now, if we substitute these values in Eq. (15.3), we see that the fourth term is

$$\text{Fourth term} = \frac{9 \cdot 8 \cdot 7}{3 \cdot 2 \cdot 1} (2a)^6(-b)^3$$

$$= 84(64)a^6(-b^3)$$

$$= -5376a^6b^3$$

The first form $(9 \cdot 8 \cdot 7)/(3 \cdot 2 \cdot 1)$ of the coefficient checks with the statement in the note just before Example 4.

● **Example 5** What is the sixth term of $(3x - 4y)^8$?

Solution The number of the desired term is $r = 6$; hence, the coefficient is

$$C(8,5) = \frac{8 \cdot 7 \cdot 6}{3 \cdot 2 \cdot 1} = 56$$

Consequently, the sixth term is

$$56(3x)^3(-4y)^5 = 56(27x^3)(-1024y^5) = -1,548,288x^3y^5$$

Exercise 15.1   Find the expansion of the binomial in each of Probs. 1 to 12 by use of the binomial formula.

1 $(x + y)^4$	**2** $(s + t)^7$	**3** $(b - y)^5$	**4** $(a - w)^6$
5 $(3x - y)^5$	**6** $(2a - b)^6$	**7** $(x + 3w)^4$	**8** $(a + 4x)^3$
9 $(2x + 3y)^4$	**10** $(3s - 4t)^4$	**11** $(2x - 5y)^3$	**12** $(5x + 3b)^5$

Find the first four terms of the expansion of the binomial in each of Probs. 13 to 16.

13 $(a + y)^{33}$ **14** $(x - y)^{51}$ **15** $(m - 2y)^{101}$ **16** $(b + 3c)^{42}$

Find the indicated power of the number in each of Probs. 17 to 20, and round off to four decimal places.

17 $(1 + 0.04)^5$ **18** 1.05^4 **19** 1.03^6 **20** 1.06^3

Find the specified term of the expansion in each of Probs. 21 to 32.

21 Fifth term of $(x - 2y)^7$ **22** Fourth term of $(2a - c)^6$
23 Sixth term of $(3x + y)^9$ **24** Third term of $(x + 4y)^8$
25 Fourth term of $(a - a^{-1})^7$ **26** Sixth term of $(2x - x^{-2})^9$
27 Seventh term of $(x^2 + 2y)^{11}$ **28** Fifth term of $(3x + y^3)^8$
29 Middle term of $(x + 2y^{1/2})^6$ **30** Middle term of $(x - 3y^{1/4})^8$
31 The term in $(x + 2y)^{10}$ that involves x^7.
32 The term in $(3x - y^{1/2})^{13}$ that involves y^4.

Problems 33 to 36 give the outline of an alternative proof of the binomial theorem which is based on mathematical induction.

33 Show that (15.2) holds for $n = 1$.
34 Assuming (15.2) holds for $n = k$, multiply both sides of the equation by $x + y$. Why is the rth term of $(x + y)^{k+1}$ equal to

$$C(k + 1, r - 1)x^{k-r+2}y^{r-1}$$

35 Continuing Prob. 34, show that after the right member of (15.2) is multiplied by $x + y$, the terms involving y^{r-1} are $C(k, r - 1)x^{k-r+2}y^{r-1} + C(k, r - 2)x^{k-r+2}y^{r-1}$.
36 Show that $C(k + 1, r - 1) = C(k, r - 1)) + C(k, r - 2)$. How does this allow you to complete the proof by induction?
37 Show that $C(n, 0) + C(n, 1) + C(n, 2) + \cdots + C(n, n) = 2^n$ for every positive integer n. *Hint:* Use (15.2) with $x = y = 1$.
38 Show that $C(n, 0) - C(n, 1) + C(n, 2) - C(n, 3) + \cdots + (-1)^n C(n, n) = 0$ for every positive integer n.
39 Show that $C(n, 0) + 2C(n, 1) + 2^2C(n, 2) + 2^3C(n, 3) + \cdots + 2^n C(n, n) = 3^n$ for every positive integer n.
40 Show that $C(2n, n) = C^2(n, 0) + C^2(n, 1) + C^2(n, 2) + \cdots + C^2(n, n)$ for every positive integer n. *Hint:* Look at the middle term in

$$(x + y)^{2n} = (x + y)^n(x + y)^n$$

15.2 Binomial theorem for fractional and negative exponents

The proof of the binomial formula for fractional and negative exponents is beyond the scope of this book; however, we will point out some elementary applications of it. It should be noted that the expansion of $(x + y)^n$ for n not a positive integer has no last term since the coefficient never becomes zero; hence it is impossible to complete the series, and we must be content with any desired or indicated number of terms. The following fact can be established, although the proof will not be given.

The binomial expansion of $x + y$ for fractional and negative exponents is valid if and only if the value of y is between the values of x and $-x$.

● **Example 1** What are the first four terms in the expansion of $(2 + x)^{1/2}$? In what interval is the expansion valid?

Solution The expansion is

$$(2 + x)^{1/2} = 2^{1/2} + (\tfrac{1}{2} \times 2^{-1/2}x) + \frac{\tfrac{1}{2} \times (-\tfrac{1}{2}) \times 2^{-3/2}x^2}{2!}$$

$$+ \frac{\tfrac{1}{2} \times (-\tfrac{1}{2}) \times (-\tfrac{3}{2}) \times 2^{-5/2}x^3}{3!} + \cdots$$

$$= \sqrt{2} + \frac{x}{2\sqrt{2}} - \frac{x^2}{16\sqrt{2}} + \frac{x^3}{64\sqrt{2}} - \cdots$$

$$= \sqrt{2}\left(1 + \frac{x}{4} - \frac{x^2}{32} + \frac{x^3}{128} - \cdots\right)$$

It is valid if and only if $-2 < x < 2$.

● **Example 2** Determine an approximation to the square root of 10.

Solution $\sqrt{10} = 10^{1/2} = (9 + 1)^{1/2} = (3^2 + 1)^{1/2}$

$$= (3^2)^{1/2} + \tfrac{1}{2} \times (3^2)^{-1/2} \times 1 + \frac{\tfrac{1}{2} \times (-\tfrac{1}{2}) \times (3^2)^{-3/2} \times 1^2}{2}$$

$$+ \frac{\tfrac{1}{2} \times (-\tfrac{1}{2}) \times (-\tfrac{3}{2}) \times (3^2)^{-5/2} \times 1^3}{2 \times 3} + \cdots$$

$$= 3 + (\tfrac{1}{2} \times 3^{-1}) - (\tfrac{1}{8} \times 3^{-3}) + (\tfrac{1}{16} \times 3^{-5}) - \cdots$$

$$= 3 + \tfrac{1}{6} - \tfrac{1}{216} + \tfrac{1}{3888} + \cdots$$

$$= 3 + 0.16667 - 0.00463 + 0.00026 + \cdots$$

$$= 3.16230 \qquad \text{to 5 decimal places}$$

By comparing the four terms in the above expansion, we see that their values decrease very rapidly. The rate of this decrease increases as the expansion is carried further. In fact, the fifth term is -0.0000178, and when this is combined with the other four terms, we obtain $\sqrt{10} = 3.1622822$, or, rounded to four decimal places, 3.1623. Hence we conclude that this is the correct value of $\sqrt{10}$ to five figures. Obviously, the expansion can be extended until we obtain any desired degree of accuracy.

Calculators   All calculators use only addition, subtraction, multiplication, and division. They give *approximations* to other function values such as $\log x$, y^x by using only the four fundamental operations.

The approximations can be programmed to be as accurate as desired, often to six or eight or ten decimal places. One way these approximations are found is illustrated by Prob. 33, where $\sqrt{1 + x}$ is first approximated by the first-degree polynomial $1 + (x/2)$, then by the second degree-polynomial $1 + (x/2) - (x^2/8)$. Higher-degree polynomials, obtained from the binomial theorem for $\sqrt{1 + x}$, are used if better approximations are needed.

Problem 35 shows that the calculator needs to store approximations only for $\sqrt{1 + x}$, not $\sqrt{2 + x}$ or $\sqrt{25 + x}$, etc. This is true since, as indicated in Prob. 35,

$$\sqrt{25 + t} = \sqrt{25\left(1 + \frac{t}{25}\right)} = 5\sqrt{1 + \frac{t}{25}}$$

and we may now replace x by $t/25$ in $\sqrt{1 + x}$, then multiply by 5.

Exercise 15.2
binomial theorem
for fractional
and negative
exponents

Find the first four terms of the expansion of the binomial in each of Probs. 1 to 20. Find the range of the variable for which the expansion is valid in Probs. 13 to 20.

1 $(x + y)^{-2}$ **2** $(a + b)^{-4}$ **3** $(a - x)^{-5}$ **4** $(b - y)^{-3}$

5 $(2a + x)^{-1}$ **6** $(x + 2a)^{-1}$ **7** $(a^2 - y)^{-3}$ **8** $(x + a^2)^{-2}$

9 $\left(x - \dfrac{1}{x}\right)^{-3}$ **10** $(x + x^{-2})^{-3}$ **11** $(y + 1)^{-4}$ **12** $(1 + y)^{-4}$

13 $(8 - x)^{1/3}$ **14** $(25 + x)^{1/2}$ **15** $(4 - x)^{-1/2}$ **16** $(9 + x)^{-1/4}$

17 $(1 + x)^{1/2}$ **18** $(1 + x)^{-1/2}$ **19** $(1 - x)^{1/2}$ **20** $(1 - x)^{-3/4}$

In Probs. 21 to 32, find the number to three decimal places by using a binomial expansion. Usually the first three terms of the expansion are enough.

21 $\sqrt{25.5} = (25 + 0.5)^{1/2}$ **22** $(7.9)^{1/3}$ **23** $(3.8)^{-1/2}$ **24** $(9.1)^{1/4}$

25 $\sqrt[5]{31}$ **26** $\sqrt[3]{28}$ **27** $\sqrt[4]{83}$ **28** $\sqrt[3]{123}$

29 1.03^{-4} **30** 1.02^{-5} **31** 0.97^{-3} **32** 1.05^{-6}

33 In Prob. 17 we found that the first four terms of the binomial expansion of $\sqrt{1 + x}$ are $1 + x/2 - x^2/8 + x^3/16$. We may thus get a linear approximation to $\sqrt{1 + x}$ by using $l(x) = 1 + x/2$, and a quadratic one by using $q(x) = 1 + x/2 - x^2/8$. Make a table of values of $y = \sqrt{1 + x}$, $y = l(x)$, and $y = q(x)$. Use $x = 0.1, 0.2, 0.3, 0.4,$ and 0.5. Notice that the closer x is to 0, the better the approximations are, especially $q(x)$.

34 Repeat Prob. 33 for $\sqrt{1 - x}$ (see Prob. 19).

35 Find $\sqrt{26}$ to three decimal places by writing it as

$$(25 + 1)^{1/2} = [25(1 + \tfrac{1}{25})]^{1/2} = 5(1 + \tfrac{1}{25})^{1/2}$$

and using Prob. 17.

36 Repeat Prob. 35 for $\sqrt{62} = (64 - 2)^{1/2} = [64(1 - \tfrac{1}{32})]^{1/2} = 8(1 - \tfrac{1}{32})^{1/2}$.

Psychology **37** The law of comparative judgment, which is basic for all experimental work on Weber's law and Fechner's law, is

$$S_1 - S_2 = x\sqrt{p^2 + q^2 - 2rpq}$$

where r is between 0 and 0.05 and p and q are the percent dispersion of two stimuli.

(a) Show that if $d = q - p$, and d^2 is small, then we have approximately

$$S_1 - S_2 = x\sqrt{2p}\sqrt{p + d}$$

(b) Still assuming that d^2 is small, expand $\sqrt{p + d}$ and show that approximately:

$$S_1 - S_2 = x\sqrt{2}\left(p + \frac{d}{2}\right) = \frac{x}{\sqrt{2}}(p + q)$$

CHAPTER 16 · PROBABILITY

Probability had its beginnings in the middle of the seventeenth century as an outcome of a disagreement over a dice game. It has grown into a varied discipline with applications in the social and natural sciences and is now used by the gambler, statistician, economist, engineer, and others. The word "probability" is used loosely and vaguely by the layman to indicate the likelihood that something will happen under conditions that are not always well-defined. It is the purpose of this chapter to attempt to bring order from the chaos caused by the loose use of the word probability.

16.1   Some basic concepts

There are 52 possible results if a card is drawn from a standard deck. There are $C(52, 5) = 52!/5!47! = 2,598,960$ in which five cards can be drawn from a deck. If a basketball game is played, there are three possible outcomes at the end of regulation time and two at the end of the game. In general, there is a set of possible results if an event or experiment is held. The set of all possible results is called the **sample space.** Each element of the sample space is called a sample point or **outcome.** Any subset of a sample space is called an **event.** If a die is cast, it may stop with 1, 2, 3, 4, 5, or 6 on top; hence, {1, 2, 3, 4, 5, 6} is the sample space, and any one of these elements is an outcome or sample point. Furthermore, any combination of one or more of them is an event.

Sample space
Outcome
Event

    If a die is made accurately and rolled honestly, it is as likely to stop with one number up as with another. Thus, each of the outcomes is **equally likely,** and we say the outcome is **random.** We will assume that

Equally likely
Random

All outcomes of experiments in this chapter are equally likely.

We will use the following notation in this chapter.

Symbol	Meaning
S	The sample space of all possible outcomes
$n(S)$	The number of elements in S
E	A set of specified outcomes in S; hence $E \subseteq S$
$n(E)$	The number of elements in E
$p(E)$	The probability that E will happen or more briefly, the probability of E

Under the assumption that each *outcome* is equally likely, we define the probability of an *event* $E \subseteq S$ as

Probability

$$p(E) = \frac{n(E)}{n(S)}$$

This is equivalent to saying that the probability of each *outcome* is $1/n(S)$. Since $0 \leq n(E) \leq n(S)$, it follows that $0 \leq p(E) \leq 1$.

● **Example 1**   Find the probability of throwing a prime number total with (a) 1 die and (b) 2 dice.

Solution   (a)   We have $n(S) = 6$, $E = \{2, 3, 5\}$, $n(E) = 3$, and so $p = \frac{3}{6} = \frac{1}{2}$.

(b)   The chart below lists all the possibilities. We have $n(S) = 6 \times 6 = 36$. Each prime sum in the chart is underlined, and the primes from 2 to 12 are $E = \{2, 3, 5, 7, 11\}$, hence $n(E) = 15$. It follows that $p = \frac{15}{36} = \frac{5}{12}$.

	1	2	3	4	5	6
1	2	3	4	5	6	7
2	3	4	5	6	7	8
3	4	5	6	7	8	9
4	5	6	7	8	9	10
5	6	7	8	9	10	11
6	7	8	9	10	11	12

● **Example 2**   If one card is drawn from a standard deck of 52 cards, find the probability that card will be a jack.

Solution   Here $S = \{x \mid x$ is a card in a deck of 52 cards$\}$. Hence $n(S) = 52$. Furthermore, $E = \{$club jack, diamond jack, heart jack, spade jack$\}$; so $n(E) = 4$. Thus

$$p(E) = \frac{n(E)}{n(S)} = \frac{4}{52} = \frac{1}{13}$$

● **Example 3**   Each of the three-digit numbers that can be formed using the integers from 1 to 9, with no digit repeated, is written on a card. The cards are then stacked and shuffled. If one card is drawn from the stack, find the probability that the sum of the digits in the number on it will be 10.

Solution   Here $S = \{x \mid x$ is a card bearing a three-digit number formed with the digits from 1 to 9 with no digit repeated$\}$. The number of cards in S is $P(9, 3) = 9 \cdot 8 \cdot 7 = 504$. Therefore, $n(S) = 504$. Furthermore,

$$E = \{x \mid x \text{ is a card in } S \text{ whose digit sum is } 10\}$$

The sets of three *different* nonzero digits whose sum is 10 are $\{1, 2, 7\}$, $\{1, 3, 6\}$, $\{1, 4, 5\}$, and $\{2, 3, 5\}$. Since the digits in each of these sets can be arranged in $P(3, 3) = 3! = 6$ ways and there are four sets, there are $4 \cdot 6 = 24$ cards in E. Hence $n(E) = 24$. Therefore,

$$p(E) = \frac{n(E)}{n(S)} = \frac{24}{504} = \frac{1}{21}$$

Note   Counting techniques, such as permutations in Example 3 and combinations in Example 4, are valuable tools in calculating probabilities. Note that a combination is a subset without regard to order.

● **Example 4**   Two cards are withdrawn from a deck of 52 playing cards. If the first is not replaced before the second is drawn, what is the probability that both will be kings?

Solution   Two cards can be chosen in $C(52, 2)$ ways and two kings can be chosen from the four kings in $C(4, 2)$ ways. Consequently, the probability called for is

$$\frac{C(4, 2)}{C(52, 2)} = \frac{4 \cdot 3}{52 \cdot 51} = \frac{1}{221}$$

● **Example 5**   In Example 4 find the probability if the first card is replaced.

Solution   There are $4 \cdot 4$ ways of drawing two kings and $52 \cdot 52$ ways of drawing two cards; so the probability is

$$\frac{4 \cdot 4}{52 \cdot 52} = \frac{1}{169}$$

● **Example 6**   If 6 balls are drawn from a bag that contains 7 black and 5 white balls, what is the probability that 4 will be black and 2 white?

Solution   There are $C(12, 6)$ ways in which 6 balls can be drawn from a bag that contains 12; furthermore, 4 black balls can be drawn from the 7 in $C(7, 4)$

ways, and 2 white balls can be drawn from 5 in $C(5, 2)$ ways. Therefore the probability of drawing the stated combination is

$$\frac{C(7, 4) \cdot C(5, 2)}{C(12, 6)} = \frac{\dfrac{7!}{3!4!} \cdot \dfrac{5!}{3!2!}}{\dfrac{12!}{6!6!}} = \frac{7!5!6!6!}{3!4!3!2!12!}$$

$$= \tfrac{25}{66} = 0.37878 \cdots$$

Basic theorems

We will now derive and explain the use of several theorems on probability. If E_1 and E_2 represent two events, and if $n(E_1)$ and $n(E_2)$ denote the number of elements in sets E_1 and E_2, respectively, then

$$n(E_1) \geq 0 \qquad \text{and} \qquad n(E_2) \geq 0 \tag{1}$$
$$n(\varnothing) = 0 \tag{2}$$
$$n(E_1 \cup E_2) = n(E_1) + n(E_2) - n(E_1 \cap E_2) \tag{3}$$

A picture illustrating Eq. (3) is shown in Fig. 16.1 provided E_1 and E_2 have some elements in common, and in Fig. 16.2 for the case in which $E_1 \cap E_2 = \varnothing$. Equation (3) can be rewritten as

$$n(E_1 \cup E_2) + n(E_1 \cap E_2) = n(E_1) + n(E_2) \tag{4}$$

In this form, any element in $E_1 \cap E_2$ is counted twice in the left member of (4) and also twice in the right member. Furthermore, an element in E_1 but not E_2 is counted once by each member, and similarly for any element in E_2 but not E_1.

Mutually exclusive · · · Quite often, we say that two events are *mutually exclusive* or *disjoint* to
Disjoint · · · indicate that $E_1 \cap E_2 = \varnothing$. If two events are mutually exclusive, then $n(E_1 \cap E_2) = 0$ and (3) becomes

$$n(E_1 \cup E_2) = n(E_1) + n(E_2) \tag{5}$$

We will now state and prove a theorem concerning the probability of two events.

FIGURE 16.1

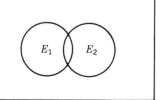

FIGURE 16.2

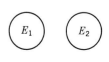

○ **Theorem** *If E_1 and E_2 are any two events, mutually exclusive or not, with probabilities $p(E_1)$ and $p(E_2)$, then the probability of E_1 or E_2 is*

$$p(E_1 \cup E_2) = p(E_1) + p(E_2) - p(E_1 \cap E_2) \qquad (16.1)$$

Proof In the proof of this theorem, we begin by noting that

$$p(E_1 \cup E_2) = \frac{n(E_1 \cup E_2)}{n(S)} \qquad \text{definition of probability}$$

$$= \frac{n(E_1) + n(E_2) - n(E_1 \cap E_2)}{n(S)} \qquad \text{by (3)}$$

$$= \frac{n(E_1)}{n(S)} + \frac{n(E_2)}{n(S)} - \frac{n(E_1 \cap E_2)}{n(S)}$$

by dividing each term of the numerator separately by the denominator. Consequently, by use of the definition of probability, we have

$$p(E_1 \cup E_2) = p(E_1) + p(E_2) - p(E_1 \cap E_2)$$

as stated in the theorem.   ○

● **Example 7** If one card is drawn from a deck of 52 playing cards, find the probability that it will be red or a king.

Solution If E_1 represents the set of red cards and E_2 represents the set of kings, then $n(E_1) = n(\text{reds}) = 26$, $n(E_2) = n(\text{kings}) = 4$, and

$$n(E_1 \cap E_2) = n(\text{red} \cap \text{king}) = 2$$

Therefore, $p(E_1) = \frac{26}{52} = \frac{1}{2}$, $p(E_2) = \frac{4}{52} = \frac{1}{13}$, and $p(E_1 \cap E_2) = \frac{2}{52} = \frac{1}{26}$. Consequently, the probability that the card drawn will be red or a king is

$$p(E_1 \cup E_2) = p(E_1) + p(E_2) - p(E_1 \cap E_2)$$
$$= \tfrac{1}{2} + \tfrac{1}{13} - \tfrac{1}{26}$$
$$= \tfrac{14}{26} = \tfrac{7}{13}$$ ●

If the events considered in Eq. (16.1) are mutually exclusive, then $n(E_1 \cap E_2) = 0$, and by (5)

$$p(E_1 \cup E_2) = \frac{n(E_1 \cup E_2)}{n(S)}$$

$$= \frac{n(E_1) + n(E_2)}{n(S)}$$

$$= p(E_1) + p(E_2)$$

Consequently, we have shown:

If E_1 and E_2 are mutually exclusive events, then the probability that one of them will occur in a single trial is

$$p(E_1 \cup E_2) = p(E_1) + p(E_2) \qquad (16.2)$$

● **Example 8**   If the probability of team A winning its conference championship in football is $\frac{1}{8}$ and the probability of team B of the same conference winning the championship is $\frac{1}{3}$, what is the probability that team A or team B will be the champion?

Solution   Since the events are mutually exclusive, the probability of A or B becoming the champion is the sum of the separate probabilities, as given by Eq. (16.2). Consequently,

$$p(A \cup B) = p(A) + p(B)$$
$$= \tfrac{1}{8} + \tfrac{1}{3} = \tfrac{11}{24}$$ ●

The theorem on mutually exclusive events can be extended to any finite number of events by considering $(E_1 \cup E_2)$ as an event and adding a third event E_3, then considering $(E_1 \cup E_2 \cup E_3)$ as an event and adding another. If this is continued, we reach the following conclusion:

The probability that some one of a set of mutually exclusive events will occur in a single trial is the sum of the probabilities of the separate events.

Symbolically,

$$p(E_1 \cup E_2 \cup \cdots \cup E_n) = p(E_1) + p(E_2) + \cdots + p(E_n) \qquad (16.3)$$

● **Example 9**   In a race for mayor, four candidates A, B, C, and D have probabilities of 0.15, 0.18, 0.24, and 0.42 of winning. What is the probability that A, B, C, or D will win?

Solution   Only one of the candidates will win; so Eq. (16.3) is applicable and gives

$$P(A \cup B \cup C \cup D) = P(A) + P(B) + P(C) + P(D)$$
$$= 0.15 + 0.18 + 0.24 + 0.42 = 0.99$$

Notice that this probability is less than 1; so there is at least one other candidate in the race (a decided dark horse).   ●

Both Eqs. (16.2) and (16.3) follow from (16.1), as does (16.4) below.
Complementary event   If E is an event and E' the *complementary event* (E' happens precisely when E does not happen), then E and E' are mutually exclusive, so (16.2) gives

$$P(E \cup E') = P(E) + P(E') \qquad (6)$$

If, however, an experiment is performed, either E or E' must occur; so $E \cup E' = S$ and $P(E \cup E') = P(S) = 1$. Thus (6) becomes

$$1 = P(E) + P(E')$$

or $$P(E') = 1 - P(E) \qquad (16.4)$$

In Example 9, if we let $E = A \cup B \cup C$, then $E' = D \cup$ dark horse, and

$$P(E') = 1 - P(E) = 1 - (0.15 + 0.18 + 0.24) = 0.43$$

We define the odds in favor of E to be

$$\frac{P(E)}{P(E')} \quad \text{or} \quad \frac{P(E)}{1 - P(E)} \tag{7}$$

Odds   If the **odds in favor** of an event E are $\frac{7}{5}$, we read this as "7 to 5" or 7 : 5. The odds against this event E, or in favor of E', are $\frac{5}{7}$ or 5 : 7.

● **Example 10**   A tote board at a race track gives the odds against a horse winning. If the board shows that the odds against Pace Maker are $\frac{7}{2}$ or 7 : 2, what is the probability that Pace Maker will win the race in the collective judgment of the public?

Solution   The odds against are $\frac{7}{2}$; so the odds for Pace Maker are $\frac{2}{7}$. If p is the probability of winning, then by Eq. (7) we have

$$\frac{2}{7} = \frac{p}{1 - p}$$

Solving for p gives first $7p = 2(1 - p) = 2 - 2p$. Thus $9p = 2$ and $p = \frac{2}{9}$.

Empirical probability

In interpreting results of certain experiments and in the analysis of statistical data, the following ratio is often used: If out of n trials an event has occurred
Relative frequency h times, the *relative frequency* of its occurrence is h/n.

It is assumed, and the assumption is justified by experience, that the larger n becomes, the more nearly the relative frequency approaches the mathematical probability. We make use of this assumption to define a ratio,
Empirical probability   known as the *empirical probability*, that is used extensively in the interpretation of statistics and in insurance. If out of n trials, where n is a large number, an event has occurred h times, then the probability that it will happen in any one trial is defined as h/n.

Life-insurance companies make use of the ratio h/n applied to information
Mortality table   tabulated in a *mortality table* in order to determine their rates. One of these tables is the American Experience Mortality Table, which was compiled from data gathered by several large life-insurance companies. Starting with 100,000 people 10 years of age, it shows the number alive at the age of 11 years, 12 years, 13 years, and so on to 95 years.

● **Example 11**   According to the American Experience Mortality Table, what is the probability that a man 20 years of age will be alive at 30, if the table shows that, out of 92,637 people alive at 20 years of age, 85,441 are living at the age of 30.

Solution   We have $h = 85,441$, $n = 92,637$, and the desired probability is 85,441/92,637 = 0.9223.   ●

**Exercise 16.1
basic theorems**

1 If a card is drawn from a standard deck, what is the probability it will be a heart? A red card?

2 If a card is drawn from a standard deck, what is the probability it will be a king? An even number?

3 If a card is drawn from the face cards, what is the probability it is a jack? A black jack?

4 If a card is drawn from the red cards, what is the probability it is a king? The heart king?

5 Find the probability of throwing a 5 in one toss of a die.

6 Find the probability of throwing a number divisible by 4 in one toss of a die.

7 Find the probability of throwing a number with four letters in it in one toss of a die.

8 Find the probability of throwing a number x with $x^2 < 11$ in one toss of a die.

9 If two dice are tossed, what is the probability of getting a sum of 2? Of 5?

10 If two dice are tossed, what is the probability of getting a sum of 9? Of 12?

11 If two dice are tossed, what is the probability the sum will be 2, 3, or 4?

12 If two dice are tossed, how many ways are there for the sum to be 7?

13 A bag has 3 red, 4 green, and 6 yellow balls. If one ball is drawn, what is the probability it is green? The probability it is not red?

14 In Prob. 13, what is the probability it is not green? The probability it is red or not yellow?

15 If three dice are tossed, what is the probability the sum will be 5?

16 A box contains seven $1 bills, six $5 bills, and eight $10 bills. If two bills are drawn simultaneously at random, find the probability that the sum drawn is $2. That the sum is $6.

17 A bridge hand of 13 cards is drawn from a deck. Find the probability of getting a hand with every card 8 or lower.

18 A poker hand of five cards is drawn from a deck. What is the probability that every card will be a 10, jack, queen, king, or ace?

19 Three people are to be chosen randomly from a group of 12 men and 8 women. What is the probability all three will be men?

20 Five dogs will be in the finals of a dog show containing 8 poodles, 10 terriers, and 14 bulldogs. What is the probability none of the finalists is a poodle?

21 A die has been loaded so that each odd number is three times as likely to occur as each even number. What is $P(4)$?

22 In Prob. 21, what is the probability that the square of the number showing has one digit?

23 In Prob. 21, what is the probability that the square of the number showing is odd?

24 If E_1 is the event described in Prob. 22 and if E_2 is the event in Prob. 23, show that $P(E_1) + P(E_2) > 1$. Is the event $E_1 \cup E_2$ certain to happen?

25 In a town of 8600 there were 43 millionaires. What is the (empirical) probability of any one person in that town being a millionaire?

26 What is the probability in Prob. 25 if 100 people on welfare move into the city?

27 In a college of 16,000 students, each student takes five courses that meet three times per week, and there are usually about 15,000 absences per week. What is the probability of a student being absent a certain day from a certain class?

28 In Prob. 27, how many students could be expected to be absent on a given day in a class of 48 students?

29 In a survey of 35 people, 17 enjoyed classical music, 19 enjoyed popular music, and 9 enjoyed both. What is the probability a person in this survey enjoys at least one of the types of music?

30 In a roomful of people, 260 have attended a Methodist service, 370 have attended a Catholic service, 50 have attended both, and 240 have attended neither. What is the probability a person in that room has attended either type of service? Only a Methodist service?

31 At a college reunion it was discovered that 16 percent of the people had climbed up a mountain, 51 percent had skied down a mountain, and 6 percent had done both. What is the probability a person has done neither?

32 In an election, 59 percent of the people voted for Proposition 1, 43 percent voted for Proposition 2, and 13 percent voted against both. What is the probability a person voted for both?

33 If the probabilities of getting an A, B, C, D, or F in a certain college are 0.26, 0.23, 0.29, 0.18, and 0.04, respectively, what is the probability of getting an A or a B? Of not failing?

34 A women's club is selecting officers. If there are 6 former presidents, 5 former vice-presidents, and 3 former treasurers among the 30 women, what is the probability the new secretary is a former president, vice-president, or treasurer?

35 The probabilities that teams L, T, and A will win a conference championship are $\frac{1}{5}$, $\frac{1}{6}$, and $\frac{1}{10}$. What is the probability one of the three will win?

36 Suppose 20 percent of the population gets exactly one cold in January, $\frac{5}{8}$ as many get exactly two colds as get one, and half as many get exactly three as get one. What is the probability of a person getting one, two, or three colds in January?

37 Albert's probability of vacationing this summer in Yellowstone is 0.22, in Cape Cod is 0.32, and in the Smoky Mountains is 0.42. What is his probability of doing none of the three?

38 What is the probability of getting a one-digit sum if two dice are tossed?

39 What is the probability that an integer between 1 and 75 inclusive has more than 3 letters in its English spelling?

40 What is the probability that the English spelling of a month will contain the letter R?

41 In choosing 5 cards from a deck of 52, what is the probability of getting

3 cards of one kind and 2 of another (this is called a full house in poker)—for example, three sevens and two queens?

42 In choosing 5 cards from 52, what is the probability of getting a pair—that is, 2 of the same kind and 3 different kinds—for example, two threes, a four, a six, and a king?

43 Find the probability of getting a bridge hand (13 cards) with all cards aces, kings, queens, jacks, or tens.

44 In a bridge game two players (13 cards each) are known to have all four queens. What is the probability that each has two queens?

45 What is the probability of choosing 3 Indians and 2 Afghans out of 5 Indians, 4 Pakistanis, and 5 Afghans?

46 What is the probability of choosing 2 Chileans and 4 Argentines out of 8 Chileans, 6 Argentines, and 3 Bolivians?

47 If 5 balls are drawn from a bag containing 8 green and 7 yellow balls, what is the probability that 3 will be green and 2 yellow?

48 Out of 15 people, 8 have disease A and 7 have disease B. If 5 are chosen, what is the probability that 3 will have B and 2 will have A?

49 Suppose dice are loaded so that $P(n) = kn$, for some constant k. Find $P(2)$.

16.2   Dependent and independent events

In this section we will consider the probability of the occurrence of two events E and F. The event $E \cap F$ indicates that both E and F occur. The event $E \mid F$ (read "E given F") is the event E happening assuming that F has already happened.

In finding $P(E \mid F)$ we may use either of two viewpoints: we may consider the original sample space S or the *reduced sample space* F. From the reduced-sample-space point of view and the definition of probability we have

Reduced sample space

$$P(E \mid F) = \frac{n(E \cap F)}{n(F)} = \frac{n(E \cap F)/n(S)}{n(F)/n(S)} = \frac{P(E \cap F)}{P(F)} \qquad (16.5)$$

Thus,
$$P(E \cap F) = P(F)P(E \mid F) \qquad (16.5a)$$

We may interchange E and F, obtaining

$$P(E \cap F) = P(E)P(F \mid E) \qquad (16.5b)$$

This may also be extended to more than two events.

● **Example 1**   If two balls are drawn without replacement from a bag containing five white and three black balls, what is the probability that both will be white? That one will be black and one white?

FIGURE 16.3

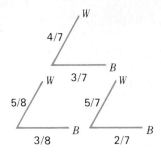

Solution A convenient way of listing all possibilities is a tree diagram (Fig. 16.3). On the first draw there are five white and three black balls, so the probability of a white ball on the first draw is $P(W_1) = \frac{5}{8}$. Now if a white ball was drawn first, there are four white and three black left. The probability of white on the second draw, given that white was drawn first, is $P(W_2 \mid W_1) = \frac{4}{7}$. Thus the probability both are white is

$$P(W_1 \cap W_2) = P(W_1)P(W_2 \mid W_1) = \frac{5}{8} \cdot \frac{4}{7} = \frac{5}{14}$$

The answer to the second question is $P(W_1 \cap B_2) + P(B_1 \cap W_2)$, since $W_1 \cap B_2$ and $B_1 \cap W_2$ are mutually exclusive events. Each can be calculated from the tree, giving

$$P(W_1 \cap B_2) + P(B_1 \cap W_2) = P(W_1)P(B_2 \mid W_1) + P(B_1)P(W_2 \mid B_1)$$

$$= \frac{5}{8} \cdot \frac{3}{7} + \frac{3}{8} \cdot \frac{5}{7} = \frac{15}{28} \qquad \bullet$$

● **Example 2** If two cards are drawn without replacement from a deck, what is the probability the first is a spade and the second is a club?

Solution The probability of a spade on the first draw is $\frac{13}{52}$, and then the probability of a club on the second is $\frac{13}{51}$; so the answer is

$$P(S_1 \cap C_2) = P(S_1)P(C_2 \mid S_1) = \frac{13}{52} \cdot \frac{13}{51} = \frac{13}{204} \qquad \bullet$$

Independent Two events E and F are *independent* if $P(E) = P(E \mid F)$. That is, the probability of E is unaffected by the occurrence or nonoccurrence of F. Using Eq. (16.5a) and this definition of independence gives

$$P(E \cap F) = P(F)P(E) \qquad (16.6)$$

for independent events E and F.

Dependent If two events are not independent, they are called *dependent*.

● **Example 3** Repeat Example 1 except that the first ball is replaced.

Solution   The answer to the first question is

$$P(W_1 \cap W_2) = P(W_1)P(W_2) = \frac{5}{8} \cdot \frac{5}{8} = \frac{25}{64}$$

and the answer to the second is

$$P(W_1 \cap B_2) + P(B_1 \cap W_2) = P(W_1)P(B_2) + P(B_1)P(W_2)$$

$$= \frac{5}{8} \cdot \frac{3}{8} + \frac{3}{8} \cdot \frac{5}{8} = \frac{15}{32}$$ ●

● **Example 4**   Find the probability that all four women in a golf foursome will cook supper that night if their respective probabilities are $\frac{2}{3}$, $\frac{3}{5}$, $\frac{5}{8}$, and $\frac{4}{7}$ and the events are independent.

Solution   The answer is the product $\dfrac{2}{3} \cdot \dfrac{3}{5} \cdot \dfrac{5}{8} \cdot \dfrac{4}{7} = \dfrac{1}{7}$.   ●

Exercise 16.2 dependent and independent events

1   The probability Mr. Tomkins will plant tomatoes is $\frac{3}{4}$, that they will grow if planted is $\frac{2}{3}$, and that they will produce bountifully if they grow is $\frac{8}{9}$. What is the probability of all three things happening?

2   What is the probability of Ross winning an election if the probabilities of getting into the first primary and general election are $\frac{2}{3}$ and $\frac{2}{5}$ and of winning the general election is $\frac{3}{8}$?

3   The probabilities that Sarah will graduate from college, then be accepted into medical school, and then become a doctor are $\frac{9}{10}$, $\frac{2}{5}$, and $\frac{3}{5}$. Find the probability that she will be accepted into medical school. That she will become a doctor.

4   The probability that a committee will recommend Burnside for president of the university is $\frac{1}{4}$, and that of being then selected is $\frac{1}{3}$. Find the probability of his being recommended and selected. Of his being recommended and not selected.

5   If two balls are drawn without replacement from a bag with eight white and four black balls, what is the probability both are white?

6   If two balls are drawn without replacement from a bag with seven red and six green balls, find the probability the first is red and the second is green.

7   If two balls are drawn without replacement from a bag with 10 black and 6 red balls, find the probability that one is red and one is black.

8   If two balls are drawn without replacement from a bag with eight red and six white balls, find the probability that both are the same color.

9   If three balls are drawn without replacement from a bag with 12 white and 6 black balls, what is the probability all three are white?

10   If three balls are drawn without replacement from a bag with eight white and four black balls, what is the probability that all three are white? Is this larger than the answer to Prob. 9?

11 A bag has six red and four white balls, and a box has eight red and five white balls. If one ball is drawn from each, find the probability that both balls are of the same color.

12 A bag has seven red and three white balls, and a box has six red and five black balls. If exactly two balls are drawn from the bag without replacement, and one is drawn from the box, what is the probability that exactly two of the three balls are red?

13 If two cards are drawn without replacement from a standard deck, find the probability that both are sevens.

14 If two cards are drawn from a standard deck without replacement, find the probability that both are of the same rank (2 sevens, 2 kings, etc.).

15 If three cards are drawn without replacement from a standard deck, find the probability that all three are of the same suit.

16 If four cards are drawn without replacement from a standard deck, find the probability that they are all of different suits.

17 The probabilities that Gene will marry and that Jean will win at Wimbledon are $\frac{7}{8}$ and $\frac{2}{9}$. What is the probability that both will happen?

18 Find the probability of throwing a 5 with a pair of dice and choosing a heart from a deck of cards.

19 Three friends agree to meet in one year if all are alive. What is the probability of their meeting if their probabilities of being alive are 0.95, 0.88, and 0.96?

20 What is the probability of throwing a 10 on one toss of a pair of dice and following it with a 4 on the next toss?

21 The probability that Webster, Hill, and Newhouse will all eat at a certain restaurant one night is $\frac{3}{8}$. If the probability that Webster will is $\frac{3}{4}$ and that Hill will is $\frac{2}{3}$, what is the probability that Newhouse will?

22 If a coin is tossed three times, what is the probability of three heads?

23 If a coin is tossed three times, what is the probability of three heads if the first two tosses were heads?

24 Find the probability of throwing a 9 with a pair of dice and two heads in two tosses of a coin.

25 If two balls are drawn with replacement from a bag with eight white and four black balls, what is the probability both are white?

26 If two balls are drawn with replacement from a bag with seven red and six green balls, find the probability the first is red and the second is green.

27 A cup has four butterbeans and three red beans. If two beans are drawn with replacement, what is the probability of getting one of each kind?

28 In Prob. 27, what is the probability both are the same type of bean?

29 If a family is known to have two children, not both boys, what is the probability they have a boy?

30 If a family has three children not all boys, what is the probability they have exactly one boy?

31 A bag has six red and four white balls, and three balls are chosen without replacement. What is the probability the third ball is red if the first ball is white?

32 A pair of dice are thrown, and one of the dice shows a 4, 5, or 6. What is the probability the total is 8? Is 7?

33 Show that if E and F are independent events, then so are E and F'.

34 Show that E and S are always independent, if $S =$ sample space.

35 Show that $P(E)P(F \mid E) = P(F)P(E \mid F)$.

36 Show that if $P(E \mid F) = P(F \mid E)$, then $P(E) = P(F)$, assuming all four quantities are not zero.

In Probs. 37 to 40, assume that E_1, E_2, and E_3 are independent events with $P(E_1) = \frac{1}{7}$, $P(E_2) = \frac{1}{5}$, and $P(E_3) = \frac{1}{2}$.

37 Find $P(E_1 \cap E_2)$ **38** Find $P(E_1 \cap E_3)$

39 Find $P(E_2 \cap E_3)$ **40** Find $P[E_1 \cap (E_2 \cap E_3)]$

16.3   Repeated trials of an event

If we know the probability of an event occurring in one trial, then the probability of its happening a given number of times in n trials is given by the following theorem:

Theorem   *If p is the probability that an event will occur in one trial, then the probability that it will occur exactly r times in n trials is equal to*

$$C(n, r)p^r(1 - p)^{n-r} \tag{16.7}$$

Proof   The r trials can be selected from the n trials in $C(n, r)$ ways, by Sec. 14.3. The probability that the event will occur r times and fail the remaining $n - r$ times is $p^r(1 - p)^{n-r}$, by Sec. 16.2, since the trials are independent and $1 - p$ is the probability of the event failing in any trial. By (16.1), the desired probability is therefore $C(n, r)p^r(1 - p)^{n-r}$.

The theorem has the following corollary.

Theorem   *If p is the probability that an event will occur in one trial, then the probability that it will occur at least r times in n trials is equal to*

$$p^n + C(n, n - 1)p^{n-1}(1 - p) + C(n, n - 2)p^{n-2}(1 - p)^2 + \cdots$$
$$+ C(n, r + 1)p^{r+1}(1 - p)^{n-r-1} + C(n, r)p^r(1 - p)^{n-r} \tag{16.8}$$

Proof   We prove the corollary as follows: The terms of the sum are the probabilities that the event will occur exactly n times, exactly $n - 1$ times, . . . , exactly $r + 1$ times, and exactly r times in n trials, and the events are mutually exclusive.

The reader should notice that the expression in this corollary is the first $n - r + 1$ terms of the expansion of the binomial $(p + q)^n$, where $q = 1 - p$. For this reason, it is often called the *binomial probability*.

● **Example 1**   A bag contains three white and four red balls. The balls are drawn from the bag one at a time and are replaced after each drawing. What is the probability of drawing exactly three red balls in five trials?

Solution   The probability of drawing a red ball in one trial is $\frac{4}{7}$. Therefore, by (16.7) the desired probability is

$$C(n, r)p^r(1 - p)^{n-r} = C(5, 3)\left(\frac{4}{7}\right)^3\left(1 - \frac{4}{7}\right)^{5-3}$$

$$= \left(\frac{5!}{3!2!}\right)\left(\frac{4^3}{7^3}\right)\left(\frac{3^2}{7^2}\right)$$

$$= \frac{5760}{16,807} \approx 0.343 \qquad ●$$

● **Example 2**   Find the probability of throwing at least one 5 in three tosses of a die.

Solution   Since $p(5) = \frac{1}{6}$, the answer, by Eq. (16.8), is

$$C(3, 1)\left(\frac{1}{6}\right)\left(\frac{5}{6}\right)^2 + C(3, 2)\left(\frac{1}{6}\right)^2\left(\frac{5}{6}\right) + C(3, 3)\left(\frac{1}{6}\right)^3 = \frac{(3)(25)}{216} + \frac{(3)(5)}{216} + \frac{1}{216}$$

$$= \frac{91}{216} \qquad ●$$

The answer could also have been obtained by use of Eq. (16.4). The probability of throwing no fives is $C(3, 0)(\frac{1}{6})^0(\frac{5}{6})^3 = (\frac{5}{6})^3$; so the answer is

$$1 - \left(\frac{5}{6}\right)^3 = 1 - \frac{125}{216} = \frac{91}{216}$$

● **Example 3**   If a certain binomial experiment is performed five times, the probability of 3 successes and 2 failures is twice the probability of 2 successes and 3 failures. Determine the probability p of success as a constant times the probability q of failure. Then find p and q.

Solution   The probability of 3 successes and 2 failures is $C(5, 3)p^3q^2$ and that of 2 successes and 3 failures is $C(5, 2)p^2q^3$. Therefore, we have

$$C(5, 3)p^3q^2 = 2C(5, 2)p^2q^3$$

$$C(5, 3)p = 2C(5, 2)q$$

$$p = \frac{2C(5, 2)}{C(5, 3)} \cdot q$$

$$p = 2q \qquad \text{since } C(5, 2) = C(5, 3)$$

$$p = 2(1 - p) \qquad \text{since } q = 1 - p$$

$$3p = 2$$

Consequently, $p = \frac{2}{3}$ and $q = 1 - p = \frac{1}{3}$. ●

**Exercise 16.3
repeated trials**

1 Find the probability of throwing, in five tosses of a coin, (a) exactly three heads, (b) at least three heads.

2 Find the probability of throwing, in three tosses of a die, (a) exactly two 5s, (b) at least two 5s.

3 The probability that a boy will be on time for a meal is 0.2. Find the probability that he will be on time (a) exactly four times in two days, (b) at least four times.

4 Each of six boys tosses a die. Find the probability that at least four of them will throw a 2.

5 If each boy in Prob. 4 tosses the die twice, find the probability that exactly four of them will throw a 6 first and a 5 second.

6 A bag contains three white, four red, and five black balls. Five withdrawals of one ball each are made, and the ball is replaced after each. Find the probability that all five will be red.

7 In Prob. 6, find the probability that exactly three balls will be red.

8 In Prob. 6, find the probability that at least three balls will be red.

9 If the probability that a certain basketball team will win the conference championship is $\frac{2}{3}$, find the probability that it will win exactly three championships in 5 years.

10 John and Tom are members of a Sunday-school class composed of five boys. The members sit on a bench. If they take seats at random and all five are present each of the four Sundays in June, find the probability that John and Tom will sit together (a) exactly three times, (b) at least three times.

11 Six history books, four mathematics books, and two civics books are on a table. If a book is removed and replaced, then another removed and replaced, and so on until six removals and replacements have been made, find the probability that a history book was removed and replaced (a) three times, (b) at least three times.

12 A student has made an A in 8 of the last 20 courses he has taken. Find the probability that he will make an A five times in the next seven courses he takes.

13 Each of three men has a box containing six balls numbered from 1 to 6. Each man draws a ball from his box, replaces it, and then draws two balls simultaneously. Find the probability that as many as two men will draw a 5 the first time, and balls whose sum is 5 the second time.

14 Six mathematics classes and four other classes meet at 7:30 A.M. in a certain building six times a week. There are six parking places near the entrance reserved for teachers. Find the probability that the mathematics teachers will park in the reserved places exactly four times in a week.

15 If the probability that a candidate will be elected to an office is $\frac{2}{3}$, find the probability that she will be elected for four successive terms and then defeated for the fifth term.

16 If the probability that a certain man will tell the truth is $\frac{1}{3}$, find the probability that he will tell 7 lies in answering 10 questions.

17 The probability of an event happening exactly twice in four trials is 18

times the probability of it happening exactly five times in six trials. Find the probability that it will occur in one trial.

18 If the probability of an event happening exactly four times in five trials is $\frac{10}{243}$, find the probability that it will happen in one trial.

19 If the probability that an event will happen exactly three times in five trials is equal to the probability that it will happen exactly two times in six trials, find the probability that it will happen in one trial.

20 Find the probability that an event will occur in one trial if the probability of it happening exactly three times in six trials is equal to the probability that it will happen exactly twice in five trials.

21 Find p if the probability of 4 successes and 1 failure is $\frac{2}{3}$ the probability of at least 4 successes in 5 trials.

22 Find p if the probability of exactly 5 successes in 6 trials is $\frac{3}{4}$ of the probability of at least 5 successes.

23 Find p if, in 7 trials, the probability of exactly 5 successes is equal to the probability of at least 6 successes.

24 In 8 trials, can the probability of 7 successes be equal to the sum of the probabilities of 6 successes and 8 successes? Why?

Key concepts

Be certain that you understand and can use the following concepts and formulas.

Sample space	Empirical probability	Bernoulli probability
Sample point	Mutually exclusive events	Dependent events
Event	Repeated trials	Independent events
Random experiment	Binomial probability	

(16.1) $P(E_1 \cup E_2) = P(E_1) + P(E_2) - P(E_1 \cap E_2)$

(16.2) $P(E_1 \cup E_2) = P(E_1) + P(E_2)$

(16.3) $P(E_1 \cup E_2 \cup \cdots \cup E_n) = P(E_1) + P(E_2) + \cdots + P(E_n)$

(16.4) $P(E') = 1 - P(E)$

(16.5) $P(E \mid F) = \dfrac{P(E \cap F)}{P(F)}$

(16.6) $P(E \cap F) = P(E)P(F)$

(16.7) $C(n, r)p^r(1 - p)^{n-r}$

(16.8) $p^n + C(n, n - 1)p^{n-1}(1 - p) + C(n, n - 2)p^{n-2}(1 - p)^2 + \cdots$
$$+ C(n, r)p^r(1 - p)^{n-r}$$

Exercise 16.4 review

1 If a ball is drawn twice with replacement from a bag with 10 red and 4 white balls, what is the probability both balls are red?

2 Repeat Prob. 1 without replacement.

3 If the probability that a person will catch a certain disease is $\frac{5}{7}$, what is the probability that both a man and his wife will catch the disease?

4 If the probability that a person will win something on a TV games show is $\frac{3}{4}$, what is the probability that exactly three of four contestants will win something?

5 If a card is drawn from a standard deck, what is the probability it is either red or a three?

6 Repeat Prob. 5 with the condition that we know it is not a face card.

7 What is the probability of throwing a 6 with one die?

8 What is the probability of throwing a 6 with two dice?

9 What is the probability of throwing a 6 with three dice?

10 If an integer from 1 to 300 is chosen, what is the probability it is divisible by 2 or 3?

11 What is the probability that an integer between 1 and 100 inclusive is divisible by 3 or 4?

12 If four cards are drawn from a deck without replacement, what is the probability that they are all of the same suit?

13 What is the probability of hitting a target exactly three times out of four if the probability of missing it in one shot is $\frac{1}{3}$?

14 What is the probability of at least two out of four people being on welfare in a city of 10,000 if 4000 are on welfare?

15 What is the probability of at least three successes out of six trials if the probability of an individual success is $\frac{2}{5}$? Is this larger than the answer to Prob. 14?

16 What is the probability of picking a committee of five people consisting of three men and two women, out of a group of 12 men and 8 women?

17 On a cruise ship 39 percent of the passengers enjoyed playing canasta, 52 percent enjoyed playing bridge, and 23 percent enjoyed neither. What is the probability a passenger enjoyed bridge but not canasta?

18 If the probability that a contract will be awarded to a firm in Kansas is 0.14, in Massachusetts is 0.18, in Tennessee is 0.21, and in Utah is 0.23, what is the probability it will not be awarded to a Tennessee or Kansas firm?

19 If a coin and two dice are thrown, what is the probability of a head and a product of 12?

20 If B is half as likely to happen as A and $P(A \mid B) = \frac{2}{5}$, find $P(B \mid A)$.

21 Nine men and six women apply for jobs at a business concern. If eight people are hired, what is the probability it is three men and five women?

22 6 tenth-graders, 10 eleventh-graders, and 9 twelfth-graders are in a piano contest with five prizes. What is the probability that tenth-graders win two prizes and twelfth-grades win three?

23 A poker hand of five cards is drawn from a standard deck of 52 cards. What is the probability that all five cards are of the same suit (a flush)?

APPENDIX

TABLE A.1
COMMON LOGARITHMS

N	0	1	2	3	4	5	6	7	8	9
10	0000	0043	0086	0128	0170	0212	0253	0294	0334	0374
11	0414	0453	0492	0531	0569	0607	0645	0682	0719	0755
12	0792	0828	0864	0899	0934	0969	1004	1038	1072	1106
13	1139	1173	1206	1239	1271	1303	1335	1367	1399	1430
14	1461	1492	1523	1553	1584	1614	1644	1673	1703	1732
15	1761	1790	1818	1847	1875	1903	1931	1959	1987	2014
16	2041	2068	2095	2122	2148	2175	2201	2227	2253	2279
17	2304	2330	2355	2380	2405	2430	2455	2480	2504	2529
18	2553	2577	2601	2625	2648	2672	2695	2718	2742	2765
19	2788	2810	2833	2856	2878	2900	2923	2945	2967	2989
20	3010	3032	3054	3075	3096	3118	3139	3160	3181	3201
21	3222	3243	3263	3284	3304	3324	3345	3365	3385	3404
22	3424	3444	3464	3483	3502	3522	3541	3560	3579	3598
23	3617	3636	3655	3674	3692	3711	3729	3747	3766	3784
24	3802	3820	3838	3856	3874	3892	3909	3927	3945	3962
25	3979	3997	4014	4031	4048	4065	4082	4099	4116	4133
26	4150	4166	4183	4200	4216	4232	4249	4265	4281	4298
27	4314	4330	4346	4362	4378	4393	4409	4425	4440	4456
28	4472	4487	4502	4518	4533	4548	4564	4579	4594	4609
29	4624	4639	4654	4669	4683	4698	4713	4728	4742	4757
30	4771	4786	4800	4814	4829	4843	5857	4871	4886	4900
31	4914	4928	4942	4955	4969	4983	4997	5011	5024	5038
32	5051	5065	5079	5092	5105	5119	5132	5145	5159	5172
33	5185	5198	5211	5224	5237	5250	5263	5276	5289	5302
34	5315	5328	5340	5353	5366	5378	5391	5403	5416	5428
35	5441	5453	5465	5478	5490	5502	5514	5527	5539	5551
36	5563	5575	5587	5599	5611	5623	5635	5647	5658	5670
37	5682	5694	5705	5717	5729	5740	5752	5763	5775	5786
38	5798	5809	5821	5832	5843	5855	5866	5877	5888	5899
39	5911	5922	5933	5944	5955	5966	5977	5988	5999	6010
40	6021	6031	6042	6053	6064	6075	6085	6096	6107	6117
41	6128	6138	6149	6160	6170	6180	6191	6201	6212	6222
42	6232	6243	6253	6263	6274	6284	6294	6304	6314	6325
43	6335	6345	6355	6365	6375	6385	6395	6405	6415	6425
44	6435	6444	6454	6464	6474	6484	6493	6503	6513	6522
45	6532	6542	6551	6561	6571	6580	6590	6599	6609	6618
46	6628	6637	6646	6656	6665	6675	6684	6693	6702	6712
47	6721	6730	6739	6749	6758	6767	6776	6785	6794	6803
48	6812	6821	6830	6839	6849	6857	6866	6875	6884	6893
49	6902	6911	6920	6928	6937	6946	6955	6964	6972	6981
50	6990	6998	7007	7016	7024	7033	7042	7050	7059	7067
51	7076	7084	7093	7101	7110	7118	7126	7135	7143	7152
52	7160	7168	7177	7185	7193	7202	7210	7218	7226	7235
53	7243	7251	7259	7267	7275	7284	7292	7300	7308	7316
54	7324	7332	7340	7348	7356	7364	7372	7380	7388	7396
N	0	1	2	3	4	5	6	7	8	9

TABLE A.1

COMMON LOGARITHMS
(*Continued*)

N	0	1	2	3	4	5	6	7	8	9
55	7404	7412	7419	7427	7435	7443	7451	7459	7466	7474
56	7482	7490	7497	7505	7513	7520	7528	7536	7543	7551
57	7559	7566	7574	7582	7589	7597	7604	7612	7619	7627
58	7634	7642	7649	7657	7664	7672	7679	7686	7694	7701
59	7709	7716	7723	7731	7738	7745	7752	7760	7767	7774
60	7782	7789	7796	7803	7810	7818	7825	7832	7839	7846
61	7853	7860	7868	7875	7882	7889	7896	7903	7910	7917
62	7924	7931	7938	7945	7952	7959	7966	7973	7980	7987
63	7993	8000	8007	8014	8021	8028	8035	8041	8048	8055
64	8062	8069	8075	8082	8089	8096	8102	8109	8116	8122
65	8129	8136	8142	8149	8156	8162	8169	8176	8182	8189
66	8195	8202	8209	8215	8222	8228	8235	8241	8248	8254
67	8261	8267	8274	8280	8287	8293	8299	8306	8312	8319
68	8325	8331	8338	8344	8351	8357	8363	8370	8376	8382
69	8388	8395	8401	8407	8414	8420	8426	8432	8439	8445
70	8451	8457	8463	8470	8476	8482	8488	8494	8500	8506
71	8513	8519	8525	8531	8537	8543	8549	8555	8561	8567
72	8573	8579	8585	8591	8597	8603	8609	8615	8621	8627
73	8633	8639	8645	8651	8657	8663	8669	8675	8681	8686
74	8692	8698	8704	8710	8716	8722	8727	8733	8739	8745
75	8751	8756	8762	8768	8774	8779	8785	8791	8797	8802
76	8808	8814	8820	8825	8831	8837	8842	8848	8854	8859
77	8865	8871	8876	8882	8887	8893	8899	8904	8910	8915
78	8921	8927	8932	8938	8943	8949	8954	8960	8965	8971
79	8976	8982	8987	8993	8998	9004	9009	9015	9020	9025
80	9031	9036	9042	9047	9053	9058	9063	9069	9074	9079
81	9085	9090	9096	9101	9106	9112	9117	9122	9128	9133
82	9138	9143	9149	9154	9159	9165	9170	9175	9180	9186
83	9191	9196	9201	9206	9212	9217	9222	9227	9232	9238
84	9243	9248	9253	9258	9263	9269	9274	9279	9284	9289
85	9294	9299	9304	9309	9315	9320	9325	9330	9335	9340
86	9345	9350	9355	9360	9365	9370	9375	9380	9385	9390
87	9395	9400	9405	9410	9415	9420	9425	9430	9435	9440
88	9445	9450	9455	9460	9465	9469	9474	9479	9484	9489
89	9494	9499	9504	9509	9513	9518	9523	9528	9533	9538
90	9542	9547	9552	9557	9562	9566	9571	9576	9581	9586
91	9590	9595	9600	9605	9609	9614	9619	9624	9628	9633
92	9638	9643	9647	9652	9657	9661	9666	9671	9675	9680
93	9685	8689	9594	9699	9703	9708	9713	9717	9722	9227
94	9731	9736	9741	9745	9750	9754	9759	9763	9768	9773
95	9777	9782	9786	9791	9795	9800	9805	9809	9814	9818
96	9823	9827	9832	9836	9841	9845	9850	9854	9859	9863
97	9868	9872	9877	9881	9886	9890	9894	9899	9903	9908
98	9912	9917	9921	9926	9930	9934	9939	9943	9948	9952
99	9956	9961	9965	9969	9974	9978	9983	9987	9991	9996
N	0	1	2	3	4	5	6	7	8	9

TABLE A.2
ACCUMULATED
VALUE: $(1 + i)^n$

$n \backslash i$	$1\frac{1}{2}\%$	2%	$2\frac{1}{2}\%$	3%	4%	5%	6%
1	1.0150	1.0200	1.0250	1.0300	1.0400	1.0500	1.0600
2	1.0302	1.0404	1.0506	1.0609	1.0816	1.1025	1.1236
3	1.0457	1.0612	1.0769	1.0927	1.1249	1.1576	1.1910
4	1.0614	1.0824	1.1038	1.1255	1.1699	1.2155	1.2625
5	1.0773	1.1041	1.1314	1.1593	1.2167	1.2763	1.3382
6	1.0934	1.1262	1.1597	1.1941	1.2653	1.3401	1.4185
7	1.1098	1.1487	1.1887	1.2299	1.3159	1.4071	1.5036
8	1.1265	1.1717	1.2184	1.2668	1.3686	1.4775	1.5938
9	1.1434	1.1951	1.2489	1.3048	1.4233	1.5513	1.6895
10	1.1605	1.2190	1.2801	1.3439	1.4802	1.6289	1.7908
11	1.1779	1.2434	1.3121	1.3842	1.5395	1.7103	1.8983
12	1.1956	1.2682	1.3449	1.4258	1.6010	1.7959	2.0122
13	1.2136	1.2936	1.3785	1.4685	1.6651	1.8856	2.1329
14	1.2318	1.3195	1.4130	1.5126	1.7317	1.9799	2.2609
15	1.2502	1.3459	1.4483	1.5580	1.8009	2.0789	2.3966
16	1.2690	1.3728	1.4845	1.6047	1.8730	2.1829	2.5404
17	1.2880	1.4002	1.5216	1.6528	1.9479	2.2920	2.6928
18	1.3073	1.4282	1.5597	1.7024	2.0258	2.4066	2.8543
19	1.3270	1.4568	1.5987	1.7535	2.1068	2.5270	3.0256
20	1.3469	1.4859	1.6386	1.8061	2.1911	2.6533	3.2071
21	1.3671	1.5157	1.6796	1.8603	2.2788	2.7860	3.3996
22	1.3876	1.5460	1.7216	1.9161	2.3699	2.9253	3.6035
23	1.4084	1.5769	1.7646	1.9736	2.4647	3.0715	3.8197
24	1.4295	1.6084.	1.8087	2.0328	2.5633	3.2251	4.0489
25	1.4509	1.6406	1.8539	2.0938	2.6658	3.3864	4.2919
26	1.4727	1.6734	1.9003	2.1566	2.7725	3.5557	4.5494
27	1.4948	1.7069	1.9478	2.2213	2.8834	3.7335	4.8223
28	1.5172	1.7410	1.9965	2.2879	2.9987	3.9201	5.1117
29	1.5400	1.7758	2.0464	2.3566	3.1187	4.1161	5.4184
30	1.5631	1.8114	2.0976	2.4273	3.2434	4.3219	5.7435
31	1.5865	1.8476	2.1500	2.5001	3.3731	4.5380	6.0881
32	1.6103	1.8845	2.2038	2.5751	3.5081	4.7649	6.4534
33	1.6345	1.9222	2.2589	2.6523	3.6484	5.0032	6.8406
34	1.6590	1.9607	2.3153	2.7319	3.7943	5.2533	7.2510
35	1.6839	1.9999	2.3732	2.8139	3.9461	5.5160	7.6861
36	1.7091	2.0399	2.4325	2.8983	4.1039	5.7918	8.1473
37	1.7348	2.0807	2.4933	2.9852	4.2681	6.0814	8.6361
38	1.7608	2.1223	2.5557	3.0748	4.4388	6.3855	9.1543
39	1.7872	2.1647	2.6196	3.1670	4.6164	6.7048	9.7035
40	1.8140	2.2080	2.6851	3.2620	3.8010	7.0400	10.2857
41	1.8412	2.2522	2.7522	3.3599	4.9931	7.3920	10.9029
42	1.8688	2.2972	2.8210	3.4607	5.1928	7.7616	11.5570
43	1.8969	2.3432	2.8915	3.5645	5.4005	8.1497	12.2505
44	1.9253	2.3901	2.9638	3.6715	5.6165	8.5572	12.9855
45	1.9542	2.4379	3.0379	3.7816	5.8412	8.9850	13.7646 < n
46	1.9835	2.4866	3.1139	3.8950	6.0748	9.4343	14.5905
47	2.0133	2.5363	3.1917	4.0119	6.3178	9.9060	15.4659
48	2.0435	2.5871	3.2715	4.1323	6.5705	10.4013	16.3939
49	2.0741	2.6388	3.3533	4.2562	6.8333	10.9213	17.3775
50	2.1052	2.6916	3.4371	4.3839	7.1067	11.4674	18.4202

TABLE A.3
PRESENT VALUE:
$(1 + i)^{-n}$

$n \backslash i$	$1\frac{1}{2}\%$	2%	$2\frac{1}{2}\%$	3%	4%	5%	6%
1	0.98522	0.98039	0.97561	0.97087	0.96154	0.95238	0.94340
2	0.97066	0.96117	0.95181	0.94260	0.92456	0.90703	0.89000
3	0.95632	0.94232	0.92860	0.91514	0.88900	0.86384	0.83962
4	0.94218	0.92385	0.90595	0.88849	0.85480	0.82270	0.79209
5	0.92826	0.90573	0.88385	0.86261	0.82193	0.78353	0.74726
6	0.91454	0.88797	0.86230	0.83748	0.79031	0.74622	0.70496
7	0.90103	0.87056	0.84127	0.81309	0.75992	0.71068	0.66506
8	0.88771	0.85349	0.82075	0.78941	0.73069	0.67684	0.62741
9	0.87459	0.83676	0.80073	0.76642	0.70259	0.64461	0.59190
10	0.86167	0.82035	0.78120	0.74409	0.67556	0.61391	0.55839
11	0.84893	0.80426	0.76214	0.72242	0.64958	0.58468	0.52679
12	0.83639	0.78849	0.74356	0.70138	0.62460	0.55684	0.49697
13	0.82403	0.77303	0.72542	0.68095	0.60057	0.53032	0.46884
14	0.81185	0.75788	0.70773	0.66112	0.57748	0.50507	0.44230
15	0.79985	0.74301	0.69047	0.64186	0.55526	0.48102	0.41727
16	0.78803	0.72845	0.67362	0.62317	0.53391	0.45811	0.39365
17	0.77639	0.71416	0.65720	0.60502	0.51337	0.43630	0.37136
18	0.76491	0.70016	0.64117	0.58739	0.49363	0.41552	0.35034
19	0.75361	0.68643	0.62553	0.57029	0.47464	0.39573	0.33051
20	0.74247	0.67297	0.61027	0.55368	0.45639	0.37689	0.31180
21	0.73150	0.65978	0.59539	0.53755	0.43883	0.35894	0.29416
22	0.72069	0.64684	0.58086	0.52189	0.42196	0.34185	0.27751
23	0.71004	0.63416	0.56670	0.50669	0.40573	0.32557	0.26180
24	0.69954	0.62172	0.55288	0.49193	0.39012	0.31007	0.24698
25	0.68921	0.60953	0.53939	0.47761	0.37512	0.29530	0.23300
26	0.67902	0.59758	0.52623	0.46369	0.36065	0.28124	0.21981
27	0.66899	0.58586	0.51340	0.45019	0.34682	0.26785	0.20737
28	0.65910	0.57437	0.50088	0.43708	0.33348	0.25509	0.19563
29	0.64936	0.56311	0.48866	0.42435	0.32069	0.24295	0.18456
30	0.63976	0.55207	0.47674	0.41199	0.30832	0.23138	0.17411
31	0.63031	0.54125	0.46511	0.39999	0.29646	0.22036	0.16425
32	0.62099	0.53063	0.45377	0.38834	0.28506	0.20987	0.15496
33	0.61182	0.52023	0.44270	0.37703	0.27409	0.19987	0.14619
34	0.60277	0.51003	0.43191	0.36604	0.26355	0.19035	0.13791
35	0.59387	0.50003	0.42137	0.35538	0.25342	0.18129	0.13011
36	0.58509	0.49022	0.41109	0.34503	0.24367	0.17266	0.12274
37	0.57644	0.48061	0.40107	0.33498	0.23430	0.16444	0.11579
38	0.56792	0.47119	0.39128	0.32523	0.22529	0.15661	0.10924
39	0.55953	0.46195	0.38174	0.31575	0.21662	0.14915	0.10306
40	0.55126	0.45289	0.37243	0.30656	0.20829	0.14205	0.09722
41	0.54312	0.44401	0.36335	0.29763	0.20028	0.13528	0.09172
42	0.53509	0.43530	0.35448	0.28896	0.19257	0.12884	0.08653
43	0.52718	0.42677	0.34584	0.28054	0.18517	0.12270	0.08163
44	0.51939	0.41840	0.33740	0.27237	0.17805	0.11686	0.07701
45	0.51171	0.41020	0.32917	0.26444	0.17120	0.11130	0.07265
46	0.50415	0.40215	0.32115	0.25674	0.16461	0.10600	0.06854
47	0.49670	0.39427	0.31331	0.24926	0.15828	0.10095	0.06466
48	0.48936	0.38654	0.30567	0.24200	0.15219	0.09614	0.06100
49	0.48213	0.37896	0.29822	0.23495	0.14634	0.09156	0.05755
50	0.47500	0.37153	0.29094	0.22811	0.14071	0.08720	0.05429

TABLE A.4 AMOUNT
OF AN ANNUITY
$$\frac{(1 + i)^n - 1}{i}$$

$n \setminus i$	$1\frac{1}{2}\%$	2%	$2\frac{1}{2}\%$	3%	4%	5%	6%
1	1.0000	1.0000	1.0000	1.0000	1.0000	1.0000	1.0000
2	2.0150	2.0200	2.0250	2.0300	2.0400	2.0500	2.0600
3	3.0452	3.0604	3.0756	3.0909	3.1216	3.1525	3.1836
4	4.0909	4.1216	4.1525	4.1836	4.2465	4.3101	4.3746
5	5.1523	5.2040	5.2563	5.3091	5.4163	5.5256	5.6371
6	6.2296	6.3081	6.3877	6.4684	6.6330	6.8019	6.9753
7	7.3230	7.4343	7.5474	7.6625	7.8983	8.1420	8.3938
8	8.4328	8.5830	8.7361	8.8923	9.2142	9.5491	9.8975
9	9.5593	9.7546	9.9545	10.1591	10.5828	11.0266	11.4913
10	10.7027	10.9497	11.2034	11.4639	12.0061	12.5779	13.1808
11	11.8633	12.1687	12.4835	12.8078	13.4864	14.2068	14.9716
12	13.0412	13.4121	13.7956	14.1920	15.0258	15.9171	16.8699
13	14.2368	14.6803	15.1404	15.6178	16.6268	17.7130	18.8821
14	15.4504	15.9739	16.5190	17.0863	18.2919	19.5986	21.0151
15	16.6821	17.2934	17.9319	18.5989	20.0236	21.5786	23.2760
16	17.9324	18.6393	19.3802	20.1569	21.8245	23.6575	25.6725
17	19.2014	20.0121	20.8647	21.7616	23.6975	25.8404	28.2129
18	20.4894	21.4123	22.3863	23.4144	25.6454	28.1324	30.9057
19	21.7967	22.8406	23.9460	25.1169	27.6712	30.5390	33.7600
20	23.1237	24.2974	25.5447	26.8704	29.7781	33.0660	36.7853
21	24.4705	25.7833	27.1833	28.6765	31.9692	35.7193	39.9927
22	25.8376	27.2990	28.8629	30.5368	34.2480	38.5052	43.3923
23	27.2251	28.8450	30.5844	32.4529	36.6179	41.4305	46.9958
24	28.6335	30.4219	32.3490	34.4265	39.0826	44.5020	50.8156
25	30.0630	32.0303	34.1578	36.4593	41.6459	47.7271	54.8645
26	31.5140	33.6709	36.0117	38.5530	44.3117	51.1135	59.1564
27	32.9867	35.3443	37.9120	40.7096	47.0842	54.6691	63.7058
28	34.4815	37.0512	39.8598	42.9309	49.9676	58.4026	68.5281
29	35.9987	38.7922	41.8563	45.2189	52.9663	62.3227	73.6398
30	37.5387	40.5681	43.9027	47.5754	56.0849	66.4388	79.0582
31	39.1018	42.3794	46.0003	50.0027	59.3283	70.7608	84.8017
32	40.6883	44.2270	48.1503	52.5028	62.7015	75.2988	90.8898
33	42.2986	46.1116	50.3540	55.0778	66.2095	80.0638	97.3432
34	43.9331	48.0338	52.6129	57.7302	69.8579	85.0670	104.1838
35	45.5921	49.9945	54.9282	60.4621	73.6522	90.3203	111.4348
36	47.2760	51.9944	57.3014	63.2759	77.5983	95.8363	119.1209
37	48.9851	54.0343	59.7339	66.1742	81.7022	101.6281	127.2681
38	50.7199	56.1149	62.2273	69.1594	85.9703	107.7095	135.9042
39	52.4807	58.2372	64.7830	72.2342	90.4091	114.0950	145.0585
40	54.2679	60.4020	67.4026	75.4013	95.0255	120.7998	154.7620
41	56.0819	62.6100	70.0876	78.6633	99.8265	127.8398	165.0477
42	57.9231	64.8622	72.8398	82.0232	104.8196	135.2318	175.9505
43	59.7920	67.1595	75.6608	85.4839	110.0124	142.9933	187.5076
44	61.6889	69.5027	78.5523	89.0484	115.4129	151.1430	199.7580
45	63.6142	71.8927	81.5161	92.7199	121.0294	159.7002	212.7435
46	65.5684	74.3306	84.5540	96.5015	126.8706	168.6852	226.5081
47	67.5519	76.8172	87.6679	100.3965	132.9454	178.1194	241.0986
48	69.5652	79.3535	90.8596	104.4084	139.2632	188.0254	256.5645
49	71.6087	81.9406	94.1311	108.5406	145.8337	198.4267	272.9584
50	73.6828	84.5794	97.4843	112.7969	152.6671	209.3480	290.3359

TABLE A.5 PRESENT
VALUE OF AN
ANNUITY

$$\frac{1 - (1 + i)^{-n}}{i}$$

$n \backslash i$	$1\frac{1}{2}\%$	2%	$2\frac{1}{2}\%$	3%	4%	5%	6%
1	0.9852	0.9804	0.9756	0.9709	0.9615	0.9524	0.9434
2	1.9559	1.9416	1.9274	1.9135	1.8861	1.8594	1.8334
3	2.9122	2.8839	2.8560	2.8286	2.7751	2.7232	2.6730
4	3.8544	3.8077	3.7620	3.7171	3.6299	3.5460	3.4651
5	4.7826	4.7135	4.6458	4.5797	4.4518	4.3295	4.2124
6	5.6972	5.6014	5.5081	5.4172	5.2421	5.0757	4.9173
7	6.5982	6.4720	6.3494	6.2303	6.0021	5.7864	5.5824
8	7.4859	7.3255	7.1701	7.0197	6.7327	6.4632	6.2098
9	8.3605	8.1622	7.9709	7.7861	7.4353	7.1078	6.8017
10	9.2222	8.9826	8.7521	8.5302	8.1109	7.7217	7.3601
11	10.0711	9.7868	9.5142	9.2526	8.7605	8.3064	7.8869
12	10.9075	10.5753	10.2578	9.9540	9.3851	8.8633	8.3838
13	11.7315	11.3484	10.9832	10.6350	9.9856	9.3936	8.8527
14	12.5434	12.1062	11.6909	11.2961	10.5631	9.8986	9.2950
15	13.3432	12.8493	12.3814	11.9379	11.1184	10.3797	9.7122
16	14.1313	13.5777	13.0550	12.5611	11.6523	10.8378	10.1059
17	14.9076	14.2919	13.7122	13.1661	12.1657	11.2741	10.4773
18	15.6726	14.9920	14.3534	13.7535	12.6593	11.6896	10.8276
19	16.4262	15.6785	14.9789	14.3238	13.1339	12.0853	11.1581
20	17.1686	16.3514	15.5892	14.8775	13.5903	12.4622	11.4699
21	17.9001	17.0112	16.1845	15.4150	14.0292	12.8212	11.7641
22	18.6208	17.6580	16.7654	15.9369	14.4511	13.1630	12.0416
23	19.3309	18.2922	17.3321	16.4436	14.8568	13.4886	12.3034
24	20.0304	18.9139	17.8850	16.9355	15.2470	13.7986	12.5504
25	20.7196	19.5235	18.4244	17.4131	15.6221	14.0939	12.7834
26	21.3986	20.1210	18.9506	17.8768	15.9828	14.3752	13.0032
27	22.0676	20.7069	19.4640	18.3270	16.3296	14.6430	13.2105
28	22.7267	21.2813	19.9649	18.7641	16.6631	14.8981	13.4062
29	23.3761	21.8444	20.4535	19.1885	16.9837	15.1411	13.5907
30	24.0158	22.3965	20.9303	19.6004	17.2920	15.3725	13.7648
31	24.6461	22.9377	21.3954	20.0004	17.5885	15.5928	13.9291
32	25.2671	23.4683	21.8492	20.3888	17.8736	15.8027	14.0840
33	25.8790	23.9886	22.2919	20.7658	18.1476	16.0025	14.2302
34	26.4817	24.4986	22.7238	21.1318	18.4112	16.1929	14.3681
35	27.0756	24.9986	23.1452	21.4872	18.6646	16.3742	14.4982
36	27.6607	25.4888	23.5563	21.8323	18.9083	16.5469	14.6210
37	28.2371	25.9695	23.9573	22.1672	19.1426	16.7113	14.7368
38	28.8051	26.4406	24.3486	22.4925	19.3679	16.8679	14.8460
39	29.3646	26.9026	24.7303	22.8082	19.5845	17.0170	14.9491
40	29.9158	27.3555	25.1028	23.1148	19.7928	17.1591	15.0463
41	30.4590	27.7995	25.4661	23.4124	19.9931	17.2944	15.1380
42	30.9941	28.2348	25.8206	23.7014	20.1856	17.4232	15.2245
43	31.5212	28.6616	26.1664	23.9819	20.3708	17.5459	15.3062
44	32.0406	29.0800	26.5038	24.2543	20.5488	17.6628	15.3832
45	32.5523	29.4902	26.8330	24.5187	20.7200	17.7741	15.4558
46	33.0565	29.8923	27.1542	24.7754	20.8847	17.8801	15.5244
47	33.5532	30.2866	27.4675	25.0247	21.0429	17.9810	15.5890
48	34.0426	30.6731	27.7732	25.2667	21.1951	18.0772	15.6500
49	34.5247	31.0521	28.0714	25.5017	21.3415	18.1687	15.7076
50	34.9997	31.4236	28.3623	25.7298	21.4822	18.2559	15.7619

ANSWERS TO SELECTED PROBLEMS

Exercise 1.1 **17** $3 \times 7 \times 11$ **18** $2 \times 3 \times 29$ **19** $2 \times 2 \times 3 \times 3 \times 11$
21 (a) Yes (b) no (c) no **22** (a) No (b) no (c) yes
23 (a) No (b) yes (c) yes **25** 3 divides $2! + 1 = 3$
26 4 does not divide $3! + 1 = 7$ **27** 6 does not divide $5! + 1 = 121$
29 $4(56) = 8(28) = 224$ **30** $3(165) = 15(33) = 495$ **31** $9(189) = 27(63) = 1701$
33 67, 6 **34** 2273, 2 **35** 339, 12 **37** 13, 279 **38** 1832, 38
39 14, 596 **41** $-\pi$, $\sqrt{18}$ **42** 1, -4, 0 **43** $-\frac{5}{8}$, -4, $-\pi$ **45** 25
46 25 **47** 17 **49** 5 **50** -8 **51** 2 **57** 8.91 and 15.246

58 11.54 and 30.504 **59** 62.8 and 139.15 **61** $\dfrac{8!}{4!4!} = 70$, $\dfrac{10!}{5!5!} = 252$

62 4 divides 8, 10 divides 30, and 11 divides 33

63 $\frac{81}{16} = 5 + \frac{1}{16}$, $\frac{256}{16} = 16$, $\frac{625}{16} = 39 + \frac{1}{16}$ **65** $7.34 \geq 7.2$

66 $16.354 \geq 13.68$ **67** $6237 \geq 5760$ **69** $841 = 841$ and $17261 < 24389$

Exercise 1.2 **1** True **2** True **3** True **5** False **6** True **7** False
9 True **10** False **11** False **13** True **14** True **15** True
17 True **18** True **19** False **21** False **22** True **23** True
25 22 **26** -5 **27** 0 **29** 0 **30** 9 **31** Not defined
41 $\sqrt{2}$, $\frac{12}{5}$, 2.6, $\frac{6}{2}$, 6.2 **42** -8, -6, -4, 5, 7
43 $-|-4| = -4$, $-1 - 2 = -3$, $|4 - 2| = 2$, $1 - (-2) = 3$, $|3 - 8| = 5$
53 1 **54** 4 **55** 4 **57** 2, 3 **58** 1, 6 **59** 5, 5
61 2 **62** 0 **63** 1 **65** 1 **66** 2 **67** 1 **69** 1 and $1 + 0 + 0$

Exercise 1.3 **1** 729 **2** 128 **3** 3125 **5** 343 **6** 25
7 64 **9** 81 **10** 64 **11** 729 **13** $\frac{32}{243}$ **14** $\frac{125}{8}$
15 $\frac{4}{49}$ **17** 5184 **18** 8,000,000 **19** 4,100,625 **21** $-x^5$ **22** x^7

23 $-x^9$ **25** $-6x^5$ **26** $-12x^7$ **27** $10x^8$ **29** $6x^7y^4$ **30** $-15x^2y^5$

31 $18x^3y^5$ **33** $-x^6$ **34** x^6 **35** x^{12} **37** $16x^4$ **38** $-27x^9$

39 $-125x^{12}$ **41** $32x^{10}y^5$ **42** $-27x^9y^3$ **43** $25x^8y^4$ **45** $6x^3y^4$

46 $-14x^5y^7$ **47** $-6x^7y^7$ **49** $4x^3y$ **50** $9x^3y^3$ **51** $6xy$

53 $8a^6b^3c^9$ **54** $9a^6b^4c^8$ **55** $625a^4b^8c^{16}$ **57** $a^6b^4/4c^6d^8$

58 $a^{12}b^9c^3/8d^6$ **59** $4x^8y^6/9w^4z^2$ **61** $324a^8b^6$ **62** $108a^{13}b^9$

63 $200c^8d^{13}$ **65** $uv/2w^2$ **66** $3b^6/4c$ **67** $4y^6/11x^6z^5$ **69** ac^2d^3

70 c^4/b^{10} **71** $b^7c^6/108a^5$ **73** $x^{2a-1}y^4$ **74** a^9d^{3b}

75 $a^{6n}b^{6n+1}$ **77** 2.9 **78** $1 - (\frac{2}{11})^8$ **79** $3/50a^2$

81 741 years

Exercise 1.4 **1** 33 **2** $-\frac{11}{57}$ **3** $\frac{9}{5}$ **5** $\frac{17}{5}$ **6** 26 **7** $81\frac{1}{2}$ **9** $16ab$ **10** $5LS$ **11** $-4alg$

13 $4a + b$ **14** $-6y$ **15** $11y - 4x$ **17** $60xy$ **18** $-48xy$ **19** -32

21 $x + 5y$ **22** $-2a + 29b$ **23** $-11ax + 22bx$ **25** $-25x + 13t$

26 $n + b + a$ **27** $5ax - 4bx - 3ay - 2by$ **29** $-5y + 3z$ **30** $-3y + 2z$

31 $2x - 5b$ **33** $-3x - y + 2$ **34** $-x + 3z - 4$ **35** $-a + 3b - 9$

37 $4b$ **38** $6a + 4b$ **39** $6a + b + c$ **41** $4a - 4ab - 2bc - 2c$ **42** $2a^3 + 10a^2$

43 $4a - 7b - 3ab$ **45** $-6x + 4y + 4z$ **46** $-3a - 5b + 6c$ **47** $12d - 6e - 6f$

49 $x^3 + 4x^2 - 6x + 3$ **50** $-x^3 + 7x^2 + 5x - 2$ **51** $x^4 + 3x^3 - 4x^2 + 6x - 1$

Exercise 1.5 **1** True **2** True **3** False **4** True **5** False **6** True

12 $a + bc = (a + b)(a + c) = a^2 + ac + ba + bc$; canceling bc (cancellation law of addition) gives $a = a^2 + ac + ab = a(a + b + c)$; if $a = 0$, this is true, and if $a \neq 0$, cancel a (cancellation law of multiplication), giving $1 = a + b + c$

13 $2 \times 3^3 \times 5 \times 11$ and $2^4 \times 13 \times 17$ **14** 14 and 8820 **15** 20 and 0; 20 and 20

16 $7.891011121314\cdots$ **17** $-\pi/\pi = -1$ **18** $-\pi/\pi = -1$ **19** 7.9

20 5.5 and 7.14 **21** $5.5111\cdots$ and $7.1777\cdots$ **22** True **23** False **24** True

25 False **26** True: multiply both members of $\frac{1}{3} = 0.333\cdots$ by 3. **27** False

28 True **29** True **30** True **31** False **32** $5art$ **33** $32x^{11}$ **34** $4x^2$ **35** x^7

36 $x - y$ **37** $13x - 4y$ **38** -6

Exercise 2.1 **1** $a - 4b + 4c$ **2** $6p - r$ **3** $2x + 5z$ **5** $ab - ac$ **6** $4xy + 2xz - yz$

7 $2pq$ **9** $4a - 7b + 9c$ **10** $3x + y + 3z$ **11** $p + 2d$ **13** $9ab^2 + 6ab - 6a^2b$

14 $-xy + 2xy^2 + 3xy^3$ **15** $p^2q + 2pq + pq^2$ **17** 7 **18** 44 **19** 40

21 $-a - b - 7c$ **22** $a + b - c$ **23** $-2x + 4y - 29z$ **25** $a^2 + b^2 + c^2$

26 $3x^2 + y^2 - z^2$ **27** $7ax + ax^2 + a^2x$ **29** $3a - 6b$ **30** $x - 3y + 5z$

31 $a - p + 2x$ **33** $-6x^3y^5$ **34** $-15x^6y^5$ **35** $14x^3y^5$ **37** $4x^6$ **38** $81x^8$

39 $64x^{15}$ **41** $6x^3y^4 - 10x^4y^5$ **42** $6x^4y^3 - 12x^5y^2$ **43** $8x^3y^6 - 12x^4y^4$

45 $-6x^3y^5 + 4x^4y^4$ **46** $-6x^3y^5 + 15x^4y^3$ **47** $10x^4y^6 - 20x^6y^5$

49 $-4x^3y + 3xy^3$ **50** $-2x^2y + 11x^3y^2$ **51** $2x^3y^4 + 13x^2y^6$

53 $6x^2 - 5xy - 6y^2$ **54** $8x^2 + 6xy - 9y^2$ **55** $15x^2 - 19xy + 6y^2$

57 $12x^2 - 37x + 28$ **58** $18x^2 - 63x + 40$ **59** $20x^2 - 9x - 20$

61 $6x^3 + x^2 - 30x - 25$ **62** $12x^3 + 19x^2 - 1$ **63** $8x^4 + 8x^3 + 25x - 14$

65 $2x^4 - 3x^3y - 8x^2y^2 + 15xy^3 - 6y^4$ **66** $6x^4 - 5x^3y + 6x^2y^2 - xy^3 - 6y^4$

67 $10x^4 - x^3y + 15x^2y^2 + 5xy^3 + 3y^4$

Exercise 2.2

1 x^3 **2** x^4 **3** $1/y^7$ **5** y^2/x^9 **6** x^3y **7** x^4/y^4 **9** $3a^3/b^2$ **10** $5a^4b^7$ **11** $19a^4$

13 $7a^5 - 5a^2 - 2a$ **14** $8a^3 - 12a + \dfrac{5}{a^2}$ **15** $-13x^5 - 11x^4 + 9x^2$

17 $-5xy + 6x^3 - 2x^2y^2$ **18** $\dfrac{3x^4}{y^5} - \dfrac{2x}{y^4} - \dfrac{7x^2}{y^6}$ **19** $\dfrac{3}{x} - 4y^3 - \dfrac{5y}{x^2}$ **21** $3x - 1, 3$

22 $2x + 3, -11$ **23** $5x - 2, -4$ **25** $3x + 2, -2$ **26** $2x - 1, 3$ **27** $5x + 3, -6$
29 $2x^2 - x + 3, 7$ **30** $5x^2 - 2x - 1, 0$ **31** $3x^2 + 2x + 1, -2$ **33** $3x - 2, 2$
34 $2x + 5, 4$ **35** $3x - 1, 3$ **37** $3x^2 + 4x + 5, -10$ **38** $2x^2 - 3x + 1, -5$
39 $4x^2 - 2x - 10, -48$ **41** $2x^3 - 3x^2 + 4x - 5, -7$ **42** $4x^3 + x^2 - 2x + 3, -5$
43 $3x^3 - 4x^2 + 5x + 3, 8$

Exercise 2.3

1 $2x^2 + 5x - 3$ **2** $3b^2 + 5b - 2$ **3** $5a^2 + 9a - 2$
5 $6x^2 + 5x - 6$ **6** $10h^2 - 19h - 15$ **7** $21i^2 + 13i - 20$
9 $6r^2 + 13rs + 6s^2$ **10** $42a^2 + 47ab + 10b^2$ **11** $8x^2 + 22xy + 15y^2$
13 $4x^2 - 4x + 1$ **14** $9i^2 - 6ij + j^2$ **15** $9a^2 - 24ab + 16b^2$
17 $25a^2 - 20ab + 4b^2$ **18** $16x^2 + 40xy + 25y^2$ **19** $49r^2 - 42rs + 9s^2$
21 $4a^4 + 12a^2b^3 + 9b^6$ **22** $16a^8 + 40a^4b^3 + 25b^6$
23 $9x^6 + 12x^3y^4 + 4y^8$ **25** $x^3 + 6x^2y + 12xy^2 + 8y^3$
26 $27x^3 + 135x^2y + 225xy^2 + 125y^3$ **27** $8x^3 + 12x^4 + 6x^5 + x^6$
29 $125x^3 - 75x^2y + 15xy^2 - y^3$ **30** $216x^3 - 540x^2y + 450xy^2 - 125y^3$
31 $343x^3 - 294x^2 + 84x - 8$ **33** $x^2 - 16$ **34** $x^2 - 49$
35 $9x^2 - 25$ **37** $4x^2 - 25y^2$ **38** $9x^2 - 16y^2$ **39** $49x^2 - 36y^2$ **41** $4a^4 - 9b^4$
42 $4a^4 - 25b^4$ **43** $25a^4 - 49b^6$ **45** $a^2/9 - b^2/4$ **46** $9a^2/16 - 16b^2/9$
47 $4a^2/9b^2 - 9c^2/16d^2$ **49** $x^2 + y^2 + z^2 + 2xy + 2xz + 2yz$
50 $4x^2 + y^2 + z^2 - 4xy + 4xz - 2yz$ **51** $x^2 + y^2 + 9z^2 - 2xy + 6xz - 6yz$
53 $4x^2 + y^2 + z^2 + 9w^2 + 4xy + 4xz - 12xw + 2yz - 6yw - 6zw$
54 $x^2 + 4y^2 + 4z^2 + w^2 - 4xy + 4xz + 2xw - 8yz - 4yw + 4zw$
55 $x^6 + 4x^5 - 2x^3 + 16x^2 - 12x + 9$ **57** $6x^2 + 12xy + 6y^2 + 13x + 13y + 6$
58 $40x^2 - 40xy + 10y^2 - 26x + 13y - 3$
59 $108x^2 - 144xy + 48y^2 + 21x - 14y - 10$ **61** $x^4 - 4x^3 + 4x^2 - 9$
62 $x^4 - 4x^2 + 12x - 9$ **63** $x^4 + 6x^3 + 9x^2 - 16$ **65** $x^6 + x^4 + 3x^2 - 1$
66 $x^4 - x^6 + x^2 - 1$ **67** $4x^8 + 8x^5 + x^2 - x^6 - 4x^4$

Exercise 2.4

1 $3(x + 3)$ **2** $7(x - 4)$ **3** $5(x + 6)$ **5** $x(x + 3)$ **6** $x(x - 4)$ **7** $x^2(x + 6)$
9 $x(x^2 - 5x - 4)$ **10** $2(2x^2 - 5x + 1)$ **11** $2x(3x^2 + 2x + 4)$ **13** $(a + x)(a - x)$
14 $(3 + y)(3 - y)$ **15** $(x + 5)(x - 5)$ **17** $(4y + 5a)(4y - 5a)$
18 $(8x + 7y)(8x - 7y)$ **19** $(4x + 9y)(4x - 9y)$ **21** $(p - q)(p^2 + pq + q^2)$
22 $(a + b)(a^2 - ab + b^2)$ **23** $(m + n)(m^2 - mn + n^2)$
25 $(x - 2y)(x^2 + 2xy + 4y^2)$ **26** $(3a + b)(9a^2 - 3ab + b^2)$
27 $(2a + 5b)(4a^2 - 10ab + 25b^2)$ **29** $(ab + 3c)(ab - 3c)$
30 $(pq + 2r)(pq - 2r)$ **31** $(3a + 2bc)(3a - 2bc)$ **33** $(x + y^2)(x - y^2)$
34 $(x^3 + y)(x^3 - y)$ **35** $(x^3 + y^4)(x^3 - y^4)$ **37** $(x^2 - y)(x^4 + x^2y + y^2)$
38 $(x^2 - y^3)(x^4 + x^2y^3 + y^6)$ **39** $(x - y^3)(x^2 + xy^3 + y^6)$
41 $(x - 2y + 3z)(x - 2y - 3z)$ **42** $(3x - y + 2z)(3x - y - 2z)$
43 $(9x + 4y - 5z)(9x - 4y + 5z)$
45 $(x + y - 1)(x^2 + 2xy + y^2 + x + y + 1)$

46 $(2x - y - 2)(4x^2 - 4xy + y^2 + 4x - 2y + 4)$
47 $(3x + 2y + 3)(9x^2 + 12xy + 4y^2 - 9x - 6y + 9)$
49 $(4x^2 + y^2)(2x + y)(2x - y)$ **50** $(x^2 + 9y^2)(x + 3y)(x - 3y)$
51 $(9x^2 + 25y^2)(3x + 5y)(3x - 5y)$ **53** $(x^2 + y^4)(x + y^2)(x - y^2)$
54 $(x^4 + 9)(x^2 + 3)(x^2 - 3)$ **55** $(x^6 + y^4)(x^3 + y^2)(x^3 - y^2)$
57 $(x^2 + y^2)(x + y)(x - y)$ **58** $(x^4 + y^4)(x^2 + y^2)(x + y)(x - y)$
59 $(x^8 + y^2)(x^4 + y)(x^4 - y)$
61 $(x + y)(x - y)(x^2 - xy + y^2)(x^2 + xy + y^2)$
62 $(x + y)(x^2 - xy + y^2)(x^6 - x^3y^3 + y^6)$ **63** $(x^2 - 3)(x^4 + 3x^2 + 9)$
65 $(2x - 1)(2x + 1)(4x^2 + 1)$ **66** $(4 + 9x^2)(2 + 3x)(2 - 3x)$
67 $(25x^2 + 9y^4)(5x + 3y^2)(5x - 3y^2)$
69 $(x + y)(x^4 - x^3y + x^2y^2 - xy^3 + y^4)$
70 $(x + y)(x^6 - x^5y + x^4y^2 - x^3y^3 + x^2y^4 - xy^5 + y^6)$
71 $(x - y)(x^4 + x^3y + x^2y^2 + xy^3 + y^4)$

Exercise 2.5   **1** $(x - 1)^2$ **2** $(x - 3)^2$ **3** $(x - 2)^2$ **5** $(2y - 1)(y + 2)$
6 $(3x + 1)(x - 2)$ **7** $(5h - 3)(h + 1)$ **9** $(3r - 1)(2r + 5)$
10 $(8y - 5)(y + 1)$ **11** $(3y - 7)(3y + 1)$ **13** $3(a + 2b)(a + 2b)$
14 $(5x + 8y)(x + 2y)$ **15** $(9u - 7v)(2u - v)$ **17** $(8h - 3k)(h + k)$
18 $(2c + 9d)(c - d)$ **19** $(3a - 2b)(a + 5b)$ **21** $(3p - 2q)(2p - 3q)$
22 $(4y + 3z)(3y + 2z)$ **23** $(5b - 4c)(3b - c)$ **25** $(3x + 1)(x + 2)$
26 $(2x + 1)(x + 2)$ **27** $(2x + 3)(3x - 1)$ **29** $(7x - 3)(x - 1)$
30 $(3x - 2)(x + 5)$ **31** $(2x + 3)(2x - 1)$ **33** 65   **34** -4 **35** -104 **37** 153
38 40   **39** 161   **41** $(x + 5)(x - 1)$ **42** $(5x - 1)(x + 1)$ **43** $(2x + 3)(x + 3)$
45 $(2x + 7y)(x - 4y)$ **46** $(3x + 8y)(x - 3y)$ **47** $(5x + 6y)(x - 3y)$
49 $(5x + 7y)(4x - 5y)$ **50** $(2x + 3y)(12x - 5y)$ **51** $(7x - 5y)(6x + 5y)$
53 $(6x - 7y)(5x + 7y)$ **54** $(6x - 5y)(9x + 7y)$ **55** $(2x + 3y)(15x - 4y)$
57 $(7x + 16y)(11x - 3y)$ **58** $(7x - 6y)(12x + 7y)$ **59** $(7x - 2y)(11x + 5y)$
61 $(15x - 20y + 2)(6x - 8y + 3)$ **62** $(4x - 6y + 9z)(3x + 4y - 6z)$
63 $(15x - 25y - 3z)(9x - 15y + 8z)$

65 (a) 15 (b) $\dfrac{(n + 1)(n + 2)}{2}$ **67** (a) $\dfrac{B - A}{B + A}$ (b) -0.16

Exercise 2.6   **1** $(a + 1)(b + 1)$ **2** $(x + 1)(y + 2)$ **3** $(u + 2)(v + 3)$
5 $(a - b)(a + 1)$ **6** $(x - 2y)(y + 1)$ **7** $(c + 2d)(2c - 3)$
9 $(a + b)(c + d)$ **10** $(x - y)(y + z)$ **11** $(2r - 5s)(3t + u)$
13 $(2a - 5b)(3a - 2c)$ **14** $(4x - 3y)(x - 6z)$ **15** $(5h - 3k)(3h + 7j)$
17 $(a - b + c)(a - 1)$ **18** $(x - y - z)(x + 1)$ **19** $(r - 2s - t)(r - 3u)$
21 $(x + y)(x - y - 1)$ **22** $(a - b)(a + b - 1)$ **23** $(a - b)(a - b - c)$
25 $(x - y)(x^2 + xy + y^2 - x + y)$ **26** $(a + b)(a + b - a^2 + ab - b^2)$
27 $(x - y)(x + 2y - x^2 - xy - y^2)$ **29** $(h - 2k + 4)(h - 2k - 4)$
30 $(r + 3s + 2t)(r + 3s - 2t)$ **31** $(5a + b - 2d)(5a - b + 2d)$
33 $(x - 2y + 2z + 3w)(x - 2y - 2z - 3w)$
34 $(2a - b + c - 3d)(2a - b - c + 3d)$
35 $(2r + 3s + t + 4u)(2r + 3s - t - 4u)$ **37** $(a - b)(a - 2b + 3c)$
38 $(x - y)(x - 2y - z)$ **39** $(2s - 3t)(2r - 3s + 4t)$
41 $(x^2 + 2x + 3)(x - 3)(x + 1)$ **42** $(x + 2)(x - 3)(x^2 + x + 6)$

43 $(x^2 + 3x - 2)(x - 2)(x - 1)$ **45** $(x - 3)(2x - 3)(2x + 1)$
46 $(x - 5)(3x - 1)(2x + 1)$ **47** $(x - 1)(x - 4)(x^2 + x + 1)$

Exercise 2.7   **1** $5a - 8b - 2c$ **2** $4a - 9c$ **3** $-27x - 2y$ **4** $-39a + 12b$
5 $4x^3 + 8x^2 + 7x + 2$ **6** $-15x^3 - 18x^2 + 8x + 4$ **7** $-22a + 90$

8 $\dfrac{1250}{x^8}$ **9** $2/x$ **10** $12x^5y - 24x^4y^4$ **11** $-x^{15}y^{18} + x^{14}y^{16}$ **12** $7x^4 - 5x^2y$

13 $8x^2 + 14xy - 15y^2$ **14** $6x^2 - 5xy - 6y^2$ **15** $9x^2 + 30xy + 25y^2$
16 $49x^2 + 56xy + 16y^2$ **17** $4x^2 - 49y^2$ **18** $9x^4 - 4y^6$
19 $4x^4 - x^2 + 6x - 9$ **20** $x^6 - 6x^5 + 9x^4 - 4x^2 + 4x - 1$
21 $6x^4 - 5x^3y - 5x^2y^2 + 5xy^3 - y^4$ **22** $5x^4 + 8x^3y - 12x^2y^2 - 17xy^3 - 2y^4$
23 $2x - 1, 4$ **24** $3x^2 - 2x + 4, 5$ **25** 2   **26** rem. $= 0$ **27** $7(x - 3y)$
28 $3x(x^2 - 2x + 3)$ **29** $(3 + 4x)(3 - 4x)$ **30** $(2x - 3 + 5y)(2x - 3 - 5y)$
31 $(x^2 + 9y^2)(x + 3y)(x - 3y)$ **32** $(x - y^2)(x^4 + x^3y^2 + x^2y^4 + xy^6 + y^8)$
33 $(x - 7)^2$ **34** $(3x - 4)^2$ **35** $(4x - 3)(3x - 1)$ **36** $(3x - 4)(4x + 1)$
37 313 is not a perfect square   **38** $x(12x - 13)$ **39** $(x + 2)(y - 3)$
40 $(x - y)(x^2 + xy + y^2 - x - y)$
41 $a^2 + b^2 + c^2 + d^2 + 2ab + 2ac + 2ad + 2bc + 2bd + 2cd$ **42** $22^2 + 29^2$
43 $3x^2 + 3xh + h^2$ **44** $1 - r^7$ **45** $(x + 5)(x - 1)(x^2 - 4x + 5)$
46 7225, 6351   **48** 0.997, 4.08   **49** $(x^2 + 3x + 1)^2$

Exercise 3.1   **1** $\dfrac{2}{x - y}$ **2** $\dfrac{-a - b}{a - b}$ or $-\dfrac{a + b}{a - b}$ **3** $\dfrac{3y - 2x - z}{z + y - x}$ **5** $\dfrac{a^2}{2ab}$

6 $\dfrac{6x^2}{9xy}$ **7** $\dfrac{a^2 + 2ab + b^2}{a^2 - b^2}$ **9** $\dfrac{y}{3x}$ **10** $\dfrac{3a}{4b}$ **11** $\dfrac{a - b}{a + b}$ **13** $\dfrac{3a^2 - 6ab}{6a^2}$

14 $\dfrac{8xy - 12y^2}{12xy}$ **15** $\dfrac{4v^2 - 16uv}{20v^2}$ **17** $\dfrac{x^2 - xy - 2y^2}{x^2 - y^2}$ **18** $\dfrac{-2a^2 - ab + b^2}{a^2 - b^2}$

19 $\dfrac{3h^2 + 10hk + 8k^2}{2h^2 + 3hk - 2k^2}$ **21** $\dfrac{5x^2 - 3xy - 2y^2}{x^3 - y^3}$ **22** $\dfrac{a^4 + a^2b^2 + b^4}{a^3 - b^3}$

23 $\dfrac{b^3 - 2bc^2 - c^3}{b^3 + c^3}$ **25** $\dfrac{x - 2}{x + 2}$ **26** $\dfrac{a + 3}{a - 2}$ **27** $\dfrac{2h - 1}{3h + 1}$ **29** $\dfrac{2x - 3y}{3x + 2y}$

30 $\dfrac{a - 3b}{a + 2b}$ **31** $\dfrac{w + 2z}{w - 2z}$ **33** $\dfrac{s - 2t}{2s - t}$ **34** $\dfrac{x - y}{x + y}$ **35** $\dfrac{3h - 2k}{2h + k}$ **37** $\dfrac{a^2 - ab + b^2}{a - b}$

38 $\dfrac{x^2 + xy + y^2}{x + y}$ **39** $\dfrac{1}{u^2 + v^2}$ **41** $\dfrac{x + y}{x^2 + xy + y^2}$ **42** $\dfrac{x - y}{x^2 - xy + y^2}$

43 $\dfrac{(x^2 + xy + y^2)(x^2 - xy + y^2)}{x^2 + y^2}$ **45** $\dfrac{1}{x + 1}$ **46** $\dfrac{1}{x + 1}$ **47** $\dfrac{1}{2x + 1}$

49 $\dfrac{1}{2x + 1}$ **50** $\dfrac{1}{x + 1}$ **51** $\dfrac{1}{x + 1}$ **53** $\left\{\dfrac{3xy}{x^2y}, \dfrac{2x}{x^2y}, \dfrac{4y}{x^2y}\right\}$

54 $\left\{\dfrac{y^2}{x^2y^2}, \dfrac{2xy}{x^2y^2}, \dfrac{3x^2}{x^2y^2}\right\}$ **55** $\left\{\dfrac{3y}{x^2y^2}, \dfrac{-2x}{x^2y^2}, \dfrac{xy}{x^2y^2}\right\}$

57 $\left\{\dfrac{(2x - y)(x + y)}{x^2 - y^2}, \dfrac{(x - 2y)(x - y)}{x^2 - y^2}, \dfrac{x + 2y}{x^2 - y^2}\right\}$

58 $\left\{\dfrac{x^2 - y^2}{x^3 + y^3}, \dfrac{x^4 + x^2y^2 + y^4}{x^3 + y^3}\right\}$

59 $\left\{\dfrac{(x - y)^2}{(x - 2y)(x^2 - y^2)}, \dfrac{(x + y)^2}{(x - 2y)(x^2 - y^2)}, \dfrac{(x - 2y)^2}{(x - 2y)(x^2 - y^2)}\right\}$

61 9.8   **62** $\frac{2}{9}, \frac{7}{36}, \frac{1}{6}, \frac{5}{36}, \frac{1}{9}, \frac{1}{12}, \frac{1}{18}, \frac{1}{36}$ **63** (a) $\frac{1}{9}$ (b) $\frac{1}{2}$

Exercise 3.2   **1** $\dfrac{c}{2}$ **2** $\dfrac{7y^2z}{3}$ **3** $\dfrac{3w}{2v}$ **5** $7c^2$ **6** $90q^2r$ **7** $\dfrac{8c}{7d^3}$ **9** $\dfrac{3x^2y}{4abz}$

10 $\dfrac{4mnp^2t^5}{5s^2u}$ **11** $\dfrac{2b^4}{ac^4d}$ **13** $\frac{5}{3}$ **14** $\dfrac{2x}{5y}$ **15** $\dfrac{2k}{3h}$ **17** $\dfrac{2a + b}{2a^2 + 4ab}$

18 $\dfrac{x - 3y}{6xy}$ **19** $\dfrac{2c(b + c)}{3b + 2c}$ **21** b **22** $\dfrac{2x - y}{2}$ **23** $\dfrac{c(3c + 5d)}{8d}$

25 $\dfrac{1}{x^2y}$ **26** $\dfrac{(a - b)b}{a}$ **27** q **29** 1 **30** y/x **31** 1 **33** $2x$

34 $\dfrac{1}{x + 3y}$ **35** x **37** $\dfrac{x - 1}{x + 6}$ **38** $\dfrac{x}{2x + 3}$ **39** 1 **41** $x + 2$

42 $\dfrac{(x - 1)(3x - 1)}{(x + 1)(3x + 1)}$ **43** $\dfrac{x + 1}{x + 3}$

Exercise 3.3   **1** $\dfrac{18x - 7}{6x + 1}$ **2** $\dfrac{-8x - 1}{x^2 + 11}$ **3** $\dfrac{6a + 7b - 9c}{d - 1}$ **5** $\dfrac{7x + 19}{18}$

6 $\dfrac{4x + 27}{12}$ **7** $\dfrac{-9x + 14}{12}$ **9** $\dfrac{4b^2c^2 - 9a^2c^2 + 5a^2b^2}{6abc}$

10 $\dfrac{15b^2c^2 - 16a^2c^2 + 6a^2b^2}{12abc}$ **11** $\dfrac{4a^2 - 6c^2 + 27b^2}{18abc}$ **13** $\dfrac{-a}{b}$ **14** 0

15 $\dfrac{a^2 + b^2}{5ab}$ **17** $\dfrac{15x^2 + 9xy - 2y^2}{2y(3x + y)}$ **18** $\dfrac{3x^2 - 2y^2}{x(2x + y)}$ **19** $\dfrac{9x^2 + 4y^2}{2y(3x + 2y)}$

21 $\dfrac{6xy}{(x - y)(x + y)}$ **22** $\dfrac{5x^2 - 9xy + 3y^2}{(2x - y)(3x - y)}$ **23** $\dfrac{19x^2 - 62xy + 34y^2}{(3x - 8y)(2x - 5y)}$

25 $\dfrac{1}{r - s}$ **26** $\dfrac{r + s}{s}$ **27** $\dfrac{r + s}{r}$ **29** $\dfrac{2x + 3y}{y(x + y)}$ **30** $\dfrac{2x - 3y}{x(3x + 2y)}$ **31** $\dfrac{2(x + 2y)}{y(x - 2y)}$

33 $\dfrac{1}{(x - y)(x - 2y)}$ **34** $\dfrac{4(a + 2b)}{(a + b)(a - 2b)(a + 3b)}$ **35** $\dfrac{7}{(a + 5b)(a - 4b)}$

37 $\dfrac{-16x^3 - 29x^2y + 25xy^2 + 33y^3}{(3x + 2y)(4x - 3y)(2x - 3y)}$ **38** $\dfrac{-5x^3 - 18x^2y - 18xy^2 + 13y^3}{(2x + 5y)(x + 3y)(x - 2y)}$

39 $\dfrac{15x^3 + 20x^2y + 40xy^2 - 5y^3}{(x + 3y)(2x + y)(3x - y)}$ **41** $\dfrac{x^2 + xy + 4y^2}{(2x + 3y)(x - y)(2x - y)}$

42 $\dfrac{2(4x^2 + xy + y^2)}{(x - 3y)(x + y)(3x + y)}$ **43** $\dfrac{x^2 - 9xy - 3y^2}{(2x - y)(x + 2y)(x - 2y)}$

49 $\dfrac{a}{\sqrt{3k^2 + 2p^2}}$ **50** $1.78 > 0.94$

51 Since $12 + 11 + 10 + \cdots + 1 = 78$, the sum of all fractions is $\frac{78}{78}$, or 100 percent.

Exercise 3.4   **1** $\frac{1}{4}$ **2** $\frac{27}{16}$ **3** $\frac{2}{23}$ **5** $\dfrac{x}{2x + 1}$ **6** $\dfrac{x - 3}{x}$ **7** $x + 4$ **9** $\dfrac{x^2 - 2}{x}$ **10** $\dfrac{x^2 - 9}{5}$ **11** $\dfrac{x}{5}$

13 $\dfrac{x - 5}{x - 4}$ **14** $\dfrac{x - 3}{x + 2}$ **15** $\dfrac{3x + 2}{x + 3}$ **17** $\dfrac{y - x}{y + x}$ **18** $\dfrac{y - x}{y + 2x}$ **19** $\dfrac{3x + y}{2x + y}$

21 $\dfrac{a - 2}{a - 4}$ **22** $\dfrac{2a + 3}{2a - 1}$ **23** $\dfrac{2a + 3}{a - 2}$ **25** $\dfrac{p}{p + 3}$ **26** $\dfrac{1}{p + q}$ **27** $\dfrac{-2p}{3(p - 1)}$

29 $\dfrac{-1}{2x + 1}$ **30** $\dfrac{-(x + 3)}{2(x + 1)}$ **31** $\dfrac{-1}{x + 1}$ **33** $\dfrac{y + 2x}{2x^2}$ **34** $\dfrac{x - 1}{2y}$ **35** $\dfrac{v}{w}$

41 (a) $\dfrac{mP - 1}{mQ - 1}$ (b) 0.205 **42** (a) $\frac{1}{3}$ (b) $\frac{1}{2}$

Exercise 3.5   **1** $\dfrac{y^2}{xy}$ **2** $\dfrac{xy}{x^2}$ **3** $\dfrac{y}{x}$ **4** $\dfrac{x}{2}$ **5** $\dfrac{-x + 2y}{x - 3y}$ **6** $\dfrac{(x - 3)(x - 4)}{x^2 - 16}$ **7** $\dfrac{(x + 4y)(x + y)}{(x - 2y)(x + y)}$

8 $\dfrac{(x - 3)(x + 3)}{(x + 1)(x + 3)}$ **9** $\dfrac{x - 2}{x - 3}$ **10** $\dfrac{2x - 1}{x + 4}$ **11** $\dfrac{x - 2}{x + 2}$ **12** $\dfrac{3x - 1}{3x + 1}$ **13** $\dfrac{x + 4y}{x - 3y}$

14 $\dfrac{x + 2y}{x + 3y}$ **15** $\dfrac{z + 2w}{2z - w}$ **16** $\dfrac{x^2 - xy + y^2}{x - y}$ **17** 1 **18** $\dfrac{1}{2x + 1}$

19 $\left\{ \dfrac{(2x - 1)(x + 1)}{x^2 - 1} ; \dfrac{(x - 2)(x - 1)}{x^2 - 1}, \dfrac{x + 2}{x^2 - 1} \right\}$ **20** $\left\{ \dfrac{x^2 - y^2}{x^3 + y^3}, \dfrac{x^4 + x^2y^2 + y^4}{x^3 + y^3} \right\}$

21 $\dfrac{y^2z^2w}{x^2}$ **22** $\dfrac{4y^2zw}{3}$ **23** $\dfrac{5x(x + 7)}{x + 2}$ **24** $xy^2(x + y)$ **25** $\dfrac{x - 3y}{x + 2y}$ **26** $x + 2$

27 $\frac{44}{45}$ **28** $\dfrac{25z^2 - 3y^2 - 27x^2}{45xyz}$ **29** $\dfrac{2x + 1}{x + 3}$ **30** $\dfrac{1}{(x - y)(x - 2y)}$

31 $\dfrac{y}{(x - y)(x - 3y)}$ **32** $\frac{1}{13}$ **33** -4 **34** $x + 2$ **35** $\dfrac{x - 3}{x + 2}$ **36** $\dfrac{y + 2x}{y - 3x}$

37 $\dfrac{1}{x + y}$ **38** $\dfrac{2(x + 1)(x - 2)}{x^2}$ **39** $\dfrac{1}{x(1 - x)}$ **42** 171 and 57   **43** 0.00696

44 (a) 1311 (b) 1185 (c) 1218

Exercise 4.1

1 $\frac{1}{16}$ **2** $\frac{1}{27}$ **3** $\frac{1}{5}$ **5** $\frac{1}{27}$ **6** $\frac{1}{5}$ **7** 1 **9** 7 **10** $\frac{1}{9}$ **11** 11 **13** $\frac{1}{9}$ **14** 64 **15** $\frac{1}{625}$
17 $\frac{3}{2}$ **18** $\frac{81}{625}$ **19** $\frac{1}{9604}$ **21** $\frac{1}{5184}$ **22** $\frac{729}{512}$ **23** $1/91{,}125$ **25** $a^{-2}b^{-2}$ **26** a^2b^2

27 a^2b^{-2} **29** $2ab^{-1}c^{-4}$ **30** $3ab^{-1}c$ **31** $12a^{-6}b^{-2}$ **33** $\dfrac{v^7w^2}{18u^2}$ **34** $\dfrac{4r^3t}{3s^5}$

35 $\dfrac{27g^5h^3}{2f^5}$ **37** c^9d^6 **38** $\dfrac{b^6}{a^2}$ **39** z^6 **41** $\dfrac{m^6}{n^2}$ **42** $\dfrac{d^2}{c^3}$ **43** $\dfrac{1}{b^4r^6}$ **45** $\dfrac{m^{10}}{9w^{10}}$

46 $\dfrac{512y^9}{x^9}$ **47** $\dfrac{512}{19{,}683b^{18}z^{21}}$ **49** $\dfrac{r^5}{a^5t^5}$ **50** $\dfrac{u^{12}m^{20}}{b^4}$ **51** $\dfrac{f^8}{c^{20}d^8}$ **53** $\dfrac{y+x}{y-x}$

54 $\dfrac{ay^2-1}{1-a^3}$ **55** $\dfrac{1}{a+b}$ **57** $x+y$ **58** $x-3y$ **59** $\dfrac{1}{x-2y}$ **61** $\dfrac{-x-9}{(x-3)^4}$

62 $-\dfrac{x-2}{(x-3)^3}$ **63** $\dfrac{(x-3)^3(2x-16)}{(x-5)^4}$ **65** $\dfrac{-14}{(3x+1)^2}$ **66** $\dfrac{3x+3}{(2x+5)^3}$

67 $\dfrac{13x+6}{(2x-3)^3(3x+4)^2}$ **69** 7.82×10^{11} **70** 2.64×10^{31} **71** 9.51×10^{21}

73 6.0×10^{26} **74** 7.7×10^{12} **75** 7.42×10^{58} **77** 35.98 **78** 260.684 **79** 1085.9
81 60.1 **82** 142.3 **83** 14.53

Exercise 4.2

1 $2\sqrt{3}$ **2** $7\sqrt{2}$ **3** $11\sqrt{3}$ **5** $5\sqrt[3]{2}$ **6** $2\sqrt[3]{5}$ **7** $2\sqrt[4]{3}$ **9** $4a^4b^3\sqrt{7a}$

10 $6b^3\sqrt{3b}$ **11** $2a^2b\sqrt{7}$ **13** $3xy^2\sqrt[3]{2}$ **14** $5xy^3\sqrt[3]{3x}$ **15** $2xy\sqrt[4]{5x^2y}$ **17** $6\sqrt[3]{6}$

18 $3\sqrt[4]{5}$ **19** $2\sqrt[5]{15}$ **21** $5xy^2\sqrt{15x}$ **22** $14x^2y\sqrt{3y}$ **23** $6x^2y\sqrt{3y}$ **25** $3xy\sqrt[3]{2y^2}$

26 $2xy^2\sqrt[3]{20x^2y^2}$ **27** $3xy\sqrt[4]{5x}$ **29** $\sqrt[4]{2}$ **30** $\sqrt[6]{5}$ **31** $\sqrt[8]{3}$ **33** $\sqrt[6]{a^2} = \sqrt[3]{a}$ **34** $\sqrt[8]{a}$

35 $\sqrt[20]{a^2} = \sqrt[10]{a}$ **37** $-\sqrt{3}$ **38** $7\sqrt{5}$ **39** $-\sqrt[3]{2}$ **41** $5\sqrt{2} + \sqrt[3]{2}$ **42** $\sqrt{3} + 5\sqrt[3]{2}$

43 $11\sqrt{2} - 3\sqrt[3]{3}$ **45** $3a\sqrt{b} - b\sqrt{a}$ **46** $(r + 3t^2)\sqrt{2rt} + r(r - t)\sqrt{r}$

47 $(2c + 3d)\sqrt[3]{c^2d} + cd(3 - c)\sqrt[3]{cd^2}$ **49** $\dfrac{x^2\sqrt{15xy}}{5y}$ **50** $\dfrac{\sqrt{6xy}}{4y^2}$ **51** $\dfrac{x\sqrt{xy}}{5y^3}$

53 $\dfrac{9x^2y\sqrt{y}}{7}$ **54** $\dfrac{2xy^2\sqrt{y}}{3}$ **55** $\dfrac{xy\sqrt{42y}}{6}$ **57** $\dfrac{3x\sqrt[3]{2y}}{2y}$ **58** $\dfrac{5\sqrt[3]{12x^2y^2}}{6y}$ **59** $\dfrac{x^4\sqrt{24xy^2}}{3y}$

61 $\dfrac{7x^2y^2\sqrt{xy}}{9}$ **62** $\dfrac{x^4y^3\sqrt{102y}}{6}$ **63** $\dfrac{3x^2y^5\sqrt{2}}{14}$ **65** $-1 - \sqrt{3}$ **66** $-(\sqrt{2} + 3)$

67 $3(\sqrt{5} - 2)$ **69** $2(\sqrt{5} + \sqrt{2})$ **70** $\dfrac{-(\sqrt{3} + \sqrt{7})}{2}$ **71** $\sqrt{10} + \sqrt{5}$

73 $\dfrac{2\sqrt{5} + 2\sqrt{3} + \sqrt{15} + 3}{2}$ **74** $\dfrac{\sqrt{3} - \sqrt{5} + 2\sqrt{15} - 10}{-2}$ **75** $-7 - 3\sqrt{6}$

77 1.8 s **78** 35.5 cm² **79** $2, 2.5937, 2.7048, 2.7169$ **81** $0.5\sqrt{21}$

82 $\sqrt{2 + \dfrac{L}{m}}$ **83** $\dfrac{a}{1 + a}$ **85** 10.71 **86** 6.530 **87** 5.298

89 1.85, 1.96, 1.99   **90** $\dfrac{1}{\sqrt{x + h + 3} + \sqrt{x + 3}}$

91 Using decimal approximations gives $\sqrt{3 + \sqrt{2}} = 2.101$ and $\sqrt{(3 + \sqrt{7})/2} +$ $\sqrt{(3 - \sqrt{7})/2} = 1.680 + 0.421 = 2.101$ **93** $\sqrt[3]{3} > \sqrt{2} = \sqrt[4]{4}$

95 $\sqrt{12}, \sqrt[3]{-81}, \sqrt[5]{77}$

101 (a) $\left(\sqrt[12]{2}\right)^4, \frac{5}{4}$ (b) $\left(\sqrt[12]{2}\right)^5, \frac{4}{3}$ (c) $\left(\sqrt[12]{2}\right)^7, \frac{3}{2}$

Exercise 4.3

1 9   **2** 6   **3** 5   **5** 8   **6** 32   **7** 32   **9** $\frac{1}{3}$ **10** $\frac{1}{3}$ **11** $\frac{1}{16}$ **13** $\frac{27}{64}$ **14** $\frac{81}{256}$ **15** $\frac{8}{27}$ **17** $\frac{5}{4}$ **18** $\frac{9}{16}$ **19** $\frac{25}{4}$ **21** 27   **22** 9   **23** 4   **25** $5bc^2$ **26** $4a^2b^3$ **27** $6c^3d^4$

29 $2m^2n^3$ **30** $3rs^3$ **31** $2x^2y$ **33** $6a^2/b^3c^4$ **34** $m^2n^4/2p$ **35** $3s^2/t^3$ **37** $\sqrt[3]{ab^2}$

38 $\sqrt[4]{a^3b^2}$ **39** $\sqrt[15]{a^{10}b^9}$ **41** $\sqrt[15]{a^{10}/y^3}$ **42** $\sqrt[6]{b^2/a^3}$ **43** $\sqrt[15]{1/a^6b^5}$ **45** $6x^{5/6}$ **46** $10x^{7/12}$ **47** $12x^{7/10}$ **49** $4x^{1/3}/y^{1/4}$ **50** $3x^{1/5}/y^{5/6}$ **51** $5x^{1/2}/y^{3/2}$ **53** $4x^{1/8}/y^{1/5}$ **54** $a/3b^{1/3}$ **55** $a^{1/3}/3b^{1/7}$ **57** $9y^{14/9}/4x^6$ **58** $3/x^{1/6}y^{1/12}$ **59** $x^{1/6}/2y^{3/4}$ **61** $\frac{4}{3}$ **62** $3/5a^3b$ **63** $1/4^{2/3}a^{8/3}b^4$ **65** $(3x + 4)/(2x + 3)^{1/2}$ **66** $(-3x + 2)/(5 - 2x)^{1/2}$ **67** $(21x - 24)/(2x - 5)^{1/4}$ **69** $(14x + 5)/(2x - 1)^{1/3}(x + 1)^{1/2}$ **70** $(13x + 10)/(3x + 2)^{1/3}(x + 1)^{1/2}$

71 $(9x + 8)/(3x - 1)^{1/2}(2x + 3)^{3/4}$ **73** $x^{4/b}$ **74** x^a **75** x^{4ab} **77** $\dfrac{5\sqrt{5}}{2}$ **78** 21.1

79 $S = S_0 + \sqrt[c]{\dfrac{R}{k}}$ **81** (a) 19,280 (b)12,929   **82** 8176

83 (a) Decreased to 90 percent (b) \$9989

Exercise 4.4

1 $10 + 7i$ **2** $10 + 3i$ **3** $9 + 7i$ **5** $-3 - 3i$ **6** 4   **7** 0   **9** $12 + 5i$ **10** $13i$

11 $14 + 48i$ **13** 25   **14** $18 + i$ **15** $21 - 20i$ **17** $\dfrac{1 - i}{2}$ **18** $\dfrac{7 + 23i}{17}$

19 $\dfrac{12 + 5i}{13}$ **21** $3 - 2i$ **22** $2 + 3i$ **23** $1 + 7i$ **25** $\dfrac{-34 - 12i}{325}$ **26** $\dfrac{32i}{25}$

27 $\frac{4}{5}$ **29** 5   **30** 5   **31** 13   **49** $x = 3, y = 2$ **50** $x = 4, y = 7$ **51** $x = 6, y = 2$ **53** $x = 2, y = 1$ **54** $x = 6, y = 1$ **55** $x = 3, y = 3$

Exercise 4.5

1 729   **2** 36   **3** $\frac{16}{81}$ **4** 64   **5** 1,259,712   **6** $15x^2y^7$ **7** $4x^2y^3$ **8** a^4b^8

9 $8a^9b^6$ **10** a^2/b^4 **11** $a^{12}bc^6$ **12** $\dfrac{32a}{c^2d}$ **13** $a^{3n}b^{n+5}$ **14** $\frac{1}{9}$ **15** 9   **16** $\frac{1}{81}$

17 $\frac{1}{2}$ **18** a^3b^2 **19** $1/a^3b^4c^3$ **20** $6b^3/a^4$ **21** a^6b^9 **22** 5   **23** 0.8   **24** 2   **25** 3   **26** $5ab^3$ **27** $2a/b^2$ **28** $26a^2/b$ **29** $6x^{7/12}$ **30** $15x^{8/15}$ **31** $5xy^2$

32 $4a^{1/5}/b$ **33** $2/3y$ **34** x **35** $\dfrac{30x - 19}{(2x - 3)^{2/3}(3x + 2)^{1/2}}$ **36** $3\sqrt{3}$ **37** $4\sqrt{5}$

38 $2\sqrt[3]{5}$ **39** $2\sqrt[5]{3}$ **40** 4 **41** 6 **42** $a\sqrt[5]{a}$ **43** $10xy^2$ **44** $3x^3y^2\sqrt{10}$

45 $\dfrac{3x^4}{4y^2}$ **46** $4\sqrt{2} - 5$ **47** $\sqrt[3]{3}$ **48** $\sqrt[3]{4}$ **49** $\sqrt[3]{x^4}$ **50** 2 **51** 0

52 $3\sqrt{a} + 2\sqrt{b} + 3\sqrt[3]{a}$ **53** $b\sqrt[3]{a} + \sqrt{ab}(2a + 3ab)$
54 $\sqrt{3ab}$ **55** -4 **56** $7\sqrt{10} - 3$ **57** $u^{3/2} + v^{3/2}$ **58** $a - b + c + 2\sqrt{ac}$
59 2.91, 2.997 **62** $0.902I_0$, $0.814I_0$, $(0.902)^nI_0$ **63** 0.040 **64** y approaches 1
65 788,000

Cumulative
review
exercise for
Chaps. 1 to 4

1 $2^2 \cdot 5 \cdot 23$ **2** Distributive **3** It has no integral factor except itself and 1.
4 $\frac{3}{2}$, -5 **5** 8 **6** It is, it is **7** $A - B = 54 - 36 = 18$
8 472 and 853 **9** True, false **10** $\frac{2}{11}$, $3.141414\cdots$ **11** -81, 81
12 1296, $\frac{125}{343}$ **13** $\frac{1}{9}$, $\frac{1}{8}$, $-\frac{1}{8}$ **14** $16a^2c^4d^3/3$ **15** $2c^2/a^3t^3$
16 $5.83(10^2)$, $5.83(10^{-2})$, $5.083(10)$, $2.03(10^{-3})$ **17** 73.6, 73.7, 0.0737, $7.37(10^5)$
18 $1.14(10^3)$, 0.725 **19** 8.539, 5.860 **20** $12a^3 - 24a^2 - 13a$
21 $15x^2 + 2xy - 8y^2$ **22** $9x^2 - 25y^2$ **23** $27x^3 - 135x^2y + 225xy^2 - 125y^3$
24 $x^2 + 4y^2 + 9z^2 - 4xy + 6xz - 12yz$ **25** $4x^2 - 25y^2 + 70yz - 49z^2$
26 $3x^2 - 5x + 4$, 19 **27** $2x^4 + x^3y - 18x^2y^2 + 13xy^3 - 2y^4$
28 $(3a - 2b)(9a^2 + 6ab + 4b^2)$ **29** $(2a - b + 3)(2a - b - 3)$
30 $(2a + 3b)(3a - 5b)$ **31** $D = 37$ is not a positive perfect square.
32 $(2x + 2y - 3)(2x - 2y + 3)$ **33** $(x^2 + 2 + x)(x^2 + 2 - x)$
34 $8x^3/3y$ **35** $2x + 3$ **36** $10x^3 - 9x^2 - 13x - 15/[(3x - 1)(x + 3)(2x + 3)]$
37 $-(7x + 9)/(8x^2 + 18x + 9)$ **38** 1 **39** 3 **40** -3

41 $6\sqrt{2}$ **42** $0.6\sqrt[4]{2}$ **43** $(2xy)\sqrt[3]{9x^2z^2}$ **44** $(3xy^2)\sqrt[5]{7y}$

45 $(-1-\sqrt{15})/2$ **46** $\sqrt[6]{6a}$ **47** $\sqrt[3]{6}$ **48** $(2ab - 3a^2 - 2b)\sqrt{2a}$

49 $\sqrt[3]{3} + 13\sqrt{2}$ **50** 37 **51** 2.27 **52** 25, -5, -5, -5, $5i$

53 $\sqrt[5]{a^2b^3}$, $\sqrt[6]{b^2/a^3}$ **54** $a/5b^{1/3}$ **55** $(8x - 9)/[(2x - 1)^{1/3}(x + 2)^{1/2}]$

56 x^a **57** $\sqrt[5]{\frac{10}{3}} \approx 1.27$ **58** $5 + 6i$ **59** $1 + 6i$ **60** $11 + 7i$
61 $(1 + 13i)/5$ **62** $\sqrt{34}$ **63** $29(1 + 13i)/170$ **64** $\frac{154}{69}$

Exercise 5.1

1 Identity **2** Conditional equation **3** Conditional equation
5 Conditional equation **6** Identity **7** Identity **9** No **10** Yes
11 No **13** -2 **14** -1 **15** 3 **17** 1 **18** -1 **19** 1 **21** 10 **22** -8
23 12 **25** 6 **26** 4 **27** 5 **29** 4 **30** 6 **31** 12 **33** 1 **34** -2 **35** -3
37 -1 **38** 3 **39** 9 **41** 5 **42** 8 **43** 4 **45** 2 **46** -2 **47** 7 **49** -2 **50** -4
51 -3 **53** 4 **54** 3 **55** 7 **57** 1 **58** 3 **59** 3 **61** 5 **62** -4 **63** -7 **65** 6

66 8 **67** 5 **69** $F = 1.8C + 32$ **70** $d = \dfrac{Ak}{4\pi C}$ **71** $f = \dfrac{pq}{p + q}$

73 $p = \dfrac{c - m(1 - d)}{c}$ **74** $r = \dfrac{Ne - IR}{IN}$ **75** $a = \dfrac{S(1 - r)}{1 - r^n}$ **77** $-\frac{5}{2}$ **78** $-\frac{11}{2}$

79 (a) $H = \dfrac{z(x^2 + W^2)}{Wx}$ (b) $H = \dfrac{2zW^2}{2Wx - x^2}$ (c) $H = \dfrac{W(y + z)}{x}$

Exercise 5.2 **1** $x + (x + 1) + (x + 2) = 75$, where x is the smallest integer; 24, 25, 26
2 $x + 4x/3 = 224$; \$96, \$128 **3** $41,209 = 2x - 5015$, where x is the 1970
population; 18,097 increase in population **5** $x + (x + 195) = 625$; \$215
and \$410 **6** $2898 = 210 + x/2$; 5376 **7** $x/3 + 2 = x/2$; 12 problems
9 Fred 1526, Bruce 1210 **10** 532 **11** \$126 **13** 680 mi
14 \$15,500 at 9 percent, \$16,250 at 10 percent
15 \$728 on living-room carpet; \$760 on bedroom carpet **17** \$400, \$350
18 78 **19** 11, 14 **21** 10 mi **22** 46 mi/h **23** 48 mi/h **25** 10 mi **26** $1\frac{1}{2}$ h
27 18 min **29** $3\frac{1}{4}$ h **30** $3\frac{3}{4}$ h **31** $17\frac{1}{2}$ min **33** 10 min **34** 10 h **35** $19\frac{1}{5}$ h
37 4 at \$3.94; 3 at \$1.89 **38** 1125 lb of 9.3 percent; 3375 lb of 11.3 percent
39 24 yd³ **41** 50 ml **42** 112 mi **43** Bus, 40 mi/h; plane, 660 mi/h
45 20 km **46** 1 h and 24 min **47** \$80

Exercise 5.3 **1** 2, -2 **2** $\frac{1}{3}$, $-\frac{1}{3}$ **3** $\frac{5}{3}$, $-\frac{5}{3}$ **5** 0, -6 **6** 0, -5 **7** 0, 2 **9** $2i$, $-2i$
10 $3i$, $-3i$ **11** $5i/2$, $-5i/2$ **13** 4, 1 **14** 4, 5 **15** 4, -3 **17** 3, 2 **18** 3, -1
19 -3, -4 **21** $\frac{2}{3}$, $\frac{3}{2}$ **22** $\frac{4}{3}$, $-\frac{3}{4}$ **23** $\frac{2}{3}$, $\frac{3}{2}$ **25** $\frac{3}{2}$, $-\frac{1}{5}$ **26** $\frac{1}{6}$, $-\frac{3}{2}$ **27** $\frac{5}{3}$, $-\frac{2}{7}$
29 3, $\frac{2}{5}$ **30** 2, $\frac{1}{3}$ **31** 2, $\frac{4}{7}$ **33** $1 \pm \sqrt{3}$ **34** $2 \pm \sqrt{5}$ **35** $3 \pm \sqrt{2}$

37 $\dfrac{2 \pm \sqrt{3}}{2}$ **38** $\dfrac{3 \pm \sqrt{2}}{3}$ **39** $\dfrac{2 \pm \sqrt{3}}{3}$ **41** $4 \pm \sqrt{21}$ **42** $-2 \pm \sqrt{10}$

43 $\dfrac{3 \pm \sqrt{23}}{2}$ **45** $2 \pm 3i$ **46** $3 \pm 2i$ **47** $4 \pm 2i$ **49** $\dfrac{3 \pm i}{2}$ **50** $\dfrac{2 \pm i}{3}$

51 $\dfrac{3 \pm 2i}{5}$ **53** 1; rational, unequal; 5, 6 **54** 121; rational, unequal; $\frac{5}{12}$, $-\frac{1}{6}$

55 zero; rational, equal; 6, 9 **57** 72; real, unequal; 6, -9
58 84; real, unequal; $-\frac{1}{2}$, $-\frac{5}{4}$ **59** -76; conjugate imaginary; $-\frac{1}{2}$, $\frac{5}{4}$
61 $x^2 - 3x + 2 = 0$ **62** $15x^2 - 2x - 8 = 0$ **63** $x^2 - 4x - 1 = 0$ **65** $-\frac{89}{8}$
66 $\frac{161}{20}$ **67** $-\frac{13}{3}$ **69** $-\frac{11}{2}$ **70** $\frac{17}{4}$ **71** $\frac{41}{4}$ **73** 5400 **74** 16
75 11,647 **77** (a) $\frac{1}{2}[B + C \pm \sqrt{(B - C)^2 + 4AD}]$ (b) Discriminant is ≥ 0
78 $(1 \pm \sqrt{5})/4$ **79** 0.766

Exercise 5.4 **1** 3, -3, 2, -2 **2** 4, -4, 2, -2 **3** 4, -4, 3, -3 **5** $\frac{2}{3}$, $-\frac{2}{3}$, $\frac{1}{2}$, $-\frac{1}{2}$
6 $\frac{3}{4}$, $-\frac{3}{4}$, $\frac{2}{5}$, $-\frac{2}{5}$ **7** $\frac{2}{3}$, $-\frac{2}{3}$, 1, -1 **9** $\pm\frac{27}{1000}$, ± 8 **10** ± 27, ± 8 **11** ± 8, $\pm\frac{27}{8}$
13 3, -3, 2, -2 **14** 2, -2, 1, -1 **15** 2, -3, 3, -4 **17** -10, 4

18 0, -1 **19** -5, 1 **21** $(5 \pm \sqrt{21})/2$, -1 **22** $-2 \pm \sqrt{5}$, $\dfrac{3 \pm \sqrt{13}}{2}$

23 1, 9 **25** 3 **26** -1 **27** 6 **29** $\frac{1}{2}$ **30** 0 **31** 4 **33** 3, -2 **34** 0, $-\frac{1}{3}$ **35** 2
37 0, 3 **38** 3, 4 **39** $\frac{1}{6}$, -2 **41** 2, $-\frac{3}{4}$ **42** 0, 3 **43** 3, 5 **45** $\frac{2}{3}$ **46** 4, $\frac{2}{3}$ **47** 5

Exercise 5.5 **1** $x(x + 1) - 2x - 1 = 19$; 5, 6; -4, -3 **2** $(4 + x)^2 + x^2 = 26$; 51
3 $x(x - 8) = 273$; 21, 13; -13, -21 **5** $(x + 6)^2 = 4x^2$; 6

6 $\dfrac{x}{x + 2} = \dfrac{1}{x}$; 2 **7** $x + \dfrac{1}{x} = \frac{25}{12}$; $\frac{4}{3}$, $\frac{3}{4}$ **9** $\dfrac{1520}{s - 10} - \dfrac{1560}{s} = 24$; 30

10 $\dfrac{10}{s} + \dfrac{25}{s+10} = \dfrac{3}{4}$; 40 mi/h   **11** $\dfrac{1}{m} + \dfrac{1}{m+20} = \dfrac{1}{24}$; 40   **13** 4, 5 or -3, -2

14 9 km   **15** 96   **17** 21, 25 or -21, -25 **18** 12, 19 or -12, -19
19 5 ft, 12ft   **21** 3 ft by 24 ft   **22** 9 ft by 14 ft   **23** 1 mi and $\frac{7}{8}$ mi
25 $\frac{141}{5}$ m   **26** 28th   **27** 10,000 m²   **29** 138,384 cm³   **30** 21,760 cm³
31 \$4.50   **33** 8th day   **34** 2.75   **35** 5.25

Exercise 5.6

1 $x > 2$ **2** $x < 4$ **3** $x > -2$ **5** $x > 2$ **6** $x > 3$ **7** $x > -2$ **9** $x < 3$
10 $x < 12$ **11** $x < -3$ **13** $x < 2$ **14** $x > 2$ **15** $x > 4$ **17** $x < -6$ **18** $x > 1$
19 $x > 1$ **21** $x < -6$ **22** $x < -1$ **23** $x > -7$ **25** $x < -1$ or $x > 2$
26 $x < 2$ or $x > 4$ **27** $x < -3$ or $x > -2$ **29** $-\frac{3}{2} < x < \frac{2}{5}$ **30** $-\frac{2}{7} < x < \frac{7}{2}$
31 $-\frac{7}{2} < x < \frac{3}{5}$ **33** $x < 1$ or $x > 3$ **34** $-3 < x < 2$ **35** Empty set   **37** All x
38 $-4 < x < -1.5$ **39** $-\frac{2}{3} < x < \frac{3}{2}$ **41** $x < -3$ or $x > 1$ **42** $-4 < x < 2$
43 $-3 < x < -\frac{1}{2}$ **45** $1 < x < 7$ **46** $x < -3$ or $x > -\frac{11}{5}$ **47** $x < \frac{4}{13}$ or $x > \frac{5}{8}$
49 $x < 1$ or $x > \frac{5}{3}$ **50** $x < -1$ or $x > 1$ **51** $-\frac{1}{2} < x < \frac{1}{2}$
53 $-\frac{4}{3} < x < -1$ or $x > \frac{1}{2}$ **54** $-\frac{3}{2} < x < -\frac{1}{3}$ or $x > 3$
55 $-2 < x < -\frac{3}{7}$ or $x > \frac{7}{3}$ **57** $-\frac{1}{2} < x < \frac{5}{3}$ or $x < -2$
58 $x < -2$ or $\frac{1}{5} < x < \frac{8}{3}$ **59** $x < -\frac{2}{7}$ or $\frac{7}{2} < x < 4$
61 $x < -4 - \sqrt{13}$ or $x > -4 + \sqrt{13}$ **62** Empty set

63 $x < \dfrac{-11 - \sqrt{141}}{2}$ or $x > \dfrac{-11 + \sqrt{141}}{2}$ **65** $k < \dfrac{3 - \sqrt{21}}{2}$ or $k > \dfrac{3 + \sqrt{21}}{2}$

66 $\dfrac{-3 - 5\sqrt{3}}{4} < x < \dfrac{-3 + 5\sqrt{3}}{4}$ **67** $7 - 4\sqrt{3} < x < 7 + 4\sqrt{3}$

69 $10 - 5\sqrt{2} < x < 10 + 5\sqrt{2}$ **70** $x > 2 + \sqrt{29}$ **71** More than 145
73 $(N/2) + 1 \leq t < N + 1$

Exercise 5.7

1 $-2, 2$ **2** 0   **3** $-3, 3$ **5** $0, 2$ **6** $-10, 4$ **7** $-10, 2$ **9** $-2, 3$ **10** $-\frac{14}{3}, 2$
11 $-3, \frac{9}{5}$ **13** $2, 5$ **14** $-2, \frac{8}{3}$ **15** $-\frac{5}{2}, 0$ **17** $-\frac{1}{2}$ **18** 2   **19** All x **21** $\frac{1}{3}, 9$
22 $-10, -\frac{4}{5}$ **23** $-1, 1$ **25** $-2 < x < 2$ **26** $-5 < x < 5$ **27** $x < -3$ or $x > 3$
29 $-7 < x < 1$ **30** $-4 < x < 8$ **31** $2 < x < 8$ **33** $x < -2$ or $x > 3$
34 $x < \frac{4}{3}$ or $x > 2$ **35** $x < -2$ or $x > -\frac{2}{3}$ **37** $-1 < x < 5$ **38** $-3 < x < 2$
39 $x < 1$ or $x > \frac{7}{3}$ **41** $x < 1$ **42** $x < 2$ **43** $\frac{2}{3} < x < 4$ **45** $x < -\frac{1}{2}$ or $x > 5$
46 Empty set   **47** $x < -2$ or $x > \frac{1}{3}$

Exercise 5.8

1 No   **2** No   **3** No   **4** No   **5** Yes   **6** Yes   **7** 3   **8** -2 **9** 2   **10** 5   **11** 6
12 8   **13** 15   **14** 18   **15** $\frac{5}{44}$ **16** 24   **17** 4   **18** 3   **19** 6   **20** 12   **21** $\pm\frac{1}{4}$ **22** $0, \frac{1}{16}$
23 $\frac{4}{3}, -\frac{1}{2}$ **24** $(-3 \pm \sqrt{7})/3$ **25** $2 \pm \sqrt{3}$ **26** $(-3 \pm \sqrt{6})/2$ **27** $2 \pm i\sqrt{3}$
28 $(5 \pm 3i)/2$ **29** $(-1 \pm 2i)/4$ **30** $\frac{5}{8}, -\frac{3}{4}$ **31** $5 \pm 3i\sqrt{3}$ **32** $\pm 3, \pm 1$

33 $2 \pm \sqrt{7}, \dfrac{4 \pm \sqrt{14}}{2}$ **34** 16   **35** 5   **36** 3   **37** $4, -\frac{3}{4}$ **38** $5, \frac{4}{3}$ **39** $1, -7$

40 $-3, 4$ **41** $-1, 1$ **42** $-\frac{4}{7}, 4$ **43** $5(F - 32)/9$ **44** $\dfrac{2S - na}{n}$ **45** $x > 3$

46 $x < 2$ **47** $x < 1$ **48** $x < 4$ **49** $-\frac{7}{3} < x < \frac{1}{2}$ **50** $x < -\frac{1}{3}$ or $x > \frac{8}{5}$
51 $x < -\frac{9}{5}$ or $x > \frac{5}{8}$ **52** $-\frac{3}{7} < x < \frac{7}{3}$ **53** $x < -\frac{9}{2}$ or $-2 < x < 4$
54 $-\frac{8}{3} < x < -\frac{3}{5}$ or $x > \frac{5}{2}$ **55** $-7 < x < 1$ **56** $x < -3$ or $x > 4$
57 $-1 < x < 1$ **58** $-1 < x < 5$ **60** $7, 41$ **61** $24, 24$ **62** $\frac{57}{8}$
63 $20{,}000$ m²   **64** $-\frac{6}{5}, -\frac{9}{5}$ **65** All k **66** $\frac{2}{5} < x < \frac{5}{2}$ **67** All x
68 $15x^2 - 11x - 12 = 0$ **69** $x^2 - 4x + 13 = 0$ **70** Yes, $D = 146^2$.
71 (a) At $\dfrac{9 + 2\sqrt{14}}{5}$ s (b) $\frac{181}{5}$ m at $\frac{9}{5}$ s

82 (a) $\sqrt{\dfrac{1 + 2rd}{1 - rd}}$ (b) 1   **83** (a) 5 (b) 6, 1

Exercise 6.1   **1** IV   **2** I   **3** II   **5** III   **6** IV   **7** III

9, 10, 11   **13, 14, 15**

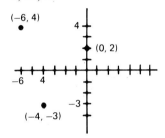   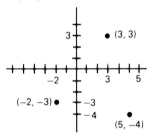

17 The x axis   **18** The negative y axis   **19** The fourth quadrant
21 All points above the horizontal line 5 units above X axis
22 First and third quadrants   **23** Vertical line 2 units to left of Y axis
25 $10\sqrt{2}, (-1, 3)$ **26** $4\sqrt{5}, (6, -1)$ **27** $4\sqrt{10}, (3, 6)$ **29** $13, (\frac{11}{2}, 1)$
30 $17, (-8, -\frac{1}{2})$ **31** $25, (\frac{5}{2}, 10)$ **33** $\sqrt{61}, (-8, -\frac{1}{2})$ **34** $\sqrt{58}, (\frac{13}{2}, \frac{3}{2})$
35 $\sqrt{173}, (-\frac{5}{2}, 5)$ **37** $(11, -4)$ **38** $(6, -5)$ **39** $(5, -5)$ **45** $(8, \frac{15}{2})$
46 $\sqrt{50} = \sqrt{50}$
47 Locate the triangle so its vertices are $(0, 0)$, $(a, 0)$, and (a, b). Now each distance from vertex to $(a/2, b/2)$ is $\sqrt{(a/2)^2 + (b/2)^2}$.
49 $\sqrt{5} + \sqrt{20} = \sqrt{45}$ **50** $\sqrt{20} + \sqrt{45} = \sqrt{125}$ **51** $\sqrt{13} + \sqrt{52} = \sqrt{117}$

Exercise 6.2   **1**   **2**   **3**

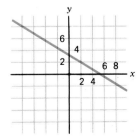   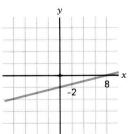   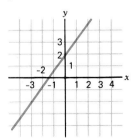

5

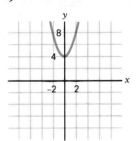

6

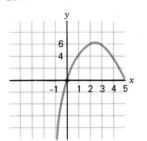

7

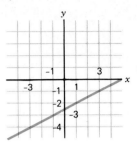

9

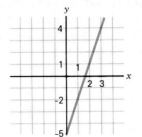

10

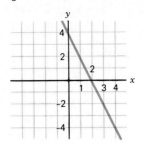

11

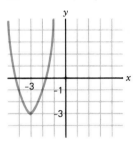

13

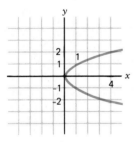

14

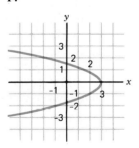

15

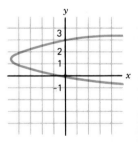

17

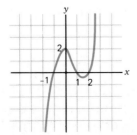

18

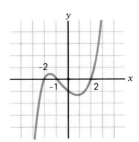

19

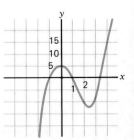

21

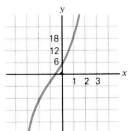

22

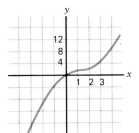

23

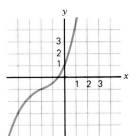

25

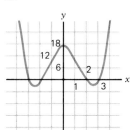

26

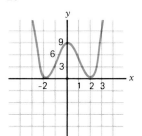

27

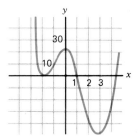

29

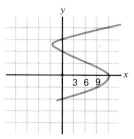

30

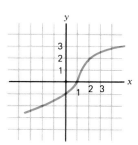

31

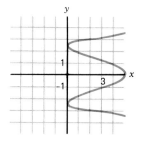

33

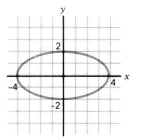

34

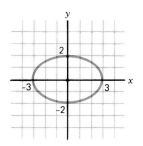

35

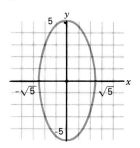

37

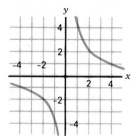

38

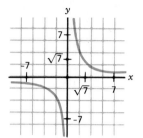

39

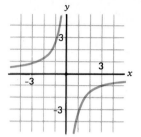

41

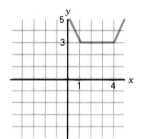

42

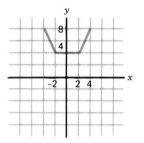

43

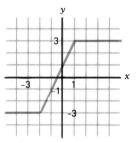

45

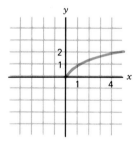

46

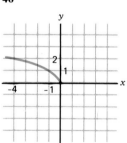

47

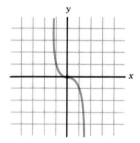

49

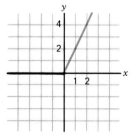

50

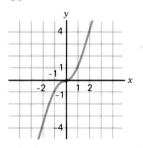

51

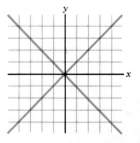

53

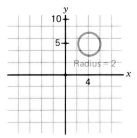

54

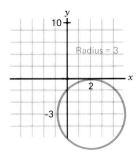

55

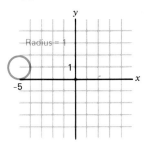

57

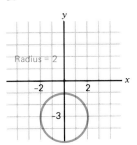

58

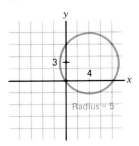

59

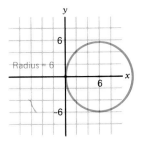

61

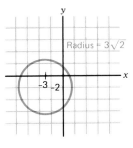

62

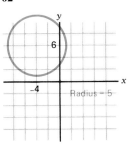

63

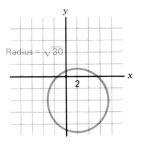

65

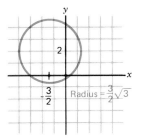

66

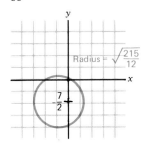

67

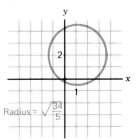

69

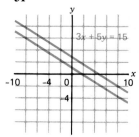

Exercise 6.3   **1** $(3, 5)$, $(-3, 5)$ **2** $(-2, -1)$, $(2, -1)$ **3** $(-3, 8)$, $(3, 8)$
5 $(2, -7)$, $(-7, -2)$ **6** $(-6, -1)$, $(-1, 6)$ **7** $(-11, 4)$, $(4, 11)$
9 x axis, y axis, origin   **10** y axis   **11** x axis   **13** Origin, $y = x$
14 x axis, y axis, origin   **15** y axis   **17** $-xy(x - y) = 5$
18 $-xy(-x + y) = 5$ **19** $-xy(x + y) = 5$ **21** $(x - 3)^2 + (y - 5)^2 = 4$
22 $(x + 2)^2 - (y + 3)^2 = 11$ **23** $xy = 5$ **25** $7x - 2y = 28$
26 $xy = 5x + 3y$ **27** $x^2 + 7x + 2y = 0$

29

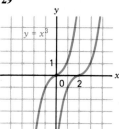

30

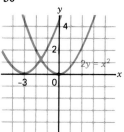

31

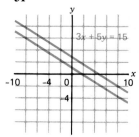

33

34

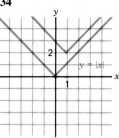

35

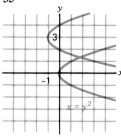

37

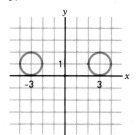

38

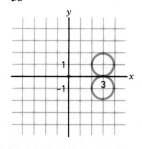

39

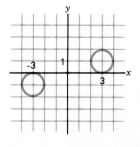

41

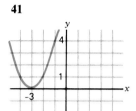

42

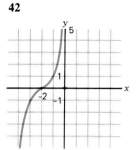

43

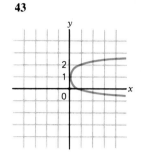

45

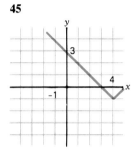

46

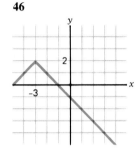

47

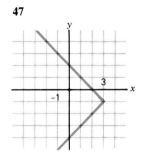

49

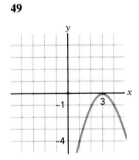

50

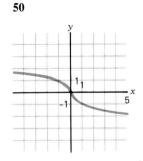

51

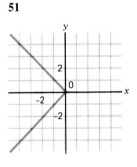

53 (a) $y = -(x + 2)^3$ (b) $y = -(x - 2)^3$

Exercise 6.4   **1** $x \geq -\frac{1}{5}$ **2** all x **3** All x **5** $x \neq \pm 2$ **6** $x \neq 0$ **7** all x **9** $\frac{2}{3}$
10 $\pm\sqrt{5}$ **11** 25   **13** Yes   **14** Yes   **15** No   **17** No   **18** Yes   **19** Yes
21 3, 9   **22** 1, 10   **23** -4, 14   **25** -4, 8   **26** -4, 23   **27** 111, -6
29 $3h - 5, 3h + 2$ **30** $22 + 2h, 2h + 13$ **31** $2h^2 - 3h + 7, 2h^2 + h + 3$
33 $2xh + h^2 + h$ **34** $4xh + 2h^2 - 5h$ **35** $10xh + 5h^2 - 5h$
37 -1 **38** -8 **39** $\frac{7}{2}$ **41** 9   **42** 21   **43** $\frac{19}{11}$ **45** -26 **46** 25   **47** -5 **49** 4
50 4   **51** $\frac{1}{10}$ **53** -6 **54** -5 **55** 6   **61** 33, 28   **62** -27, 27   **63** 7   **65** 2

66 $\frac{2}{3}$ **67** $-\frac{1}{5}$ **69** $2x + h + 3$ **70** $\dfrac{-1}{x(x + h)}$ **71** $10x + 5h - 2$

73

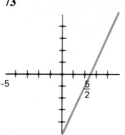

74

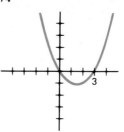

75

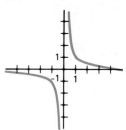

77 (a) 74 (b) $\frac{4}{5}$ **78** (a) 420 (b) 54   **79** (a) $A = 2T$ (b) 122

81 (a) 355.61 (b) 3236.84   **82** $y = \dfrac{120(175 - x)}{10 + x}$ **83** 276   **85** 225

86 3700; 4700; 5700; 10,700   **87** $(5t^2 + u^2 + 8)/2$ **89** 2.55
90 (a) 9.24 (b) 9 (c) $(4m + 3)/3$ **91** 16.72

Exercise 6.5   **1** 6   **2** $\frac{1}{3}$ **3** -1 **5** $y = x - 2$ **6** $y = x - 3$ **7** $3x + 4y + 2 = 0$
9 $2x - 3y = 5$ **10** $5x - 2y + 7 = 0$ **11** $5x + 3y = 3 \cdot$ **13** $y - 1 = 2(x - 4)$
14 $y - 2 = -3(x + 2)$ **15** $y - 3 = 3x/4$ **17** $y = 5x + 3$ **18** $y = -4x - 1$
19 $y = -3x + \frac{2}{3}$ **21** $m = 3, b = -7$ **22** $m = -5, b = 1$ **23** $m = -\frac{5}{2}, b = 2$
25 $a = 3, b = 5$ **26** $a = 4, b = -7$ **27** $a = \frac{3}{2}, b = 8$ **29** Parallel
30 Perpendicular   **31** Perpendicular   **33** Perpendicular   **34** Neither
35 Parallel   **41** $x + y = 5$ **42** $6x - 7y = -11$ **43** $y = 6x - 5$
45 $25/\sqrt{20} = 5\sqrt{5}/2$ **46** $7/\sqrt{58}$ **47** $20/\sqrt{26}$

49

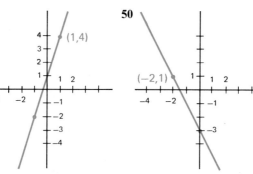

50

51

53

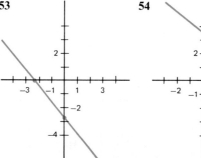

54

55

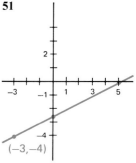

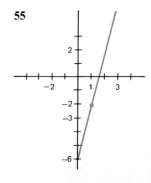

57 $f(x) = 3x - 5$ **58** $x + 2y = 8$ **59** $y - 5 = -3(x - 1)$ **61** $x + y = 7$
62 $14x - 20y - 65 = 0$ **63** $4x + 38y + 183 = 0$

Exercise 6.6   **1** Yes **2** No **3** Yes **5** No **6** No **7** Yes **9** $f^{-1}(x) = (x - 2)/3$
10 $f^{-1}(x) = (x + 1)/5$ **11** $f^{-1}(x) = (x - 7)/(-2)$ **13** $f^{-1}(x) = 3x - 4$

14 $f^{-1}(x) = (1 + 7x)/2$ **15** $f^{-1}(x) = (4x + 1)/12$ **17** $f^{-1}(x) = \sqrt[3]{x} - 1$

18 $f^{-1}(x) = \sqrt[3]{x + 3}$ **19** $f^{-1}(x) = \sqrt[5]{x - 4}$ **21** $f^{-1}(x) = 1/x$ **22** $f^{-1}(x) = 3x/4$
23 $f^{-1}(x) = (1 + 3x)/2x$ **25** 5, 1 **26** 0, $-\frac{21}{4}$ **27** 1, 1 **29** 1 **30** 4 **31** -2

33

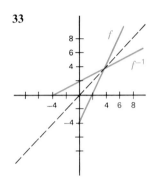

34

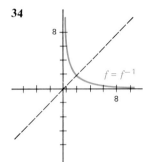

35

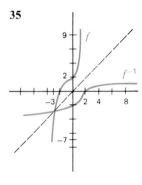

41 $f(x) = (x + 1)^2 + 4$, so $-1 < a < b$ implies $0 < a + 1 < b + 1$,
hence $(a + 1)^2 < (b + 1)^2$ **42** $a < b$ implies $3a < 3b$ implies $3a + 5 < 3b + 5$
43 x^3 is increasing, and $2x$ is also, hence so is their sum.

45 $f(x) = 1 + \dfrac{2}{x - 1}$ which gets smaller as x gets larger

46 $g(g(x)) = g(-x + b) = -(-x + b) + b = x - b + b = x$

47 $k = 3$ **49** 3 **50** 0, 1 **51** $-\frac{5}{2}$, 4

Exercise 6.7   **1** $m = kn^2$ **2** $s = k/t$ **3** $p = kqr^3$ **5** 14 **6** 6 **7** 2 **9** 100 **10** 40 **11** $\frac{7}{2}$
13 2816 in³ **14** $\frac{340}{9}$ **15** Intensity at 5 ft is $\frac{49}{25}$ of the intensity at 7 ft.
17 600 barrels **18** 900 barrels **19** 1280 lb **21** First is $\frac{9}{4}$ as strong as the second
22 $\frac{8}{1}$ **23** $\frac{4}{25}$ **25** 145 lb **26** 30 A
27 The first produces $\frac{4}{9}$ as much illumination as the second **29** 54 **30** $62.50
31 8 in³ **33** 128 **34** 1875 lb **35** 0.054 in **37** (a) $(w - 10)/3$ (b) $-\frac{2}{3}$
38 I is multiplied by 8 **39** $V_1/V_2 = \frac{64}{343}$

Exercise 6.8   **1** (a) No (b) No (c) Yes (d) Yes **2** 2, -10, $6t - 4$ **3** -2, -2, -2 **4** 3, $\frac{1}{2}$
review   **5** 29 **6** No **7** $5x - 2y = 13$ **8** $y = -3x + 2$ **9** $2x + 7y = 10$ **10** 3

11 No **12** $-7, \frac{1}{2}$ **13** $f^{-1}(x) = \dfrac{1 + 3x}{2 - x}$ **16** $(\frac{7}{30}, \frac{8}{15})$

18

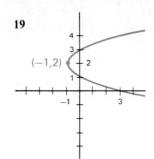

19

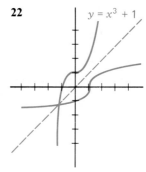

20

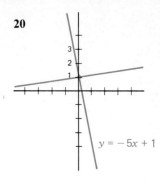

$y = -5x + 1$

21

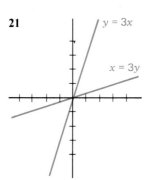

$y = 3x$

$x = 3y$

22

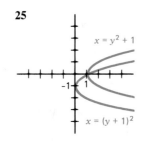

$y = x^3 + 1$

23

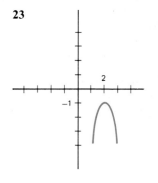

2

-1

24

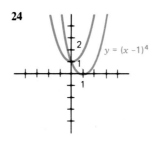

2

1

1

$y = (x - 1)^4$

25

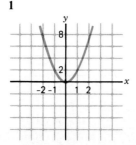

$x = y^2 + 1$

-1 1

$x = (y + 1)^2$

26 *O*   **27** *I*   **28** *U*   **29** *W*   **30** *V*   **31** *X*   **32** *M*   **33** *E*   **34** *C*   **35** *N*   **36** *S*
37 *Z*   **40** 8.618   **41** 47.1   **42** 800   **43** $\frac{2}{9}$ h   **44** 0.01 s   **45** They are the same.
46 Energy at 50 mi/h is 25 times that at 10 mi/h.   **47** 671 days   **48** 64 lb

Exercise 7.1   **1**

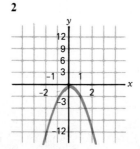

2

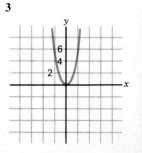

3

5

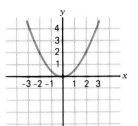

6

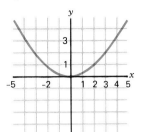

7

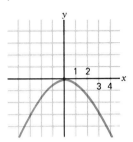

9

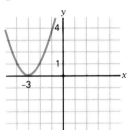

10

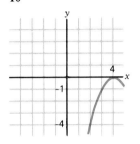

11

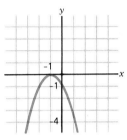

13

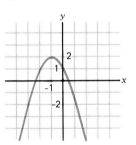

14

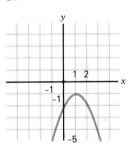

15

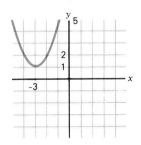

17

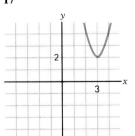

18

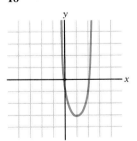

19

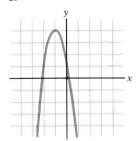

21 $x = -1, (-1, -8)$ **22** $x = 2, (2, 5)$ **23** $x = 2, (2, 5)$

25 $y = -1, (2, -1)$ **26** $y = -3, (-11, -3)$ **27** $y = -2, (3, -2)$

29 $y - 3 = 2(x + 2)^2$ **30** $y + 4 = 3(x + 5)^2$ **31** $y + 3 = 5(x - 1)^2$

33

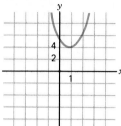

34

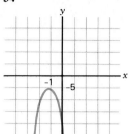

35

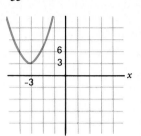

37

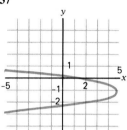

38

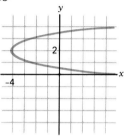

39

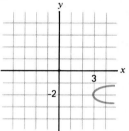

41 $f(2) = 65, \dfrac{4 + \sqrt{26}}{2}$ **42** $f(2.6) = 84.6, \dfrac{26 + \sqrt{846}}{10}$

43 $f(1.8) = 74.4, \dfrac{9 + \sqrt{186}}{5}$ **45** $x \geq -\dfrac{b}{2a}$

Exercise 7.2   **1**

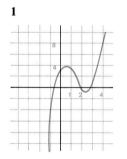

2

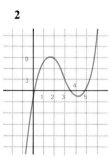

3

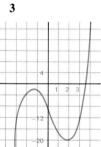

5

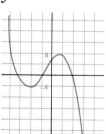

6

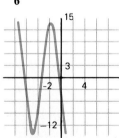

7

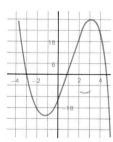

9

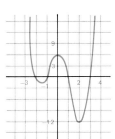

10

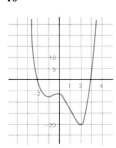

11

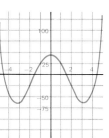

13

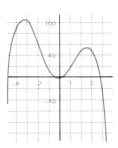

14

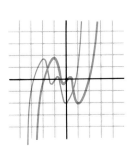

15

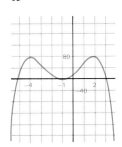

17

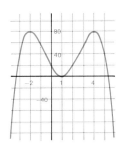

18

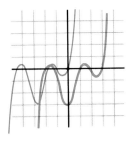

19

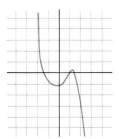

21

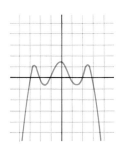

22

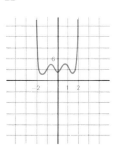

23

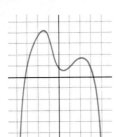

25

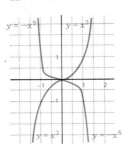

26

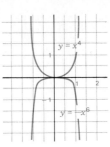

27

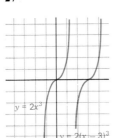

29

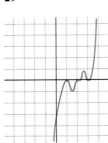

30

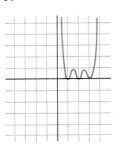

31

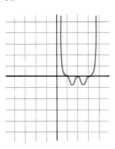

33

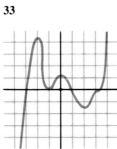

34

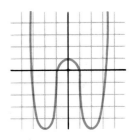

35

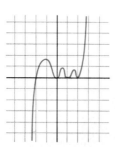

Exercise 7.3

1

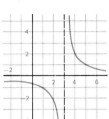

2

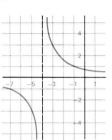

3

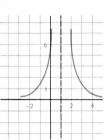

5

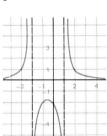

6

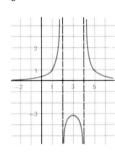

7

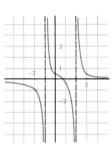

9

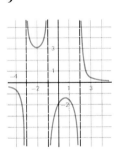

10

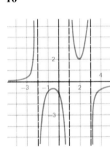

11

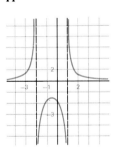

13

14

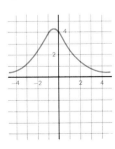

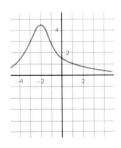

15

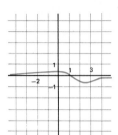

17

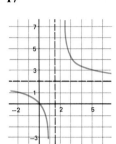

18

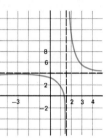

19

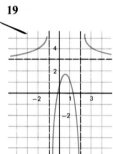

21

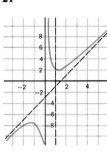

22

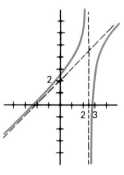

23

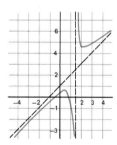

Exercise 7.4   **1**

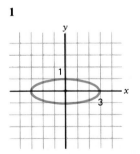

2

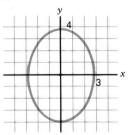

3

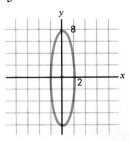

5

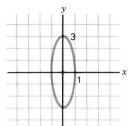

6

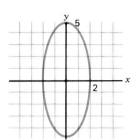

7

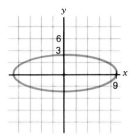

9

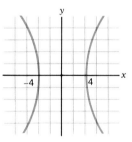

10

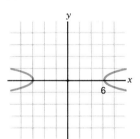

11

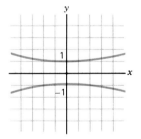

13

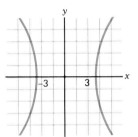

14

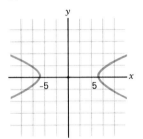

15

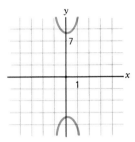

17

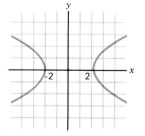

18

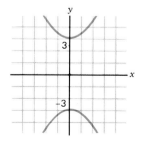

19

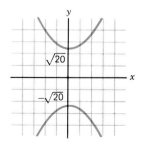

21

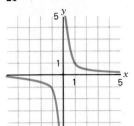

22

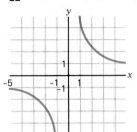

23

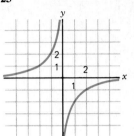

25

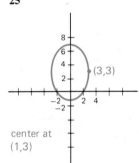

center at
(1,3)

26

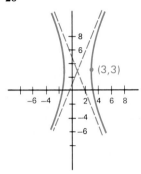

27

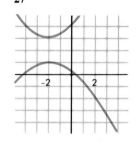

29

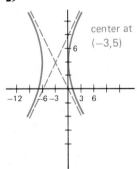

center at
(−3,5)

30

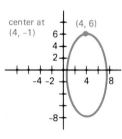

center at
(4, −1)

31

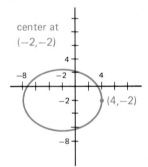

center at
(−2,−2)

33

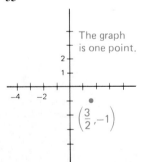

The graph
is one point.

34

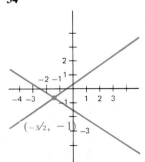

35 Same as Prob. 27.

37

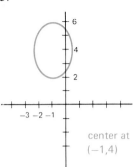

center at
$(-1,4)$

38

center at
$\left(-2,-\dfrac{3}{2}\right)$

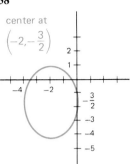

39

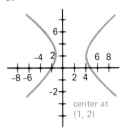

center at
$(1, 2)$

50 A circle

51

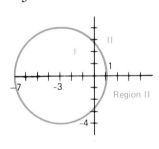

$x^2 - \dfrac{y^2}{9} = 1$

$y^2 - \dfrac{x^2}{9} = 1$

Exercise 7.5   **1**

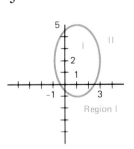

Region I

2

Region I

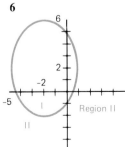

3

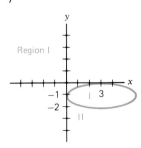

Region II

5

Region I

6

Region II

7

Region I

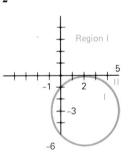

9

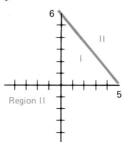

10

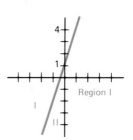

11

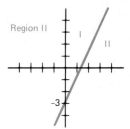

13

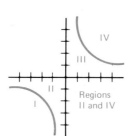

14

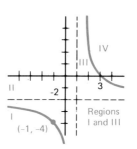

15

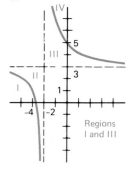

17

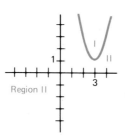

18

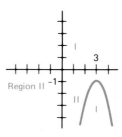

19

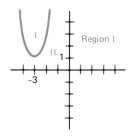

21

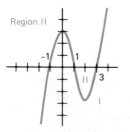

22

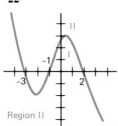

23

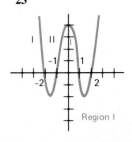

25

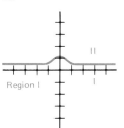

26

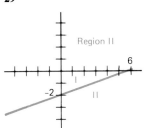

27

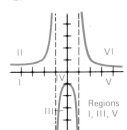

29

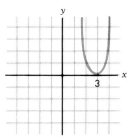

30

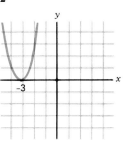

31

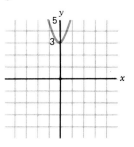

Exercise 7.6   **1**

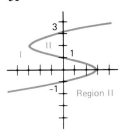

2

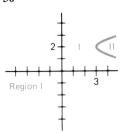

3

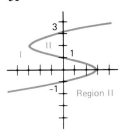

4

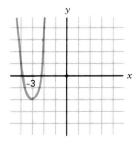

5

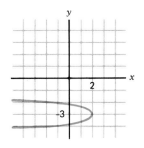

6

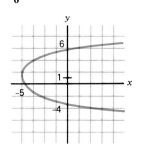

7

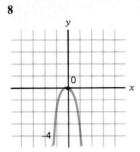

8

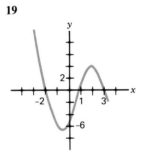

9 $(-\frac{3}{2}; 8)$, minimum   **10** $(2, 17)$, maximum   **11** $(\frac{1}{2}, 0.8125)$, minimum   **12** Ellipse
13 Parabola   **14** Parabola   **15** Hyperbola   **16** Hyperbola   **17** Ellipse

18

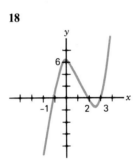

19

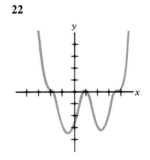

20

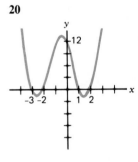

21

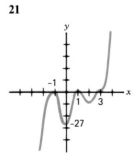

22

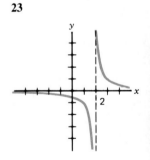

23

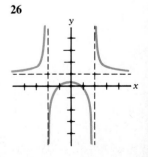

24

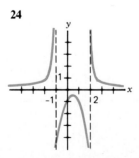

25

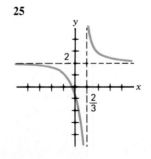

26

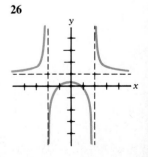

27

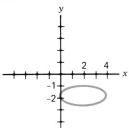

28

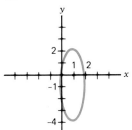

29

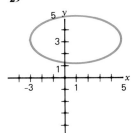

30

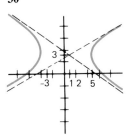

31

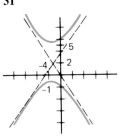

32

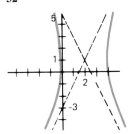

33

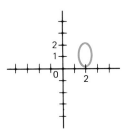

34

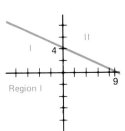

35

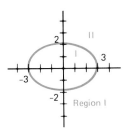

36

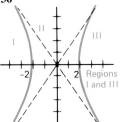

37

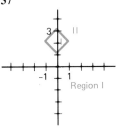

38

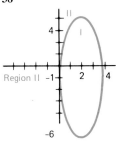

39

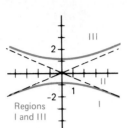

40

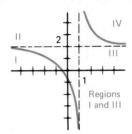

Exercise 8.1   **1**

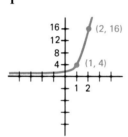

2

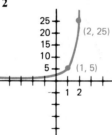

3

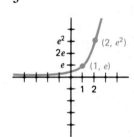

5

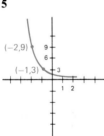

6

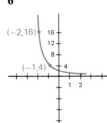

7

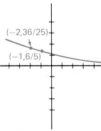

9

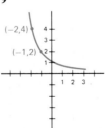

10

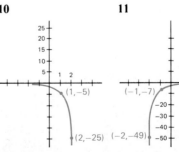

11

13 **14** **15**

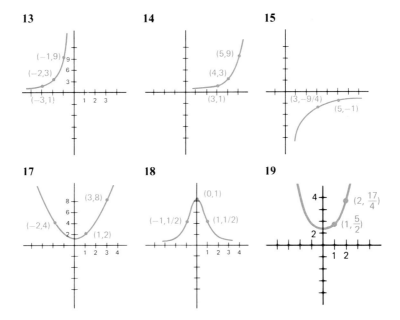

17 **18** **19**

21 $-\frac{3}{4}$ **22** -3 **23** $-\frac{3}{2}$ **25** $-\frac{1}{2}$, 1 **26** -8 **27** -6 **29** 2, 4 **30** 0, 2 **31** 0, 1

33 $6(2^{15}) = 196,608$ **34** 1000 to 1 **35** Better at $6\frac{1}{4}$ percent simple interest

37 7.36 lb/in² **38** \$3555 **39** Barely less than 4 min **41** False **42** False

43 True **45** 2401 **46** $\frac{1}{1024}$ **47** 6 **49** 16 **50** $\frac{1}{4}$ **51** $2^8 - 2^{-8} = 255\frac{255}{256}$

57 15.43, 16.189, 16.241 **58** 16.242 **59** 1.848 and 1.990 **62** y approaches b/c

63 p approaches $1 - \dfrac{b}{a}$ **65** 2.70833, 2.71806, 2.71828

Exercise 8.2   **1** $\log_2 8 = 3$ **2** $\log_3 243 = 5$ **3** $\log_5(1/25) = -2$ **5** $\log_{32} 8 = 3/5$

6 $\log_{27} 9 = 2/3$ **7** $\log_{\frac{1}{2}} 16 = -4$ **9** $3^4 = 81$ **10** $7^2 = 49$ **11** $5^{-3} = 1/125$

13 $8^{5/3} = 32$ **14** $32^{3/5} = 8$ **15** $64^{2/3} = 16$ **17** 5 **18** 3 **19** 4

21 9 **22** 32 **23** 81 **25** 2 **26** 3 **27** 25

29 **30** **31**

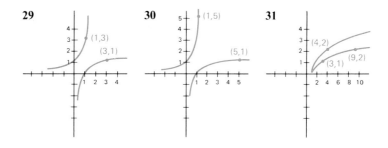

33   **34**   **35**

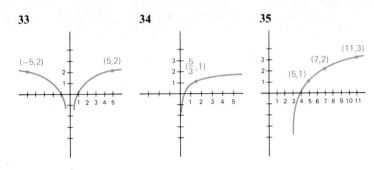

37   **38**   **39**

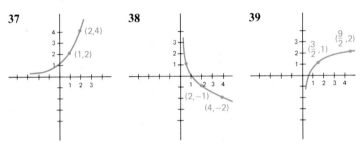

41 $\log_e x + \log_e(x^3 - 1) - \log_e(5x + 3)$ **42** $4\log_e x + \frac{2}{3}\log_e(x + 1) - \log_e(x + 23)$
43 $\log_e(x - 4) + 5\log_e(2x + 7) - \log_e x - 3\log_e(3x + 8)$ **45** $\log_b 3x^2 y^9$
46 $\log_b x^2 z^{1/3}/y^{1/2}$ **47** $\log_b x^3$ **49** 1.45   **50** 2.81   **51** 1.32   **53** True   **54** True
55 False   **57** False   **58** True   **59** False   **69** $-2N \log_e(1 - f)$

70 (a) $\log(1/x) = -\log x$ (b) $\frac{7}{4}$ **71** $N\left[1 - \left(1 - \frac{1}{N}\right)^F\right]$

73 $\dfrac{\log(S/S_0)}{\log(1 + c)}$ **74** (a) $10^{-d/c}$ (b) $x\left[c \log_{10}\left(\dfrac{A}{x}\right) + d\right]$

75 $|W|\sqrt{e^{2A/WH} - 1}$

Exercise 8.3   **9** 0.5752   **10** 0.9926   **11** 0.3118   **13** 1.5899   **14** 1.7007   **15** 2.8280
17 $9.5092 - 10$ **18** $9.8519 - 10$ **19** $8.4871 - 10$ **21** $7.7160 - 10$
22 $6.3385 - 10$ **23** $7.9652 - 10$ **25** 3.4412   **26** 3.7663   **27** 1.8911
29 $9.7576 - 10$ **30** $9.9396 - 10$ **31** $8.6995 - 10$ **33** 2.13   **34** 30.7   **35** 579
37 $4.62(10^3)$ **38** $3.71(10^4)$ **39** $8.14(10^6)$ **41** 0.925   **42** 0.0178   **43** 0.00294
45 0.00907   **46** 0.000118   **47** 0.663   **49** 22.0   **50** 314   **51** 0.0523   **53** 196.5
54 30.85   **55** 1.242   **57** 0.5027   **58** 0.005984   **59** 0.05826   **61** 16   **62** 57
63 120   **65** 91,701,120, within 1 percent of the actual 1910 population
66 3.32, 9   **67** 4

Exercise 8.4   **1** 81.9   **2** 809   **3** 61.2   **5** 285   **6** 1.95   **7** 0.0695   **9** 2.18   **10** 60.3   **11** 5.18
13 213   **14** $4.92(10^6)$ **15** 0.0104   **17** 8.73   **18** 17.1   **19** 1.19   **21** 0.903
22 0.335   **23** 0.976   **25** 0.466   **26** 1.98   **27** 61.8   **29** $5.91 < 6.12 < 6.20 < 6.35$
30 $2.76 < 2.79 < 2.81 < 2.83$ **31** $10 > 9.5$ **33** 175.0   **34** 931.3

35 0.006385   **37** 1.781   **38** 1.248   **39** 0.3993   **41** 3.53   **42** 2.92   **43** 1.85
49 3.00   **50** 4.17   **51** 4.71   **53** 35 years, 70 years   **54** $P(t) = A(2^{t/35})$
55 $P(t) = 6.72(2^{-0.2304t})$, $P(t) = 6.72e^{-0.1597t}$
57 $S = 1000(1.0175)^{4n}$, 9.99 years   **58** $j = 5.8$ percent
59 The first is 15.8 times as intense as the second
61 58   **62** 33 months   **63** 0.98 min   **65** 6.4   **66** $4.0(10^{-12})$
67 $(14.7)10^{-0.0882m}$ **69** 0.257, 0.214   **70** 128,000; 165,000; 272,000
71 3, 54, 6878   **73** By Eq. (8.7), each equals $\log_b a$.

Exercise 8.5   **1** $x = \log y$ **2** $x = -\log y$ **3** $x = \frac{1}{2}(\log 5 - \log y)$ **5** $x = \ln(y + \sqrt{y^2 - 1})$

6 $x = \ln(y + \sqrt{y^2 + 1})$ **7** $x = \frac{1}{2} \ln \dfrac{y + 1}{y - 1}$ **9** 4   **10** 6   **11** ± 2 **13** 0.637

14 0.869   **15** 0.617   **17** 2   **18** 2   **19** 6   **21** 3   **22** 5   **23** -2 **25** 3   **26** 4   **27** 6
29 $x = 1.97, y = 1.01$ **30** $x = 0.97, y = 3.01$ **31** $x = 1.98, y = 2.01$
33 $x = 0.31, y = 0.88$ **34** $x = 0.42, y = 1.58$ **35** $x = 0.23, y = 0.035$
37 $x < 1.51$ **38** $x < 0.389$ **39** $x > -1.17$
41 $x < 1.55$ **42** $x < 1.08$ **43** $x < 24.3$

45   **46**   **47**

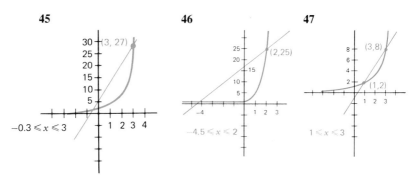

49 3.5000   **50** 1.7550   **51** 1.3713

Exercise 8.6   **1**   **2**   **3**

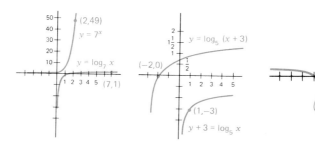

4a   **4b**

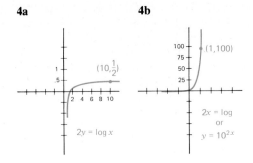

16 $3x$ **17** $1 + y - z - 2x$ **18** $(z + 2y)/2$ **19** 2.6857   **20** 1.8274
21 $9.9886 - 10$ **22** 0.3347   **23** $5.52(10^4)$ **24** 0.00000845   **25** 0.08035
26 0.216   **27** $93^{95} > 95^{93}$ **28** 173   **29** 0.576   **30** 1.44   **31** $2.78(10^4)$
32 0.3335   **33** 1.387   **34** 4.32   **35** 2.08   **36** 0.929   **37** 1.21   **38** 3   **39** $1, -4$
40 2.82   **41** 8   **42** 8   **43** -4.61 **44** 3^{16} **45** $x = 2.37, y = 3.58$
46 $x < 0.435$ **47** $x < 0.885$ **48**

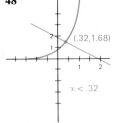

49 3h, 4.75 h   **50** $10^{3.7}/1$ **51** 2220 years
52 $t \geq 14.3$, v approaches 120 ft/s as t gets larger and larger
53

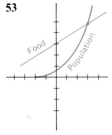

54 $4.53 < 4.55 < 4.58$ **55** 915
56 $1.01(10^{29})$ **58** 6.514   **59** 32
61 $0.1333 < 0.1335 < 0.1336$

Exercise 9.1   **1** (2, 1)   **2** (1, −2)   **3** (3, −1)   **5** (3, −2)   **6** (1, −1)   **7** (2, −1)
9 (−1, 2)   **10** (−3, 4)   **11** (−2, 3)   **13** (2, 1, 1)   **14** (1, 0, −2)
15 (−4, −2, 3)   **17** (2, 1, 3)   **18** (−2, 1, −1)   **19** (3, −2, −1)
21 (−1, 0, 1, 2)   **22** (−2, −1, 1, 3)   **23** (−3, −1, 0, 2, 4)   **25** (2, 1)
26 Inconsistent   **27** Dependent   **29** Dependent   **30** (1, −2)
31 Dependent   **33** (1, 2)   **34** (2, 0)   **35** (0, 1)

Exercise 9.2

1 $(1, 0), (-1, 0)$ **2** $(1, 1), (-1, 1), (-1, -1), (1, -1)$
3 $(1, 1), (1, -1), (-1, 1), (-1, -1)$
5 $(2, 1), (2, -1), (-\frac{11}{3}, \frac{1}{3}\sqrt{26}), (-\frac{11}{3}, -\frac{1}{3}\sqrt{26})$
6 $(3, 1), (3, -1), (-5.75, i\sqrt{58.0625}), (-5.75, -i\sqrt{58.0625})$
7 $(2, 3), (-2, 3), (0, -1)$ **9** $(-1, 1), (-\frac{4}{5}, \frac{9}{5})$ **10** $(2, 5), (-7, -\frac{10}{7})$
11 $(-1, 1), (-\frac{1}{13}, \frac{109}{13})$ **13** $(7, 6), (3, 4)$ **14** $(\frac{1}{3}, 0), (\frac{1}{27}, -\frac{4}{9})$
15 $(19, -6), (3, 2)$ **17** $(10, 2), (-2, -2), (5, 1.5), (-4, -1.5)$
18 $(1, -2), (-1, -4), (\sqrt{7}, 3 + \sqrt{7}), (-\sqrt{7}, 3 - \sqrt{7})$
19 $(20, 2), (-4, -2), (9 + 9\sqrt{2}, 1.5\sqrt{2}), (9 - 9\sqrt{2}, -1.5\sqrt{2})$
21 $(2, 1), (-2, -1), (0.5\sqrt{2}, 2\sqrt{2}), (-0.5\sqrt{2}, -2\sqrt{2})$
22 $(1, 3), (-1, -3), (1.5\sqrt{6}, \frac{1}{3}\sqrt{6}), (-1.5\sqrt{6}, -\frac{1}{3}\sqrt{6})$
23 $(5, -1), (-5, 1), (\sqrt{2}, -2.5\sqrt{2}), (-\sqrt{2}, 2.5\sqrt{2})$
33 $(3, -1), (-3, 1), (4, 2), (-4, -2)$
34 $(1, 1), (1, -2), (-1, -1), (-1, 2)$ **35** $(0, i), (0, -i), (2i, i), (-2i, -1)$
37 $(4, -1), (-4, 1), (\sqrt{2}, \sqrt{2}), (-\sqrt{2}, -\sqrt{2})$
38 $(2, -\frac{1}{3}), (-2, \frac{1}{3}), (1, -1), (-1, 1)$
39 $(3, -0.5), (-3, 0.5), (2, 1), (-2, -1)$ **41** $(2, 1), (-\frac{2}{5}, \frac{1}{5})$
42 $(-1, 2), (2.5, -1.5)$ **43** $(-2, -1), (0.5, 1.5)$ **45** $(\frac{2}{3}, -\frac{1}{3}), (\frac{1}{2}, -\frac{1}{2})$
46 $(-1, -2), (-\frac{9}{13}, -\frac{6}{13})$ **47** $(3, -1), (\frac{1}{2}, \frac{3}{2})$ **49** $(2, 1), (3.4, -1.8)$
50 $(3, 2), (-1, -1)$ **51** $(4, 1), (3, 2)$ **53** $(3, 3)$ **54** $(2, -1), (-1, 2)$
55 $(2, 3), (3, 2)$ **57** $(2, 1), (-1.3, -\frac{8}{3})$ **58** $(1, -1)$
59 $(0, 1), (23, -1.3)$

Exercise 9.3

1 Sucker, \$0.10; apple, \$0.15 **2** 2 qt whipping cream, $\frac{1}{2}$ qt half-and-half
3 4, 2 **5** 54, 36 **6** 21 mi **7** 10 one-bedroom, 7 two-bedroom **9** 9, 5
10 1880, 500 **11** 200 mi, 220 mi **13** \$1, \$2 **14** 18, shorter hike; 17, longer hike
15 6, 3 **16** \$65, \$80 **17** 333-6483 **18** Terry 49 days; Elaine 38 days
19 Second, $2\frac{1}{2}$ h; third, $1\frac{1}{2}$ h **21** Square, 15 by 15 ft; rectangle, 15 by 10 ft
22 40 rods, 40 rods, 70 rods **23** Base, 6 by 6 ft; depth, 5 ft
25 160 by 40 rods, or 80 by 80 rods **26** 800 by 300 rods, or 600 by 400 rods
27 80, \$12 **29** 14 by 20 ft, or $\frac{126}{47}$ by $\frac{2308}{47}$ ft **30** \$4, \$6
31 16 in, 14 in; or 26.56 in, 7.28 in

Exercise 9.4   **1**   **2**   **3**

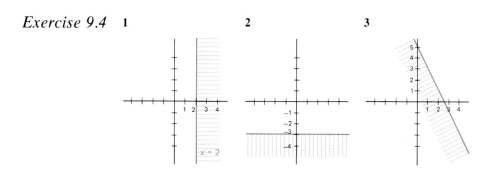

5

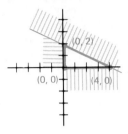

6

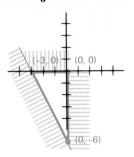

7

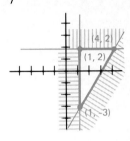

9

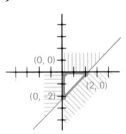

10

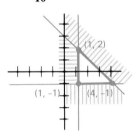

11

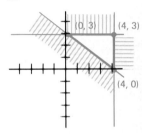

13

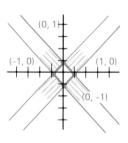

14

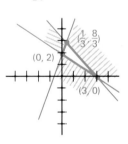

15

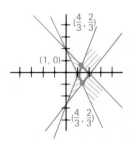

17

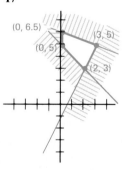

18

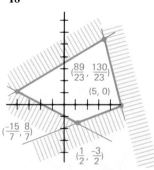

19

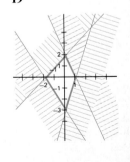

21   **22**   **23**

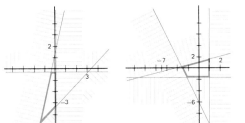

25   **26**   **27**

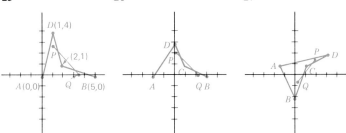

29   **30**   **31**

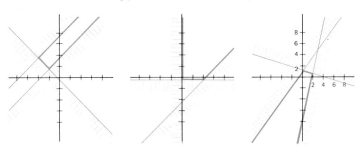

33 $(2, -1)$, $(1, 3)$, $(-3, 1)$ **34** $(3, 0)$, $(-1, 4)$, $(1, -1)$
35 $(-3, -4)$, $(-4, 3)$, $(1, 2)$, $(2, -3)$

37   **38**   **39**

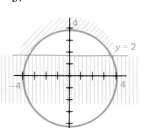
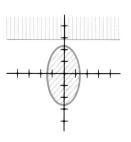
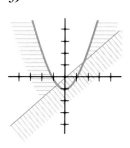

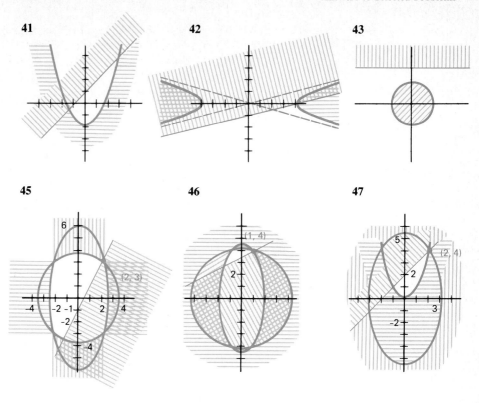

Exercise 9.5   **1** min 0 at (3, 5), max 11 at (4, 1)   **2** min 3 at (0, 0), max 11 at (3, 2)
3 min 1 at $(-1, -2)$, max 20 at $(3, -1)$
5 min -8 at (4, 3), max 16 at $(-1, -1)$
6 min -51 at (2, 7), max 16 at $(-3, -2)$
7 min -11 at $(-3, 0)$ and $(0, -2)$, max 13 at (0, 6)
9 min -5 at (0, 2), max 19 at $(3, -1)$ **10** min -4 at (1, 5), max 23 at (4, 2)
11 min -16 at $(2, -2)$, max 24 at $(-3, 2)$
13 min -11 at $(2, -3)$, max 24 at (3, 5)
14 min -25 at $(-1, 3)$, max 29 at $(1, -3)$ **15** min 9 at (0, 1), max 41 at (6, 2)
17 $M = 15$ at (0, 2), $m = 1$ at (0, 0)   **18** $M = 13$ at (3, 0), $m = -2$ at (0, 0)
19 $M = 18$ at (2, 2), $m = 2$ at (0, 0)
21 $M = 2$ at (1, 0) and (0, 1), $m = 0$ at $(-1, 0)$ and $(0, -1)$
22 $M = 13$ at (3, 2), $m = 0$ at $(-1, 1)$ **23** $m = -5$ at (0, 2), $M = 5$ at $(1, -1)$
25 $m = 8$ at (0, 0), $M = 19$ at $(1, -2)$ **26** $m = -13$ at $(-2, 2)$, $M = -1$ at (0, 0)
27 $m = 5$ at (0, 1), $M = 17$ at $(2, -2)$
29 $m = -\frac{40}{7}$ at $(\frac{23}{7}, -\frac{13}{7})$, $M = 16$ at (2, 3)
30 $m = -4$ at $(-1, 1)$, $M = 2$ at $(3, -1)$ **31** $m = -13$ at $(0, -3)$, $M = 4$ at (2, 2)
33 X, 44 units; y, 60 units; Z, 100 units; profit, \$11,240
34 X, 62 units; y, 47 units; Z, 100 units; profit, \$11,525
35 X, 38 units; y, 47 units; Z, 140 units; profit, \$11,595
37 X, 4 kg; Y, 2 kg   **38** X, 4 kg; Y, 2 kg   **39** X, 4 kg; Y, 2 kg

41 Oxfords, 200 pair; boots, 180 pair; high-top, 120 pair; $2290

42 Oxfords, 200 pair; boots, 80 pair; high-top, 120 pair; $1690

43 Oxfords, 250 pair; boots, 130 pair; high-top, 120 pair; $2165

45 500 acres of corn, 300 acres of wheat

46 200 acres of corn, 400 acres of wheat, $50,000

47 200 units of day lillies, no amaryllis, $60,000

Exercise 9.6

1 $\dfrac{1}{x-1} - \dfrac{3}{3x-2}$ **2** $\dfrac{6}{2x+3} - \dfrac{3}{x+3}$ **3** $\dfrac{3}{3x+7} - \dfrac{2}{2x+5}$

5 $\dfrac{1}{2x+1} + \dfrac{3}{x-3}$ **6** $\dfrac{10}{3x+5} - \dfrac{6}{x+3}$ **7** $\dfrac{5}{x-3} - \dfrac{4}{x-4}$

9 $\dfrac{4}{2x+1} - \dfrac{1}{x+2} - \dfrac{1}{x-3}$ **10** $\dfrac{3}{3x-2} - \dfrac{2}{x+5} + \dfrac{2}{2x+1}$

11 $\dfrac{2}{2x-3} + \dfrac{1}{x+4} - \dfrac{3}{3x-1}$ **13** $\dfrac{2}{2x-1} - \dfrac{5}{x-4} + \dfrac{3}{(x-4)^2}$

14 $\dfrac{2}{2x+3} - \dfrac{1}{3x-1} + \dfrac{5}{(3x-1)^2}$ **15** $\dfrac{2}{5x+2} - \dfrac{3}{2x-7} - \dfrac{1}{(2x-7)^2}$

17 $\dfrac{1}{x-1} - \dfrac{x+3}{x^2+2}$ **18** $\dfrac{3}{x+1} + \dfrac{2x-1}{x^2+1}$ **19** $\dfrac{1}{x-3} + \dfrac{2x-5}{x^2+x+1}$

21 $\dfrac{2}{x-1} - \dfrac{1}{2x+1} + \dfrac{x-4}{x^2+3}$ **22** $\dfrac{2}{x+5} + \dfrac{3}{2x-1} - \dfrac{x-5}{x^2+3}$

23 $\dfrac{2}{x-3} + \dfrac{1}{3x+1} - \dfrac{2x-1}{x^2-x+2}$ **25** $\dfrac{1}{x-1} - \dfrac{2}{(x-1)^2} + \dfrac{2x-3}{x^2+1}$

26 $\dfrac{2}{x-2} + \dfrac{1}{(x-2)^2} - \dfrac{3x+1}{x^2+5}$ **27** $\dfrac{3}{x-2} - \dfrac{1}{(x-2)^2} + \dfrac{x-3}{x^2+x+1}$

29 $\dfrac{3x-1}{x^2+2} - \dfrac{x-1}{x^2-2}$ **30** $\dfrac{2x-5}{x^2-3x-1} + \dfrac{x-2}{x^2+1}$ **31** $\dfrac{x-3}{x^2-5} + \dfrac{2x+1}{x^2-x+3}$

33 $\dfrac{x+2}{x^2-2} + \dfrac{2x+1}{(x^2-2)^2}$ **34** $\dfrac{2x-3}{x^2+3} - \dfrac{3x+1}{(x^2+3)^2}$ **35** $\dfrac{2x+1}{x^2-x-3} + \dfrac{x-5}{(x^2-x-3)^2}$

37 $\dfrac{2}{x^2} + \dfrac{x-1}{(x^2+3x-1)^2}$ **38** $\dfrac{1}{x^2} + \dfrac{x+1}{x^2-x+2} + \dfrac{x-7}{(x^2-x+2)^2}$

39 $\dfrac{1}{x^2} - \dfrac{x+2}{(x^2+1)^2}$ **41** $A = \dfrac{cr+d}{r-s}, B = \dfrac{cs+d}{s-r}$

Exercise 9.7

1 Dependent   **2** Inconsistent   **3** Independent, (2, 1)

4 (2, 1), $(\frac{14}{23}, -\frac{41}{23})$ **5** (1, 1), (1, −1), (−1, 1), (−1, −1)

6 (1, −2), (−1, 2), $(\dfrac{4}{\sqrt{3}}, \dfrac{-\sqrt{3}}{2})$, $(\dfrac{-4}{\sqrt{3}}, \dfrac{\sqrt{3}}{2})$

7   **8**   **9**

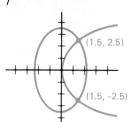

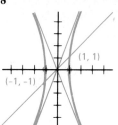

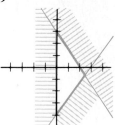

10 $(1, 1)$, $(5, 2)$, $(3, -2)$ **11** max of 29 at $(3, -2)$, min of -7 at $(-1, 2)$
12 min of -4 at $(0, 0)$, max of 10 at $(4, 2)$
13 Profit of \$21,800 for 100 acres of maize, 400 acres of wheat, 200 acres of corn

14 $\dfrac{0.4}{x} + \dfrac{2.4}{4x + 5}$ **15** $\dfrac{2}{x - 3} + \dfrac{5}{(x - 3)^2}$

16 $\dfrac{2}{x + 1} + \dfrac{3x - 1}{x^2 + x + 3}$ **17** $\dfrac{1}{x - 2} + \dfrac{2x - 1}{x^2 - x + 3} + \dfrac{3x + 2}{(x^2 - x + 3)^2}$

Exercise 10.1   **1** $a = 1$, $b = 2$, $c = 3$, $d = -2$ **2** $a = -5$, $b = 1$, $c = 0$, $d = 3$
3 These two matrices can not be equal.

5 2×3, $\begin{bmatrix} 2 & 3 \\ 1 & -6 \\ 5 & 2 \end{bmatrix}$ **6** 3×3, $\begin{bmatrix} -1 & 4 & 0 \\ 4 & 5 & 2 \\ 0 & 2 & 3 \end{bmatrix}$ **7** 3×2, $\begin{bmatrix} 3 & 1 & 0 \\ -2 & 5 & 4 \end{bmatrix}$

9 $\begin{bmatrix} 6 & -2 & 4 \\ 8 & 0 & 10 \end{bmatrix}$ **10** Not defined   **11** $\begin{bmatrix} 2 & -3 & 0 \\ -1 & -5 & 4 \end{bmatrix}$

13 $\begin{bmatrix} 1 & 2 & 2 \\ 5 & 5 & 1 \end{bmatrix}$ **14** $\begin{bmatrix} -11 & 13 & -2 \\ 0 & 20 & -21 \end{bmatrix}$ **15** $\begin{bmatrix} 5 & 3 & 6 \\ 14 & 10 & 7 \end{bmatrix}$

17 $\begin{bmatrix} 7 & 2 \\ 0 & 1 \end{bmatrix}$ **18** $\begin{bmatrix} 7 & 4 \\ 0 & 1 \end{bmatrix}$ **19** $\begin{bmatrix} 0 & 5 \\ -1 & 0 \end{bmatrix}$

37 $\begin{bmatrix} 1 \\ 7 \\ 1 \end{bmatrix}$ **38** $\begin{bmatrix} -1 \\ 18 \\ 19 \end{bmatrix}$ **39** $\begin{bmatrix} 21 & 1 \\ 1 & 10 \end{bmatrix}$

Exercise 10.2   **1** $x = 2$, $y = 1$ **2** $x = 3$, $y = -2$ **3** $x = -2$, $y = -5$ **5** $x = 1$, $y = 2$, $z = 1$
6 $x = 1$, $y = 0$, $z = 3$ **7** $x = 2$, $y = -1$, $z = -2$ **9** $x = 0$, $y = 1$, $z = -1$
10 $x = 2$, $y = 2$, $z = -2$ **11** $x = \frac{1}{2}$, $y = \frac{1}{3}$, $z = \frac{5}{6}$ **13** $x = 1$, $y = 2$, $z = 1$, $w = 2$
14 $x = 3$, $y = 0$, $z = 1$, $w = -1$ **15** $x = 0$, $y = 2$, $z = 3$, $w = -4$
21 $x = -6 + 4z$, $y = \frac{1}{2}(5 - 3z)$, $z = z$
22 $x = \frac{1}{10}(-z - 3)$, $y = \frac{1}{10}(13z - 31)$, $z = z$
23 $x = -2z - 10$, $y = \frac{1}{2}(z + 16)$, $z = z$
25 $x = \frac{1}{5}(4 - 3z)$, $y = \frac{1}{5}(3 - z)$, $z = z$
26 $x = \frac{1}{7}(1 + z)$, $y = \frac{1}{7}(4 - 10z)$, $z = z$
27 $x = 5w - 8$, $y = 23 - 13w$, $z = 10w - 17$, $w = w$
29 $a = 2$, $b = -3$, $c = -5$

Exercise 10.3   **1** 2   **2** 7   **3** −22   **5** 8 − 6a   **6** 6 + a   **7** 2a + 3b

9 $\begin{vmatrix} 2 & 6 \\ 5 & 2 \end{vmatrix}$, $\begin{vmatrix} 3 & 1 \\ 5 & 2 \end{vmatrix}$, $\begin{vmatrix} 3 & 1 \\ 2 & 6 \end{vmatrix}$ **10** $\begin{vmatrix} 1 & 4 \\ 4 & 1 \end{vmatrix}$, $\begin{vmatrix} 3 & 4 \\ 4 & 1 \end{vmatrix}$, $\begin{vmatrix} 3 & 4 \\ 1 & 4 \end{vmatrix}$

11 $\begin{vmatrix} 3 & -3 \\ 1 & -2 \end{vmatrix}$, $\begin{vmatrix} 2 & -1 \\ 1 & -2 \end{vmatrix}$, $\begin{vmatrix} 2 & -1 \\ 3 & -3 \end{vmatrix}$

13 −74   **14** 24   **15** 51   **17** 23   **18** −16   **19** 84
21 3   **22** $\frac{1}{3}$ **23** 1   **25** 2   **26** 2   **27** −2, 2
29 11   **30** 25.5   **31** 0   **45** Discriminant = $(a + c)^2 − 4(ac − b^2) = (a − c)^2 + 4b^2 \geq 0$.   **47** 0

Exercise 10.4   **17** 7   **18** 14   **19** 7   **21** 42   **22** −7   **23** −28   **25** −6   **26** 0
27 $a(ab + b^2 − ac)$ **29** 2   **30** 3, −3   **31** 5, −3   **33** −12   **34** 12   **35** −90
41 −28   **42** 46   **43** 0   **45** 12   **46** −116   **47** −54
49 $−SPQR$ **50** $DIAG$ **51** $(ad − bc)(eh − fg)$

Exercise 10.5   **1** (2, 1)   **2** (1, −1)   **3** (3, −1)   **5** (1, −2)   **6** (2, −1)
7 (−2, −3)   **9** (3, 0)   **10** (−1, 4)   **11** (−1, −2)   **13** (a, b)
14 (2a − b, b)   **15** (a, b)   **17** (1, 2, 1)   **18** (1, −2, −1)
19 (2, −1, 1)   **21** (1, −2, 3)   **22** (−2, −1, 5)   **23** (2, 0, −1)
25 (2, 3, −2)   **26** (4, 3, −5)   **27** (2, −1, −3)   **29** (1, −1, 0, 2)
30 (1, −1, 2, 0)   **31** (1, −4, 2, −1)   **33** x = 2, y = 1, z = 4, w = 3, t = 2
34 x = 3, y = 2, z = 1, w = 2, t = 3   **35** x = 4, y = 3, z = 2, w = 1, t = 0
37 m = 2.5, b = 1.5, y = 2.5x + 1.5   **38** m = −3.6, b = 38, y = −3.6x + 38
39 $m = \frac{24.16}{14} \approx 1.73$, $b = \frac{128.38}{14} \approx 9.17$, y = 1.73x + 9.17

Exercise 10.6   **1** $\frac{1}{2}\begin{bmatrix} 7 & -4 \\ -3 & 2 \end{bmatrix}$ **2** $\frac{-1}{3}\begin{bmatrix} 3 & 0 \\ -2 & -1 \end{bmatrix}$ **3** $\frac{1}{7}\begin{bmatrix} -1 & -5 \\ 2 & 3 \end{bmatrix}$

5 $\frac{-1}{25}\begin{bmatrix} -13 & 5 & 11 \\ -8 & 5 & 1 \\ 6 & -10 & -7 \end{bmatrix}$ **6** $\frac{-1}{3}\begin{bmatrix} 1 & 0 & 1 \\ -11 & -3 & -2 \\ -8 & -3 & -2 \end{bmatrix}$ **7** $\frac{-1}{70}\begin{bmatrix} -3 & -23 & 7 \\ -7 & -7 & -7 \\ 6 & -24 & -14 \end{bmatrix}$

9 $\frac{1}{2}\begin{bmatrix} 2 & 2 & -4 & 2 \\ 0 & 1 & 0 & 1 \\ 0 & 0 & 0 & -2 \\ 0 & -1 & 2 & -1 \end{bmatrix}$ **10** $\frac{1}{4}\begin{bmatrix} -8 & 8 & 4 & -2 \\ 8 & -8 & -4 & 4 \\ -4 & 4 & 4 & -2 \\ 4 & -2 & -2 & 1 \end{bmatrix}$ **11** $\begin{bmatrix} 1 & 1 & -1 & 0 \\ 0 & 1 & -1 & 1 \\ 1 & -1 & 0 & -1 \\ 1 & 0 & -1 & -1 \end{bmatrix}$

13 (2, −3)   **14** (4, 1)   **15** (3, 2)   **17** (4, 4, 3)   **18** (3, 2, 5)
19 (1, 2, 1)   **21** (3, 0, 2)   **22** (2, 1, 0)   **23** (4, 6, 2, 8)

Exercise 10.7   **1** x = 2, y = 3   **2** x = 1, y = 2, z = −1   **3** $\begin{bmatrix} 0 & 2x & x \\ 3 & 1 & x \end{bmatrix}$

4 $\begin{bmatrix} -12 & 10 & 36 & 24 \\ -9 & 6 & 21 & 15 \\ 6 & -3 & -10 & -8 \end{bmatrix}$ **5** $\begin{bmatrix} 1 & 0 & 1 \\ 2 & 0 & a \\ 3 & -1 & -1 \end{bmatrix}$ **6** $\begin{bmatrix} -4 & 11 & 1 & 5 \\ -7 & 25 & -1 & 18 \end{bmatrix}$

7 $\begin{bmatrix} 40 & -27 & 25 \\ -23 & 16 & -14 \\ -9 & 5 & -4 \end{bmatrix}$ **8** $\begin{bmatrix} 5 & 34 & -14 \\ 1 & 23 & -9 \\ -12 & -55 & 24 \end{bmatrix}$ **9** $\begin{bmatrix} 1 & 2 & 5 \\ 2 & 3 & 8 \\ -1 & 1 & 2 \end{bmatrix}$

12 10   **13** 15   **14** -24 **15** 0   **16** 48   **17** -1 **24** 7   **25** 1   **26** -4, 3
27 $x = 2$, $y = -3$ **28** $x = 3$, $y = 5$ **29** $x = 1$, $y = -2$, $z = 5$
30 $x = 3$, $y = 0$, $z = -4$ **31** $x = 1$, $y = -2$, $z = 5$
32 $x = 3$, $y = 0$, $z = -4$ **33** Yes   **38** The constant sum is $270 = (15)(18)$.
42 $y = 3x - 4$

43 Using two decimal places gives $A^2 = \begin{bmatrix} 0.83 & 0.17 \\ 0.34 & 0.66 \end{bmatrix}$,

$A^4 = \begin{bmatrix} 0.75 & 0.25 \\ 0.51 & 0.49 \end{bmatrix}$, $A^8 = \begin{bmatrix} 0.69 & 0.31 \\ 0.63 & 0.37 \end{bmatrix}$, and $A^{16} = \begin{bmatrix} 0.67 & 0.33 \\ 0.67 & 0.33 \end{bmatrix}$

44 $\begin{bmatrix} 4 & 18 & 4 \\ 0 & 1 & 0 \\ 12 & 20 & 4 \end{bmatrix}$ **45** $I = T - TN = T(I - N)$

Exercise 11.1   **1** 5   **2** -54 **3** -3 **5** -2 **6** -108 **7** 13   **9** 22   **10** -2 **11** 7   **37** 3
38 2   **39** 3   **57** -1.862672 **58** 17.836412   **59** -2.6644

Exercise 11.2   **1** 6; 1, 3; -2, 1; 3, 2   **2** 9; -1, 5; 2, 3; -3, 1   **3** 8; -6, 2; 3, 4; 1, 1; -5, 1
5 3; $\frac{7}{2}$, 2; $-\frac{2}{3}$, 1   **6** 7; $-\frac{5}{3}$, 4; $\frac{1}{4}$, 3   **7** 9; $-\frac{2}{3}$, 5; 3, 3; $-\frac{7}{2}$, 1
9 -1, 1, 2   **10** 3, 1, -4 **11** 2, -2, 3   **13** 1, 1, -1, -2
14 -1, -1, i, $-i$ **15** 1, 2, $2i$, $-2i$ **17** -1, i, $-i$ **18** $\frac{3}{2}$, $2i$, $-2i$
19 $-\frac{5}{3}$, $2i$, $-2i$ **21** $-\frac{3}{2}$, $1 + i$, $1 - i$ **22** $-\frac{1}{4}$, $-1 + 2i$, $-1 - 2i$
23 $\frac{3}{4}$, $2 - i$, $2 + i$ **25** $2i$, $-2i$, $-1 + 2i$, $-1-2i$
26 3, $-2 + i$, $-2 + i$, $-2 - i$, $-2 - i$
27 $-1 + 2i$, $-1 + 2i$, $-1 - 2i$, $-1 - 2i$, i, $-i$
29 $3x^3 - 2x^2 - 3x + 2 = 0$ **30** $4x^3 - 13x^2 + 11x - 2 = 0$
31 $3x^3 - 2x^2 + 3x - 2 = 0$ **33** $3x^3 - 5x^2 + 13x + 5 = 0$
34 $2x^3 + 5x^2 - 22x - 70 = 0$ **35** $x^3 - 5x^2 + 9x - 5 = 0$
37 $3x^3 - 5x^2 - 5x - 1 = 0$ **38** $2x^3 + 5x^2 - 34x - 28 = 0$
39 $x^3 - 5x^2 + 6x - 2 = 0$ **41** $x^4 + 8x^3 + 26x^2 + 40x + 25 = 0$
42 $x^4 + 4x^3 + 2x^2 - 4x + 1 = 0$ **43** $x^4 - 8x^3 + 32x^2 - 64x + 64 = 0$
45 $2(x - 1)(x - 3)(x - 6)$
46 $3(x + 2)(x)(x - 3)$
47 $(x + 3)(x - 1)(x - 3)(x - 4)$
49 $8(x^2 - 2)(x - 1)$
50 $4(x^2 - 3)(x - 2)$
51 $5(x^2 - 4x - 1)(x - 4)$
53 $(x^2 + 4)(x - 1)$

54 $3(x^2 + 9)(x + 2)$
55 $2(x^2 - 4x + 13)(x^2 - 6x + 13)$
58 $x^3 + x^2 - 7x + 65 = 0$ **59** $x^4 - 10x^3 + 38x^2 - 58x + 13 = 0$

Exercise 11.3
1 $1, -2$ **2** $2, -3$ **3** $2, -1$ **5** $3, -6$ **6** $7, -1$ **7** $2, -4$ **9** $1, -1$
10 $1, -3$ **11** $1, -4$ **13** 2 or 0 positive, 1 negative   **14** 1 positive, 2 or 0
negative   **15** 0 positive, 3 or 1 negative   **17** 2 or 0 positive, 2 or 0 negative
18 2 or 0 positive, 2 or 0 negative   **19** 3 or 1 positive, 1 negative   **21** 4, 2,
or 0 positive, 1 negative   **22** 3 or 1 positive, 2 or 0 negative   **23** 1 positive,
2 or 0 negative   **25** -4 and -3, -1 and 0, 0 and 1   **26** 0 and 1, 1 and 2,
4 and 5   **27** 0 and 1, 3 and 4, 5 and 6   **29** -2 and -1, 0 and 1, 3 and 4,
5 and 6   **30** -5 and -4, -3 and -2, -2 and -1, 4 and 5   **31** -2 and -1,
-1 and 0, 1 and 2, 4 and 5   **33** 3 and 4   **34** 0 and 1   **35** -1 and 0,
2 and 3   **37** -1 and $-\frac{1}{2}$, $-\frac{1}{2}$ and 0, 2 and 3   **38** -2 and -1, 1 and $\frac{2}{3}$, $\frac{3}{2}$ and 2
39 -2 and -1, 1 and $\frac{2}{3}$, $\frac{3}{2}$ and 2   **41** No positive or negative roots
42 None positive, 1 negative   **43** 1 positive, none negative

Exercise 11.4
1 $-2, 1, 2$ **2** $-1, 2, 3$ **3** $1, 2, -3$ **5** $-3, 2, \frac{1}{2}$ **6** $-5, -2, \frac{2}{3}$
7 $-1, -\frac{1}{2}, \frac{3}{2}$ **9** $-\frac{1}{2}, \frac{1}{3}, \frac{1}{2}$ **10** $-\frac{3}{2}, \frac{3}{2}, \frac{3}{2}$ **11** $-\frac{3}{2}, \frac{1}{2}, 3$ **13** $2, 1 - \sqrt{2}$,
$1 + \sqrt{2}$ **14** $-3, 1 - \sqrt{3}, 1 + \sqrt{3}$ **15** $-1, \frac{1}{4}(-1 + \sqrt{41}), \frac{1}{4}(-1 - \sqrt{41})$
17 $\frac{1}{2}, i, -i$ **18** $-\frac{1}{3}, -1 + i, -1 - i$ **19** $\frac{3}{2}, i, -i$ **21** $-1, 2, -\frac{3}{2}, \frac{1}{3}$
22 $-3, 1, \frac{2}{3}, -\frac{3}{2}$ **23** $-5, 1, 1, 2$ **25** $-1, 2, i, -i$ **26** $-1, 1, 1 + i, 1 - i$
27 $-\frac{3}{2}, \frac{1}{2}, i, -i$ **37** $x^2 - 6 = 0$ has no rational roots.   **38** $x^3 - 5 = 0$
has no rational roots.   **39** $x^2 - 4x - 1 = 0$ has no rational roots.
41 $x^4 - 18x^2 + 25 = 0$ has no rational roots.   **42** $x^4 - 14x^2 + 9 = 0$
has no rational roots.   **43** $x^3 - 36 = 0$ has no rational roots.
45 $0, -2, -1 \pm 5i$ **46** $1, -2, \dfrac{-1 \pm i\sqrt{23}}{2}$ **47** $0, 3, \dfrac{3 \pm \sqrt{17}}{2}$

Exercise 11.5
1 3.59 **2** 2.44 **3** -0.29 **5** 1.38 **6** 2.24 **7** -1.88 **9** 0.12 **10** 0.53
11 0.66 **13** 1.18 **14** 1.42 **15** 1.67 **17** $2.732, -0.732$ **18** $6.236, 1.764$
19 $5.449, 0.551$ **21** $-3.260, -1.340, 1.602$ **22** $-4.769, -3.116, 1.884$
23 $-0.262, 1.660, 4.602$ **25** $1 + \sqrt{3}, 1 - \sqrt{3}, 1 + \sqrt{5}, 1 - \sqrt{5}; 3.236$
26 $2 + \sqrt{5}, 2 - \sqrt{5}, -1 + i\sqrt{2}, -1 - i\sqrt{2}; 4.236$
27 $3 + \sqrt{2}, 3 - \sqrt{2}, -2 + \sqrt{3}, -2 - \sqrt{3}; 4.414$
29 $i, -i, 4 - \sqrt{3}, 4 + \sqrt{3}; 5.732$ **30** $i\sqrt{3}, -i\sqrt{3}, 2 - \sqrt{6}, 2 + \sqrt{6}; 4.449$
31 $2i, -2i, -3 - 2\sqrt{2}, -3 + 2\sqrt{2}; -0.172$ **33** $\dfrac{7 \pm \sqrt{93}}{2}, \dfrac{3 \pm i\sqrt{35}}{2}; 8.322$
34 $-1 \pm i\sqrt{2}, -5 \pm 2\sqrt{7}; 0.292$ **35** $\pm 2i, -3 \pm \sqrt{13}; 0.606$
39 $t^3 - 5t + 8$

Exercise 11.6
1 14 **2** 20 **3** $d^4 - 2d^3 + 5d + 2$ **9** $2x^2 + 4x + 13; 20$
10 $3x^3 - x^2 + 4x - 2; 1$ **11** $2, 2, -3$ **12** $\pm i\sqrt{2}$ **13** $x^3 - 8x^2 + 22x - 20$
14 $x^3 - 5x^2 + 4x + 6$ **15** $x^4 + 2x^3 + 2x^2 + 10x + 25$
16 $x^4 - 8x^3 + 24x^2 - 32x + 7$ **17** 3 **18** 2 **19** sum $= 1$, product $= 1$

20 sum $= 4$, product $= 16$ **21** -1 is a lower bound, 1 is an upper bound
22 There is one change of sign in $f(x)$.   **23** 1, 2, 3
25 $x^3 + 3x^2 - 5x + 4 = x^3 + 2x^2 - 9x - 3$ is equivalent to $x^2 + 4x + 7 = 0$,
and this has only imaginary roots.   **26** 2, i, $-i$ **27** $-\frac{3}{2}$, $-\frac{3}{2}$, $2i$, $-2i$
28 $\frac{5}{2}$, -2, $\frac{1}{3}$, $1 - i$, $1 + i$ **30** $x^4 - 6 = 0$ has no rational roots.
31 $x^4 - 10x^2 + 1 = 0$ has no rational roots.   **32** No, it is 0.
33 -0.18 **34** 0.73, -2.73 **35** -0.618, 1.618, 1.732, -1.732
36 $(1 \pm \sqrt{5})/2$, $\pm\sqrt{3}$ **37** $27t^3 - 90t + 2 = 0$ **38** $z^6 + 2z^3 + 1 = 0$

Exercise 12.1   **1** 1, 4, 7, 10   **2** 4, 5, 6   **3** 13, 11, 9, 7, 5   **5** 2, 4, 6, 8
6 -3, 0, 3, 6, 9   **7** -7, -5, -3, -1, 1, 3   **9** 10, 6, 2, -2, -6
10 11, 5, -1, -7, -13, -19 **11** 2, 3, 4, 5, 6, 7   **13** 11
14 -2 **15** -11 **17** 56   **18** 7   **19** 16   **21** 7   **22** -3
23 13   **25** $l = 12$, $s = 21$ **26** $l = -6$, $s = 21$ **27** $d = -4$, $s = 48$
29 $d = -1.5$, $n = 7$ **30** $a = 1$, $d = \frac{2}{3}$ **31** $l = -5$, $n = 8$
33 $a = 8$, $n = 6$ **34** $a = -8$, $n = 4$ **35** $a = 1$, $n = 6$ **37** 204
38 315   **39** $2n(n + 1)$ **41** -3 **42** $\frac{8}{9}$ **43** $x = 2$, $y = 3$
45 156   **46** 208 ft, 784 ft   **47** 25 s   **49** \$1102   **50** \$12,700
51 802 π cm   **53** 6, 9, 12   **54** 5, 7, 9, 11, 13   **55** 6, 2, -2, -6
57 $\frac{1}{24}$ **58** $\frac{3}{19}$ **59** $\frac{1}{6}$ **61** $\frac{1}{4}$ **62** $-\frac{2}{3}$ **63** $\frac{1}{5}$

65 $\dfrac{1}{2}\left(\dfrac{1}{a} + \dfrac{1}{b}\right) = \dfrac{a + b}{2ab}$ **66** $d = \log 3$ **67** $d = \sqrt{6} - \sqrt{3}$

Exercise 12.2   **1** 1, 2, 4, 8   **2** 3, 9, 27, 81, 243   **3** 2, -4, 8, -16, 32
5 81, 27, 9, 3, 1, $\frac{1}{3}$ **6** 128, -64, 32, -16, 8, -4, 2, -1
7 $\frac{3}{4}$, 1, $\frac{4}{3}$, $\frac{16}{9}$, $\frac{64}{27}$ **9** 162   **10** 6   **11** -1 **13** 170
14 $\frac{121}{3}$ **15** 1   **17** 127   **18** $\frac{211}{3}$ **19** $\frac{1}{5}$ **21** $l = 162$, $s = 242$
22 $l = 48$, $s = 93$ **23** $n = 5$, $s = \frac{55}{2}$ **25** $s = 400$, $n = 4$
26 $a = \frac{1}{9}$, $s = \frac{40}{9}$ **27** $r = \frac{1}{2}$, $s = 511$ **29** $l = 162$, $n = 5$
30 $a = 625$, $l = 1$ **31** $a = 343$, $n = 4$ **33** $r = -2$, $a = 3$
34 $a = 8$, $n = 5$ **35** $r = 3$, $l = 45$; $r = -4$, $l = 80$ **37** 504
38 3279   **39** $2^{n+1} - 1$ **41** 2, $-\frac{2}{15}$ **42** 2, 5, 8; $r = 3$
45 2   **46** \$24,300   **47** $4096n$ **49** \$511, \$1 profit
50 \$1419.20   **51** 254   **53** 6, 18   **54** $\frac{1}{3}$, $\frac{1}{2}$, $\frac{3}{4}$, $\frac{9}{8}$
55 4, 8, 16 and -4, 8, -16 **57** Geometric, $\frac{32}{243}$, $\frac{64}{729}$
58 Harmonic, $\frac{2}{9}$, $\frac{4}{21}$ **59** Arithmetic, $-\frac{2}{9}$, $-\frac{4}{9}$ **61** Harmonic, $-\frac{1}{8}$, $-\frac{1}{11}$
62 Geometric, $\frac{81}{8}$, $\frac{243}{16}$ **63** Arithmetic, $\frac{13}{6}$, $\frac{8}{3}$ **67** 4, 12, 36 or $\frac{4}{9}$, $-\frac{20}{9}$, $\frac{100}{9}$

Exercise 12.3   **1** 49   **2** 10.5   **3** $\frac{49}{13}$ **5** $\frac{25}{9}$ **6** 9   **7** 8   **9** $\frac{66}{7}$
10 60 or $\frac{60}{7}$ **11** $\frac{64}{3}$ **13** 3   **14** 5   **15** $-\frac{3}{5}$ **17** $\frac{1}{2}$
18 2   **19** $-\frac{2}{7}$ **21** $\frac{4}{9}$ **22** $\frac{8}{33}$ **23** $\frac{232}{99}$ **25** $\frac{371}{90}$
26 $\frac{199}{90}$ **27** $\frac{2161}{330}$ **33** $\frac{3}{4}$ **34** $\frac{4}{5}$ **35** $\frac{2}{7}$ **37** 2400 cm
38 120 m

39 18 m   **41** $p < 9$ **42** 9 mg   **43** 200 in.^{2} **45** $\{x \mid x < 0\} \cup \{x \mid x > 1\}$, $\dfrac{1}{2x - 2}$

46 $\{x \mid x < 0\} \cup \{x \mid x > \frac{4}{3}\}$, $\dfrac{2}{3x - 4}$ **47** All x, $\dfrac{1}{x^2 + 1}$

Exercise 12.4   **1** $132, $1332   **2** $168, $868   **3** $432.15, $6194.15   **5** $836.47   **6** $947.87
7 8 percent   **9** $1529.86   **10** $6309.99   **11** $9400.10   **13** $3502.82   **14** $345.68
15 $581.72   **17** 3 years   **18** 5.5 years   **19** 17 years   **21** 5 percent
22 6 percent   **23** 6 percent   **25** 6.09 percent   **26** 4.04 percent   **27** 5.06 percent
29 8.24 percent   **30** 6.14 percent   **31** 10.38 percent   **33** $2039.90
34 Pay in 5 years, $20.70   **35** Pay cash, $1.16   **37** $190,099   **38** $3895.80
39 $17,879.44   **41** 14.2 years   **42** 17.5 years   **43** 0.25 years   **45** 4 percent
46 5 percent   **47** 6 percent

Exercise 12.5   **1** $3949.15, $3001.05   **2** $3315.36, $2597.70   **3** $8463.87, $3143.19
5 $13,831.88, $4498.50   **6** $60,878.88, $19,551.02   **7** $174,079, $26,493.88
9 $18,765.73, $7851.60   **10** $18,335.51, $9183.85   **11** $66,086.94, $32,340.47
13 $13,202.55   **14** $4593.04   **15** $17,589.20   **17** $22,927.80   **18** Annuity
19 $88,672.32   **21** $875.44   **22** $512.84   **23** $158.94   **25** 4 percent
26 10 percent   **27** 10 percent   **29** 21 years   **30** 24 years   **31** 11 years
33 $5537.92   **34** $80.33   **35** $536.83   **37** 4 percent   **38** 6 percent
39 10 percent   **41** $3347.19, $8021.79   **42** $18,028.26, $46,640.38
43 $23,654.17, $271,252.20

Exercise 12.6   **1** $3, \frac{7}{2}, 4, \frac{9}{2}, 5, \frac{11}{2}$, 6   **2** $\frac{3}{4}, \frac{1}{2}, \frac{1}{3}, \frac{2}{9}, \frac{4}{27}, \frac{8}{81}$ **3** 12, 3, $\frac{12}{7}, \frac{6}{5}, \frac{12}{13}, \frac{3}{4}$ **5** $\frac{2}{243}$, sum $= \frac{728}{243}$
6 $\frac{5}{8}$, sum $= \frac{77}{24}$ **7** $a = \frac{1}{4}, d = \frac{1}{2}, l = \frac{9}{4}, s = \frac{25}{4}$ **8** $a = \frac{1}{8}, r = 4, l = 32, n = 5$
9 45,878   **10** $\frac{21}{11}$, 7   **11** $1425.80   **12** $\frac{17}{33}$ **13** $-\frac{1}{7}$ **21** 2, 6, 18; 18, 6, 2
22 $\frac{1}{3}$, 0   **23** $4\sqrt{2}$, 8, $8\sqrt{2}$ **24** 7, 10, 13   **25** $\frac{64}{13}, \frac{32}{5}, \frac{64}{7}$ **28** 2, $-1, \frac{1}{2}, -\frac{1}{4}, \ldots$
29 $\frac{3}{4}, \frac{11}{12}$, arithmetic   **30** $\frac{16}{135}, \frac{32}{405}$, geometric   **31** $\frac{7}{9}, \frac{17}{18}$, arithmetic
32 $\frac{3}{4}, \frac{2}{3}$, harmonic   **33** $\frac{4}{5}, \frac{8}{11}$, harmonic   **34** 12.55 percent   **35** $22,397.38
36 $7924.12   **37** $23,279.49   **38** $43,984
39 $2240, $2247.20, $2251, $2255   **40** 9 weeks, 13.5 weeks
41 1, 1, 2, 3, 5, 8, 13, 21

Exercise 14.1   **1** 468,000   **2** 280   **3** 40   **5** 17,576,000   **6** 625   **7** 9261   **9** 248,832
10 5040   **11** 100   **13** 3584   **14** 7,529,536   **15** 18   **17** 243   **18** 36
19 140,608   **25** Selection   **26** Combination   **27** Sample   **29** Sample
30 Selection   **31** Combination   **33** Combination   **34** Permutation
35 Permutation

Exercise 14.2   **1** 60   **2** 2520   **3** 1680   **5** 9   **6** 24   **7** 42   **13** 3   **14** 8   **15** 10   **17** 10,080
18 105   **19** 5040   **21** 336   **22** 120   **23** 720   **25** 7 min   **26** 15 min
27 414,720   **29** 10^7 **30** 40,320   **31** 120   **33** $2P(25, 3) = 27,600$
34 $P(7, 5) = 2520$ **35** $P(6, 6) = 720$ **37** 3, 1, 4, 2; 2, 1, 4, 3; 4, 1, 2, 3;
2, 4, 1, 3; 3, 4, 1, 2; 4, 3, 1, 2; 2, 3, 4, 1; 3, 4, 2, 1; 4, 3, 2, 1
38 9, 44, 1854   **39** $D_6 = 265$

Exercise 14.3   **1** 45   **2** 330   **3** 56   **9** $C(9, 4) = 126$ **10** $C(10, 3) = 120$
11 $C(52, 13) = 635,013,559,600$ **13** $C(52, 5) = 2,598,960$

14 $C(18, 12) = 18,564$ **15** $C(22, 6) = 74,613$ **17** $C(5, 2)C(7, 3) = 350$

18 $C(10, 3) = 120$ **19** $[C(8, 2)]^3 = 21,952$ **21** $\dfrac{22!}{6!5!7!3!1!}$

22 $\dfrac{9!}{2!3!4!} = 1260$ **23** $\dfrac{11!}{7!2!2!} = 1980$

26 $C(5, 1)\dfrac{27!}{5!5!5!5!7!} + C(5, 2)\dfrac{27!}{5!5!5!6!6!}$

29 (a) $C(6, 3)$ (b) $C(4, 3)$ **30** $C(6, 3)$ **31** $C(11, 5)$ **33** $C(8, 4)$
34 $C(7, 3)$ **35** $C(12, 5)$

Exercise 14.4   **1** 1320   **2** 336   **3** 220   **4** 56   **9** 56   **10** 20   **11** 120
12 $C(5, 2)C(7, 4)C(9, 4) = 44,100$ **13** 5400   **14** 287,496   **15** All integers $n \geq 3$
16 415,800   **17** 5040   **18** $C(8, 5) + C(8, 6) + C(8, 7) + C(8, 8) = 93$ **20** $C(n, 3)$
21 $8^3 = 512$; $P(8,3) = 336$; $C(8, 3) = 56$; $C(10, 3) = 120$ **22** $5^3 = 125$
23 $P(5, 3) = 60$ **24** $C(5, 3) = 10$ **25** $C(7, 3) = 35$ **26** $P(4, 4) = 24$
32 $f = 23\frac{1}{3}$, $K = 24$

Exercise 15.1   **1** $x^4 + 4x^3y + 6x^2y^2 + 4xy^3 + y^4$
2 $s^7 + 7s^6t + 21s^5t^2 + 35s^4t^3 + 35s^3t^4 + 21s^2t^5 + 7st^6 + t^7$
3 $b^5 - 5b^4y + 10b^3y^2 - 10b^2y^3 + 5by^4 - y^5$
5 $243x^5 - 405x^4y + 270x^3y^2 - 90x^2y^3 + 15xy^4 - y^5$
6 $64a^6 - 192a^5b + 240a^4b^2 - 160a^3b^3 + 60a^2b^4 - 12ab^5 + b^6$
7 $x^4 + 12x^3w + 54x^2w^2 + 108xw^3 + 81w^4$
8 $a^3 - 12a^2x + 48ax^2 + 64x^3$
9 $16x^4 + 96x^3y + 216x^2y^2 + 216xy^3 + 81y^4$
10 $81s^4 - 432s^3t + 864s^2t^2 - 768st^3 + 256t^4$
11 $8x^3 - 60x^2y + 150xy^2 - 125y^3$
13 $a^{33} + 33a^{32}y + 528a^{31}y^2 + 5456a^{30}y^3$
14 $x^{51} - 51x^{50}y + 1275x^{49}y^2 - 20,825x^{48}y^3$
15 $m^{101} - 202m^{100}y + 20,200m^{99}y^2 - 1,333,200m^{98}y^3$
17 1.2167   **18** 1.2155   **19** 1.1941   **21** $560x^3y^4$ **22** $-160a^3c^3$ **23** $10,206x^4y^5$
25 $-35a$ **26** $-2,016x^{-6}$ **27** $29,568x^{10}y^6$ **29** $160x^3y^{3/2}$
30 $5670x^4y$ **31** $960x^7y^3$

Exercise 15.2   **1** $x^{-2} - 2x^{-3}y + 3x^{-4}y^2 - 4x^{-5}y^3$ **2** $a^{-4} - 4a^{-5}b + 10a^{-6}b^2 - 20a^{-7}b^3$

3 $a^{-5} + 5a^{-6}x + 15a^{-7}x^2 + 35a^{-8}x^3$ **5** $\dfrac{1}{2a} - \dfrac{x}{4a^2} + \dfrac{x^2}{8a^3} - \dfrac{x^3}{16a^4}$

6 $\dfrac{1}{x} - \dfrac{2a}{x^2} + \dfrac{4a^2}{x^3} - \dfrac{8a^3}{x^4}$ **7** $\dfrac{1}{a^6} + \dfrac{3y}{a^8} + \dfrac{6y^2}{a^{10}} + \dfrac{10y^3}{a^{12}}$ **9** $\dfrac{1}{x^3} + \dfrac{3}{x^5} + \dfrac{6}{x^7} + \dfrac{10}{x^9}$

10 $\dfrac{1}{x^3} - \dfrac{3}{x^6} + \dfrac{6}{x^9} - \dfrac{10}{x^{12}}$ **11** $\dfrac{1}{y^4} - \dfrac{4}{y^5} + \dfrac{10}{y^6} - \dfrac{20}{y^7}$

13 $2 - \dfrac{x}{12} - \dfrac{x^2}{288} - \dfrac{5x^3}{20,736}$, $-8 < x < 8$

14 $5 + \dfrac{x}{10} - \dfrac{x^2}{1000} + \dfrac{x^3}{50{,}000}$, $-25 < x < 25$

15 $\dfrac{1}{2} + \dfrac{x}{16} + \dfrac{3x^2}{256} + \dfrac{5x^3}{2048}$, $-4 < x < 4$ **17** $1 + \dfrac{x}{2} - \dfrac{x^2}{8} + \dfrac{x^3}{16}$, $-1 < x < 1$

18 $1 - \dfrac{x}{2} + \dfrac{3x^2}{8} - \dfrac{5x^3}{16}$, $-1 < x < 1$

19 $1 - \dfrac{x}{2} - \dfrac{x^2}{8} - \dfrac{x^3}{16}$, $-1 < x < 1$ **21** 5.050   **22** 1.992   **23** 0.513   **25** 1.987

26 3.037   **27** 3.018   **29** 0.888   **30** 0.906   **31** 1.096

33

x	0.1	0.2	0.3	0.4	0.5
$l(x) = 1 + \dfrac{x}{2}$	1.05	1.10	1.15	1.20	1.25
$q(x) = 1 + \dfrac{x}{2} - \dfrac{x^2}{8}$	1.04875	1.095	1.1388	1.18	1.2188
$\sqrt{1 + x}$	1.0488	1.0954	1.1402	1.1832	1.2247

34

x	0.1	0.2	0.3	0.4	0.5
$l(x) = 1 - \dfrac{x}{2}$	0.95	0.90	0.85	0.80	0.75
$q(x) = 1 - \dfrac{x}{2} - \dfrac{x^2}{8}$	0.94875	0.895	0.8388	0.78	0.7188
$\sqrt{1 - x}$	0.9487	0.8944	0.8367	0.7746	0.7071

35 5.099

Exercise 16.1   **1** $\frac{1}{4}, \frac{1}{2}$ **2** $\frac{1}{13}, \frac{5}{13}$ **3** $\frac{1}{3}, \frac{1}{6}$ **5** $\frac{1}{6}$ **6** $\frac{1}{6}$ **7** $\frac{1}{3}$ **9** $\frac{1}{36}, \frac{1}{9}$ **10** $\frac{1}{9}, \frac{1}{36}$ **11** $\frac{1}{6}$ **13** $\frac{4}{13}, \frac{10}{13}$
14 $\frac{9}{13}, \frac{7}{13}$ **15** $\frac{1}{36}$ **17** $C(28, 13)/C(52, 13)$ **18** $C(20, 5)/C(52, 5)$
19 $C(12, 3)/C(20, 3)$ **21** $\frac{1}{12}$ **22** $\frac{7}{12}$ **23** $\frac{3}{4}$ **25** $\frac{1}{200}$ **26** $\frac{43}{8700}$ **27** $\frac{1}{16}$ **29** $\frac{27}{35}$
30 $\frac{29}{41}, \frac{21}{82}$ **31** $\frac{39}{100}$ **33** 0.49, 0.96   **34** $\frac{7}{15}$ **35** $\frac{7}{15}$ **37** 0.04   **38** $\frac{5}{6}$ **39** $\frac{71}{75}$
41 $\dfrac{13 \cdot C(4, 3) \cdot 12 \cdot C(4, 2)}{C(52, 5)} = \dfrac{3744}{2{,}598{,}690} \approx 0.0014$

42 $\dfrac{C(13, 1)C(4, 2)C(12, 3)C(4, 1)C(4, 1)C(4, 1)}{C(52, 5)} = \dfrac{1,098,240}{2,598,690} \approx 0.4226$

43 $\dfrac{C(20, 13)}{C(52, 13)} \approx 1.22 \times 10^{-7}$

45 $\dfrac{C(5, 3)C(5, 2)}{C(14, 5)} = \dfrac{100}{2002} \approx 0.0500$

46 $\dfrac{C(8, 2)C(6, 4)}{C(17, 6)} = \dfrac{420}{12376} \approx 0.0339$

47 $\dfrac{C(8, 3)C(7, 2)}{C(15, 5)} = \dfrac{56}{143} \approx 0.3916$

49 $k(1 + 2 + 3 + 4 + 5 + 6) = 1$, so $k = \frac{1}{21}$ and $P(2) = \frac{2}{21}$

Exercise 16.2 **1** $\frac{2}{5}$ **2** $\frac{1}{10}$ **3** $\frac{9}{25}, \frac{27}{125}$ **5** $\frac{14}{33}$ **6** $\frac{7}{26}$ **7** $\frac{1}{2}$ **9** $\frac{55}{204}$ **10** $\frac{14}{55}$, no **11** $\frac{34}{85}$ **13** $\frac{1}{221}$
14 $\frac{1}{17}$ **15** $\frac{22}{425}$ **17** $\frac{7}{36}$ **18** $\frac{1}{36}$ **19** 0.80256 **21** $\frac{3}{4}$ **22** $\frac{1}{8}$ **23** $\frac{1}{2}$ **25** $\frac{4}{9}$ **26** $\frac{42}{169}$
27 $\frac{24}{49}$ **29** $\frac{2}{3}$ **30** $\frac{3}{7}$ **31** $\frac{2}{3}$ **37** $\frac{1}{35}$ **38** $\frac{1}{14}$ **39** $\frac{1}{10}$

Exercise 16.3 **1** $\frac{5}{16}, \frac{1}{2}$ **2** $\frac{5}{72}, \frac{2}{27}$ **3** 0.01536, 0.01696 **5** $\dfrac{6125}{725,594,112}$ **6** $\frac{7}{243}$ **7** $\frac{40}{243}$ **9** $\frac{80}{243}$

10 $\frac{96}{625}, \frac{112}{625}$ **11** $\frac{5}{16}, \frac{21}{32}$ **13** $\dfrac{133}{91,125}$ **14** $\dfrac{43,681}{5,717,741,400,000}$ **15** $\frac{16}{243}$ **17** $\frac{1}{3}$ **18** $\frac{1}{3}$

19 0.451 **21** $\frac{5}{7}$ **22** $\frac{2}{3}$ **23** $(49 - \sqrt{133})/54$

Exercise 16.4 **1** $\frac{25}{49}$ **2** $\frac{45}{91}$ **3** $\frac{25}{49}$ **4** $\frac{27}{64}$ **5** $\frac{7}{13}$ **6** $\frac{11}{20}$ **7** $\frac{1}{6}$ **8** $\frac{5}{36}$ **9** $\frac{5}{108}$ **10** $\frac{2}{3}$ **11** $\frac{1}{2}$
12 $\frac{44}{4165}$ **13** $\frac{32}{81}$ **14** $\frac{328}{625}$ **15** $\frac{1424}{3125}$, no **16** $\dfrac{C(12, 3)C(8, 2)}{C(20, 5)} \approx 0.397$ **17** $\frac{19}{50}$
18 0.65 **19** $\frac{1}{18}$ **20** $\frac{1}{5}$ **21** $C(9, 3)C(6, 5)/C(15, 8) \approx 0.078$
22 $C(6, 2)C(9, 3)/C(25, 5) \approx 0.024$ **23** $C(4, 1)C(13, 5)/C(52, 5) \approx 0.001981$

INDEX

AXIOMS, DEFINITIONS, AND THEOREMS

(4.14) $a^{m/n} = (\sqrt[n]{a})^m = (a^{1/n})^m$
$= (a^m)^{1/n} = \sqrt[n]{a^m}$

(5.1) $x = \dfrac{-b}{a}$

(5.2) $x = \dfrac{-b \pm \sqrt{b^2 - 4ac}}{2a}$

(5.3) $r + s = \dfrac{-b}{a}$ sum of the roots

(5.4) $rs = \dfrac{c}{a}$ product of the roots

$a < b$ is equivalent to:

$a + c < b + c$ for any real c
$ac < bc$ for $c > 0$
$ac > bc$ for $c < 0$

(6.1) $(a,b) = (c,d)$ if and only if $a = c$ and $b = d$

(6.2) Distance $= \sqrt{(x_2 - x_1)^2 + (y_2 - y_1)^2}$

(6.3) Midpoint $= \left(\dfrac{a + c}{2}, \dfrac{b + d}{2} \right)$

(6.4) $(x - h)^2 + (y - k)^2 = r^2$

(6.5) Slope $= m = \dfrac{y_2 - y_1}{x_2 - x_1}$

(6.6) $y - y_1 = \dfrac{y_2 - y_1}{x_2 - x_1}(x - x_1)$

(6.7) $y - y_1 = m(x - x_1)$

(6.8) $y = mx + b$

(6.9) $m_1 = m_2$ for parallel lines

(6.10) $m_1 m_2 = -1$ for perpendicular lines

(6.11) $f(f^{-1}(x)) = x = f^{-1}(f(x))$ for inverse functions

(7.1) $y - k = a(x - h)^2$

(7.2) Ellipse $\dfrac{x^2}{a^2} + \dfrac{y^2}{b^2} = 1$
or $\dfrac{y^2}{a^2} + \dfrac{x^2}{b^2} = 1$ with $a > b$

(7.3) Hyperbola $\dfrac{x^2}{a^2} - \dfrac{y^2}{b^2} = 1$
or $\dfrac{y^2}{a^2} - \dfrac{x^2}{b^2} = 1$
with a^2 in the positive term

(7.4) $Ax^2 + Cy^2 + Dx + Ey + F = 0$:
Line if $A = C = 0$
Parabola if $A = 0$ or $C = 0$, but not both.
Circle if $A = C \neq 0$
Ellipse if A and C have same sign
Hyperbola if A and C have opposite signs

(8.1) $\log_b N = L$ means $b^L = N$, and $b^{\log_b N} = N$

(8.2) $\log_b(MN) = \log_b M + \log_b N$

(8.3) $\log_b(M/N) = \log_b M - \log_b N$

(8.4) $\log_b(N^p) = p \log_b N$

(8.5) $\log_b \sqrt[r]{N} = \dfrac{1}{r} \log_b N$

(8.7) $\log_a N = \dfrac{\log_b N}{\log_b a}$

(10.3) $A_{ij} = (-1)^{i+j} M_{ij}$

(10.4) $D_n = a_{i1}A_{i1} + a_{i1}A_{i2} + \cdots + a_{ij}A_{ij}$
$+ \cdots + a_{in}A_{in}$